Learn the techniques with step-by-step surgical videos and animations on MediaCenter.Thieme.com!

Simply visit MediaCenter.thieme.com and, when prompted during the registration process, enter the code below to get started today.

5GSZ-A6J7-528W-LNC9

	WINDOWS & MAC	TABLET
Recommended Browser(s)	Recent browser versions on all major platforms and any mobile operating system that supports HTML5 video playback. *All browsers should have JavaScript enabled.*	
Flash Player Plug-in	Flash Player 9 or higher. *For Mac users, ATI Rage 128 GPU doesn't support full-screen mode with hardware scaling.*	Tablet PCs with Android OS support Flash 10.1.
Recommended for optimal usage experience	Monitor resolutions: • Normal (4:3) 1024×768 or higher • Widescreen (16:9) 1280×720 or higher • Widescreen (16:10) 1440×900 or higher A high-speed internet connection (minimum 384 Kps) is suggested.	WiFi or cellular data connection is required.

Thieme

Epiduroscopy

Atlas of Procedures

Daniel H. Kim, MD, FAANS, FACS
The Nancy, Clive, and Pierce Runnells Distinguished Chair in Neuroscience
Professor, Director of Spinal Neurosurgery
Reconstructive Peripheral Nerve Surgery
Director of Microsurgical Robotic Lab
Department of Neurosurgery
University of Texas
Houston, Texas, USA

Salahadin Abdi, MD, PhD
Professor and Chairman
Department of Pain Medicine
Helen Buchanan & Stanley Joseph Seeger Endowed Research Professor
The University of Texas M.D. Anderson Cancer Center
Houston, Texas, USA

Günter Schütze, Dr. med.
Director of Anesthesiology, Critical Care Medicine, and Pain Medicine
Marien Hospital Iserlohn
Privatpraxis für Spezielle Schmerztherapie
Dortmund, Germany

360 illustrations

Thieme
New York • Stuttgart • Delhi • Rio de Janeiro

Executive Editor: Timothy Y. Hiscock
Managing Editor: J. Owen Zurhellen IV
Editorial Assistant: Mary Wilson
Director, Editorial Services: Mary Jo Casey
Production Editor: Torsten Scheihagen
International Production Director: Andreas Schabert
International Marketing Director: Fiona Henderson
International Sales Director: Louisa Turrell
Director of Sales, North America: Mike Roseman
Senior Vice President and Chief Operating Officer: Sarah Vanderbilt
President: Brian D. Scanlan
Printer: Everbest Printing Co.

Library of Congress Cataloging-in-Publication Data

Names: Kim, Daniel H., editor. | Abdi, Salahadin, editor. | Schütze, G. (Günter), editor.
Title: Epiduroscopy : atlas of procedures / [edited by] Daniel H. Kim, Salahadin Abdi, Günter Schütze.
Description: First edition. | New York : Thieme, 2017. | Includes bibliographical references and index.
Identifiers: LCCN 2017018203 (print) | LCCN 2017019030 (ebook) | ISBN 9781626232679 (E-book) | ISBN 9781626232662 (hardcover : alk. paper)
Subjects: | MESH: Low Back Pain–surgery | Endoscopy | Epidural Space | Atlases
Classification: LCC RD771.B217 (ebook) | LCC RD771.B217 (print) | NLM WE 17 | DDC 617.5/64057–dc23
LC record available at https://lccn.loc.gov/2017018203

Thieme Publishers New York
333 Seventh Avenue, New York, NY 10001 USA
+1 800 782 3488, customerservice@thieme.com

Thieme Publishers Stuttgart
Rüdigerstrasse 14, 70469 Stuttgart, Germany
+49 [0]711 8931 421, customerservice@thieme.de

Thieme Publishers Delhi
A-12, Second Floor, Sector-2, Noida-201301
Uttar Pradesh, India
+91 120 45 566 00, customerservice@thieme.in

Thieme Publishers Rio de Janeiro, Thieme Publicações Ltda.
Edifício Rodolpho de Paoli, 25º andar
Av. Nilo Peçanha, 50 – Sala 2508
Rio de Janeiro 20020-906 Brasil
+55 21 3172-2297 / +55 21 3172-1896

Cover design: Thieme Publishing Group
Typesetting by DiTech Process Solutions

Printed in China by Everbest Printing Co. 5 4 3 2 1

ISBN 978-1-62623-266-2

Also available as an e-book:
eISBN 978-1-62623-267-9

Important note: Medicine is an ever-changing science undergoing continual development. Research and clinical experience are continually expanding our knowledge, in particular our knowledge of proper treatment and drug therapy. Insofar as this book mentions any dosage or application, readers may rest assured that the authors, editors, and publishers have made every effort to ensure that such references are in accordance with **the state of knowledge at the time of production of the book.**

Nevertheless, this does not involve, imply, or express any guarantee or responsibility on the part of the publishers in respect to any dosage instructions and forms of applications stated in the book. **Every user is requested to examine carefully** the manufacturers' leaflets accompanying each drug and to check, if necessary in consultation with a physician or specialist, whether the dosage schedules mentioned therein or the contraindications stated by the manufacturers differ from the statements made in the present book. Such examination is particularly important with drugs that are either rarely used or have been newly released on the market. Every dosage schedule or every form of application used is entirely at the user's own risk and responsibility. The authors and publishers request every user to report to the publishers any discrepancies or inaccuracies noticed. If errors in this work are found after publication, errata will be posted at www.thieme.com on the product description page.

Some of the product names, patents, and registered designs referred to in this book are in fact registered trademarks or proprietary names even though specific reference to this fact is not always made in the text. Therefore, the appearance of a name without designation as proprietary is not to be construed as a representation by the publisher that it is in the public domain.

FSC
www.fsc.org
MIX
Paper from responsible sources
FSC® C124385

To my loving family: Anslie, Elise, Rebecca, Sarah, and Isaiah

Contents

Menu of Accompanying Videos

I **Sacral Epiduroscopic Laser Decompression (SELD)**

1. SELD Technique Animation
Courtesy of Lutronic Corp.

2. SELD Technique Step-by-Step Demonstration (Sacral Hiatus Approach for Epiduroscopy)
Paulo Pereira and Pedro Monteiro

3. SELD Case A
Chan Hong Park

4. SELD Case B
Chan Hong Park

5. SELD Case C
Chan Hong Park

6. SELD Case D
Chan Hong Park

7. SELD Case E
Chan Hong Park

8. SELD Case F
Chan Hong Park

9. SELD Case G
Chan Hong Park

10. SELD Case H
Chan Hong Park

11. SELD Case I
Chan Hong Park

12. SELD Case J
Chan Hong Park

13. SELD Case K
Chan Hong Park

14. SELD Case L
Chan Hong Park

15. SELD Case M
Chan Hong Park

16. SELD Case N
Chan Hong Park

17. SELD Case O
Chan Hong Park

18. SELD Case P
Chan Hong Park

19. SELD Case Q
Chan Hong Park

II **Transforaminal Epiduroscopic Laser Annuloplasty (TELA)**

20. TELA Technique Animation
Courtesy of Lutronic Corp.

21. TELA Technique Step-by-Step Demonstration (TELA Procedure Video)
Courtesy of Lutronic Corp.

22. Transforaminal Epiduroscopic Approach for Laser Annuloplasty
Jongsun Lee

23. Probing for Concordant Discogenic Pain with Subsequent Annuloplasty (Probing of Tender Annulus during TELA)
Jongsun Lee

24. Right-sided Transforaminal Epiduroscopic Approach for Foraminoplasty, Discography, and Annuloplasty at L4-5
Jongsun Lee

25. Right-sided Transforaminal Epiduroscopic Approach for Foraminoplasty and Discectomy and L3-4 (Transforaminal Epiduroscopic Discotomy & Laser Annuloplasty)
Jongsun Lee

26. Left-sided Transforaminal Epiduroscopic Approach for Discography, Discectomy, and Annuloplasty at L2-3
Jongsun Lee

III **Pathological Findings**

27. Accidental Contrast Injection into the Spinal Fluid Space (Classical Picture of an Accidental Contrast in Spinal Space (Lateral + A.P. Fluoroscopy Sequence) Endoscopic Video—Intrathecal Space)
Günter Schütze

28. Epiduroscopic View of Epidural Fibrosis (Epidural Fibroses (Scars) Associated with Stenoses)
Günter Schütze

Preface

While endoscopy has been widely embraced in many medical and surgical disciplines, its use has lagged in the management of spine disorders. Endoscopic technology has now advanced to the point where practitioners can access, visualize, and treat spine pathologies that were previously only accessible by open surgical means. An evolution of endoscopy has led to its use to access the spinal epidural space, called "epiduroscopy." Since many of the spine pathologic clinical symptoms result from the interaction of the neural elements and its surrounding structures in the epidural space, epiduroscopy provides a means for the diagnosis and treatment through direct visualization and focused intervention.

The intending audience for our book includes all clinicians interested in furthering their knowledge of diagnostic and treatment options available for spinal disorders, including practicing physicians, fellows, and residents. This book may also be of interests to nurses, physical therapist, chiropractors, and medical device professionals. With the detailed step-by-step guides in this book, clinicians can successfully learn and use epiduroscopic techniques in managing the spine disorders of their patients.

Epiduroscopy: Atlas of Procedures contains almost four hundred images, including photos of procedures and medical illustrations. In addition, the accompanying videos show actual cases illustrating spinal pathologies and procedures. Along with case demonstrations and different approaches, the videos include easy-to-follow, step-by-step video animations of these techniques to supplement understanding.

After beginning with an introduction to the history, the relevant anatomy, and the pathophysiology of spinal pain, the book describes the equipment, tools, and consideration for this specialized procedure. This is followed by an in-depth look into different approaches to the epidural space as well as instruments used to treat various spinal diseases presented as step-by-step procedures. The text is organized in bullet points with many illustrations, making it easy to find specific information.

This book takes current and future physicians caring for spine patients through the most up-to-date epiduroscopic spinal equipment, techniques, and skills one step at a time, allowing them to glean from the years of experience of the editors and authors.

Acknowledgments

I would like to thank Ashley Brown and Victor Lo for their tireless work, as well as the others who have contributed their energy, expertise, and talents to make this book what it is.

Contributors

Salahadin Abdi, MD, PhD
Professor and Chairman
Department of Pain Medicine
Helen Buchanan & Stanley Joseph Seeger Endowed Research Professor
The University of Texas M.D. Anderson Cancer Center
Houston, Texas

Ashley E. Brown, BS
Department of Neurosurgery
University of Texas Health Science Center at Houston
Houston, Texas

Frank Calixto, MD
Senior Resident
Department of Anesthesiology
Tulane School of Medicine
New Orleans, Louisiana

Jianguo Cheng, MD, PhD, FIPP
Vice President for Scientific Affairs
American Academy of Pain Medicine
Professor of Anesthesiology
Director
Cleveland Clinic Pain Medicine Fellowship Program
Departments of Pain Management and Neurosciences
Cleveland Clinic
Cleveland, Ohio

Jatinder S. Gill, MD
Division of Pain Medicine
Department of Anesthesiology, Critical Care and Pain Medicine
Beth Israel Deaconess Medical Center
Assistant Professor
Harvard Medical School
Boston, Massachusetts

Radhika Grandhe, MD, MBBS
Assistant Professor, Pain Management
Department of Anesthesiology
University of New Mexico
Albuquerque, New Mexico

Jae Taek Hong, MD, PhD
Professor
Department of Neurosurgery
St. Vincent's Hospital
The Catholic University of Korea
Seoul, Republic of Korea

Billy K. Huh, MD, PhD
Professor and Medical Director
Department of Pain Medicine
The University of Texas M.D. Anderson Cancer Center
Houston, Texas

Dae Hyun Jo, MD, PhD
Professor
Department of Anesthesiology and Pain Medicine
Daejeon St. Mary's Hospital
The Catholic University of Korea
Daejeon, Republic of Korea

Carrie E. Johnson, MD, PhD
Attending Pain Physician
Carolinas Pain Institute
Winston-Salem, North Carolina

Mark R. Jones, BA
Senior Medical Student
Tulane School of Medicine
Research Associate
Louisiana State University School of Medicine
New Orleans, Louisiana

Alan David Kaye, MD, PhD, DABA, DABPM, DABIPP
Professor, Program Director, Director of Pain Services, and Chairman
Department of Anesthesiology
Louisiana State University School of Medicine
New Orleans, Louisiana

Daniel H. Kim, MD, FAANS, FACS
The Nancy, Clive, and Pierce Runnells Distinguished Chair in Neuroscience
Professor, Director of Spinal Neurosurgery
Reconstructive Peripheral Nerve Surgery
Director of Microsurgical Robotic Lab
Department of Neurosurgery
University of Texas
Houston, Texas

Il Sup Kim, MD, PhD
Associate Professor
Department of Neurosurgery
St. Vincent's Hospital
The Catholic University of Korca
Seoul, Republic of Korea

Jae Do Kim, MD, PhD
Emeritus Professor
Department of Orthopaedic Surgery
Kosin University Gospel Hospital
Director
Kim's Stem Orthopaedics Clinic
Busan, Republic of Korea

Jun Young Kim, MD
Fellow
Department of Neurosurgery
St. Vincent's Hospital
The Catholic University of Korea
Seoul, Republic of Korea

Gun Woo Lee, MD
Assistant Professor
Department of Orthopaedic Surgery and Spine Center
Yeungnam University College of Medicine
Yeungnam University Hospital
Daegu, Republic of Korea

Hyung Gon Lee, MD, PhD
Department of Anesthesiology and Pain Medicine
Chonnam National University Medical School
Dong-gu, Gwangju, Republic of Korea

Jongsun Lee, MD, PhD
Director of Spine Surgery
Sewoori Spine and Joint Hospital
Daejeon, Republic of Korea

Sing-Ong Lee, MD
Attending Pain Physician
Department of Anesthesiology
Cathay General Hospital
Taipei, Taiwan

Victor P. Lo, MD, MPH
Resident
Department of Neurosurgery
The University of Texas Health Science Center at Houston
Houston, Texas

Diana Mekler, MD
Pain Medicine Fellow
Louisiana State University School of Medicine
New Orleans, Louisiana

Pedro Monteiro, MD
Neurosurgery Resident
Department of Neurosurgery
Centro Hospitalar São João
Assistant, Faculty of Medicine
University of Porto
Porto, Portugal

Jee Youn Moon, MD, PhD, FIPP, CIPS
Clinical Associate Professor
Anesthesiology and Pain Medicine
Seoul National University Hospital
Cancer Pain Center
Seoul National University Cancer Hospital
Seoul, Republic of Korea

Kent H. Nouri, MD
Assistant Professor
Department of Pain Medicine
The University of Texas M.D. Anderson Cancer Center
Houston, Texas

Chan Hong Park. MD, PhD
Director, Department of Anesthesiology and Pain Medicine
Chairman
Spine Health Wooridul Hospital
Daegu, Republic of Korea

Paulo Pereira, MD, PhD
Consultant Neurosurgeon
Head of the Spine Unit
Department of Neurosurgery
Centro Hospitalar São João
Invited Assistant Professor
Faculty of Medicine
University of Porto
Coordinator of the Spine Unit
CUF Porto Instituto
Porto, Portugal

Alissa Redko, BA
Department of Neurosurgery
University of Texas Health Science Center at Houston
Houston, Texas

Carlos J. Roldan, MD FACEP, FAAEM
Associate Professor of Emergency Medicine
Department of Emergency Medicine
The University of Texas Health Science Center at Houston
Memorial Hermann Hospital
Lyndon Baines Johnson General Hospital
Fellow of Pain Medicine
Department of Pain Medicine
The University of Texas Health Science Center at Houston
Houston, Texas

Günter Schütze, Dr. med.
Director of Anesthesiology, Critical Care Medicine, and Pain Medicine
Marien Hospital Iserlohn
Privatpraxis für Spezielle Schmerztherapie
Dortmund, Germany

Adam Shomstein, DO, MBA
Pain Medicine Fellow
Louisiana State University School of Medicine
New Orleans, Louisiana

Siddarth Thakur, MD
Resident Physician
Physical Medicine and Rehabilitation
Baylor College of Medicine
Houston, Texas

Richard D. Urman, MD, MBA
Department of Anesthesiology, Perioperative and Pain Medicine
Brigham and Women's Hospital
Associate Professor of Anesthesia
Harvard Medical School
Boston, Massachusetts

Shiraz Yazdani, MD
Assistant Professor
Department of Pain Medicine & Anesthesiology
Texas Tech University Health Sciences Center
Lubbock, Texas

YiLi Zhou, MD, PhD
Medical Director
Florida Pain and Rehabilitation Center
Courtesy Research Associate Professor
University of Florida
Lady Lake, Florida

1 The History of Epiduroscopy

Hyung Gon Lee and Salahadin Abdi

The role of endoscopy is expanding in the diagnosis and treatment of many medical conditions, but the use of this technology in spine care is still in its infancy. The technology of spinal canal endoscopy has developed slowly over the past century, initially as a rigid and later as a flexible endoscope. As such, the epiduroscopy technology with flexible optics has been used for clinical application only since the early 1990s.[1]

In 1931, Michael Burman,[2] an orthopedic surgeon at the Hospital for Joint Diseases New York, described the visualization of the spinal canal and its content. For his pioneering study, he used a rigid arthroscopic instrument with incandescent light source to visualize and study cadaver vertebral columns. It is noteworthy to point out that the diameter of the trocar in which the lamp was mounted (~3/8 inch or 9.5 mm) was greater than the average width of the spinal canal itself. Thus, the viewing lens was only in the wide areas of the spinal canal where the scope could be inserted to visualize and evaluate anatomical structures such as the dura mater, blood vessels, and the cauda equina.

Five years later, in 1936, Elias Stern[3] from Columbia University developed a spinascope, which was intended to be used in vivo to examine the contents of the spinal canal during spinal anesthesia. His vision was to use his technique to perform posterior rhizotomies for intractable pain and spastic conditions under direct visualization.

In 1937, J. Lawrence Pool,[4] an American neurosurgeon, used the myeloscope to improve the preoperative diagnostic assessment of patients with lumbar–sciatic syndrome under anesthesia. His initial challenge was bleeding, which obscured his view; thus, a clear visualization of the lumbosacral nerve roots was not possible. Fortunately, he was able to improve his technique and thus in subsequent evaluation, he was able to visualize and report about cauda equina and blood vessels, and blood flow through epidural vessels in seven volunteer patients without complications. A few years later, Pool published a summary of his experience with 400 patients.[5,6] He was able to diagnose or confirm the diagnosis of various types of spine pathologies, including neuritis, herniated nucleus pulposus, hypertrophied ligamentum flavum, primary and metastatic neoplasms, varicose vessels, and arachnoid adhesions. Unfortunately, this technique was not picked up by others and it took almost 30 years to read about this technique again. One of the main drawbacks of this technique was the fact that it was not possible to capture images during the procedures.[7]

As the era of fiberoptics emerged, Ooi et al[8,9,10,11] developed an endoscope for intradural and extradural examinations. Their device, which was inserted via the lumbar interspinous processes, was much leaner compared to the device used by their predecessors.[11] Interestingly, for the first time, the device (1.8-mm rigid scope and fiberoptics were used as a light source) enabled them to record black and white photographic images of ligamentum flavum, epidural adipose tissue, the surface of the dural sac, and the cauda equine. Most importantly, they did not report any serious complications from their initial 86 cases.[12] As the authors gained more experience and performed increased number of cases, they published an interesting article on myeloscopy and blood flow changes of the cauda equina during Lasègue's test.[13] On the other hand, even though the attempt to explore the epidural space using myeloscopy with rigid optics was unsatisfactory, it was not until the late 1980s when a small caliber flexible optics was developed for clinical application of epiduroscopy.

In 1985, Rune Blomberg[14] was the first to observe and describe the content of the epidural space using a small rigid endoscope with fiberoptic light source. He observed that the content of the epidural space varied widely in terms of fat and connective tissue. Furthermore, Blomberg and Olsson[15] performed epiduroscopies in 10 patients scheduled for partial laminectomies for herniated lumbar disks and reported that the epidural space is divided into two compartments by a dorsomedian connective tissue band. Additionally, they reported that the paramedian approach to the epidural space had less bleeding complications than the midline approach.[16]

In 1991, Shimoji et al[17] published their study using small (0.5–1.4 mm) flexible fiberoptic scopes. Thanks to the availability of video chips in the late 1980s and early 1990s, it was possible to have simultaneous video images while recording all aspects of a surgical procedure. As device technology and anesthetic techniques improved, it was possible to perform provocative nerve root tests using epiduroscopy under sedation.[17]

In the early 1990s, Saberski and Kitahata[18] popularized the clinical use of flexible fiberoptic-based epiduroscopy. They also described and used flexible, steerable epiduroscope with irrigation system. The ideal fiberoptic epiduroscope device must be easily maneuverable, have a working channel, have a lens with short focal length, and/or incorporate a mechanism that prevented tissue from obstructing the lens. Despite the progress in technology, indications for clinical use of epiduroscopy remained unclear. As far as the approach to the epidural space is concerned, the caudal approach, as described by Saberski and Kitahata[18] in 1995, seems to be safer as it reduces the risk of dural puncture. Through the caudal approach, it is possible to steer a fiberoptic without difficulty to specific pathological nerve root and deliver steroid medication under direct visualization.

In 1991, Heavner et al[19] successfully used a flexible endoscope with outside and inside diameters of 2.1 and 1.4 mm, respectively, to evaluate the epidural and subarachnoid spaces in rabbits, dogs, and in human cadavers. They also reported how to access the cervical epidural space via the lumbar approach safely.[20]

In 1995, Schutze and Kurtze[21] reported their experience using epiduroscopy in 12 patients with various pain syndromes. They were able to visualize normal and abnormal anatomy such as pronounced adhesions and fibrosis in patients with failed back surgery. They also used this technique to implant permanent epidural catheters. Furthermore, Schutze was the first to report on epiduroscopy-assisted placement of spinal cord stimulation electrode in 1996.[22] Since the mid-1990s, many providers around the world used the spinal endoscopy not only for targeted epidural steroid injection but also for performing epidural adhesiolysis using mechanical, high-volume lasers and balloon techniques.

Even though the most commonly used approach is the caudal followed by interlaminar, Hammer et al[23] in 2001 modified the technique and reported on the transforaminal approach, especially to perform a ventral epidural adhesiolysis. Later, Ruetten et al[24] expanded its application by using epiduroscopy-assisted laser therapy for postnucleotomy low back pain syndrome. Finally, over the past 10 years, there have been several articles describing the role of epiduroscopy in terms of its cost-effectiveness and safety for targeted epidural steroid injection for spinal stenosis, and failed back surgery syndrome.[25,26]

References

[1] Schütze G. Epiduroscopy: Spinal Endoscopy. Heidelberg, Germany: Springer Medizin Verlag; 2008:4

[2] Burman MS. Myeloscopy or the direct visualization of the spinal cord. J Bone Joint Surg. 1931; 13:695–696

[3] Stern EL. The spinascope: a new instrument for visualizing the spinal canal and its contents. Medical Record (NY). 1936; 143:31–32

[4] Pool JL. Direct visualization of dorsal nerve roots of the cauda equina by means of a myeloscope. Arch Neurol. 1938; 39:1308–1312

[5] Pool JL. Myeloscopy: diagnostic inspection of the cauda equina by means of an endoscope. Bull Neurol Inst.. 1938; 7:178–189

[6] Pool JL. Myeloscopy: intraspinal endoscopy. Surg Clin North Am. 1957; 37 (5):1401–1402

[7] Saberski LR, Brull SJ. Spinal and epidural endoscopy: a historical review. Yale J Biol Med. 1995; 68(1–2):7–15

[8] Ooi Y, Morisaki N. Intrathecal lumbar endoscope. Clin Orthopedic Surgery (Japan). 1969; 4:295–297

[9] Ooi Y, Satoh Y, Hirose K, Mikanagi K, Morisaki N. Myeloscopy. Acta Orthop Belg. 1978; 44(6):881–894

[10] Ooi Y, Satoh Y, Morisaki N. Myeloscopy: a preliminary report. J Jpn Orthop Assoc. 1973; 47:619–627

[11] Ooi Y, Satoh Y, Morisaki N. Myeloscopy. Int Orthop. 1977; 1:107–111

[12] Satoh Y, Hirose K, Ooi Y, et al. Myeloscopy in the diagnosis of low back pain syndrome. Proceedings of Third Congress of International Rehabilitation Medicine Association, Basel, Switzerland, July 2–9, 1978

[13] Ooi Y, Satoh Y, Inoue K, Mikanagi K, Morisaki N. Myeloscopy, with special reference to blood flow changes in the cauda equina during Lasègue's test. Int Orthop. 1981; 4(4):307–311

[14] Blomberg R. A method for epiduroscopy and spinaloscopy. Presentation of preliminary results. Acta Anaesthesiol Scand. 1985; 29(1):113–116

[15] Blomberg RG, Olsson SS. The lumbar epidural space in patients examined with epiduroscopy. Anesth Analg. 1989; 68(2):157–160

[16] Blomberg RG. Technical advantages of the paramedian approach for lumbar epidural puncture and catheter introduction. A study using epiduroscopy in autopsy subjects. Anaesthesia. 1988; 43(10):837–843

[17] Shimoji K, Fujioka H, Onodera M, et al. Observation of spinal canal and cisternae with the newly developed small-diameter, flexible fiberscopes. Anesthesiology. 1991; 75(2):341–344

[18] Saberski LR, Kitahata LM. Direct visualization of the lumbosacral epidural space through the sacral hiatus. Anesth Analg. 1995; 80(4):839–840

[19] Heavner JE, Cholkhavatia S, Kizelshteyn G. Percutaneous evaluation of the epidural and subarachnoid space with the flexible fiberscope. Reg Anesth. 1991; 15:85

[20] Heavner J, Chokhavatia K, McDaniel K, et al. Diagnostic and therapeutic maneuvers in the epidural space via a flexible endoscope [Abstract 1534]. In: Abstracts of the Seventh World Congress on Pain, Paris: Raven Press; 1993

[21] Schutze G, Kurtze H. Direct observation of the epidural space with a flexible catheter-secured epiduroscopic unit. Reg Anesth. 1995; 19:85–89

[22] Schütze G. Epiduroscopically guided percutaneous implantation of spinal cord stimulation electrodes. Management of pain, a world perspective II. In: Raj P, Erdine S, Niv D, eds. Management of Pain. Bologna, Italy: Monduzzi Editore S.p.A; 1996:301–304

[23] Hammer M, Doleys DM, Chung OY. Transforaminal ventral epidural adhesiolysis. Pain Physician. 2001; 4(3):273–279

[24] Ruetten S, Meyer O, Godolias G. Application of holmium:YAG laser in epiduroscopy: extended practicabilities in the treatment of chronic back pain syndrome. J Clin Laser Med Surg. 2002; 20(4):203–206

[25] Igarashi T, Hirabayashi Y, Seo N, Saitoh K, Fukada H, Suzuki H. Lysis of adhesions and epidural injection of steroid/local anaesthesia during epiduroscopy. Br J Anaesth. 2004; 93:181–187

[26] Avellanal M, Diaz-Reganon G. Interlaminar approach for epiduroscopy in patients with failed back surgery syndrome. Br J Anaesth. 2008; 101(2):244–249

2 Applied and Radiological Anatomy of the Epidural Space for Spinal Endoscopy

Jatinder S. Gill

2.1 Introduction

The epidural space forms an interface between the thecal sac and the bony canal. Most spine pathologies involve some alteration or distortion of this space. All spinal endoscopic interventions are either launched or conducted in this space. This space offers safe passage to endoscopic instruments. It is thus clear that precise knowledge of epidural anatomy is critical not only to understand spine pathology but also for the safe and efficient performance of spinal endoscopy. Magnetic resonance imagings (MRIs) are used to describe anatomy. In this chapter, MRI and computed tomography have been shown to correlate well with anatomical dissections.[1]

2.2 Compartments of the Epidural Space

The epidural space can be divided into three compartments: the posterior epidural space, the lateral epidural space, and the anterior epidural space.[2]

2.2.1 Posterior Epidural Space

- The posterior epidural space is a triangular space that is most spacious in the midsagittal interlaminar area (▶ Fig. 2.1, ▶ Fig. 2.2, ▶ Fig. 2.3). It is restricted superiorly and inferiorly by the proximity of the dura mater to the superior edge of the lamina (▶ Fig. 2.1, ▶ Fig. 2.2, ▶ Fig. 2.4).
- The ligamentum flavum (LF) courses from the ventral part of the superior edge of lamina to the top of the lamina below, and thus forms the posterior and posterolateral boundary of the epidural space (▶ Fig. 2.2, ▶ Fig. 2.3). The two LFs meet and generally fuse in the midline.
- The posterior epidural space is filled with fat globules that form a semifluid cushion for the thecal sac and provide elasticity for spinal motion (▶ Fig. 2.2, ▶ Fig. 2.3).
- Laterally, the space becomes obscured (potential space) as the dura mater approaches the LF (▶ Fig. 2.2, ▶ Fig. 2.3) and then becomes the lateral epidural space.
- The anterior boundary of the posterior epidural space is formed by the dura mater, which is at its thickest point in this area (▶ Fig. 2.3).

Relevance of Posterior Epidural Space to Spine Pathology and Spine Endoscopy

- The posterior epidural space is very commonly accessed for epidural analgesia as well as for therapeutic interventions such as an epidural steroid injection (▶ Fig. 2.5).
- LF hypertrophy is a common cause of lumbar spinal stenosis. Endoscopic laminectomy and flavectomy require intricate knowledge of the posterior epidural space.

2.2.2 Lateral Epidural Space

- The lateral epidural space lies in the region of the intervertebral foramen, freely communicating with the posterior and the anterior epidural space (▶ Fig. 2.2, ▶ Fig. 2.4, ▶ Fig. 2.6).
- The dura mater forms the medial boundary of the epidural space (▶ Fig. 2.7). It also forms the dural nerve sheath for the exiting nerve root. The dural sleeve lies within the medial half of the foramen (▶ Fig. 2.4, ▶ Fig. 2.8).
- The dura mater is in close proximity to the pedicles, and hence the lateral epidural space above and below the intervertebral foramen is restricted (potential space; ▶ Fig. 2.2, ▶ Fig. 2.4, ▶ Fig. 2.8).

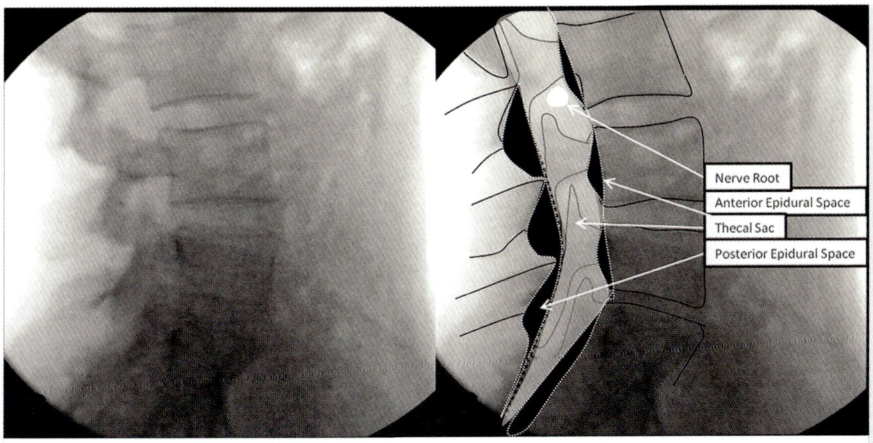

Fig. 2.1 Lateral view: The area of the epidural space (midsagittal) and the dural sac (midsagittal) is drawn, as these represent the most posterior boundaries. Given the circular nature of the epidural space, these structures will be more anterior as the sagittal plane moves laterally. The posterior epidural space is compartmentalized, as the dura mater is in close proximity to the canal at the superior edge of the lamina. The large space is visible only in the midline and is obscured paracentrally as the dura mater and the LF come together. The dura mater is adherent at the disk level, and the anterior epidural space is seen between the disks. It is much less prominent than the anterior epidural space and joins to the lateral epidural space. Note that inferiorly the thecal sac becomes smaller and more posteriorly oriented.

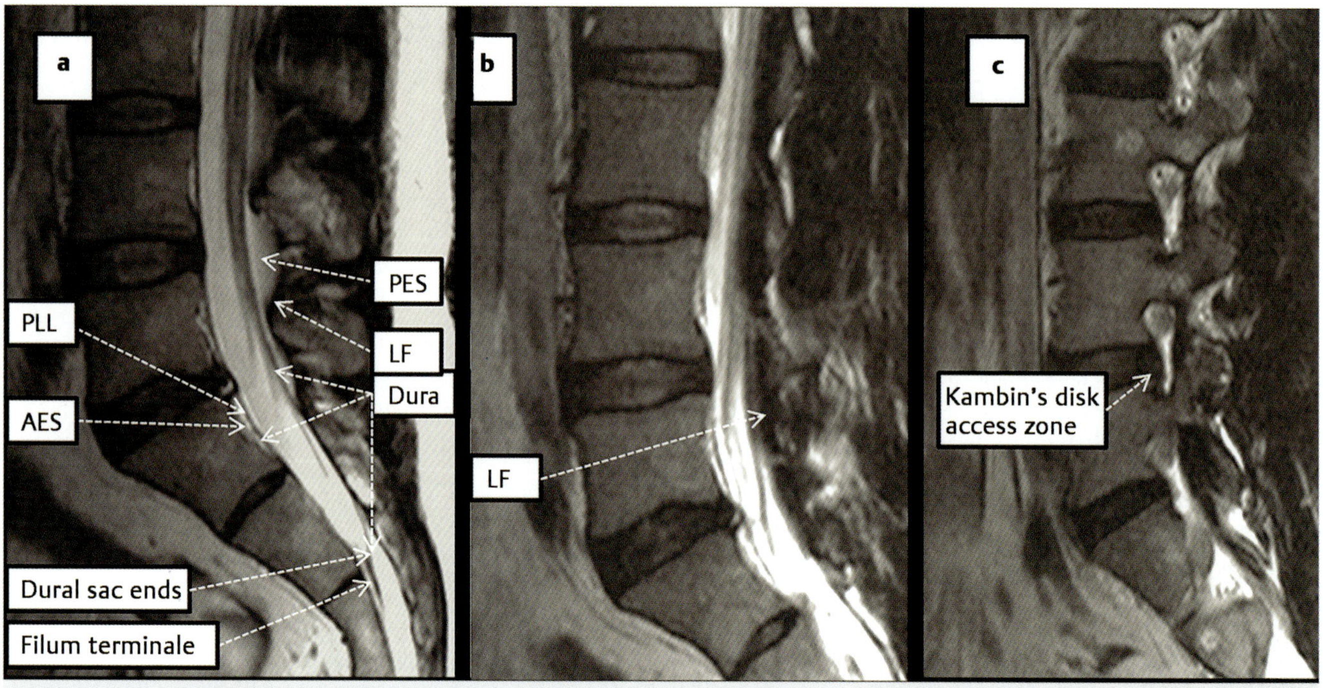

Fig. 2.2 The three sagittal views from midsagittal to foraminal show the transformation of posterior epidural space to lateral epidural space. **(a)** The LF is deficient and epidural space is substantial. **(b)** The LF is thick and the epidural space is no longer abundant. **(c)** Epidural fat surrounds the nerve root. Notice the ending of dural sac is dorsal and the caudal canal has no compartments. AES, anterior epidural space; LF, ligamentum flavum; PLL, posterior longitudinal ligament; PES, posterior epidural space.

- The lateral margin of the foramen is unrestricted and is formed by the fascia and the psoas muscle. By convention, the epidural space ends at the narrowest part of the foramen, usually midforamen.
- The contents of the lateral epidural space include the sensory and motor root, the dorsal root ganglia (DRG), venous and arterial blood vessels, and fat.
- The DRG is extradural, usually lying below the pedicle above the level of the disk.[3]

Relevance of Lateral Epidural Space to Spine Pathology and Spine Endoscopy

- Understanding the anatomy of the lateral epidural space is critical for spine endoscopy.
- Lateral epidural space is traversed during discography and intradiskal procedures.
- The neural foramen is the site for docking instruments and carrying out extradiscal interventions, such as removal of sequestrated disk, foraminoplasty, and lateral recess decompression.
- Epiduroscopy may be carried out by the transforaminal approach.

2.2.3 Anterior Epidural Space

- The posterior longitudinal ligament (PLL) is adherent to the dura mater and the disk annulus (▶ Fig. 2.3, ▶ Fig. 2.9).
- Above and below the disk, the PLL is in close proximity to the dura mater and reflects away from the vertebral body, creating the epidural space (▶ Fig. 2.1, ▶ Fig. 2.2).
- This potential space is seen mostly below L4–L5, as the thecal sac tends to translate posteriorly.[1]
- The fascial membrane extending laterally from the PLL isolates the anterior epidural space from the rest of the epidural space and also contains disk displacement.
- The anterior epidural space is a neurovascular space containing basivertebral veins and nerves.[1]
- It also harbors several pain generators including the PLL and the sensitized annulus.

Relevance of Anterior Epidural Space to Spine Pathology and Spine Endoscopy

- This is a common site for disk herniation.
- Intricate knowledge of anatomy is a must when accessing this area for the removal of herniated disk.

2.2.4 Caudal Epidural Space

- The anatomy of the caudal space is important to understand, as this is a common method of accessing the epidural space.
- Below L5, the dural sac is located more medially and posteriorly. It progressively becomes smaller as it descends inferiorly.

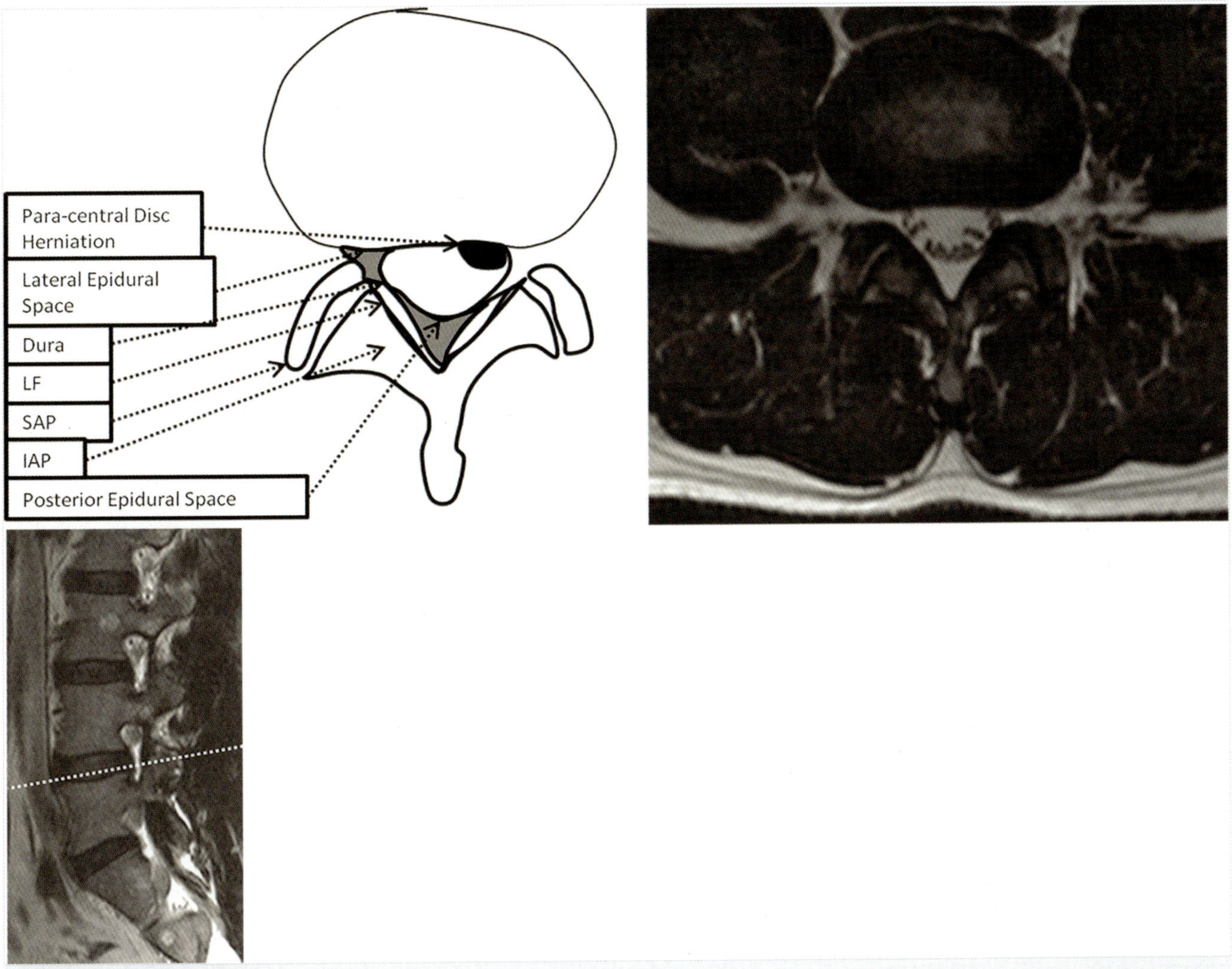

Fig. 2.3 T2 axial cut at the level of L4–L5 disk. The posterior epidural space is well formed centrally as the lamina angles posteriorly, creating a recess. On the sides, the recess is well occupied by the LF and epidural space is only a potential space. Laterally, the epidural space has fat and blood vessels with the exiting root slightly above. This region of the disk is the Kambin's triangle. This provides safe access to the disk as well as the epidural space. This is a common site of disk pathology, and disk displacement may be seen centrally, paracentrally, foraminally, and extraforaminally. Foraminal and extraforaminal disks are easy to access endoscopically but can be challenging during open surgical procedures. The posterior epidural space is obliterated because of the attachment of the disk to PLL. This area can be very sensitive to stimulation and challenging to traverse with a catheter. IAP, inferior articular process; LF, ligamentum flavum; SAP, superior articular process.

- Most common termination is upper third of S2, but dural sac can terminate as high as L5 or as low as S3.[4,5] It is extremely rare for the dural sac to extend below S3, although in one study, Tarlov cysts were found below the S3 level in 0.3% of the patients.[6]
- The space is uniform and filled with fat and nerve roots, but there are no longer any compartments (▶ Fig. 2.10, ▶ Fig. 2.11).
- It is important to recognize the fluoroscopic appearance of a spinal needle, guidewire, or contrast in the epidural space (▶ Fig. 2.12, ▶ Fig. 2.13).
- The filum terminale continues below the termination of the dural sac and attaches itself directly to the coccyx (▶ Fig. 2.3).
- The lower two sacral segments often fail to fuse in the midline and this leads to the formation of the sacrococcygeal hiatus, which is covered by the superficial and the deep sacrococcygeal ligament (▶ Fig. 2.10, ▶ Fig. 2.11a, ▶ Fig. 2.14).
- The dorsal sacrococcygeal ligament is a continuation of the LFs and attaches to the coccyx below.
- The ventral sacrococcygeal ligament is a continuation of the PLL and also attaches to the coccyx.
- Ventrally the last sacral segment directly attaches to the first coccygeal segment forming the sacrococcygeal joint.
- The sacrococcygeal hiatus may be obscured in older people.

2.3 Fluoroscopic Anatomy

Understanding fluoroscopic anatomy is the key for a spinal endoscopist. In the transforaminal approach, the needle, dilator,

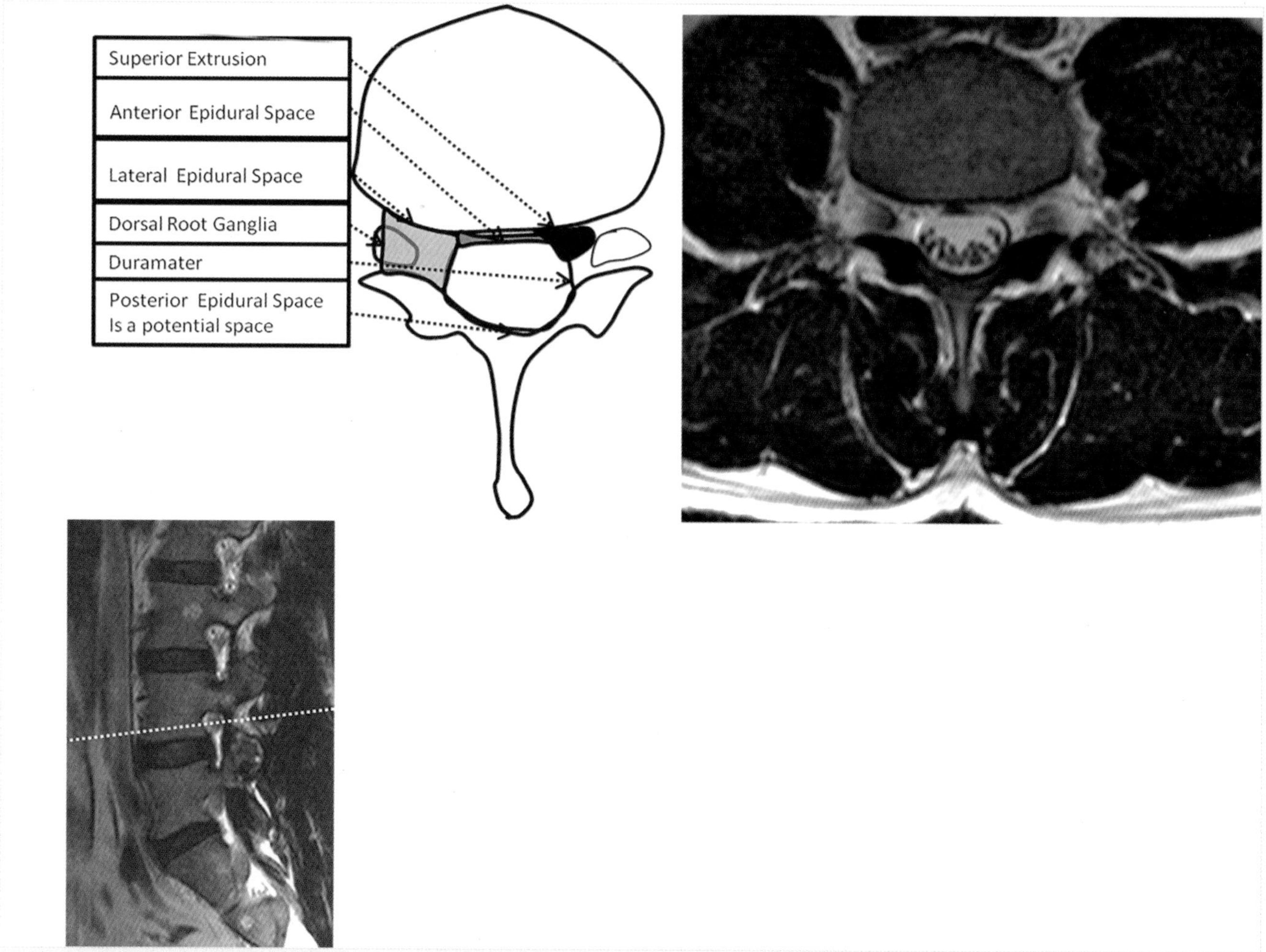

Fig. 2.4 Axial T2 cut just above the disk. The posterior epidural space is a potential space only, as the lamina projects anteriorly and in close proximity to the dura mater. The lateral epidural space is best seen along with the dorsal root ganglia. This space is bounded above and below by the pedicle, medially by the dura mater, and is continuous anteriorly and posteriorly with the anterior and posterior epidural space. This space contains nerve root, blood vessels, and fat. The lateral margin of the dural sac is medial to the midforamen (usually more medial as seen here) and to the pedicle below L5. This space is open laterally. It is an important site for cephalic sequestration and extrusion and can be accessed with special endoscopic instruments.

and the working channel are placed into the foramen and the epidural space under fluoroscopic control. This can sometimes be challenging, especially at the L5–S1 level. In the caudal approach, the needle is placed through the sacral hiatus and advanced till the S3 level. A good technique allows minimization of the radiation dose to the patient and the operator, while providing best and focused visualization. Important points include the following:

- The fluoroscope should be used in the pulse mode. This reduces radiation dose and at the same time removes motion artifact.
- Collimation should be used as much as possible. This allows for a lower dose while improving the visualization of the target anatomy.
- The image intensifier should be placed as close to the patient as possible. This provides a bigger field of view allowing for collimation while reducing the skin dose to the patient.
- For appropriate alignment of the instrument with the disk, the operative disk space should be squared, that is, the anterior and posterior margins of the end plate should be superimposed.
- Understanding the anatomy and relationship of the epiduroscope with bony, neural, and epidural structures facilitates the interpretation of fluoroscopic imaging (▶ Fig. 2.15).
- For the anteroposterior (AP) view, the spinous process should be equidistant from the pedicles so that a true AP view is obtained. This is important to know the exact mediolateral location of the instruments (▶ Fig. 2.6, ▶ Fig. 2.7, ▶ Fig. 2.16, ▶ Fig. 2.17).
- For the lateral view, the posterior vertebral body margins should overlap so that a true lateral view is obtained. This is important to know the exact AP plane of the instruments (▶ Fig. 2.1, ▶ Fig. 2.18, ▶ Fig. 2.19).

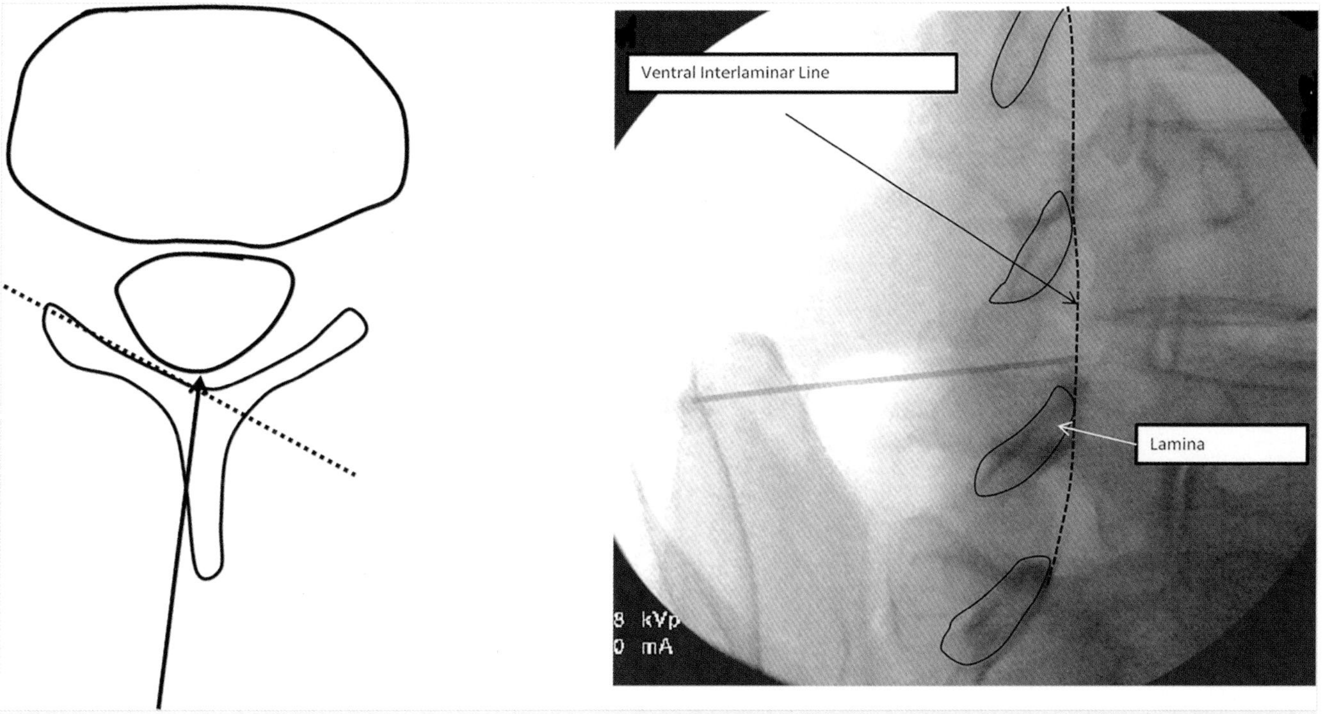

Fig. 2.5 Interlaminar access of the epidural space can be much easier and with greater precision when using the contralateral oblique view. The ventral laminar margin is visualized, and a line joining the ventral margins is the ventral interlaminar line. The epidural space is located at or just beyond that line.

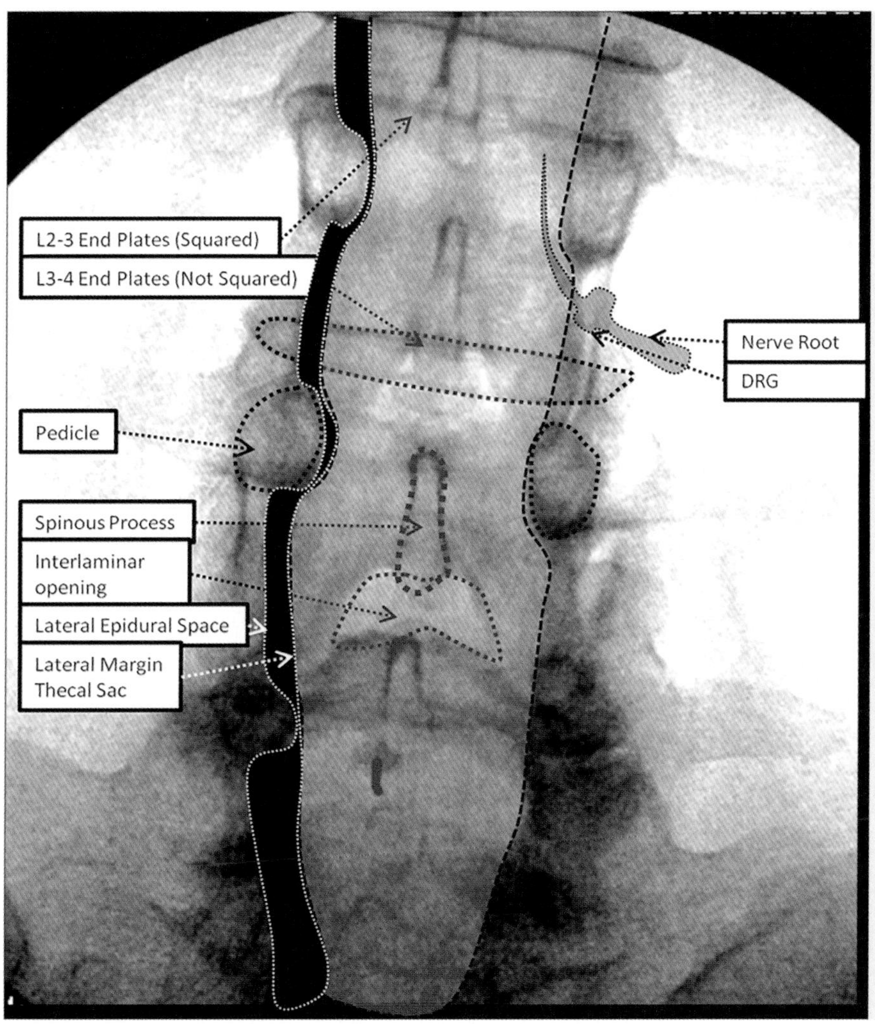

Fig. 2.6 Anteroposterior (AP) view of spinal cord with superimposed thecal sac and lateral epidural space location. Note that thecal sac flares out where the nerve root emerges, and sometimes a dural sleeve may extend further into foramen. The DRG can have variable location but commonly lies below the pedicle and is extradural. Note also that the lateral epidural space is compartmentalized by thecal sac. The narrowest part of the foramen (midforamen) by convention forms the lateral margin of the epidural space. Notice that this is a true AP view. Finally, also notice that the thecal sac starts diminishing in size below L5.

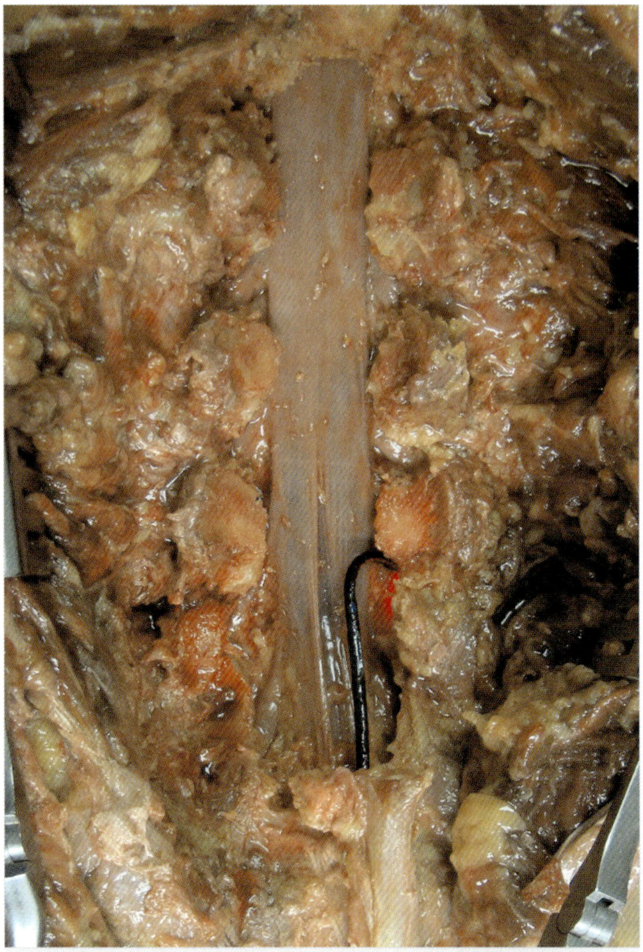

Fig. 2.7 Cadaveric dissection of the lumbar spine with an L1–L5 laminectomy and proximal sacrum unroofed. The tip of the epiduroscope can be seen in the dorsal epidural space at the right L4–L5 neural foramen.

- For the contralateral oblique view, the angle of 45 degrees has been shown to be the most accurate for predicting the posterior margin of the lumbar epidural space during interlaminar epidural access.[7]

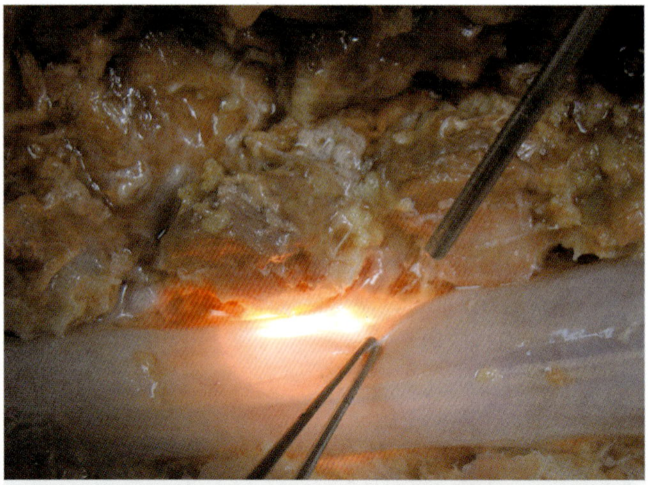

Fig. 2.8 Cadaveric dissection of the lumbar spine with an L1–L5 laminectomy and proximal sacrum unroofed. The tip of the epiduroscope can be seen in the ventral epidural space at the right L2–L3 neural foramen.

- After positioning the patient, the feasibility of obtaining the requisite views should be determined prior to prepping and draping the patient.
- For the oblique view, the angle of obliquity should be based on the location of the herniated disk. Thus, central herniations require higher degree of obliquity (from AP), whereas foraminal and extraforaminal herniations allow for a more upright (less oblique from AP) angle of placement.
- Distance of insertion from the midline or the angle of obliquity from AP can be obtained from preoperative imaging.
- Predetermined obliquity allows for a coaxial needle insertion, but AP and lateral view should be taken to gauge needle location as needed.
- The path should be carefully examined on preoperative imaging.
- The iliac crest at L5–S1 further determines the feasibility of the approach at this level.

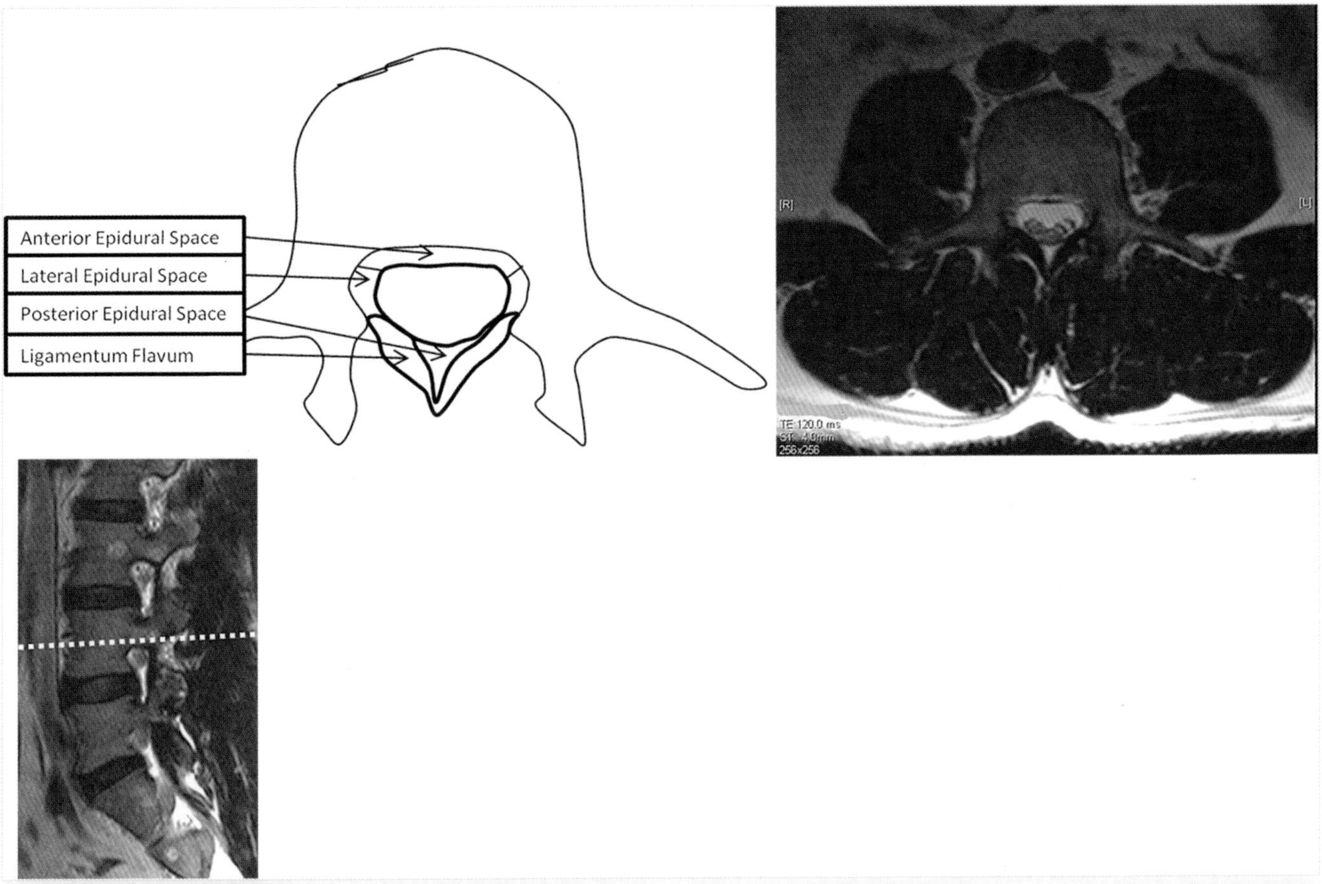

Fig. 2.9 Axial cut at the level of L3 pedicle. The posterior epidural space is present and gets obscured just below where the lamina attaches to the dura mater. The anterior epidural space is the largest, as the PLL pulls away from the vertebral body to follow the dura mater, especially below the level of the L4–L5 disk. A sequestered disk segment may lie here, but this is not the usual site for disk herniation to be observed. The basivertebral vein and nerve lie here in the posterior epidural space and penetrate the vertebral body. Lateral epidural space is small and insignificant without any important structures.

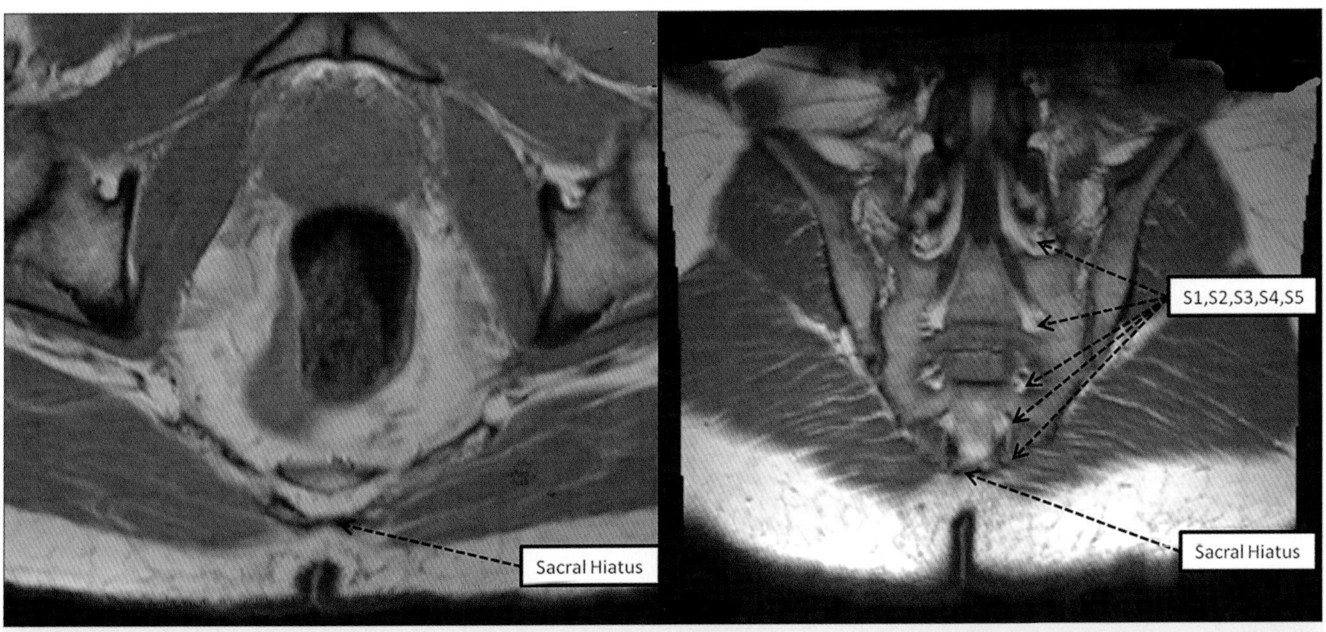

Fig. 2.10 Sacral hiatus and sacral canal at that level are seen on axial cut. Note that there is no bone dorsally. Coccygeal nerves are seen in the canal, and the coccyx is seen ventrally. On coronal section, the ventral wall of the sacrum is seen.

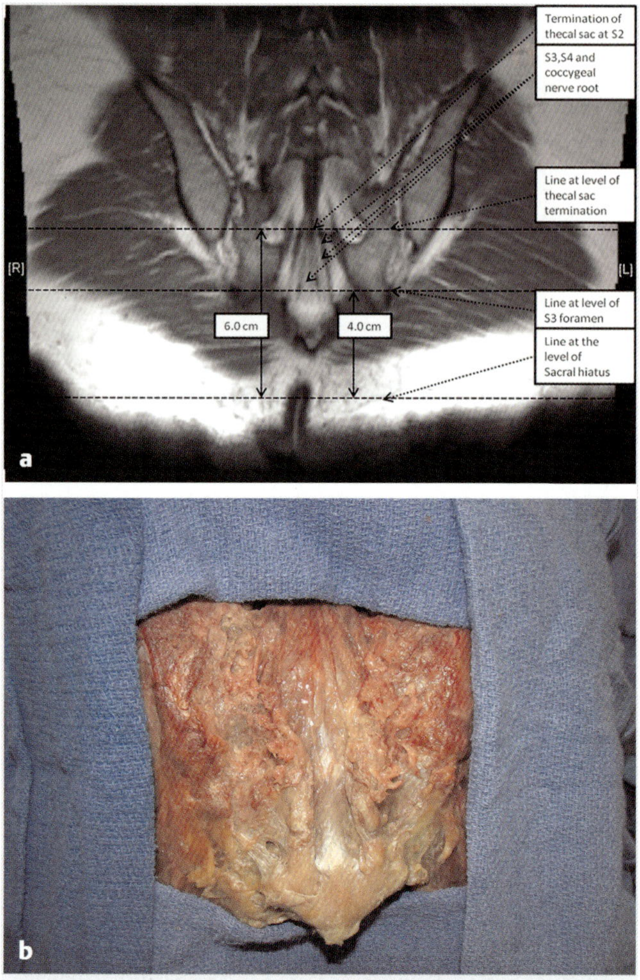

Fig. 2.11 **(a)** Coronal section through the sacrum showing the anatomy of the sacral canal. Termination of thecal sac is seen and level is determined from the corresponding axial cut. **(b)** Cadaveric dissection of the sacral hiatus. The dorsal surface of the sacrum has been removed to visualize the epidural space and sacral nerve roots.

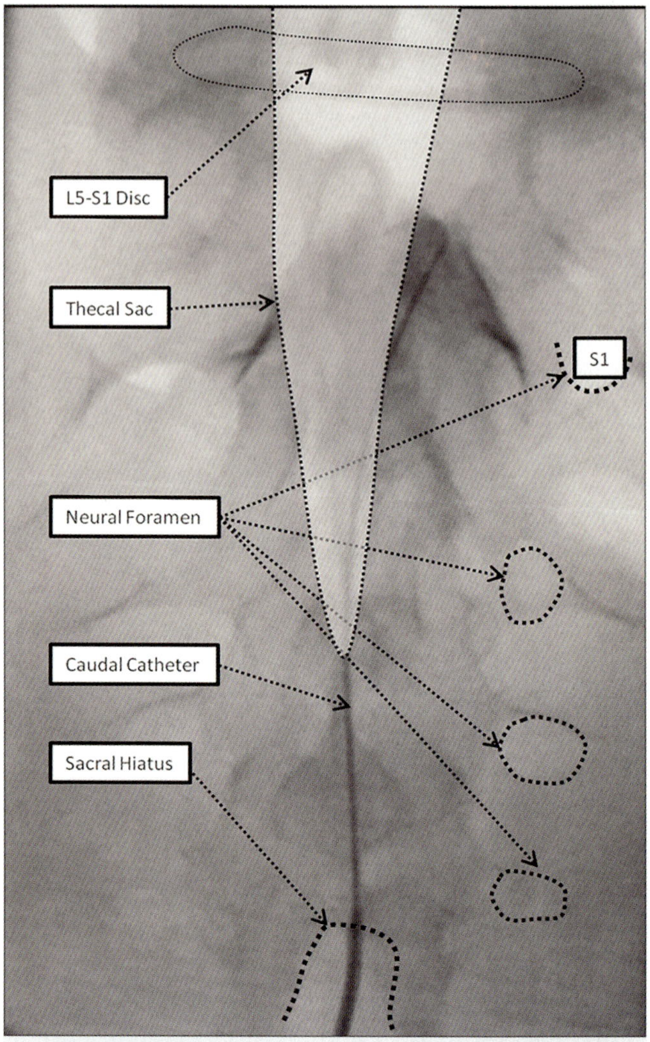

Fig. 2.12 The sacrococcygeal hiatus is formed when S4 and S5 do not fuse in the midline. The needle should not be advanced beyond S3 without checking for access. In about 1% of patients, the presence of Tarlov cysts creates the possibility of inadvertent dural access.

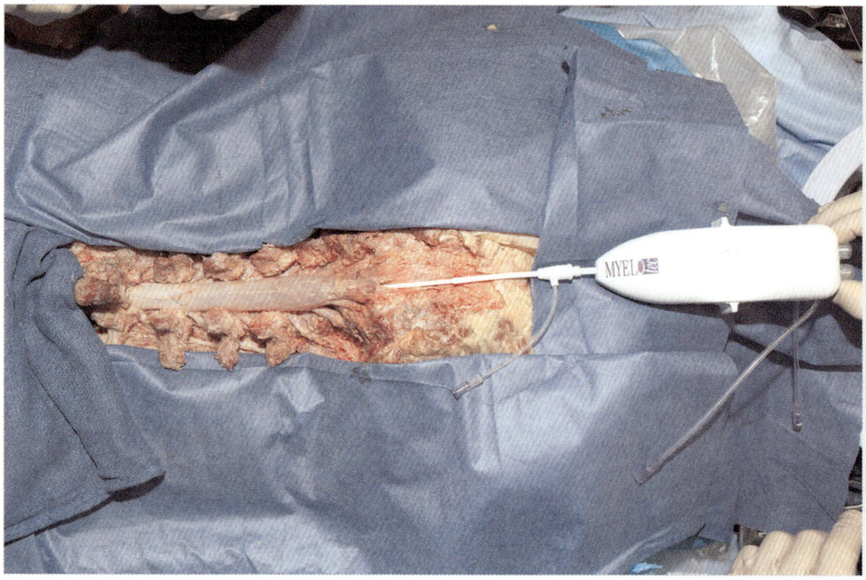

Fig. 2.13 Cadaveric dissection of the lumbosacral spine with the lamina and dorsal surface of the sacrum removed. The sheath and epiduroscope can be seen entering dorsally through the sacral hiatus. The epiduroscope can be steered into the ventral epidural space.

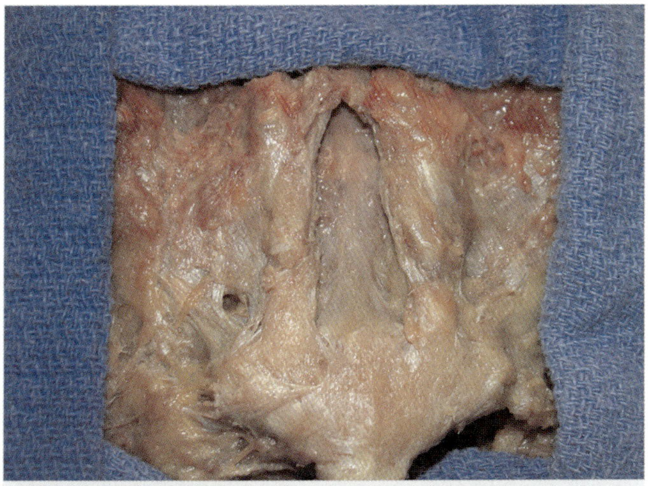

Fig. 2.14 Cadaveric dissection of the sacral hiatus. The sacrococcygeal ligament has been opened to visualize the sacral epidural space.

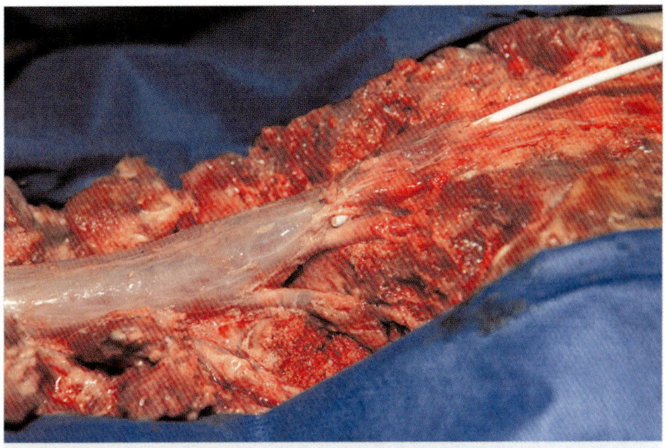

Fig. 2.15 Cadaveric dissection of the lumbosacral spine with the lamina, inferior articular facets, and dorsal surface of the sacrum removed. In this image, the epiduroscope is in the ventral epidural space, at the left S1 exiting nerve root. The location of the epiduroscope can be confirmed by fluoroscopy and direct visualization on the monitor.

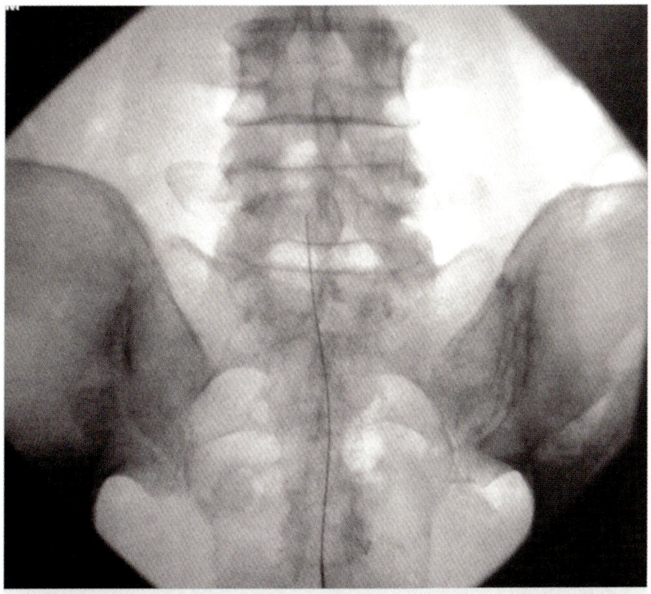

Fig. 2.16 Anteroposterior fluoroscopy demonstrating a guidewire through the sacral hiatus in the epidural space. A dilator and sheath are passed over the guidewire and visualized in the epidural space.

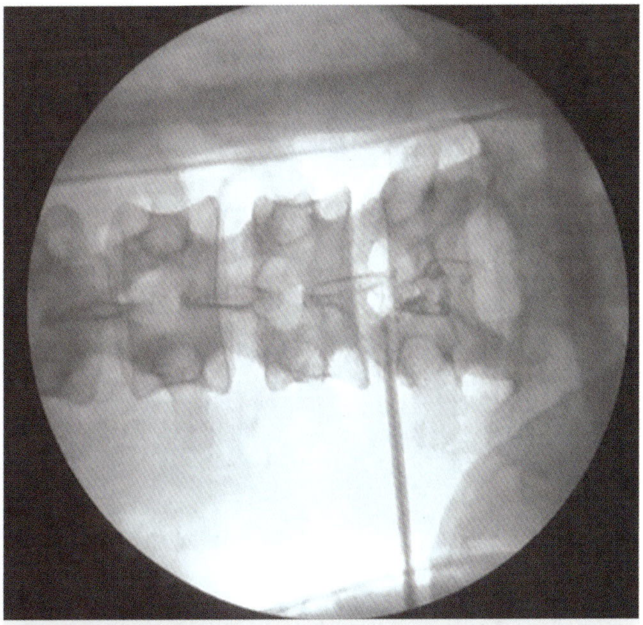

Fig. 2.17 Anteroposterior fluoroscopy demonstrating contrast in the epidural space at the foramen via a transforaminal approach.

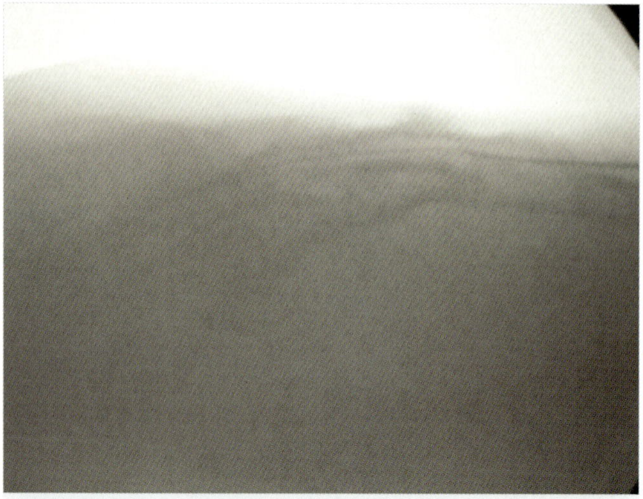

Fig. 2.18 Lateral fluoroscopy demonstrating a spinal needle through the sacral hiatus in the epidural space. A guidewire is passed through the spinal needle into the epidural space.

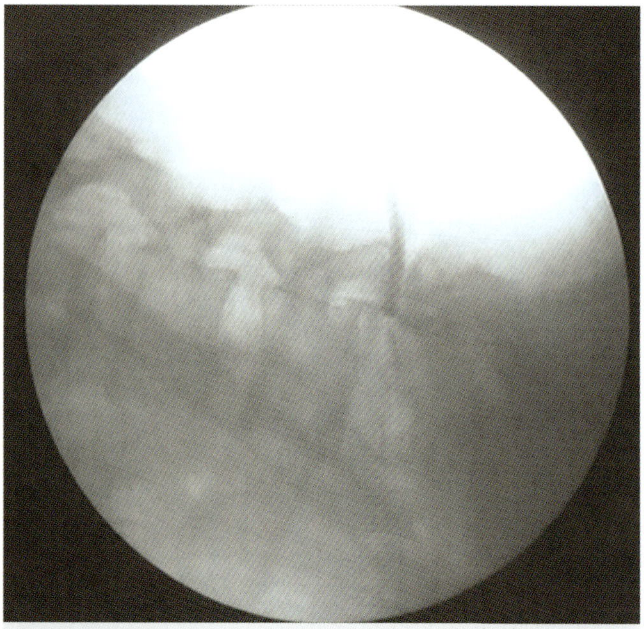

Fig. 2.19 Lateral fluoroscopy demonstrating contrast in the epidural space at the foramen via a transforaminal approach.

References

[1] Cohen MS, Wall EJ, Kerber CW, Abitbol JJ, Garfin SR. The anatomy of the cauda equina on CT scans and MRI. J Bone Joint Surg Br. 1991; 73(3):381–384

[2] Hogan QH. Lumbar epidural anatomy. A new look by cryomicrotome section. Anesthesiology. 1991; 75(5):767–775

[3] Cohen MS, Wall EJ, Brown RA, Rydevik B, Garfin SR. 1990 AcroMed Award in basic science. Cauda equina anatomy. II: extrathecal nerve roots and dorsal root ganglia. Spine. 1990; 15(12):1248–1251

[4] Binokay F, Akgul E, Bicakci K, Soyupak S, Aksungur E, Sertdemir Y. Determining the level of the dural sac tip: magnetic resonance imaging in an adult population. Acta Radiol. 2006; 47(4):397–400

[5] Phongkitkarun S, Jaovisidha S, Dhanachai M. Determination of the thecal sac ending using magnetic resonance imaging: clinical applications in craniospinal irradiation. J Med Assoc Thai. 2004; 87(11):1368–1373

[6] Senoglu N, Senoglu M, Ozkan F, Kesilmez C, Kızıldag B, Celik M. The level of termination of the dural sac by MRI and its clinical relevance in caudal epidural block in adults. Surg Radiol Anat. 2013; 35(7):579–584

[7] Gill JS, Nagda JV, Aner MM, Keel JC, Simopoulos TT. Contralateral oblique view is superior to the lateral view for lumbar epidural access. Pain Med. 2015

3 Pathophysiology of Spinal Pain and Pain Pathways

Radhika Grandhe and Jianguo Cheng

3.1 Introduction

Spinal pain or pain originating from structures of the spine is the most common chronic pain condition and has a lifetime prevalence of 54 to 80%.[1] It has significant health, economic, and societal impact. Spinal pain can arise from the myofascial layers, facet joints, intervertebral disks, vertebral bodies, or sacroiliac joint, or from compression of the nerve root or the spinal cord. In this chapter, we provide an overview of the pathophysiology of certain commonly encountered conditions in clinical practice.

3.2 Degenerative Disease of the Spine

Degenerative changes can occur in cervical, thoracic, lumbar, and lumbosacral region of the spine but are most common in the midcervical and lower lumbar regions.

- The three-joint complex between vertebral bodies consists of an intervertebral disk and two facet joints on each side. Because the three of them form "functional unit of spine," changes in one will be followed by changes in the other.[2] Disk changes usually precede changes in other structures (▶ Fig. 3.1).
- Loss of disk vascularity and hydration causes fibrosis of nucleus pulposus, and fissure formation in the annulus. Eventually, radial and circular tears occur in the annulus (▶ Fig. 3.2). Mechanical stresses such as flexion can cause herniation of the nucleus along a radial tear. The herniated disk can impinge upon the exiting nerve root in the lateral recess or far lateral zone, causing pain and weakness in nerve root distribution (▶ Fig. 3.3). Inflammatory mediators released from the herniated nucleus also contribute to pain.

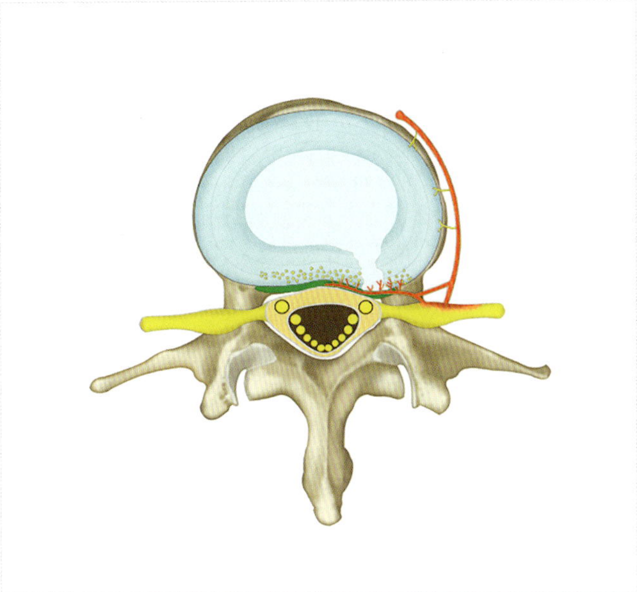

Fig. 3.2 Pain mechanism generation. Superior view at the level of the L4–L5 intervertebral disk. A posterolateral annular tear generates pain signal through the sinuvertebral nerve and gray ramus communicans. The pain signal is transmitted to the DRG.

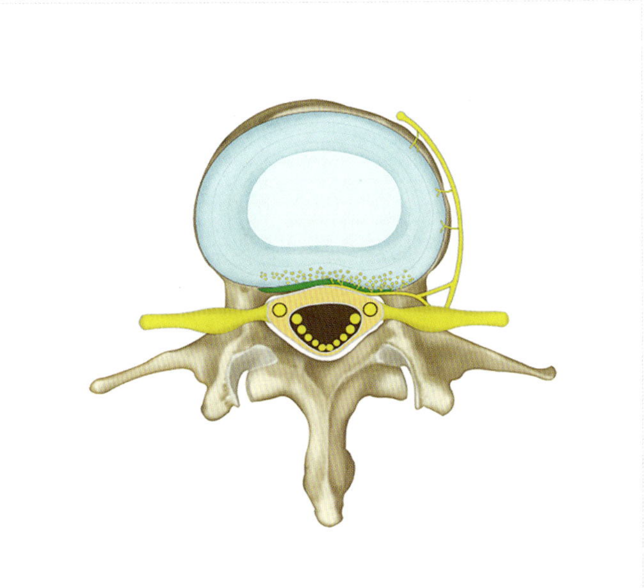

Fig. 3.1 Superior view at the level of the L4–L5 intervertebral disk. Normal relationship between the intervertebral disk and L4 exiting nerve root and L5 traversing nerve root.

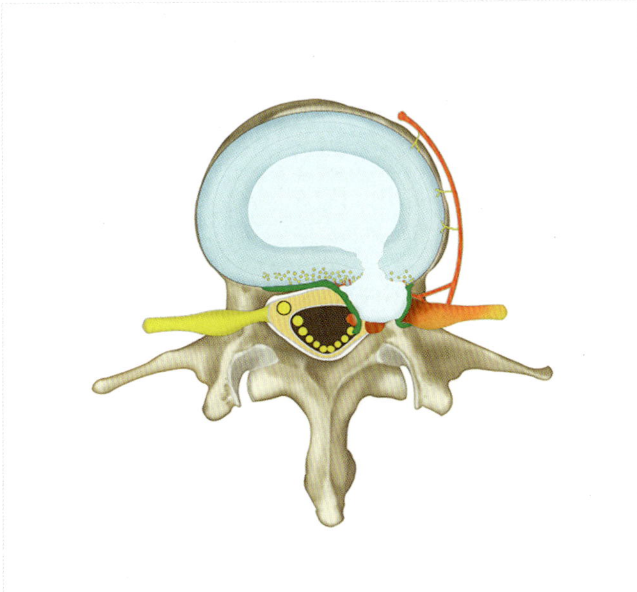

Fig. 3.3 Superior view at the level of the L4–L5 intervertebral disk. A non-contained-herniated nucleus pulposus is seen compressing the exiting L4 nerve root as well as the traversing L5 nerve.

- Progressive changes in the disk cause loss of height, compromise the disk buffering capacity, and alter biomechanical stresses on other joints in the functional unit. This leads to the formation of osteophytes at the facet joints and uncovertebral hypertrophy in the cervical region.

3.2.1 Lumbar Spinal Stenosis

- Loss of disk height, facet joint hypertrophy, buckling of ligamentum flavum, and osteophytosis result in encroachment of central canal or neural foramen, causing neurogenic claudication, myelopathy, or radiculopathy (▶ Fig. 3.4, ▶ Fig. 3.5).
- Neurogenic claudication can be attributed to either direct mechanical compression or indirect vascular insufficiency from venous congestion in the upright posture. Sitting posture or forward flexion may unbuckle the ligamentum flavum, open the canal and lateral recess, and allow adequate oxygen delivery (▶ Fig. 3.6).

3.2.2 Degenerative Spondylolisthesis

- Asymmetric degeneration of disks or facet joints can lead to varying degrees of instability and cause either anterolisthesis or retrolisthesis, resulting in stenosis or scoliosis (▶ Fig. 3.7). Once spondylolisthesis sets in, the imbalance in stress and deformity can cause self-propagating instability.

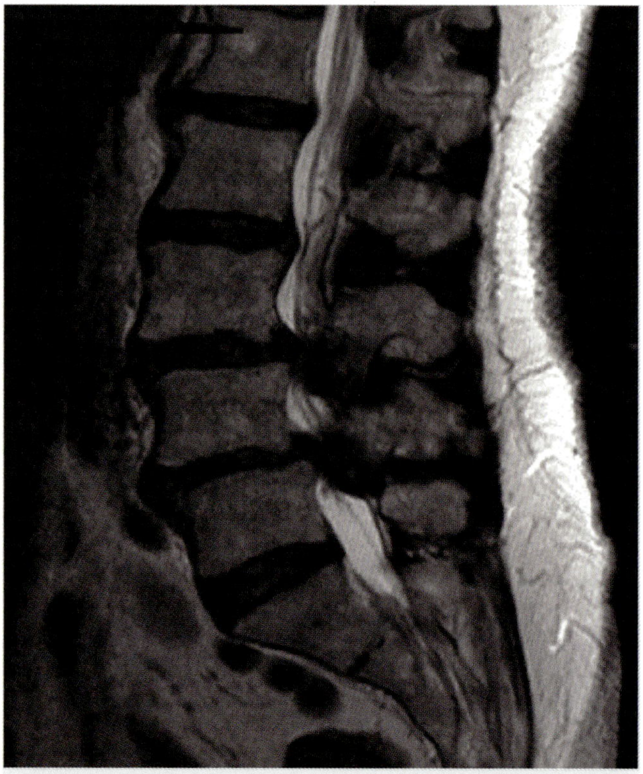

Fig. 3.4 T2-weighted sagittal magnetic resonance imaging of the lumbar spine demonstrating severe central canal stenosis at L3–L4 and L4–L5 due to ligamentum flavum hypertrophy and disk herniation.

3.2.3 Discogenic Pain

- Internal disruption of the disk (IDD) can cause axial spine pain. IDD consists of radial tears in the annulus, degradation of nucleus, and disk collapse with endplate failure in the absence of significant modifications of the external contour of disk and with no compression of neural structures.
- Within healthy disks, nerve endings are restricted to the outer 2 to 3 mm of annulus fibrosus due to high intradiskal pressures. With degeneration, nociceptor nerve fibers may reach as far inward as the nucleus and this neoinnervation may lead to discogenic pain.

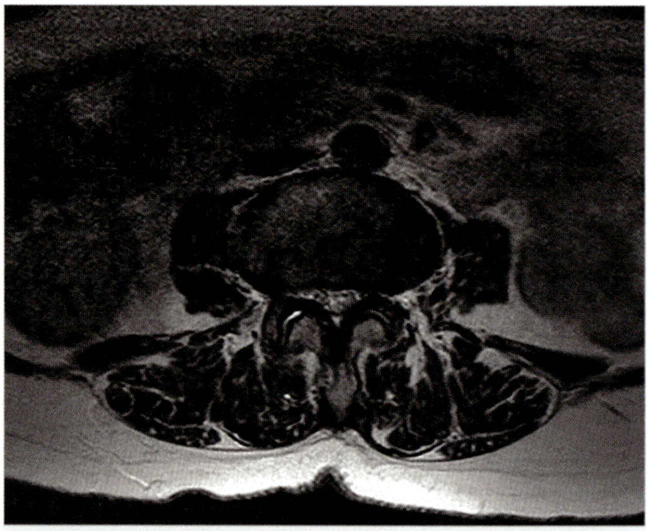

Fig. 3.5 T2-weighted axial magnetic resonance imaging of the lumbar spine at L3–L4 demonstrating severe central canal and lateral recess stenosis due to ligamentum flavum hypertrophy.

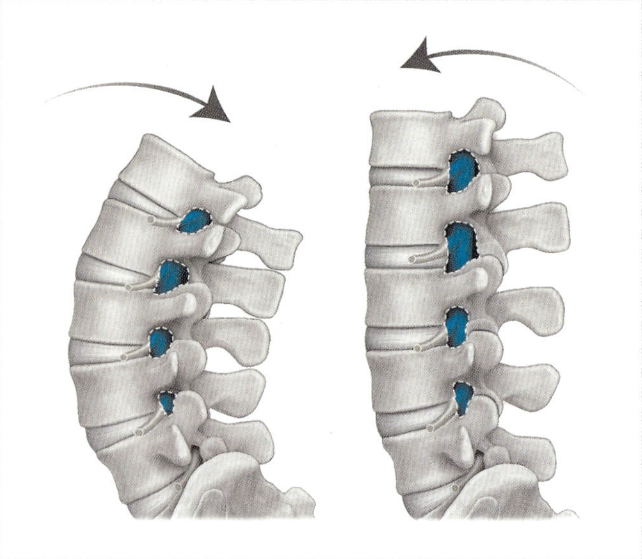

Fig. 3.6 Positional effects on the neural foramen. Extension causes narrowing of the foramen, while flexion increases the size of the foramen.

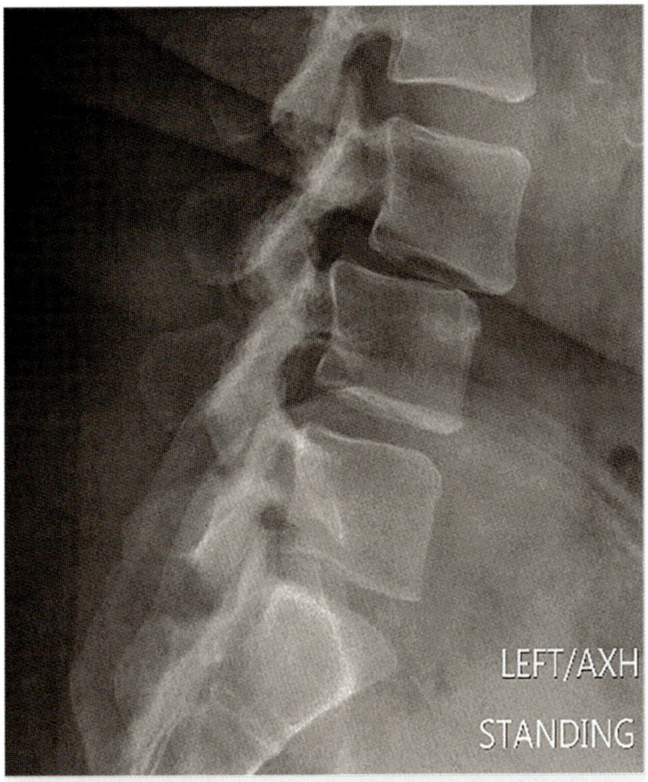

Fig. 3.7 Upright lateral lumbar spine X-ray demonstrating a grade 1 anterolisthesis of L3 on L4.

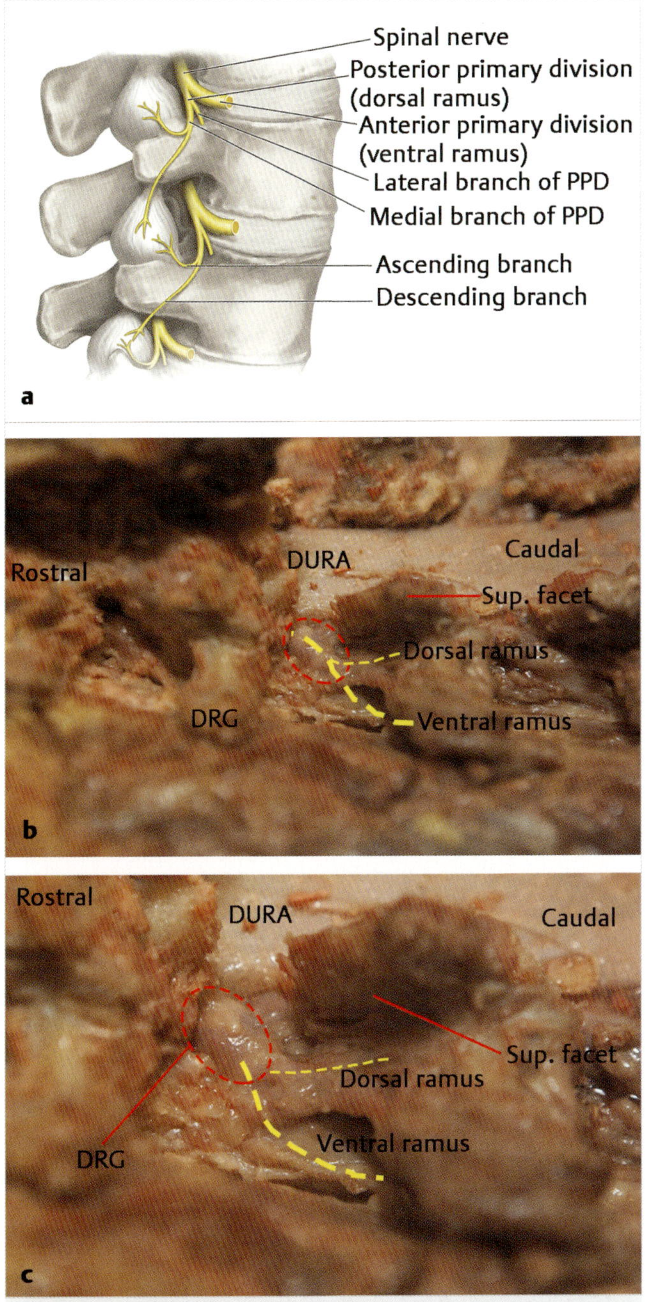

Fig. 3.8 (a) Innervation of the lumbar zygapophyseal joint. Note that each primary posterior division of the dorsal ramus divides into a lateral and medial branch. The medial branch has an ascending division that supplies the zygapophyseal joint at the corresponding vertebral level and a descending division that supplies the joint at the level below. **(b)** Cadaveric dissection of the lumbar spine with L4 and L5 laminectomy. The DRG, dorsal ramus, and ventral ramus in relation to the epidural space and facets. **(c)** A magnified view of the DRG, dorsal ramus, and ventral ramus in relation to the epidural space and facets. The thick yellow lines demonstrate the course of the ventral ramus and the thin yellow lines demonstrate the course of the dorsal ramus.

3.2.4 Facet Arthropathy

- Facet joints or zygapophyseal joints are synovial joints that allow certain ranges of motion of the spine and also support its stability by preventing excessive rotation of the spine.
- These are important pain generators in the back by virtue of abundant nociceptor innervation. Each joint is innervated by two medial branches: one from the dorsal ramus at its corresponding level and the other from the level above (▶ Fig. 3.8).[3]
- Inflammation and arthritic changes in joints, facet joint effusions/cysts, compression of nerve roots by hypertrophied joints, and entrapment of the medial branches by calcified mamilloaccessory ligament are some of the proposed mechanisms for facetogenic pain.[4]

3.2.5 Myofascial Pain Syndrome

- Myofascial pain is associated with postural derangements such as unbalanced gait, fatigue, and dystonia, as these conditions cause asymmetric tone in extensor or flexor group of paraspinal muscles. Repetitive stress on specific muscle groups can be a triggering factor.
- Other causes include spondylolisthesis with painful muscle spasms, medial branch irritation causing muscle spasm, emotional stress, and deconditioning.
- Pathophysiologic changes may begin in the motor endplate zone. Increased acetylcholine release results in sustained muscle contractions and release of vasoactive and neuroactive substances. Central sensitization with malfunction of supraspinal inhibitory pathways may also play a role.

3.2.6 Sacroiliac Joint Pain

- The sacroiliac joints are paired diarthrodial joints between the sacrum and ilium. These joints are essential for mechanical load transmission between the trunk and lower extremities and are also involved in position sense. The joint normally permits approximately 2 to 18 degrees of rotation across the transverse axis.
- Pain can originate from several joint structures such as the synovium, capsule, and multiple supporting muscles (gluteus medium, gluteus minimus, and piriformis and ligaments (ligamenta sacroiliaca), which have complex innervation from the lumbosacral plexus and dorsal rami of S1, S2, and S3.
- Common causes of sacroiliac joint pain include arthritis, trauma, inflammation, malignancies, fractures, ligamentous, enthesopathy, and spondyloarthropathies.

3.2.7 Postlaminectomy Syndrome

- Postlaminectomy syndrome is characterized by recurrent, persistent, and chronic back pain, with or without radiation to the lower extremities after surgical treatment.
- The etiology is variable and the differentials include loose disk fragments, adjacent segment disease from increased biomechanical forces on adjacent joint complex, pseudarthrosis, extensive fusion with flat back syndrome, mechanical instability caused by facet joint failure after decompression, recurrent or residual foraminal or central stenosis, pseudomeningocele, and adhesive arachnoiditis.

3.3 Nondegenerative Spinal Disease

- Primary and metastatic tumors of the spine or neural contents.
- Osteomyelitis, diskitis, and epidural abscess.
- Noninfectious disorders such as ankylosing spondylitis and psoriatic arthritis.
- Developmental or congenital conditions such as Scheuermann's disease.
- Metabolic disorders such as Paget's disease.

3.4 Referred Pain

- *Pelvis*: Endometriosis, prostatitis.
- *Renal disease*: Ureteric stones, pyelonephritis.
- *Vascular disorder*: Abdominal aortic aneurysm or dissection.
- *Gastrointestinal conditions*: pancreatitis, cholecystitis.
- *Other joints*: Sacroiliac joint, hip osteoarthritis.

3.5 Pain Pathways

How these degenerative and pathological changes of the spine lead to chronic pain is complex and variable. Here, we present a brief review of anatomical pathways for pain transmission and the common processes that are involved in transition from acute nociceptive pain to persistent pain by both peripheral and central mechanisms.

3.5.1 Anatomic Outline

- *Primary afferent neurons*: They have cell bodies that are located in the dorsal root ganglion (DRG). Their peripheral processes contain nociceptors for detection of painful stimuli and central processes extend to the spinal cord. The Aδ and C fibers are involved in nociception. The Aδ fibers mediate acute, localized pain, while the C fibers mediate slow, diffuse pain. The Aδ and C fibers have structurally and functionally diverse receptor proteins for detection of distinct pain modalities.
- *Spinal cord*: The central processes of primary afferents converge to the dorsal horn of the spinal cord. The dorsal horn is stratified into laminae, and each lamina receives input from specific class of afferent fibers. Of note, laminae I and II generally receive noxious input while laminae III and IV are responsive to innocuous stimuli. Lamina V contains neurons that are called "wide dynamic range" neurons because they response to input from both noxious and nonnoxious somatic stimuli, as well as to visceral input. These cells are considered as integrators of convergent sensory input.
- *Ascending pathways:* The second-order neurons from the dorsal horn project into thalamus and then to the somatosensory cortex through the spinothalamocortical pathway. The second-order neurons also have collateral projections to the parabrachial nucleus in the brainstem, which projects to multiple areas such as limbic system, amygdala, cingulate cortex, prefrontal cortex, and basal ganglia. The first pathway accounts for sensory–discriminative dimension and the second pathway accounts for affective–motivational dimension of pain.
- *Descending pathways*: These pathways originate from the periaqueductal gray in the midbrain and neurons in the rostral ventral medulla, project to the dorsal horn of the spinal cord, and mediate inhibitory or facilitatory influences on synaptic transmission in the spinal cord. The neurotransmitters involved in these descending modulatory systems are endogenous opioids, norepinephrine, and serotonin.

3.5.2 Molecular Mechanisms of Pain

Nociceptive Pain

- *Activation of nociceptor*: Several ion channels and metabotropic receptors such as transient receptor potential (TRP) channels, acid sensitive ion channels (ASICs), subfamily of potassium channel proteins (KCNK channels), and purine receptors are involved in transduction of thermal, chemical, and mechanical noxious stimuli into electrical energy. Noxious stimulation in the spine may be related to tissue injuries, inflammation, degeneration, and insufficiency of blood circulation (► Fig. 3.9a).
- *Transmission of pain signal*: The conduction of receptor potential from the site of nociceptor to the dorsal horn requires voltage-gated sodium and potassium channels (► Fig. 3.9a,b). Voltage-gated calcium channels enable neurotransmitter release at synaptic sites of the nervous system.
- *Clinical relevance*: Loss of function mutation of $Na_V1.7$ sodium channel would result in congenital insensitivity to pain. Gain of function mutations in $Na_V1.7$ causes erythromelalgia and paroxysmal extreme pain disorder.[5] $Na_V1.8$ is the only sodium channel that is functional at cold temperatures and is

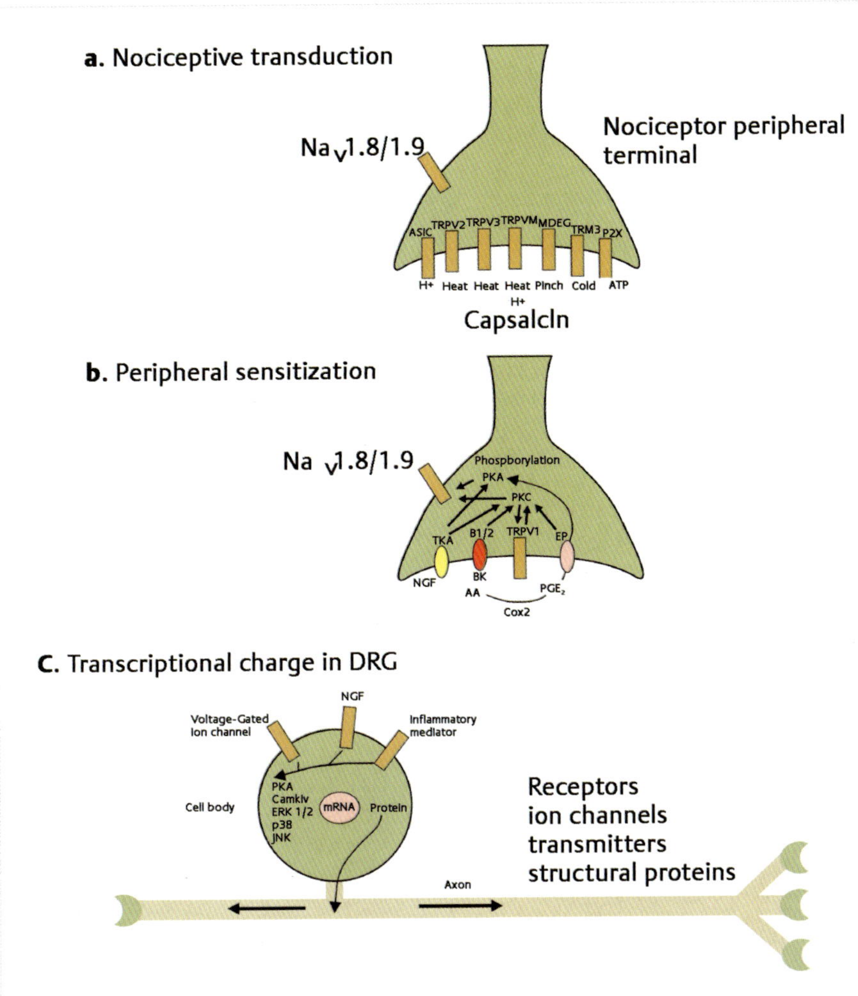

Fig. 3.9 **(a)** Noxious stimuli are transmitted through their respective receptors. **(b)** Voltage-gated sodium channels play a critical role in both nociceptive transduction and peripheral sensitization. **(c)** Transcriptional changes occur over time with continued pain signaling.

responsible for pain perception at these temperatures. Sodium channels are targets of local anesthetics and membrane stabilizers used in the treatment of pain.

- *Calcium channels*: These channels are upregulated following nerve injury and are crucial for the development of sensitization. Ziconotide, a peptide toxin, blocks N type calcium channels and is used intrathecally for chronic pain conditions. Gabapentinoid class of drugs that are extensively used for neuropathic pain act on α2δ subunit of calcium channels.

Persistent Pain

Pain that persists beyond the expected duration of initial noxious stimulus is termed "persistent" pain. Several changes occurring in the peripheral and central nervous system contribute to the development of chronic pain.

Peripheral Sensitization

Tissue injury elicits release of chemical mediators from nociceptors and nonneural cells such as mast cells, macrophages, platelets, keratinocytes, and fibroblasts. The mix of these factors at the site of injury is aptly called "inflammatory soup."[6] These mediators act on specific receptors/channels on Aδ and C nociceptors and alter the transduction and conduction properties of these primary afferents (▶ Fig. 3.9b).

TRPV1, TRPA1, and ASICs are some of the ion channels targeted by the inflammatory soup, causing a decrease in their threshold and an increase in their sensitivity to noxious stimulation. There is also spontaneous ectopic firing and increased response to regular nonnoxious and noxious stimuli. Over time, transcriptional changes occur to perpetuate pain signals (▶ Fig. 3.9c).

Central Sensitization

Pain signal transmission is augmented within the central nervous system by a variety of mechanisms such as the following (▶ Fig. 3.9b, ▶ Fig. 3.10):

- *Glutamate/NMDA receptor activation*: Upon intense stimulation or persistent injury, activated C and Aδ nociceptors release a variety of neurotransmitters including glutamate, substance P, calcitonin gene related peptide (CGRP), and adenosine triphosphate (ATP) that act on output neurons in lamina I and V of the dorsal horn. Consequently, normally silent NMDA glutamate receptors located in the output neuron become active, increase intracellular calcium, and activate a host of calcium-dependent signaling pathways and second

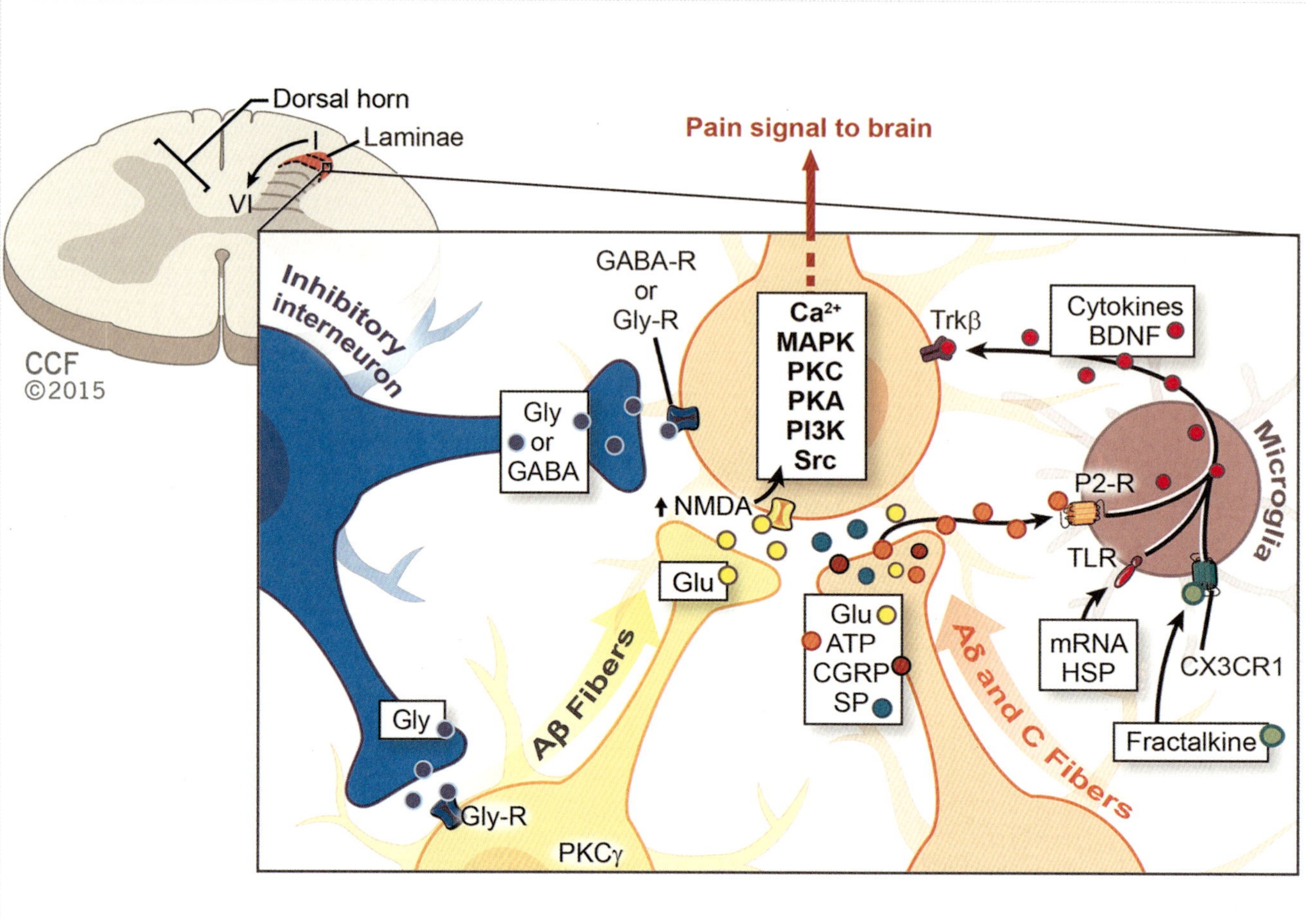

Fig. 3.10 Molecular mechanism of pain signaling to the brain. (Courtesy of the Cleveland Clinic.)

messengers including mitogen-activated protein kinase, protein kinase C, protein kinase A, phosphatidylinositol 3-kinase, and Src. This cascade of events leads to increased excitability of the output neuron, thus enhancing transmission of pain messages to the brain.

- *Loss of tonic inhibitory tone of GABAergic and glycinergic pathways*: Under normal circumstances, inhibitory interneurons (blue) maintain a tonic inhibitory effect on the lamina I output neurons by continuously releasing γ-aminobutyric acid and/or glycine. This inhibition can be lost in the setting of injury, leading to disinhibition of lamina I output neurons. In addition, decreased inhibition also allows nonnociceptive myelinated Aβ primary afferents to engage the pain transmission circuitry so that normally innocuous stimuli can be perceived as painful. This occurs, in part, through the disinhibition of excitatory PKCγ-expressing interneurons in inner lamina II.
- *Activation of microglia via purinergic and toll-like receptors*: Peripheral nerve injury promotes release of ATP and the chemokine, fractalkine, from the central endings of nociceptors. ATP and fractalkine can stimulate microglial cells through activation of purinergic P2-R receptors, CX3CR1, and toll-like receptors on microglia, which, in turn, release brain-derived neurotrophic factor (BDNF). Through activation of TrkB receptors expressed by lamina I output neurons, BDNF increases excitability of lamina I output neurons and pain transmission in response to both noxious and innocuous stimulation. In addition, activated microglia also release a host of cytokines, such as tumor necrosis factor–α, interleukin-1β and 6, and other factors that contribute to central sensitization.

References

[1] Manchikanti L, Singh V, Datta S, Cohen SP, Hirsch JA, American Society of Interventional Pain Physicians. Comprehensive review of epidemiology, scope, and impact of spinal pain. Pain Physician. 2009; 12(4):E35–E70

[2] Porensky P, Thomas NWM, Rea GL, Weinstein PR. Anatomy and pathophysiology of acquired spinal disorders. In: Benzel E, ed. Spine Surgery: Techniques, Complication Avoidance and Management. 3rd ed. Philadelphia, PA: Saunders; 2012:83-94

[3] Cohen SP, Raja SN. Pathogenesis, diagnosis, and treatment of lumbar zygapophysial (facet) joint pain. Anesthesiology. 2007; 106(3):591–614

[4] Pope J, Cheng J. Facet joint injections: cervical, lumbar and thoracic. In: Benzon H, Huntoon M and Narouze S, eds. Spinal Injections and Peripheral Nerve Blocks. 1st ed. Philadelphia, PA: Elsevier; 2010: 129-135.

[5] Basbaum AI, Bautista DM, Scherrer G, Julius D. Cellular and molecular mechanisms of pain. Cell. 2009; 139(2):267–284

[6] Wainger GB, Brenner GJ. Mechanisms of chronic pain. In: Longnecker DE, Brown DL, Newman MF, Zapol WM, eds. Anesthesiology. 2nd ed. New York, NY: McGraw-Hill; 2012.

4 Anatomic Origin of Spinal Pain and Clinical Presentations

YiLi Zhou

4.1 Introduction

Approximately 80% of adults experience back pain at some point in their lives. The following anatomic structures in the spine can cause acute and chronic back pain:

- Muscles and ligaments.
- Intervertebral disks.
- Nerve roots.
- Facet joints.
- Vertebral bodies.
- Narrowed spinal canal.
- Sacroiliac (SI) joint.
- Sacrococcygeal (SC) joint.
- Postsurgical changes, such as recurrent disk herniation, epidural abscess, adhesive arachnoiditis, and epidural scar tissues.

4.2 Pain from Spine Muscles and Ligaments

4.2.1 Muscles and Back Pain

The spine is covered by multiple layers of muscles (▶ Fig. 4.1). In the lumbar area, there are thoracolumbar fascia, erector spine, spinalis muscles, longissimus muscles, iliocostalis muscles, and quadratus lumborum.

- Twisting, improperly lifting of objects, or overstretching of the back muscles can cause muscle strains or tears in the back. Patient may experience severe pain in the lower back. Physical examination can reveal tenderness and spasm of the back muscles.
- Conditions such as myofascial pain syndrome and fibromyalgia may also cause pain, tenderness, and spasm in the spine muscles.
- Patients with other pathology in spine, such as disk herniation, radiculopathy, annular tear, facet joint syndrome, vertebral fracture, tumor, or failed back surgery syndrome (FBSS), could present with pain, tenderness, and spasm of back muscles.
- Clinicians should try to look for possible underlying pathology while treating patients with pain in the back muscles.

4.2.2 Ligaments and Back Pain

Several strong ligaments, such as anterior and posterior longitudinal ligaments, ligamentum flavum, and inter- and supraspinous ligaments, are critical to maintain the stability and flexibility of spine. Loosening, injury, hypertrophy, and calcification of these ligaments can lead to pain and instability.

- Damage of anterior longitudinal ligament in whiplash injury may lead to instability and pain.[1] Kinetic functional magnetic resonance imaging (MRI) study may help to define such ligament damage, abnormal movement, and spine instability.
- Ossification of the ligamentum flavum is mainly seen in the Japanese population, but rare cases have been reported in Caucasian and black populations. Patients may have back pain, decreased range of motion in the involved spine segment, or even paralysis.
- Loosening of anterior and posterior longitudinal ligament, ligamentum flavum, and supraspinous ligament can all contribute to the formation of spondylolisthesis.
- Hypertrophy of posterior longitudinal ligament and ligamentum flavum can cause spinal stenosis.
- Clinically, it is very seldom that ligament abnormality alone presents as a pure source of pain. More studies are needed to further define the role of the spine ligaments in pathogenesis of back pain.

4.3 Pain from Intervertebral Disks

4.3.1 Anatomy of the Intervertebral Disk

A lumbar intervertebral disk is composed of the annulus fibrosus and nucleus pulposus (▶ Fig. 4.2). Annulus fibrosus, composed of a three-dimensional network of collagen fibers, surrounds the internal gelatinous nucleus pulposus. This arrangement provides very strong bands between adjacent vertebrae while allowing some degree of movement of the vertebrae. A normal nucleus pulposus made of proteoglycans and collagens serves as a shock absorber for axial forces and as a semifluid ball during flexion, extension, rotation, and lateral flexion of the vertebral column. As the intervertebral disk ages, the number of viable cells in the inner regions of the disk decreases. Collagen cross-links formed by enzymatically mediated glycosylation diminish with age. Loss of large aggrecan and proteoglycans molecules reduces the amount of hydration in the disk, changing its shape and volume. This likely affects the ability to absorb and distribute loads effectively and makes the tissues more susceptible to mechanical failure. Common types of intervertebral disk disease include acute disk herniation, annular tears, and chronic degenerative changes such as disk dehydration, loss of height, and diffuse disk bulging.

4.3.2 Acute Disk Herniation and Lumbar Radiculopathy

Acute lumbar disk herniation (▶ Fig. 4.3) is usually a result of heavy lifting, injury, or an accident. A herniated disk may exert mechanic pressure on the nerve root(s), causing ischemia of the nerve root(s) or chemical irritation to the nerve root. Chemical substance such as phospholipase A2 leaking from the herniated disk can initiate a process of nerve root inflammation by increasing the concentration of inflammatory mediators such as prostaglandin E2 (PGE2), tumor necrosis factor (TNF)-alpha,

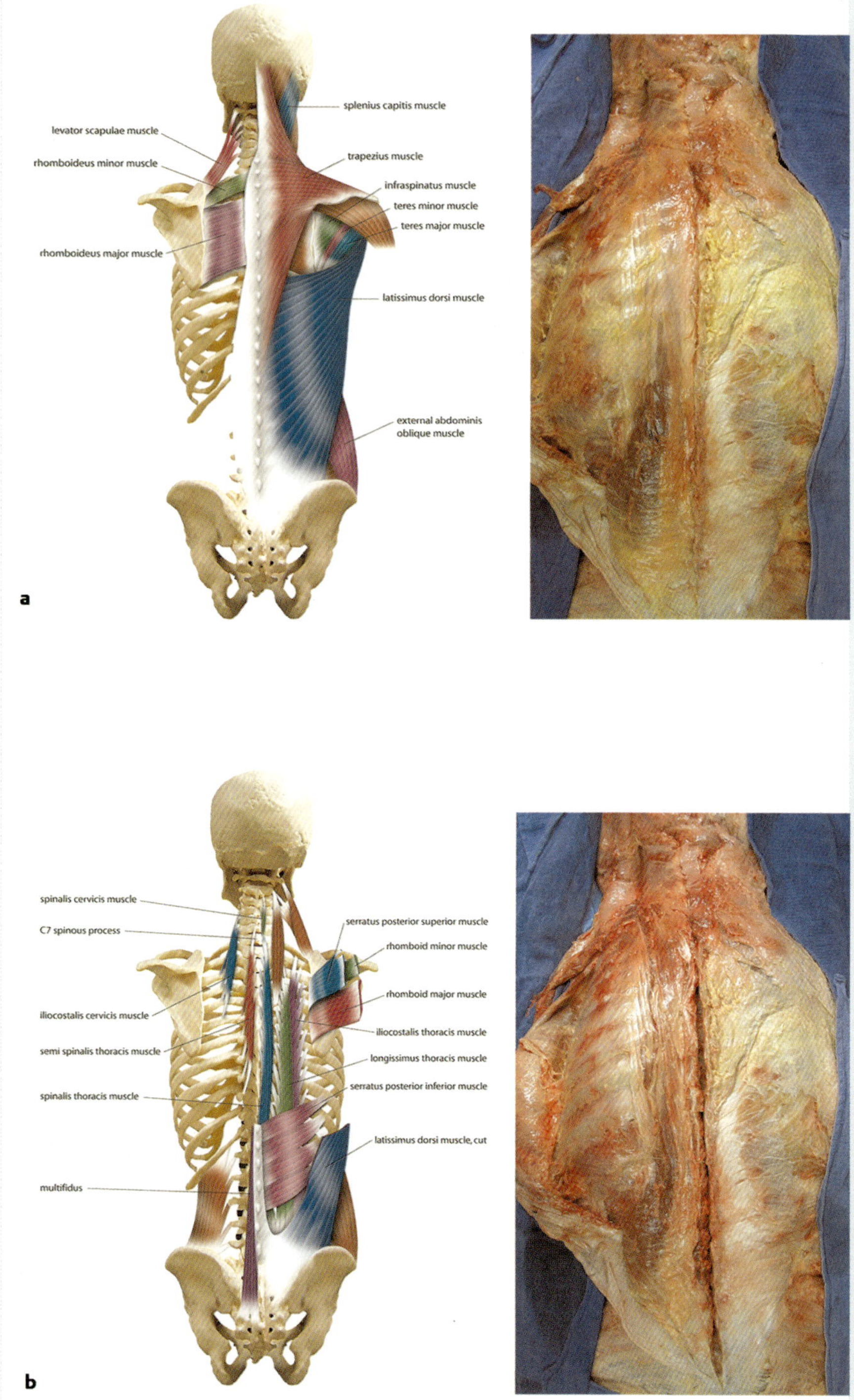

Fig. 4.1 Muscles of the back. **(a)** Superficial layer muscles: trapezius, rhomboids, levator scapulae, and latissimus dorsi muscles; cadaver dissection of superficial layer muscles. **(b)** Deep-layer muscles: erector spinae and transversospinalis muscles; cadaver dissection of deep-layer muscles.

cytokines, for example, interleukin-(IL)-1 and IL-15, and nitric oxide around the nerve root.[2]

Clinical Presentation of Acute Disk Herniation and Lumbar Radiculopathy

- History of heavy lifting or accident such as fall or auto injury.
- Acute onset of low back and leg pain.
- Radicular pain radiation depending on the level of disk herniation and nerve root irritation. An L4–L5 disk herniation may cause L4 radiculopathy. Patients could have pain in the front thigh and knee. A patient with L5–S1 disk herniation can induce L5 radiculopathy. The patient may have pain involving the anterior calf, dorsal medial foot, and big toe. A patient with S1 radiculopathy may have pain in the small toe and lateral foot.

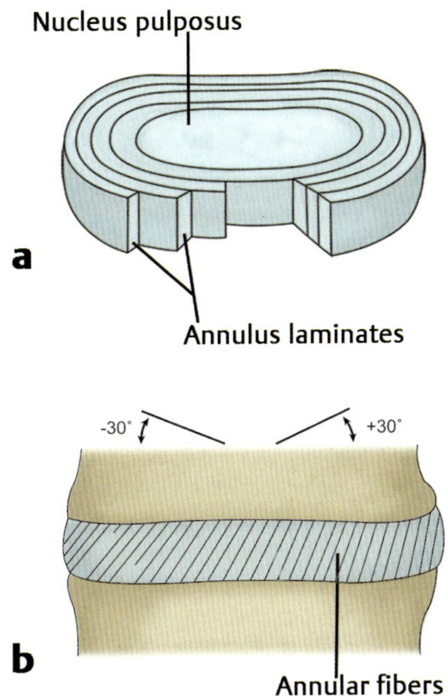

Fig. 4.2 Intervertebral disk. **(a)** The nucleus pulposus is central; annulus fibers are concentric and peripheral. **(b)** The disk lies between adjacent vertebral bodies. Each layer of annulus fibers is oriented in the same fashion, but the orientation of the fibers in adjacent layers differs by 30 degrees.

- Physical examination can find tenderness over the lumbar spine and paraspinal muscles, and straight leg raise testing on the side of irritated nerve root(s) may be positive.
- If there is mechanical damage and a loss of function of the nerve root, a neurologic examination may detect decreased sensation to pinprick and muscle weakness in the muscle innervated by the nerve root.

4.3.3 Annular Tear and Diskogenic Pain

The lumbar intervertebral disk is composed of the nucleus pulposus and annulus fibrosus (▸ Fig. 4.2). Extreme external pressure or shear force can cause rupture of the annulus fibrosus or annular tear to the disk, regardless of whether the disk has existing degenerative changes or not. There are different kinds of annular tears, including radial tear and circumferential tears. Pathogenesis of diskogenic low back pain may be different from the lumbar radicular pain caused by lumbar disk herniation. There is no direct nerve root compression or ischemia in the nerve roots in patients with back pain due to annular tear (▸ Fig. 4.4).

The mechanism of diskogenic pain due to annular tear is still unclear. However, increased expression of nerve endings sensitive to pain in the ruptured disk, epidural inflammation due to leakage of substance such as phospholipase A2, and increased concentration of the inflammatory mediators such as IL-1beta, IL-6, IL-8, PGE2, nitric oxide, monocyte chemoattractant protein-1 (MCP-1), basic fibroblast growth factor, and transforming growth factor-beta all have been involved in the pathogenesis of diskogenic pain.[3]

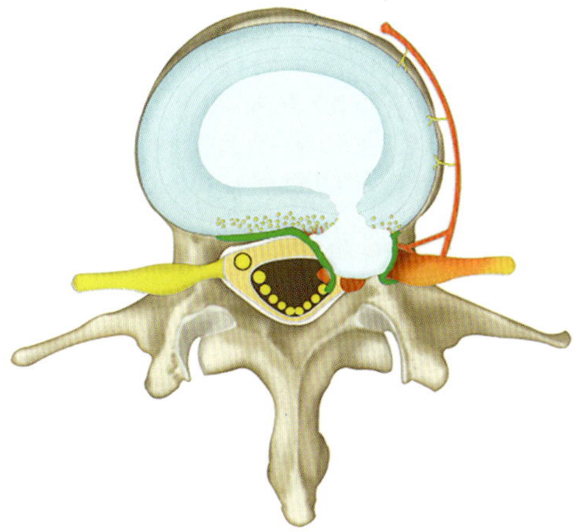

Fig. 4.3 Disk herniation. Superior view at the level of the L4–L5 intervertebral disk. A noncontained herniated nucleus pulposus is seen compressing the exiting L4 nerve root as well as the traversing L5 nerve.

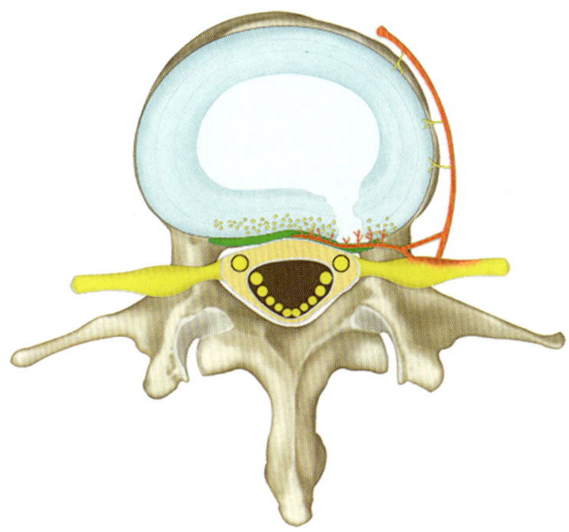

Fig. 4.4 Annular tear pain mechanism generation. Superior view at the level of the L4–L5 intervertebral disk. A posterolateral annular tear generates pain signal through the sinuvertebral nerve and gray ramus communicans. The pain signal is transmitted to the dorsal root ganglion.

Clinical Features of Lumbar Diskogenic Pain due to Annular Tear

- History of trauma with a description of their pain as burning or aching.
- Axial back pain, with less radicular symptoms.

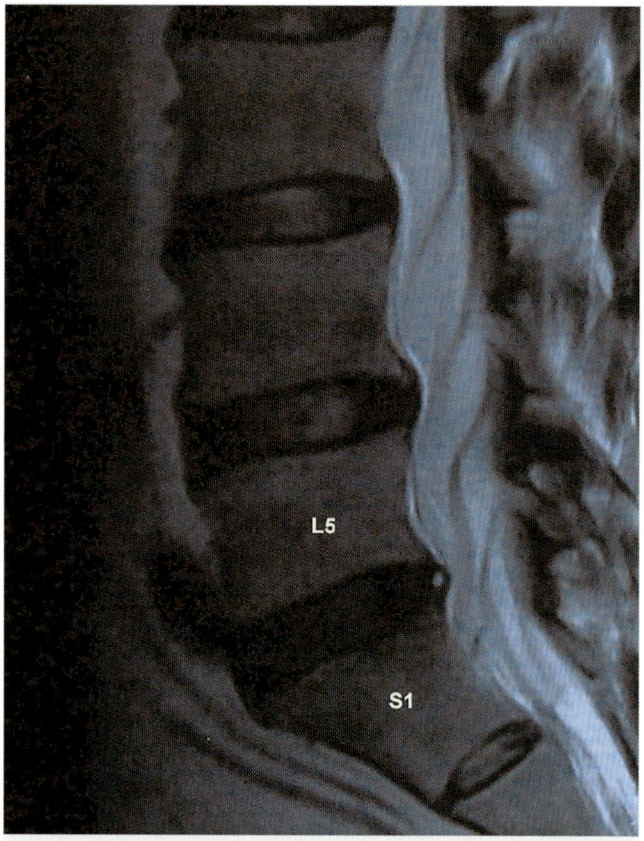

Fig. 4.5 High-intensity zone in the posterior part of the L5–S1 disk on a T2-weighted MR image.

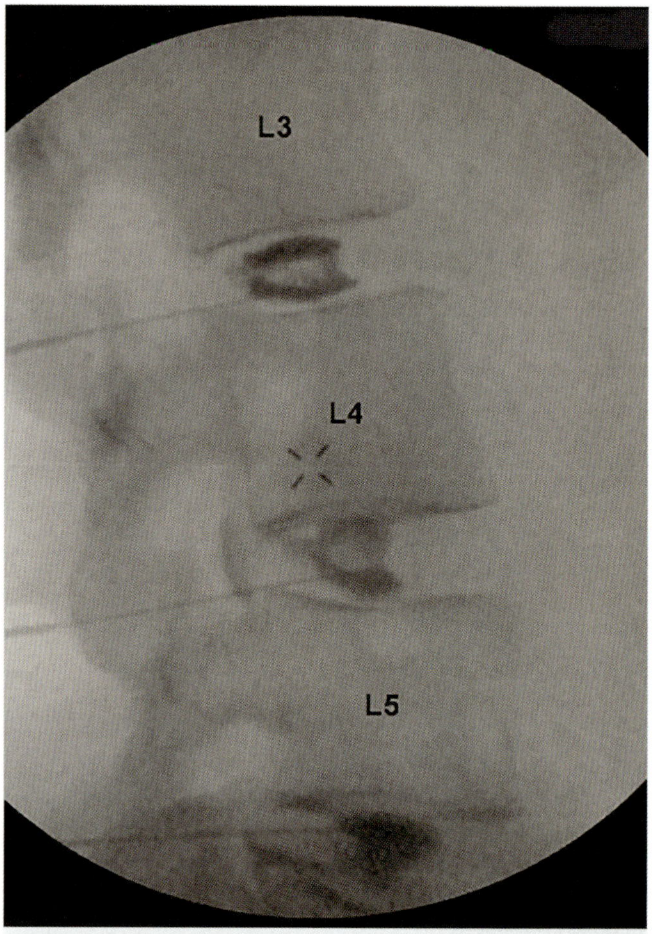

Fig. 4.6 Lumbar diskogram shows an X-ray picture of lumber diskography. There is a leakage of contrast at both L4–L5 and L5–S1 disks, indicating an annular tear.

- Sitting intolerance: The patients may have more pain when they sit for a period of time.
- Tenderness over the lumbar spine with muscle spasm. Straight leg raising test may be negative.
- No neurologic deficits.
- MRI may find a high-intensity zone in the posterior part of the disk (▸ Fig. 4.5) with approximately 50% of sensitivity and specificity.
- Definitive diagnosis of diskogenic pain with annular tear has to be made with a diskography (▸ Fig. 4.6), confirming a positive annular tear with concordant pain when the contrast is injected.

4.3.4 Chronic Degenerative Changes of the Lumbar Disks

Degenerative changes of lumbar disks such as loss of disk height, disk dehydration, and diffuse disk bulging are commonly seen in senior population. Almost 100% of normal population who are 70 years old or older have some degree of degenerative changes. ▸ Fig. 4.7 shows a typical lumbar spine MRI with diffuse degenerative changes including dehydration, bulging, and loss of heights of the lumbar intervertebral disks in a senior citizen. However, less than 30% of senior population have lower back pain, meaning that there is a 70% or more chance that these degenerative changes are not directly causing pain. In patients with low back pain, disk dehydration, loss of disk heights, and diffuse annular bulging, there are often other changes in the spine as well that may cause pain. It is still unclear whether these degenerative anatomic changes alone cause pain. When a patient with such a diffuse degenerative change starts to have pain, sometimes it is very difficult to anatomically localize the source of pain.

4.4 Pain from Lumbar Facet Joints

Facet joints or zygapophyseal joints are a set of synovial joints between the articular processes of two adjacent vertebrae (▸ Fig. 4.8). The biomechanical function of each pair of facet joints is to guide and limit the movement of the spinal motion segment. In the lumbar spine, the function of facet joint is to protect the motion segment from anterior shear forces, excessive rotation, and flexion. Degenerative changes, osteoarthritis, dislocation, fracture, injury, and instability from trauma and surgery can disrupt the function of lumbar facet joints and cause chronic low back pain or pain due to lumbar facet joint syndrome. It has been reported that lumbar facet joint syndrome accounts for more than 30% of cases of chronic low back pain.

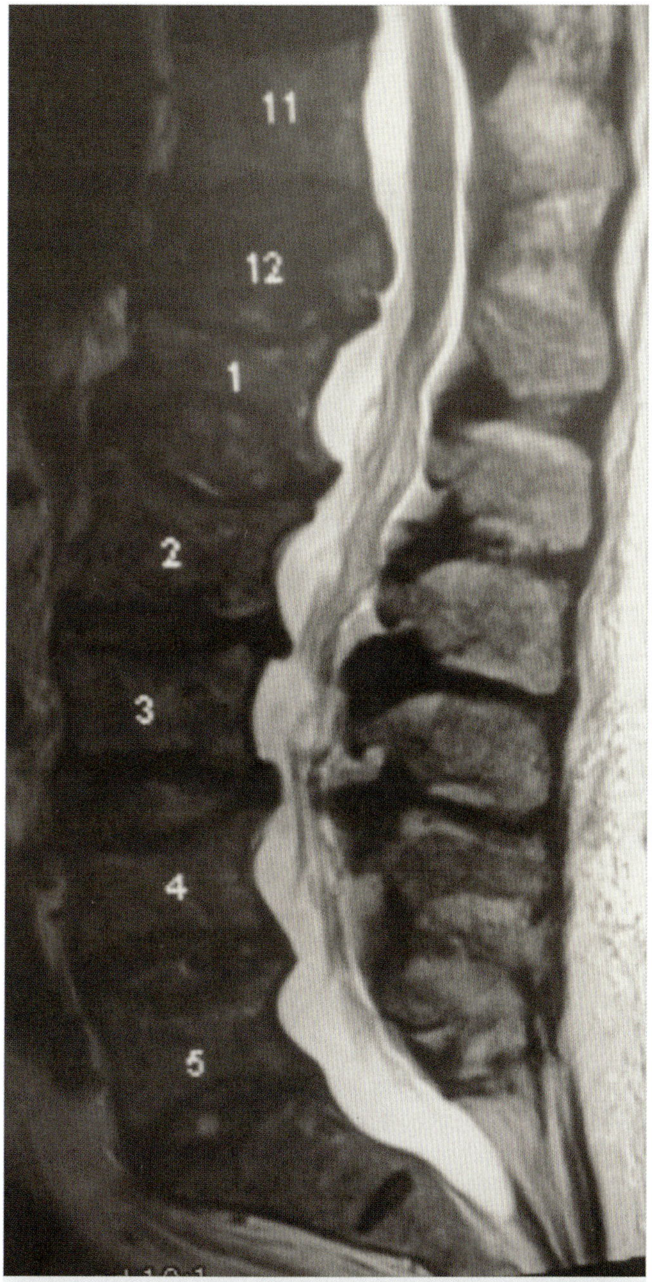

Fig. 4.7 Lumbar degenerative changes. T2-weighted MRI of the lumbar spine of a 74-year-old patient. It shows diffuse degenerative changes in the lumbar and thoracic spine. There are multiple chronic compression fractures at T12 and L1 vertebrae with retropulsion (T12). All the lumbar intervertebral disks are dark and dehydrated. There are multilevel disk bulging and osteophytes at T12–L1, L1–L2, L2–L3, L3–L4, and L4–L5. There is moderate to severe spinal stenosis at L2–L3, L3–L4, and L4–L5, and mild spinal stenosis at T12–L1 and L1–L2.

4.4.1 Clinical Features of Facet Joint Syndrome

- Some patients may have a history of injury such as being in an auto accident. Facet joint syndrome due to osteoarthritis may start slowly without a history of injury.

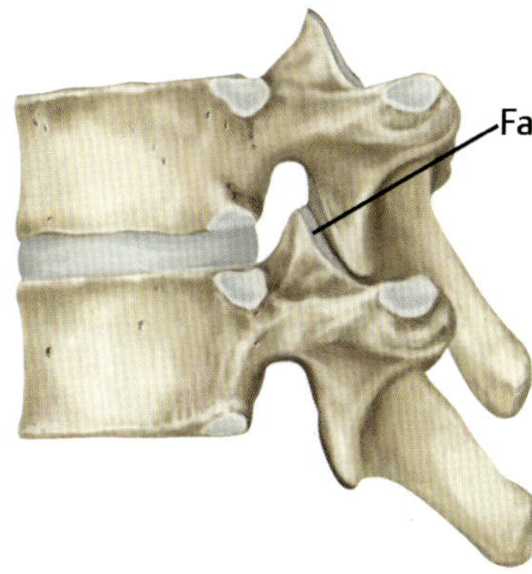

Fig. 4.8 Facet joint. Synovial joints between the articular processes of two adjacent vertebrae.

- Pain is mainly axial, without radicular symptoms.
- Patients may have tenderness over lumbar facet joints. Lateral rotation and extension of the lumbar spine (facet joint loading test) may increase the pain.
- If pain is purely from facet joints, the patient should have no neurologic deficits, such as sensory loss or leg weakness, and straight leg raising test should be negative.
- MRI and computed tomography (CT) have a very high false-positive rate for facet joint syndrome. In elderly patients, more than 90% have MRI/CT finding of facet joint hypertrophy or osteoarthritis. However, only less than 30% of these elderly patients have low back pain.
- A diagnosis of facet joint syndrome should not be made based on the MRI or CT findings. Rather, a final diagnosis of facet joint syndrome should be made by more than 50% of pain decrease after medial branch block.

4.5 Pain from Vertebral Bodies

Common pathologic changes in the vertebral bodies include compression fractures, primary or metastatic tumors, and osteomyelitis.

4.5.1 Compression Fractures of Vertebral Body

A compression fracture is a collapse of a vertebral body (▶ Fig. 4.9). It may be caused by trauma or a weakening of the vertebral body. This weakening of vertebra is often due to osteoporosis, primary or metastatic tumors, and infection. The most common cause of vertebral fracture is osteoporosis. Worldwide, osteoporosis causes more than 8.9 million fractures annually, resulting in an osteoporotic fracture every 3 seconds.

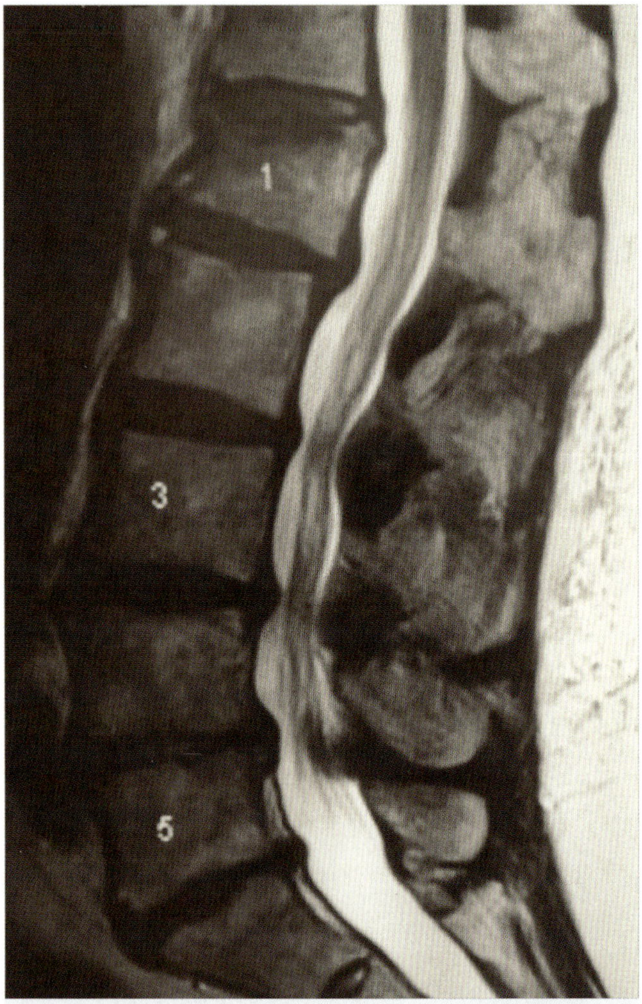

Fig. 4.9 Acute L1 compression fracture. T2-weighted MRI shows an acute L1 compression fracture with mild edema and retropulsion.

Clinical Presentation

- Acute compression fracture due to trauma is usually very painful. Patients may describe their pain as sharp and stabbing. The patient may have severe tenderness over the fractured vertebra and severe muscle spasm.
- Patients with osteoporotic compression fracture may or may not have acute pain. It is unclear what percentage of osteoporotic compression fracture is painful.
- In addition to pain, vertebral fractures can lead to loss of height, increased kyphosis, deformity, immobility, longer hospital stays, and even reduced pulmonary function. It can also significantly affect activities of daily living and cause loss of self-esteem, distorted body image, and depression.
- Diagnosis of compression fracture can be made by plain X-ray, CT, or MRI. MRI is most helpful to differentiate if the fracture is new or old. A new fracture can have edema with increased signal in T2-weighted image (▸ Fig. 4.9) and low-intensity signal in T1-weighted images.[4]
- Due to the fact that many chronic and even acute compression fractures are not painful, clinicians should be aware that an MRI finding of chronic and even acute compression may not necessarily be the source of pain.

4.5.2 Primary or Metastatic Tumors

Hemangioma

Hemangioma is the most common benign primary tumor in the vertebral bodies. It may be found in up to 26.9% of patients in MRI study, more common in females (30%) than males (23%).[5] It is reported that in 0.9 to 1.2% of patients, back pain can be caused by hemangioma.[6] The clinical presentation of hemangioma is as follows:

- The majority of patients with vertebral hemangioma are asymptomatic. No treatment is needed.
- Less than 0.9 to 1.2% of patients with hemangioma may have severe mechanical back pain. Physical exertion, collapse of the vertebral body, and encroachment into the neural canal are some of the classic causes of pain. Radicular pain may occur when the hemangioma extends into the spinal canal or neural foramen.
- Physical exam can reveal tenderness over the affected vertebral body.
- MRI and biopsy are the golden standard of diagnosis (▸ Fig. 4.10).
- Pain could be relieved with vertebroplasty or kyphoplasty.

Metastatic Tumor in the Vertebral Column

- Approximately 5 to 10% of cancer patients may develop spinal metastases. Back pain is the most common presenting symptom in patients with metastatic tumor to the vertebral column.
- There are two distinct types of back pain in patients with metastatic spine tumors: tumor-related and mechanical.
- Tumor-related pain is often nocturnal or occurs early in the morning. Pain improves with activity during the day. This pain may be caused by inflammatory mediators or tumor stretching the periosteum of the vertebral body and responds to administration of low-dose steroids.
- Mechanical pain could be the result of a structural abnormality of the spine, such as a pathologic compression fracture. This pain is related to movement and may be exacerbated by sitting, standing, or doing any activity that increases the axial load on the spine.
- Radiculopathy can occur when nerve roots are compressed, infiltrated by metastatic tumor, or chemically irritated. Radiculopathy in the cervical or lumbar spine can cause pain or weakness in the upper or lower extremity, respectively. In the thoracic spine, radiculopathy occurs as a bandlike pain at a segmental level.
- Compression or direct infiltration of spinal cord can lead to myelopathy. Patients may have progressive weakness, loss of pain and temperature below the level of the spinal cord compression, hyperreflexia, positive Babinski sign, and loss of control of bowel and bladder function.
- MRI is the most sensitive and specific modality for imaging spinal metastases. Tumor may show hypointensity on T1-weighted image and hyperintensity relative to marrow on standard T2-weighted imagines.

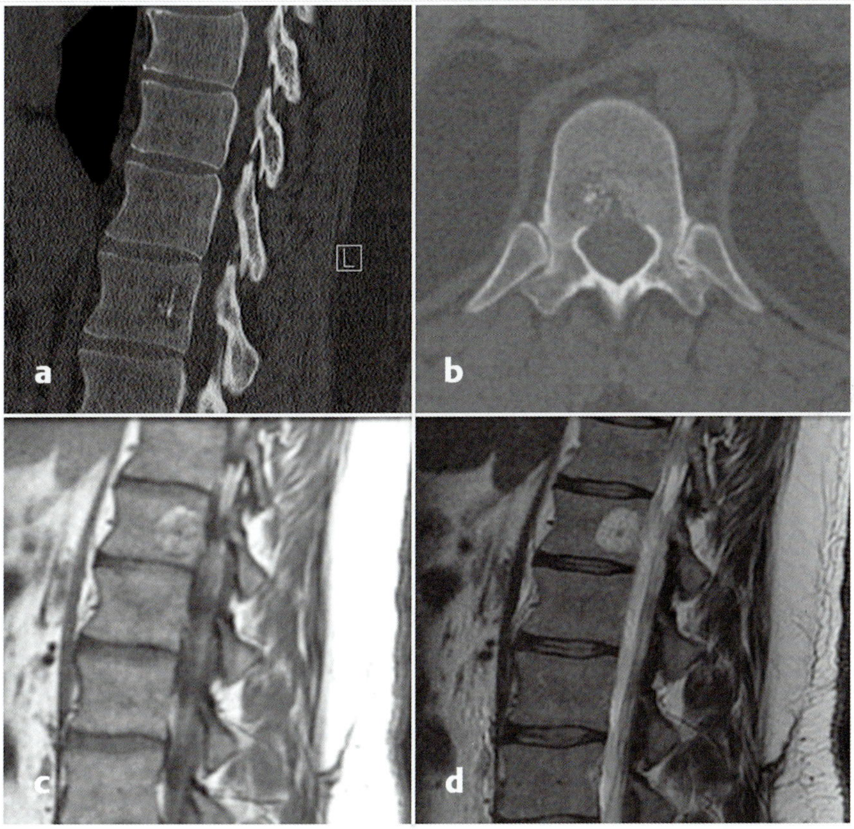

Fig. 4.10 Hemangioma. **(a)** Sagittal and **(b)** axial CT scan of thoracic spine demonstrates bony trabeculations and "polka dot" sign of a hemangioma, respectively. Hyperintense signal in both **(c)** T1-weighted and **(d)** T2-weighted sagittal MRIs of a hemangioma.

Vertebral Osteomyelitis

- Vertebral osteomyelitis is relatively rare and accounts for 2 to 4% of all bone infections.
- The most common microorganism associated with vertebral osteomyelitis is the bacteria *Staphylococcus aureus*.
- General symptoms include back pain, muscle spasm, fever, swelling at the infection site, episodes of night sweats, and sometimes difficulty transitioning from a standing to a sitting position.
- Frequently vertebral osteomyelitis may have a subtle onset. Correct diagnosis of the disease is often delayed for an average of 6 to 12 weeks due to vague and ambiguous symptoms.
- Vertebral osteomyelitis could invade the adjacent disks causing diskitis. Approximately 40% of patients with an advanced vertebral osteomyelitis could experience some type of neurologic deficiency, including myelopathy, which places the patient at risk for paralysis.
- Laboratory tests can find increased C-reactive protein, white blood cells (WBC), and the erythrocyte sedimentation rate (ESR), indicating for inflammation in the body.
- MRI scans are highly sensitive; they can show hypointensity damages on T1-weighted images, with enhancement with gadolinium (▶ Fig. 4.11).

4.6 Spinal Stenosis and Pain

Abnormal narrowing of the spinal canal can lead to spinal stenosis. It may occur in any of the regions of the spine, more common in lumbar and cervical spine. Spinal stenosis can be congenital, traumatic, or degenerative. Degenerative changes such as chronic disk bulging, growth of osteophytes, and ligament and facet joint hypertrophy are among the most common causes for spinal stenosis in elderly patients (▶ Fig. 4.7). The relationship of spinal stenosis and pain is not clear. Pain is not proportional to the severity of stenosis. Direct impingement of the nerve roots, ischemia, or increase of inflammatory mediators such as TNF have been implicated in the pathogenesis of pain in spinal stenosis.[7]

4.6.1 Classification of Spinal Stenosis

- *Mild stenosis*: Narrowing of less than 25% of the spinal canal.
- *Moderate stenosis*: Narrowing of 25 to 50% of the spinal canal.
- *Severe spinal stenosis*: Narrowing of more than 50% of the spinal canal.

4.6.2 Clinical Features of Lumbar Spinal Stenosis

- Intermittent neurogenic claudication is the typical feature in chronic lumbar spinal stenosis.
- It is characterized by lower back and limb pain, weakness, and numbness associated with paresthesia. Prolonged standing, walking, or spine extension could increase the symptoms. Sitting, lying supine, or flexion of the spin can relieve the symptoms.
- Physical examination may be completely negative without any tenderness over the lumbar spine, negative straight leg lifting test, and no leg weakness.

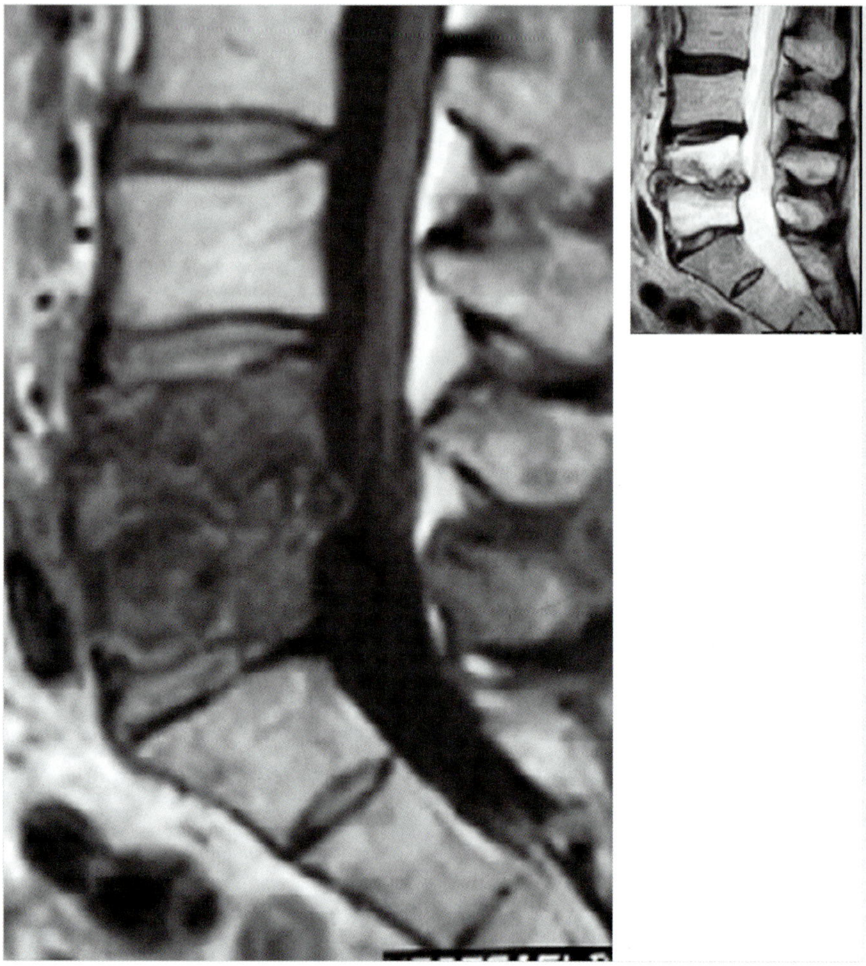

Fig. 4.11 Osteomyelitis. **(a)** A 68-year-old man developed severe, unrelenting back pain several months after experiencing bacteremia from a chronic lower extremity infection. Sagittal T1-weighted MRI demonstrates diskitis and osteomyelitis involving L4–L5. **(b)** Sagittal T2-weighted MRI demonstrates bony involvement with pathologic fracture of L4.

4.7 Sacroiliac Joint and Pain

The sacroiliac (SI) joints are paired C-shaped or L-shaped joints formed between the articular surfaces of the sacrum and the ilium bones. Strong anterior, posterior, interosseous SI joint ligaments, and sacrotuberous and sacrospinous ligaments hold the joint together with very limited range of motion (▶ Fig. 4.12). Sacroiliitis or inflammation of one or both SI joints and SI joint dysfunction can lead to chronic pain.

4.7.1 Clinical Features of Sacroiliac Joint Pain

- Patients may report a mild to moderate achy pain around the dimple or posterior superior iliac spine region.
- The pain may be increased when standing up from a sitting position or lifting the knee toward the chest during stair climbing.
- Pain could be referred to the buttock, thigh, hip, and groin, but rarely below the knee.
- Physical examination could find positive tenderness over the SI joint area.
- There are numerous provocative and nonprovocative maneuvers for the diagnosis of the SI joint dysfunction and pain. However, no single test is very reliable in the diagnosis of SI joint dysfunction.

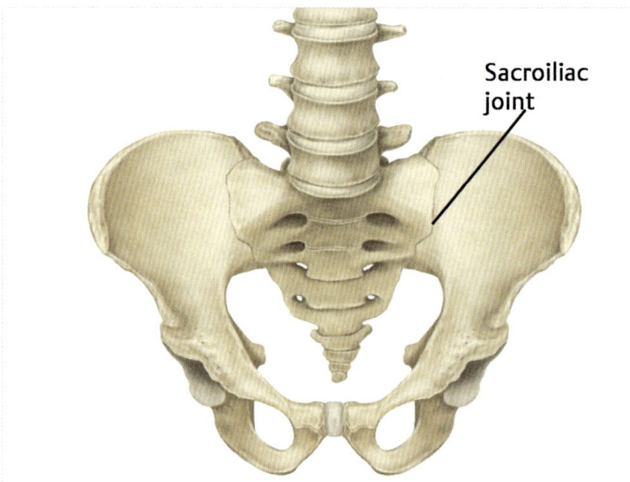

Fig. 4.12 Sacroiliac joint. Paired **C**-shaped or **L**-shaped joints, formed between the articular surfaces of the sacrum and the ilium bones.

- The current gold standard for diagnosis of SI joint pain is more than 75% pain relief with two consecutive CT- or fluoroscope-guided SI joint injections of local anesthetics.[8]

4.8 Sacrococcygeal Joint Pain (Coccydynia)

The SC joint is an amphiarthrodial joint formed between the oval surface at the apex of the sacrum and the base of the coccyx (tail bone). The coccyx itself is made up of three to five vertebrae, some of which may be fused together. The coccyx attaches the sacrum through a symphysis or as a true synovial joint, and also to the gluteus maximus muscle, the coccygeal muscle, and the anococcygeal ligament. The SC joint is slightly movable. Postacchini and Massobrio[9] described the coccyx in four types of orientation:

- Type I: The coccyx is curved anteriorly with its apex facing downward and caudally.
- Type II: The coccyx pointing forward more dramatically.
- Type III: The coccyx angles forward sharply.
- Type IV: The coccyx being subluxated at the SC joint.
- Common causes for coccydynia include the following:
- Sudden impact due to fall.
- Childbirth pressure in women.
- Excessive sitting and repetitive trauma of the surrounding ligaments and muscles, or leisure activities such as cycling and skateboarding.
- Idiopathic.
- SC teratoma or other tumor in the vicinity of the coccyx.

4.8.1 Clinical Presentation of Coccydynia

- Coccydynia is rare. However, obesity and female gender have higher risks of developing coccydynia.[10]
- Patient usually has a history of falls and injury to the tail bone.
- Pain is mainly in the tail bone, low sacrum, SI joint area, and sometime in back of the thigh.
- Activities that put pressure on the SC joint area, such as prolonged sitting, bicycling, and horseback riding, can increase coccydynia. Sometimes patient may feel more pain when changing from a sitting to a standing position.
- Physical exam could find severe tenderness over the SC joint.
- Injection of local anesthetic into the SC joint could help diagnosis. If the pain relates to the coccyx, this should produce immediate relief.
- Dynamic (sit/stand) X-ray or MRI scan may show whether the coccyx dislocates when the patient sits.

4.9 Pain in Postlaminectomy Syndrome

Postlaminectomy syndrome (also known as FBSS) refers to a condition characterized by persistent back and leg pain after back surgery. It is estimated that approximately 600,000 cases of spine surgery are performed each year in the United States. With 60 to 70% of success rate of back surgery, approximately 30 to 40% of patients continue to have pain after back surgery. Postoperative pain could be more severe than the preoperative pain.[11] Commonly, it is very difficult to define a clear source of pain for some specific patients with FBSS. However, it is always important for the treating physicians to identify the pain source in order to offer an effective treatment.

Common pathologic changes in FBSS include the following:

- Recurrent or persistent disk herniation.
- Postoperative infection.
- Nerve injury.
- Adhesive arachnoiditis.
- Epidural postoperative fibrosis.

4.9.1 Recurrent or Persistent Disk Herniation

- The most common cause of an FBBS is due to recurrent disk herniation at the same level originally operated.
- Despite the name of diskectomy, the surgery itself cannot remove the entire disk. The most complete surgical excision of the disk may still leave 30 to 40% of the disk. Percutaneous diskectomy may leave more disk tissues.
- Recurrent back and radicular pain is the main presentation.
- Patients may have a pain-free interval after the surgery. Same radicular pain may return after a short or long period of time.
- MRI could help detect the recurrent disk herniation. T1-weighted image could show peripheral enhancement for the recurrent disk herniation, while the epidural scar tends to demonstrate diffuse enhancement.

4.9.2 Postoperative Infection

- Postoperative infection such as epidural abscess has been reported in 0 to 12% of patients.
- Increasing complexity of the procedure and operating time, and usage of metal implants (instrumentation) may increase the risk of infection. Other factors such as diabetes mellitus, obesity, malnutrition, smoking, previous infection, rheumatoid arthritis, and immunodeficiency may increase the risk of postoperative infection.
- Any patient with increased pain and temperature after surgery has to be cleared for possible epidural abscess.
- WBC and ESR could be increased.
- Physical examination may reveal severe tenderness over the spine or even leaking infectious drainage.
- Early diagnosis and treatment for an epidural abscess is the key to avoid possible catastrophic consequences. If the infection is not diagnosed or treated in a timely manner, the patient may be paralyzed. Also, the localized infection could spread and cause severe meningitis, which can lead to death.
- MRI could find marginally enhancing lesions in the epidural space.

4.9.3 Nerve Injury

- Chronic compression of the nerve root by disk, osteophyte, and subsequent ischemia before the surgery can cause damage to the nerve root.
- During the surgery, laceration, cautery, and traction can injure the nerve roots and lead to chronic pain.
- After surgery, scar tissues and inappropriately placed pedicle screws and rods can cause further damage to the nerve roots.

- Theoretically, all patients with FBSS may have some degree of nerve injury or damage. It has been reported that 57% of cases were found to have pathology at the same site at the level of previous surgery.
- Persistent radicular pain, or numbness or weakness of muscles innervated by a specific nerve root may indicate the possibility of nerve root damage. However, clinically it is difficult to make a definite diagnosis of nerve root damage.
- Current MRI/CT tests do not have definition high enough to visualize nerve root damages. At most, MRI/CT may find some pathology such as herniated disk, osteophyte, scar tissue, or a pedicle screw next to a nerve root.
- Eletromyographic/nerve conduction velocity studies mainly test the large fiber function and are not very sensitive to detect the pathologic and functional damage to small fibers, which conduct the pain sensation.
- A complete pain relief after a selective nerve root block may indicate that the pain may be due to the dysfunction of the specific nerve blocked during the procedure.

4.9.4 Adhesive Arachnoiditis

Arachnoiditis is characterized by inflammation of the meninges and subarachnoid space. Fitt and Stevens[12] studied 129 symptomatic patients at least 1 year after extradural lumbar disk surgery using high-resolution MRI. They reported 20% cases of adhesive arachnoiditis, which dropped to 3% when patients who had undergone oil-based myelography were excluded. Arachnoiditis was diffuse in 88% and focal in 12%. It is now clear that even though oil-based myelography before surgery is a major cause of adhesive arachnoiditis, extradural lumbar disk surgery or lumbar laminectomy without direct access to intrathecal space can also cause adhesive arachnoiditis independently of the use of some myelographic contrast media.

Clinical Presentation of Adhesive Arachnoiditis

- Arachnoiditis has no consistent pattern of symptoms; it frequently affects the nerves that supply the legs and lower back. Chronic debilitating pain is most common.
- Spinal cord involvement and damage of nerve roots can lead to numbness and tingling.
- Bowel, bladder, and sexual functioning can be affected if the lower part of the spinal cord is affected.
- Final diagnosis of adhesive arachnoiditis can be made by MRI or myelogram.

4.9.5 Epidural Postoperative Fibrosis (Scarring)

- Epidural scarring may bind the dura mater and nerve roots to the surrounding structures. The probability of recurrent pain increases when scar score increases.
- The chief clinical presentation of epidural scar is continuous or recurrent low back and radicular pain.
- Patients with extensive peridural scar are more likely to experience recurrent radicular pain than patients with less extensive peridural scarring.[13]
- Epidural scar tends to show higher signal intensity on T2-weighted MR images than disk material or annulus. Epidural scars enhance consistently and intensely on T1 fat-saturated weighted images if the MRI is performed within 30 minutes after the gadolinium injection, whereas disk material usually does not enhance or only has peripheral enhancing.

4.10 Pain That Is Not Structure Specific

Traditionally, clinicians are trained to make a diagnosis of low back pain while trying to identify the anatomic sources of pain, such as a herniation disk, an irritated nerve root, or a fractured vertebra. However, quite often, the anatomic source(s) cannot be clearly identified. Recently, increasing evidence has suggested that back pain is not just mechanical or due to anatomic abnormality, but rather pain can also be caused by biochemical abnormality in the spine. Many neuropeptides and inflammatory mediators such as TNF, nerve growth factor, IL-1beta, IL-6, calcitonin gene-related peptide, ionized calcium-binding adaptor molecule-1, and glial fibrillary acidic protein may all contribute to the pathogenesis of back pain. TNF is released not only by the herniated or protruding disk, but also, in cases of disk tear (annular tear), by facet joints,[14] as well as in spinal stenosis.[15] In addition to causing pain and inflammation, TNF may also contribute to disk degeneration. This may well explain why surgical removal of a herniated disk or fusion of multiple vertebral bodies may not relieve the pain, and might even exacerbate it, resulting in FBSS, while treatments aiming to decrease the inflammatory mediators such as nonsteroidal anti-inflammatory drugs, steroid injection, or epidural application of TNF inhibitors may relief the back pain.

Physicians should try to identify the anatomic as well as chemical abnormalities while trying to make a diagnosis and formulating a treatment plan for the back pain.

References

[1] Shuang F, Hou SX, Zhu JL, et al. Establishment of a rat model of lumbar facet joint osteoarthritis using intraarticular injection of urinary plasminogen activator. Sci Rep. 2015; 5:9828

[2] Moen A, Schistad EI, Rygh LJ, Røe C, Gjerstad J. Role of IL1A rs1800587, IL1B rs1143627 and IL1RN rs2234677 genotype regarding development of chronic lumbar radicular pain; a prospective one-year study. PLoS ONE. 2014; 9(9): e107301

[3] Yang H, Liu H, Li Z, et al. Low back pain associated with lumbar disc herniation: role of moderately degenerative disc and annulus fibrous tears. Int J Clin Exp Med. 2015; 8(2):1634–1644

[4] Dewar C. Diagnosis and treatment of vertebral compression fractures. Radiol Technol. 2015; 86(3):301–320, quiz 321–323

[5] Barzin M, Maleki I. Incidence of vertebral hemangioma on spinal magnetic resonance imaging in Northern Iran. Pak J Biol Sci. 2009; 12(6):542–544

[6] Jones JO, Bruel BM, Vattam SR. Management of painful vertebral hemangiomas with kyphoplasty: a report of two cases and a literature review. Pain Physician. 2009; 12(4):E297–E303

[7] Ohtori S, Miyagi M, Eguchi Y, et al. Epidural administration of spinal nerves with the tumor necrosis factor-alpha inhibitor, etanercept, compared with dexamethasone for treatment of sciatica in patients with lumbar spinal stenosis: a prospective randomized study. Spine. 2012; 37(6):439–444

[8] Laux CJ, Osterhoff G, Werner CM. Sacroiliac joint pain - diagnostic algorithm and therapeutic approaches [in German]. Praxis (Bern 1994). 2015; 104 (1):33–39

[9] Postacchini F, Massobrio M. Idiopathic coccygodynia. Analysis of fifty-one operative cases and a radiographic study of the normal coccyx. J Bone Joint Surg Am. 1983; 65(8):1116–1124

[10] Lirette LS, Chaiban G, Tolba R, Eissa H. Coccydynia: an overview of the anatomy, etiology, and treatment of coccyx pain. Ochsner J. 2014; 14(1):84–87

[11] Teixeira MJ, Yeng LT, Garcia OG, Fonoff ET, Paiva WS, Araujo JO. Failed back surgery pain syndrome: therapeutic approach descriptive study in 56 patients. Rev Assoc Med Bras. 2011; 57(3):282–287

[12] Fitt GJ, Stevens JM. Postoperative arachnoiditis diagnosed by high resolution fast spin-echo MRI of the lumbar spine. Neuroradiology. 1995; 37 (2):139–145

[13] Ross JS, Robertson JT, Frederickson RC, et al. ADCON-L European Study Group. Association between peridural scar and recurrent radicular pain after lumbar discectomy: magnetic resonance evaluation. Neurosurgery. 1996; 38(4):855–861, discussion 861–863

[14] Stemper BD, Yoganandan N, Pintar FA, Rao RD. Anterior longitudinal ligament injuries in whiplash may lead to cervical instability. Med Eng Phys. 2006; 28 (6):515–524

[15] Wang YF, Chen PY, Chang W, et al. Clinical significance of tumor necrosis factor-α inhibitors in the treatment of sciatica: a systematic review and meta-analysis. PLoS ONE. 2014; 9(7):e103147

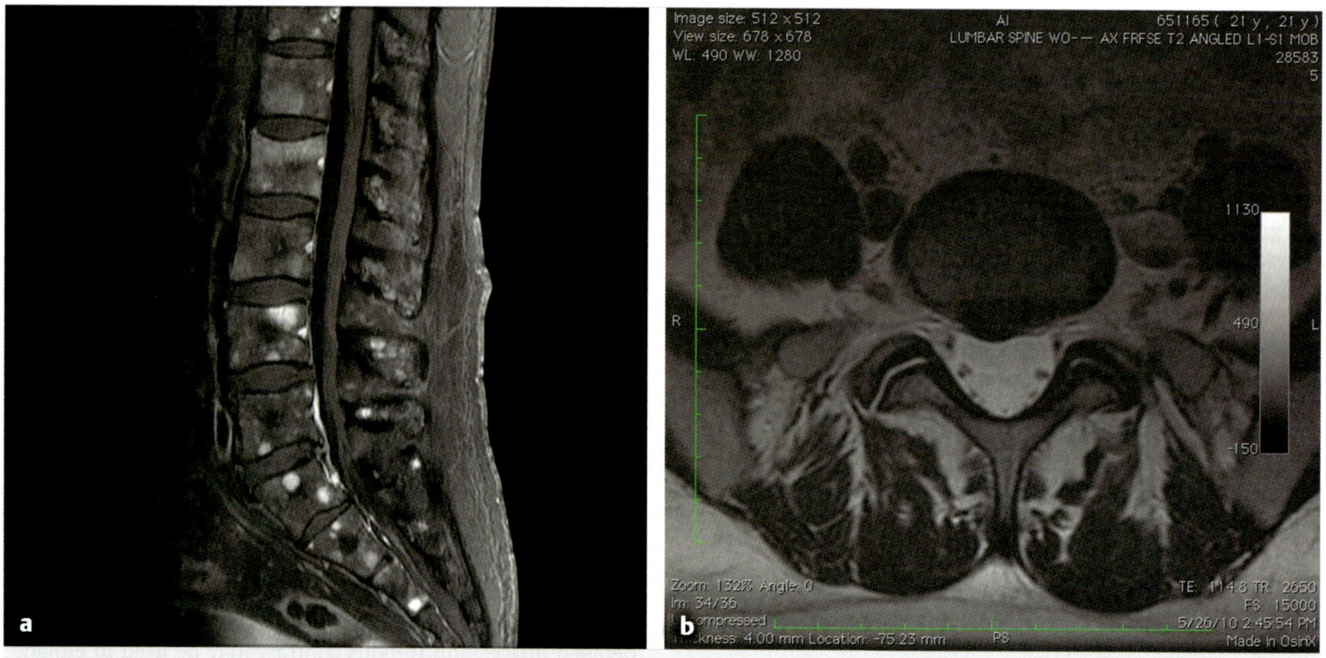

Fig. 5.6 (a,b) MRI of spine showing details of soft tissue and bony pathologies.

Advantages

- MRI does not involve the use of ionizing radiation.
- Useful for showing soft tissue structures, such as ligaments and cartilage, and organs.
- Sensitive for detecting all spinal cord and disk pathologies including narrowing and compression of the spinal cord and spinal nerve roots.
- Allows visualization of soft tissue inflammation and inflammatory lesions within bone.
- Diffusion tensor imaging technique allows detecting degenerative changes in the spinal cord in the course of chronic compression by disk herniation or osteophytes.[8]

Disadvantages

- MRI scanners are very expensive.
- Allergic reactions to contrast dye.
- Patients need to lie still in a tubular structure for 25 to 45 minutes, which can cause claustrophobia for patient.
- May cause heart pacemakers, intrathecal pump, spinal cord stimulator, and other implanted devices to malfunction or may cause tissue damage from overheating.
- May cause metallic devices such as some artificial joints, spinal hardware, certain metal clips in the brain, and other hardware to move or shift.
- Limited sensitivity for the detection of spinal inflammation.[9]
- Poor correlation between the image findings and symptoms among patients with lateral lumbar spinal canal stenosis.[10]

5.1.7 Electromyograms and Nerve Conduction Studies

An EMG measures the electrical activity of muscles at rest and during contraction. Nerve conduction studies measure how well and how fast the nerves can send electrical signals. An EMG can assess the electrical activity of a nerve root and is sometimes recommended for patients with back pain. A prolonged pressure on a nerve root can cause the muscles innervated by the nerve to contract spontaneously. Compression of a nerve will also slow electrical conduction along that nerve. EMGs are also sometimes useful in distinguishing nerve degeneration (neuropathy) from nerve root compression (radiculopathy).

The test involves placing small needles into the muscles; therefore, there is some discomfort for the patient. There are no major risks. However, the tests are not highly reliable in determining which nerve is compressed.[11]

An EMG is helpful in:

- Finding pathologic changes in muscle tissue, nerves, or the junctions between nerve and muscle.
- Evaluating the cause of weakness, paralysis, or muscle spasm.

5.1.8 Diskograms

While MRI and CT scans can provide detailed information about disk pathology, a diskogram can be diagnostic and sometimes therapeutic. It involves placing a needle into the disk and injecting a contrast in the center of the disk. Pain might be reproduced as the pressure in the affected disk is increased by injecting contrast (▶ Fig. 5.7).

Advantages

- A good diagnostic and confirmatory test for disk pathology.
- A dynamic test able to provide live information about the integrity of the disk.

Disadvantages

- Inherent risk of infection and neural injury due to its invasiveness.

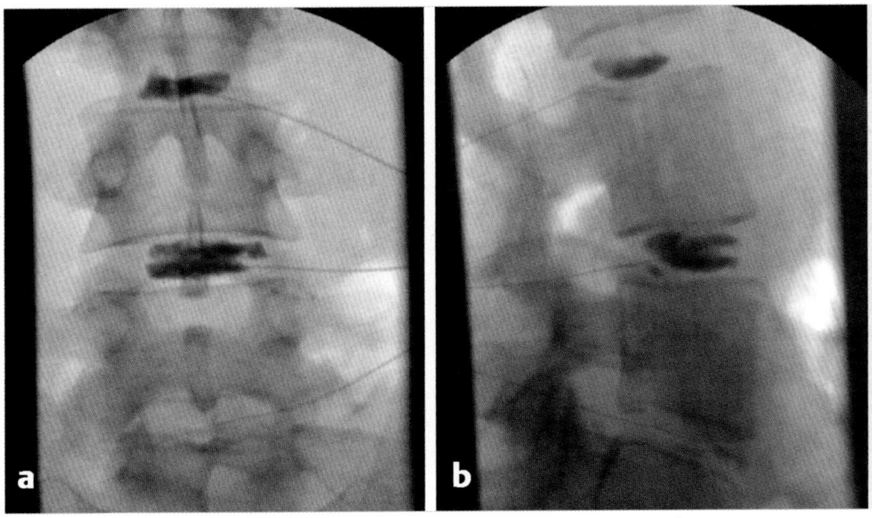

Fig. 5.7 Pictures of diskograms: (a) AP view and (b) lateral view.

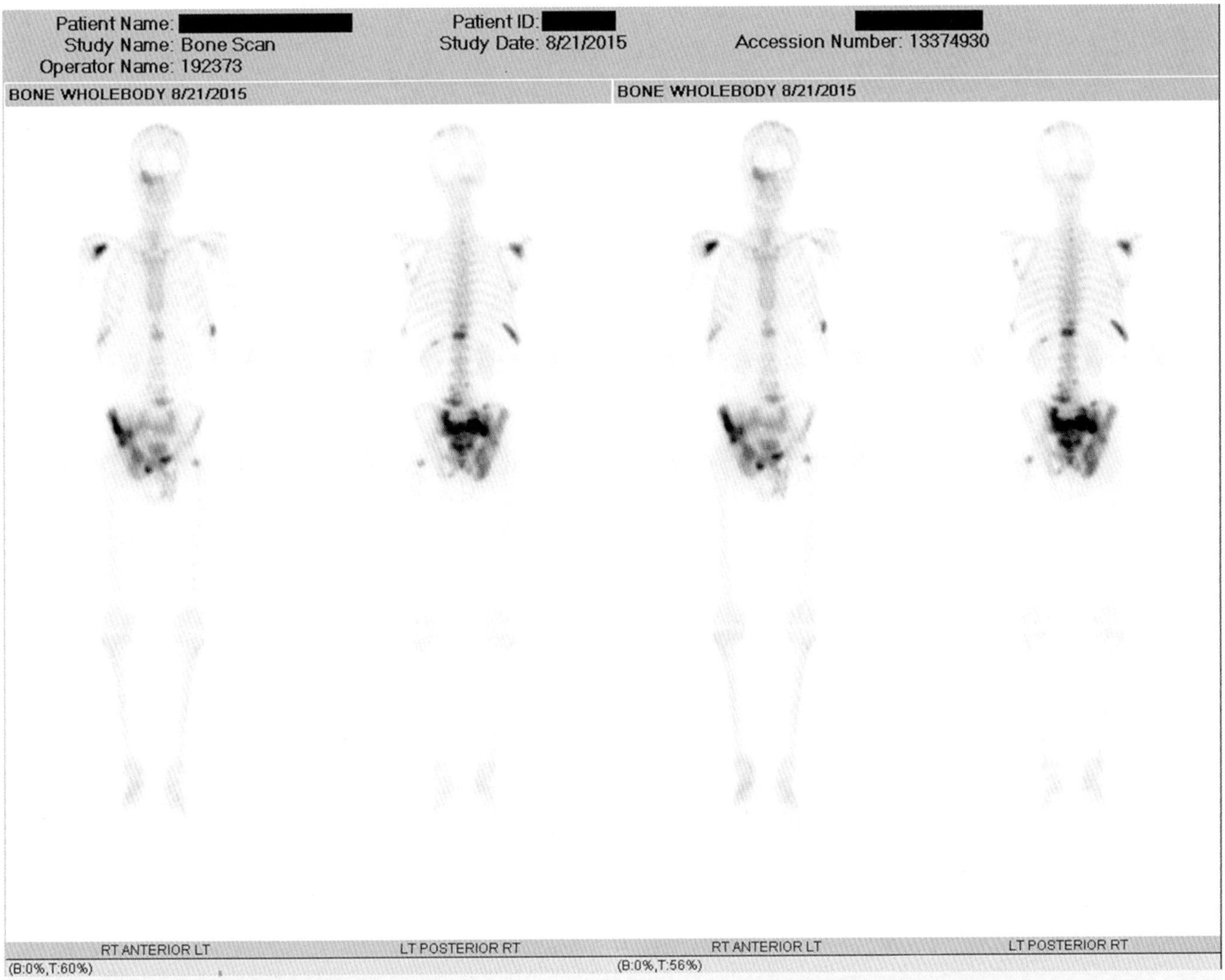

Fig. 5.8 Bone scan of whole body.

- Needs CT scan or X-ray assistance to confirm the integrity of the disk.
- Requires patient self-report of pain, hence prone to false-positives.

5.1.9 Nuclear Medicine

A bone scan is the most commonly used nuclear scanning test to detect bone abnormalities, including cancer of the bone or metastasized to the bone, inflammation, fractures, and infections which may not be visible in traditional X-ray images. It detects biochemical changes by mapping the presence of radiographic compounds incorporated into hydroxyapatite crystals during new bone formation or bone turnover, which is common in bone metastasis, spine tumors, fractures, infarction, infection, and metabolic bone diseases.

A bone scan is performed by injecting a small amount of radioactive marker into an intravenous line. Three hours later, the patient is placed through a scanner, and the radioactive marker will be concentrated in any region where there is high bone turnover.

Single-photon emission computed tomography is another nuclear medicine technique which can provide a three-dimensional image that can significantly enhance the detection of pathologic lesions (▶ Fig. 5.8).

Advantages

- Noninvasive.
- A good screening imaging for unsuspected bone lesions.
- When using gallium-67, it is the most effective radioactive tracer in assessing infectious spondylitis.[12]

Disadvantages

- Expensive.
- Limited to specialized centers and to specific pathologies.
- Not useful for procedural techniques.
- Needs specialized reader.

References

[1] Plastaras C, McCormick Z, Macron D, et al. Adverse events associated with fluoroscopically guided zygapophyseal joint injections. Pain Physician. 2014; 17 (4):297–304

[2] Provenzano DA, Narouze S. Sonographically guided lumbar spine procedures. J Ultrasound Med. 2013; 32(7):1109–1116

[3] Shaikh F, Brzezinski J, Alexander S, et al. Ultrasound imaging for lumbar punctures and epidural catheterisations: systematic review and meta-analysis. BMJ. 2013; 346:f1720

[4] Chin KJ, Ramlogan R, Arzola C, Singh M, Chan V. The utility of ultrasound imaging in predicting ease of performance of spinal anesthesia in an orthopedic patient population. Reg Anesth Pain Med. 2013; 38(1):34–38

[5] Bondár A, Szucs S, Iohom G. Thoracic paravertebral blockade. Med Ultrasound. 2010; 12(3):223–227

[6] Soh E, Karmakar MK. Assessment of the spine with CT and MRI prior to interspinous/interlaminar spinal procedures: a pictorial review. Br J Radiol. 2013; 86(1026):20130066

[7] Tepel M, Aspelin P, Lameire N. Contrast-induced nephropathy: a clinical and evidence-based approach. Circulation. 2006; 113(14):1799–1806

[8] Sąsiadek MJ, Bladowska J. Imaging of degenerative spine disease—the state of the art. Adv Clin Exp Med. 2012; 21(2):133–142

[9] Maksymowych WP. MRI and X-ray in axial spondyloarthritis: the relationship between inflammatory and structural changes. Arthritis Res Ther. 2012; 14 (2):207

[10] Kuittinen P, Sipola P, Aalto TJ, et al. Correlation of lateral stenosis in MRI with symptoms, walking capacity and EMG findings in patients with surgically confirmed lateral lumbar spinal canal stenosis. BMC Musculoskelet Disord. 2014; 15:247

[11] WebMD. Electromyogram (EMG) and nerve conduction studies. Available online at: http://www.webmd.com/brain/electromyogram-emg-and-nerve-conduction-studies

[12] Humphreys SC, Eck JC, Hodges SD. Neuroimaging in low back pain. Am Fam Physician. 2002; 65(11):2299–2306

6 Epiduroscopy Equipment

Günter Schütze

6.1 Introduction

The technical requirements for spinal endoscopy systems for clinical use in the diagnosis of and therapy for neuraxial pain syndromes are regulated, for example, in the United States by the Food and Drug Administration (FDA). A large number of endoscopic instruments, including the necessary tools for epiduroscopy, are licensed under the FDA 510(k) process and under country-specific or European medical product licensing statutes, respectively. For example, the steerable, flexible Karl Storz 2.8-mm outside-diameter endoscope with effective length 70 cm or 40 cm, equipped with a working channel, has had an 501k license since 1992; the Myelotec endoscope (Myeloscope) received one (K960194) in 1996 as did the Racz catheter (K954584).[1,2]

The complex endoscopic diagnostic and therapeutic systems for spinal endoscopy are benefiting from the newest technologies, such as high-definition endoscopy, digital chromoendoscopy (with FICE or SPIES), image-enhanced endomicroscopy (with NBI or i-SCAN), chip-on-the-tip technology, and autofluorescence and spectroscopy. With the new flexible video endoscopes, for example, the image information collection and transfer occurs through chip-on-the-tip technology from an image sensor in the tip of the endoscope.

In addition to having a digital endocamera system, a fluoroscopic unit (an X-ray image converter or C-arm), an infusion system, and a sophisticated integrated epiduroscopic-sonographic-laser system, sterile microsurgical and microinvasive instruments should be provided for performing spinal endoscopy and microsurgical endoscopic procedures.

Regardless of the invasive intervention planned, spinal endoscopy (epiduroscopy) must be performed under adequate, continuous monitoring of vital functions with an anesthesiologist in attendance and only under absolutely sterile conditions in a suitable operating room (OR). There is no distinction in hygienic requirements between inpatient and outpatient spinal endoscopies. Along with surgical hand disinfection, the wearing of sterile gloves, smocks, masks, and caps, and the use of sterile drapes and of sterile single-use materials (covering of light guide, laser, image converter, etc.) is essential. High standards for the preparation (disinfection and sterilization) of reusable endoscopes have been established and must be met.[3]

A basic prerequisite for the use of the flourscopic unit (X-ray image converter or C-arm) for spinal endoscopy is the observance of the respective ordinances about protection against injury from ionizing radiation. When using a laser, the observance of nationally established laser operator instructions to prevent laser complications is essential as is the observance of general accident prevention prescriptions.

6.2 Equipment for Endoscopic Spinal Diagnostics

The complexity of epiduroscopic diagnosis requires the option of simultaneous co-observation and complete documentation, which can be accomplished above all through high-quality information and communication technology, including high-quality digital video technology.

By exploiting the technical possibilities of telemedicine, epiduroscopic diagnosis and therapy could be optimized and further developed. Considering that the interventional pain therapist could use telemedicine in the future as a significant instrument for achieving more and ensuring quality in pain medicine, the development of a structural recommendation for an international telemedicine network for interventional pain therapy would be a worthwhile goal.

6.2.1 Epiduroscopes

The first clinically useful flexible epiduroscope was described by Schütze and Kurtze in the 1993 publication "Percutaneous investigation of the epidural space using a flexible endoscope."[4] The first black-and-white endoscopic images from the epidural space of patients with pain syndromes were produced with this percutaneous flexible endoscopic system for the neuraxial region (▶ Fig. 6.1).[5]

In the flexible epiduroscopes currently available, optical lenses conduct the image from the epidural space to a video

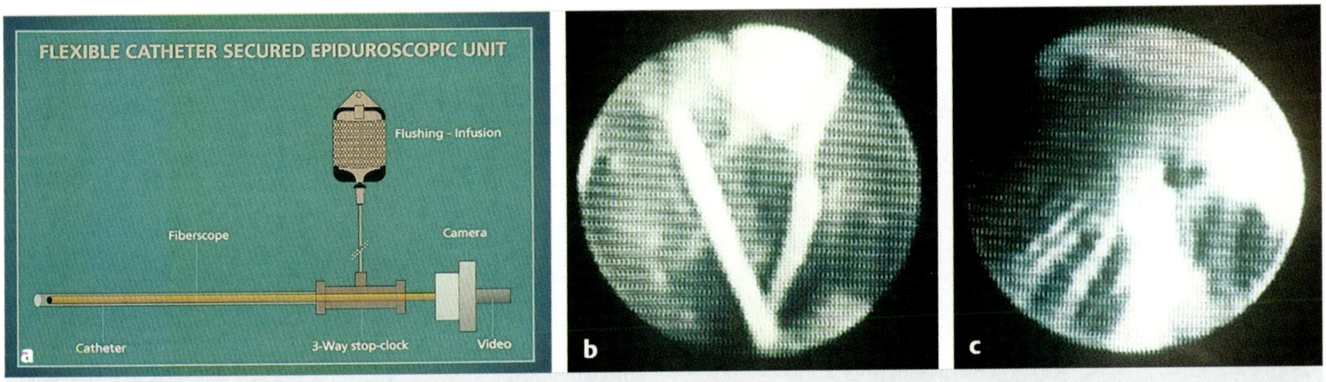

Fig. 6.1 (a) An illustration representing the principle of a percutaneous flexible epiduroscopic unit (1994) for epidural diagnostics in pain patients.[5] (b, c) Endoscopic representation of epidural adhesions and fibroses from the epidural space with the percutaneous flexible epiduroscopic unit (1990).

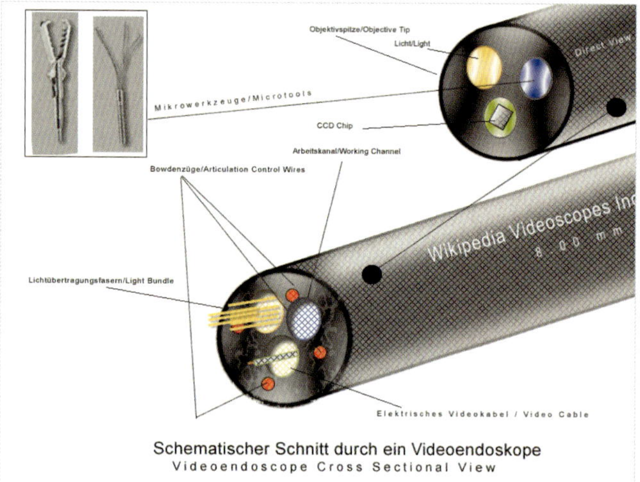

Fig. 6.2 Schematic view of a modern endoscope tip. (Courtesy of Wikipedia.)

camera, which is located at the end of the endoscope. Up until recently the entire spinal canal, from the sacral to the cervical region, could be reached only with flexible epiduroscopes equipped with fiber optic technology (a technique employing an organized bundle of glass fibers for image transport). Now some epiduroscopes available on the international market are equipped with tiny charge-coupled devices (CCD) chip or complementary metal-oxide–semiconductor (CMOS) chips integrated into the very tip of the endoscope and the image information is converted into a video signal there and reaches the monitor over a thin cable. A distinction must be made between the two different types of sensor technology (fiber optic versus chip-based) according to their capabilities. The image quality of the fiber optic endoscopes is not equal to that from the high pixel count CCD or CMOS sensors. As a rule, the CCD chip endoscope is superior to the fiber optic (glass fiber) endoscope and CMOS chip endoscope on account of the optical image quality. The image quality is dependent on the combination of video sensor, optical lens system, luminance, and image monitor.

In some videoscopes, external light projectors have been replaced by light-emitting diode (LED) illuminators integrated into the endoscope tip or even into the handgrip. In the latter, the light is conducted from the handgrip over built-in glass fibers to the tip. Very uniform illumination occurs with LEDs built into the tip of the instrument (▶ Fig. 6.2). Admittedly, the luminance of an external xenon light projector has not yet been matched by LEDs.

On the basis of their differing technical quality, endoscope length, reusability, external diameter, working channel, flexibility, image quality, ease of repair, etc., most endoscopes do not meet the high technical requirements for spinal endoscopy.

Technical Note

The flexible epiduroscope, a complex and delicate precision technical instrument for medicine, is built with various materials, for example, plastic, metal, and glass, bonded together in a watertight manner.

Functional Properties of Epiduroscopes

To fulfill the requirements of the various certifications for clinical use in spinal endoscopy, epiduroscopes (flexible, reusable endoscopes) should meet the following specifications for these important functional properties:[3,4,6]

- External diameter, effective length, and working channel.
    - Since the usual sagittal diameter of the spinal canal varies between 15 and 20 mm,[7] an epiduroscope should have an external diameter between 2.4 and 3.8 mm, a sufficient effective length of 90 cm, and a working channel with a diameter of 0.9 to 1.5 mm.
- Steerability, flexibility and strength of materials, and laser protection.
    - Through a special shaft construction method, a direct transmission of torque can be attained by which a better steerability of the epiduroscope is possible within the anatomical boundaries of the epidural space. The epiduroscope should be built with a certain hardness and buckling strength for spinal navigation. To guarantee the epiduroscope protection against laser radiation, a Laserite ceramic tip, for example, should be incorporated at the distal end of the working channel.
- Image quality and image capture.
    - On a small video camera at the ocular of the epiduroscope, image capture takes place. The optical system is divided: one part conducts the image information to the investigator's eye, and the second part converts the image information into a video signal at the camera.
    - In state-of-the-art epidiuroscopes, image capture takes place at an image sensor (a chip at the tip of the epiduroscope), which decomposes the image into a raster of electronic signals. In contrast to conventional fiber-optic epiduroscopes, this chip technology yields brilliant, true-to-life image quality and significantly improved, contrast-rich representation. An integrated LED illuminator also provides improved illumination in the neuraxial region.
- Reusability, cleaning, disinfection, and sterilization.
    - In principle, cleaning, disinfection, and sterilization of operating instruments or instrument systems should be carried out according to proper, validated procedures, such that the result of these procedures is retraceably and the safety and health of patients, operators, or third parties is not endangered.

Some available epiduroscopes do not meet these high quality requirements, they are inadequate or unsuitable for spinal endoscopy.

Selection of Epiduroscopes

The following selection of epiduroscopic (EDS) systems for diagnosis of and therapy for neuraxial pain syndromes are used in our pain clinic:

- Karl Storz Epiduroscope (Karl Storz, after G. Schütze).

- This is the first flexible epiduroscope (after G. Schütze), with an outside diameter of 2.8 mm, a 40-cm or 70-cm effective length, and containing a 1.2-mm diameter working channel. The distal end can be steered upward to 120° and downward to 170°.[6,8]
- Karl Storz Flexible Epiduroscope with FLEX-X (▶ Fig. 5.1).[9]
  - At the distal end of the working channel is a Laserite ceramic tip, which guarantees protection against laser radiation.
  - The eyepiece of the endoscope is attached to a camera system.
  - A digital three-chip camera system has also proved itself in clinical practice, because the endoscopic image is transferred on three separate color chips. Thereby color, contrast, and resolution can all be optimized directly at the camera head (▶ Fig. 6.4).
- Richard Wolf BOA vision Flexible Endoscope.
  - The BOA vision endoscope for spinal endoscopy, with an effective length of 68 cm and a shaft diameter of 8.7 Fr, features, above all, an excellent image sensor at the tip of the endoscope. An oblique atraumatic cylindrical stainless steel tip has an outside diameter of 6.6 Fr (▶ Fig. 6.5). The maximum flexibility is 270°. The upward/downward deflection is possible even with suitable auxiliary instruments and laser fibers.
  - The BOA vision endoscope image-sensor system is state of the art, yielding brilliant, true-to-life image quality and significantly improved, contrast-rich representation. Manual focusing and white adjustment are not required.
  - The integrated LED technology provides homogeneous and improved illumination in the neuraxial region.
  - The endoscope possesses a very high shaft stiffness, which is significant for introduction over the introducer sheath and the sacral segments.
- Myelotec Flexible Fiber Optic Endoscope.
  - This epiduroscope, mainly used in the United States, is available in external diameters of 0.9 mm, 1.2 mm, and 1.3 mm and can be placed with the help of a steerable, video-supported 2.7-mm or 3.0-mm video-guided catheter. This catheter contains two working channels (▶ Fig. 6.6). The 3000E endoscope has 0.9-mm diameter; a 10K-pixel fiber bundle; approximately 40 × magnification, depending on the eyepiece coupler and the video system; a 70° field of view; a 0° angle of view; a 32-mm eyecup; and an 85-mm effective length. It can be used with universal camera systems (▶ Fig. 6.7).

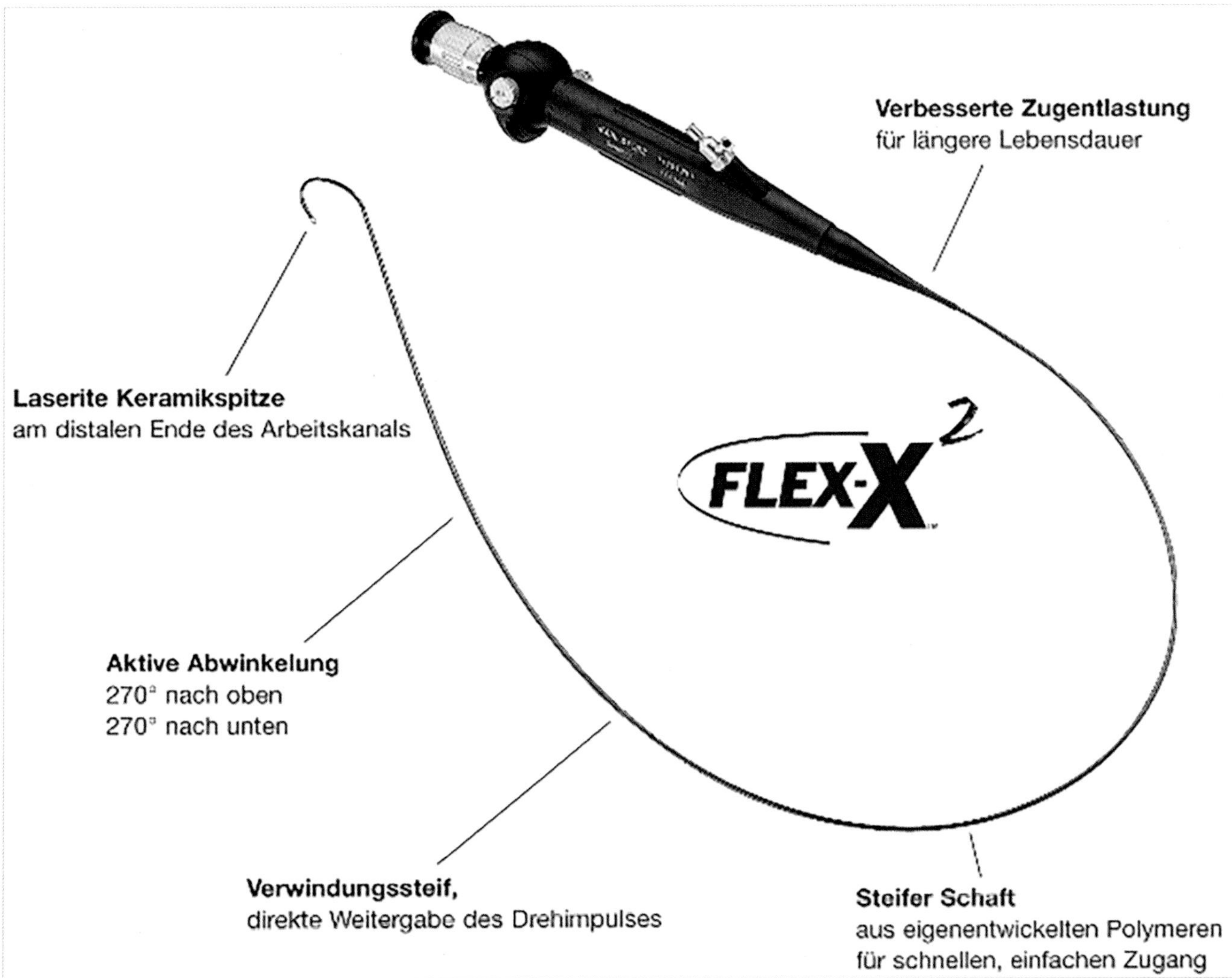

Fig. 6.3 Flexible endoscope FLEX-X.[9] (Courtesy of Karl Storz, Tuttlingen, Germany.)

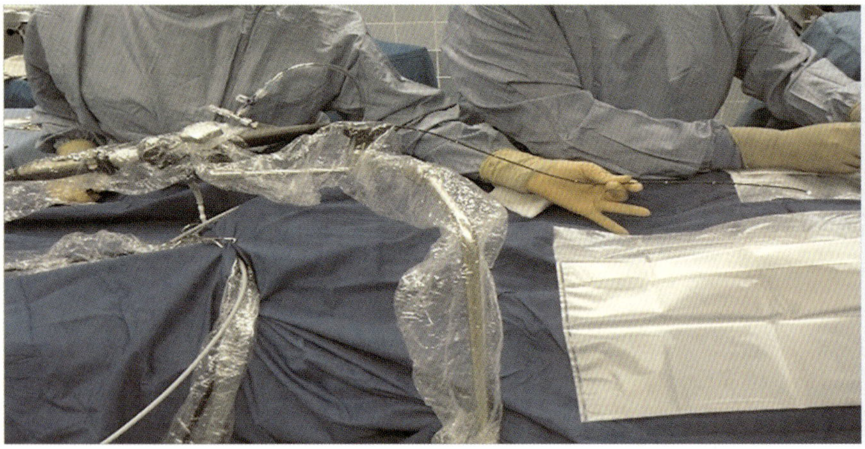

Fig. 6.4 The Karl Storz sterile, flexible epiduroscope with FLEX-X2 in clinical use.

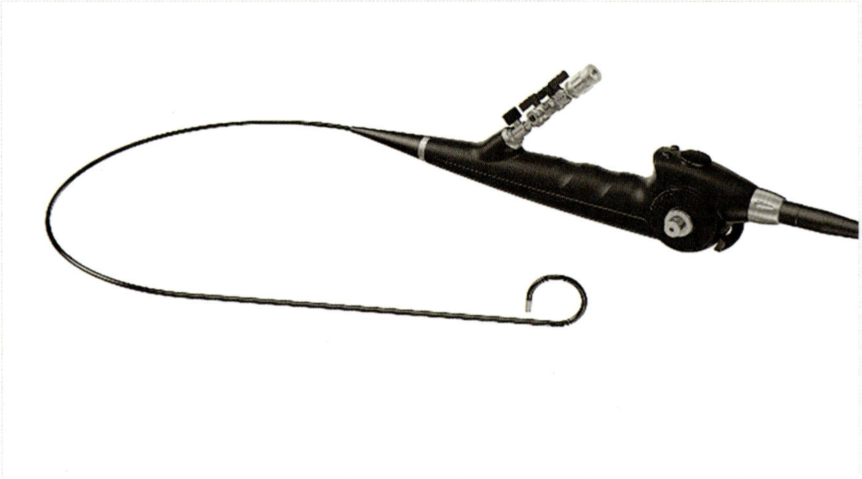

Fig. 6.5 Flexible one-channel sensor endoscope (ureterorenoscope). (Courtesy of Richard Wolf, Knittlingen, Germany.)

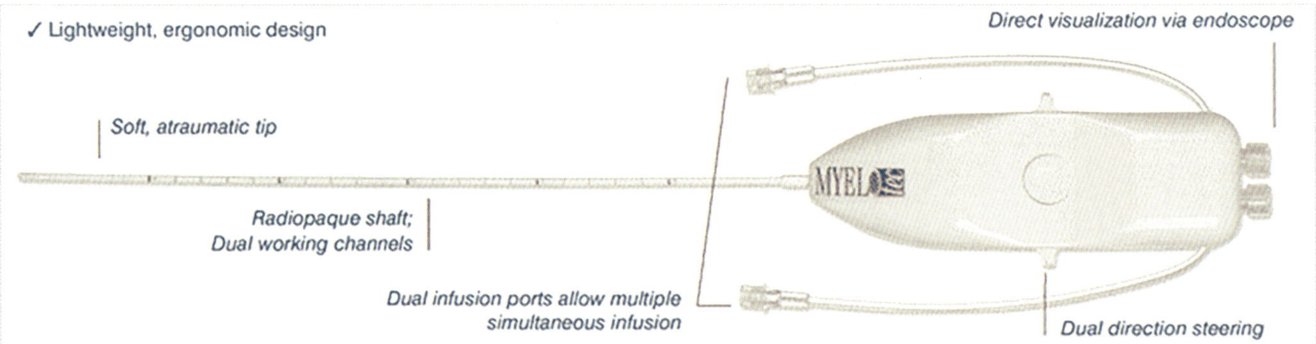

Fig. 6.6 Myelotec Video Guided Catheter. (Courtesy of Myelotec, Roswell, Georgia.)

   - Kawanishi et al[10] reported that since 1996 they have introduced Myelotec epidural endoscopes with 0.9-mm diameters through the sacral hiatus to investigate the lumbar epidural space. The authors note that some problems of these epiduroscopes must still be solved. However, they rated them as very promising diagnostic tools.
- Karl Storz Flex-XC Flexible Endoscope.
   - This new flexible endoscope (11278 VS Video-Flex-X SPIES) has an effective length of 70 cm with an outside diameter of 8.4 Fr (2.8 mm). The flexible endoscope is equipped with a CMOS chip and a Laserite ceramic liner at the distal end of a 3.6-Fr (1.2-mm) diameter working channel. The endoscope allows a maximum deflection of 270° in either direction and has automatic focus and illumination. Through the SPIES Spectra, even the finest tissue structures can be imaged by filtering out the deep red section of the visible spectrum. Through the simultaneous presentation of white light and a SPIES mode image, the tissue structures can be compared directly on the monitor intraoperatively, which makes endoscopic diagnosis easier (▶ Fig. 6.8).

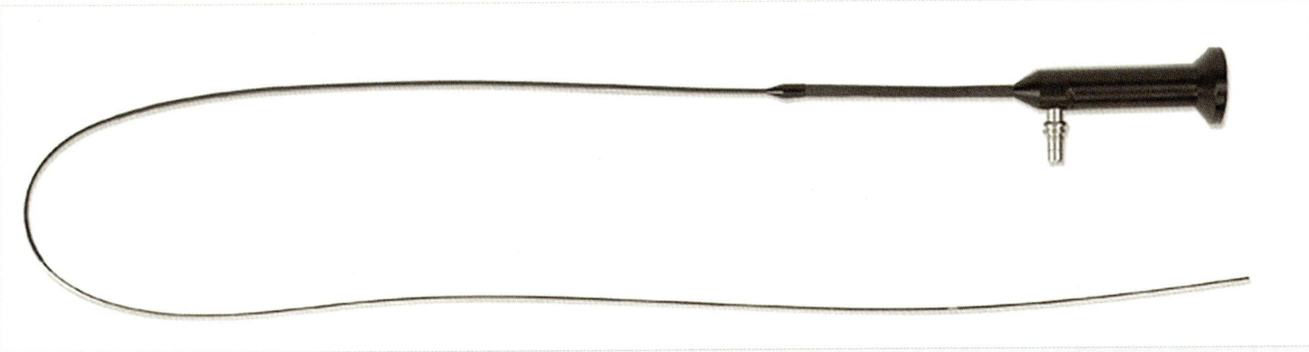

Fig. 6.7 Myelotec 3000E Flexible Fiber Optic Endoscope. (Courtesy of Myelotec, Roswell, Georgia.)

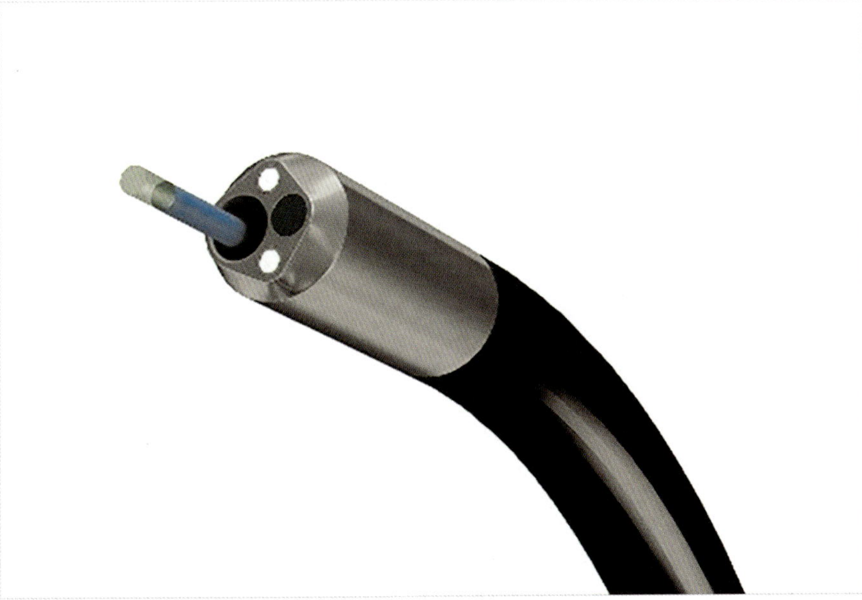

Fig. 6.8 The thinnest, flexible one-channel-sensor endoscope, the atraumatic, oblique stainless steel distal tip with 3.6 Fr. working and irrigation channel. (Courtesy of Karl Storz, Tuttlingen, Germany.)

6.2.2 Camera and Video Techniques and Equipment

Chip-on-the-Tip

At the very tip of the endoscope, tiny CCD or CMOS chips are integrated as image sensors. The image information is converted into video signals in the image sensors and conducted over a thin glass fiber to the monitor. CCD chip technology allows for the collection and analysis of more than a million pixels per image resulting in better resolution in these first-generation high-resolution video endoscopes.[11] To date, chip-on-the-tip endoscopes have, however, used fixed-focus optics, which can represent objects at only one distance optimally. If the object is located at a different distance, the image loses sharpness.

High-Definition Endoscopy

High-definition television (HDTV), whose high-resolution images surpass in quality those from the older phase-alternating line (PAL) transmission, has recently been introduced into endoscopy.

Video from the new miniaturized endoscopic color chips, which collects and analyzes more than 1 million pixels per video image, can now can be played on a standard HDTV television with up to 1080 lines per video image.[12] The currently available high-resolution HD endoscopy systems achieve a resolution of 1400 × 1080 image points.

Karl Storz Image 1 video camera

The most commonly used Image 1 digital video camera is for the PAL and NTSC color system. But with the introduction of the new digital epiduroscopic 3-chip Image 1 video camera with a digital image processor (an integrated image-processing module [IPM]), our expectations for significantly improved image quality for epiduroscopic diagnostics have been fulfilled. Images are transmitted in 16:9 format with 1080-pixel resolution, allowing for full high-definition images with 2 million image points. For medical applications, the highest possible image resolution is needed. Image 1 is a video camera system in which the optical analog images of the CCD sensor are directly converted to digital form behind the sensor while still in the camera head. This has the advantage that the image data is always in digital form and at the highest resolution throughout the entire camera system.

In addition to insensitivity to noise, such as that from high-frequency (HF) surgery, this camera and processing system also brings with it the advantages of digital image optimization. This is a capability is available in all Karl Storz video systems, the standard models through to the digital IPM models.

Karl Storz Tricam 3D

With the Endovision Tricam 3D, Karl Storz has introduced a three-dimensional 3-chip endoscopic video system for the PAL and NTSC color system. The modular construction offers the possibility of integrating central steering, data processing and conversion, and telemedicine in the pain operating room (OR). Because all camera heads from Karl Storz work with one camera control unit (CCU), the operators can use each Image 1 camera head for all endoscopic interventions, not just those in the spinal region. Moreover, all new Image 1 products are fully compatible with extant Image 1 systems.

With the optical parfocal zoom (up to 2 ×) and digital endoscope filtration, the autoclavable Image 1 cameras automatically achieve image optimization on all epiduroscopes. The use in the sterile OR is made easy by push-button operation of all camera functions.

Along with being able to output analog signals, the Image 1 video camera system is also able to output digital signals such as digital video (DV) and serial digital interface (SDI) without restriction, which makes possible high-resolution, true-to-life digital images.

Richard Wolf Endocam Logic HD

The BOA vision flexible endoscope (Richard Wolf) contains a sensor unit that works with the current Endocam Logic HD camera platform, which consists of the Endocam Logic HD controller,the hand remote, 8-GB USB flash memory, CAN-BUS terminator, 3.0-m lockable HDMI/DVI-D cable, network cable, and DVD license.

The new Endocam Logic HD camera platform is one of the latest developments in camera systems from Richard Wolf. Endocam Logic HD'shigh-resolution image is developed within the endoscope The camera is a component of a new generation of digital camera heads, intelligent LED light sources, light guides, monitors, and the highly developed chip-on-the-tip technology.

The important parameters of dialog-capable instruments are displayed on the monitor. This dialog function also allows for automatic regulation of the light source over the camera; it is advantageous that the operator needs to make no further adjustments. Endocam Logic HD possesses a large number of rapidly selectable tested clinical application profiles, the menu for which is simple to navigate. The physicians are also able to specify a large number of settings themselves and define their own profiles.

In our pain OR, SDI is the preferred method for signal transmission, image recording, and display. This technique makes it possible for the new ergonomic thin-film-transistor (TFT) liquid-crystal–display monitors to supply true-to-life, brilliant video sequences. For a brilliant endoscopic image, the Radiance 26-in. G2 high-bright liquid-crystal–display (HB LCD) monitor by Richard Wolf is recommended.

Endomicroscopy

New techniques in endoscopy, such as high-definition endoscopy, digital chromoendoscopy (with FICE or SPIES), image-enhanced,endomicroscopy (with NBI or i-SCAN), chip-on-the-tip technology, and use of autofluorescence and spectroscopy are rapidly being developed.

The systems recommended for use for endomicroscpy are high-definition video endoscopes. These instruments are equipped with special color-encoding properties (called chromoendoscopy).

> In connection with the i-SCAN technology, high-resolution (HD +) endoscopy makes more details of the mucosal surface visible, which could lead to substantial improvements in spinal endoscopic diagnostics.

- Confocal microscope
  - Confocal endomicroscopy (confocal laser endomicroscopy [CLE]) is a special medical endoscopic investigative technique. During endoscopic examination with CLE, cellular, vascular, and connective tissue structures can be differentiated in high resolution (1000 × magnification) by means of laser-supported, confocal fluorescence technology.[13]

CLE systems are available as handheld devices that can be inserted through the working channel of an endoscope or as a miniaturized system integrated into the distal end of an endoscope. In the endoscopic image, the mucous membrane is made visible in various fluorescent colors.

- The confocal argon laser works as follows to produce an image:
  - After application of the fluorescent dye, the confocal (endomicroscopy) argon laser strikes the sample with a wavelength of 488 nm (blue laser light). The 1012 × 1012-pixel endomicroscopic image obtained can be analyzed and evaluated on a miniaturized scanner in the endoscope. The tissue remains unharmed.
  - During the endoscopic investigation, many histologic images can be recorded, which makes a more accurate assessment of the tissue possible.
  - The magnification is so high that even individual blood corpuscles in the vessels can be seen.
  - Normal, noninflamed tissue usually presents with low background fluorescence that as a rule does not suffice to produce a sufficiently strong signal for endomicroscopy. Therefore, fluorescent dyes (fluorescein sodium) must be applied a few seconds before endomicroscopy.

Virtual chromoendoscopy

Using virtual chromoendoscopy, ,a histologic analysis of tissue is possible during an endoscopic investigation.[11] Comparison of the results of CLE and standard biopsy show. However, that CLE has a high sensitivity and specificity for the correct diagnosis.[14, 15] Using this technology, tissue sampling can be better targeted and the total number of necessary biopsies can be reduced.

Autofluorescence and spectroscopy

In comparison to normal tissue, a number of pathologic processes such as inflammation, ischemia, and dysplasia show distinctive fluorescence behavior. The principle of fluorescence diagnostics rests on the recognition that light with a defined wavelength (400–500 nm) not only is absorbed and reflected in tissue but also causes autofluorescence in natural or introduced fluorophores, such as 5-aminolevulinic acid.[16,17]

> Endomicroscopy provides a magnified image of the endoscopically investigated tissue and allows for histologic evaluation during endoscopy, that is, cellular level in vivo diagnosis, which is referred to as an optical biopsy.

Light-filtering image-enhancement technologies: NBI, i-SCAN, FICE, and SPIES

In virtual chromoendoscopy, various color spectra are produced through modulation of incident light with filters (NBI) or through software-based follow-up processing of reflected light (i-SCAN, FICE, or SPIES)[15]and through this process the individual components of the mucous membrane, such as its surface or vascular structure, are more clearly represented. NBI, i-SCAN, FICE, and SPIES are light-filtering systems that alter the wavelength ranges of reflected light thereby enhancing or augmenting vessels and various tissue structures in the images as an optical follow-up.

NBI (Narrow-Band Imaging; Olympus, Tokyo, Japan) is the oldest established process of virtual chromoendoscopy and is based on narrowband incident light that consists of only two wavelengths, 415 nm (blue light) and 540 nm (green light). NBI is a technique for optical contour emphasis of blood vessels and mucous membrane structures.

"I-SCAN is a software-based, image-enhancement technology that is classified as a digital contrast method among endoscopic imaging techniques."[18] The technology (Pentax Medical, Tokyo, Japan) is based on an integrated software tool that magnifies the surfaces with the help of a surface refinement function, and the virtual chromoendoscopy is made possible by the additional application of specific color filters.

FICE (Flexible Spectral Imaging Color Enhancement; Fuji Film, Tokyo, Japan) and SPIES (Storz Professional Image Enhancement System; Karl Storz) are systems for computer-supported virtual chromoendoscopy. The SPIES CLARA and CHROMA systems improve the endoscopic representation of detail through homogeneous illumination and contrast emphasis (LED technology). While perserving natural coloring, superficial structures are crisply imaged (by the complementary metal-oxide semiconductor [CMOS] technology).

6.2.3 Equipment for Documentation

For OR data recording and transfer, we use AIDA 2.0 (Advanced Image and Data Acquisition 2.0; Karl Storz), a modular system. Image, video, and audio data of the diagnostic or therapeutic intervention are recorded and saved directly from the sterile operating room (OR) by touch screen, foot switch, or camera-head key.

The system is geared to the activities and requirements in the OR and offers step-by-step exactly the functionalities needed, from the recording of patient data to the definition of the use of that data. The documentation of an endoscopy examination includes:

- Patient identification and indication of the investigation.
- Identification of the investigator and his or her assistants.
- Endoscope type and record of instruments.
- Time record of the investigation and intervention process.
- Endoscopic diagnosis and therapy.
- Coding by ICD (International Classification of Diseases) and OPS (Operations and Procedures Manual, OPS is an adaptation of the English version of the International Classification of Procedures in Medicine).
- Report of findings with endoscopic image material (▶ Fig. 6.9).

The consensus committee of the World Initiative on Spinal Endoscopy[19] agreed in 2006 in Graz, Austria, that the data for

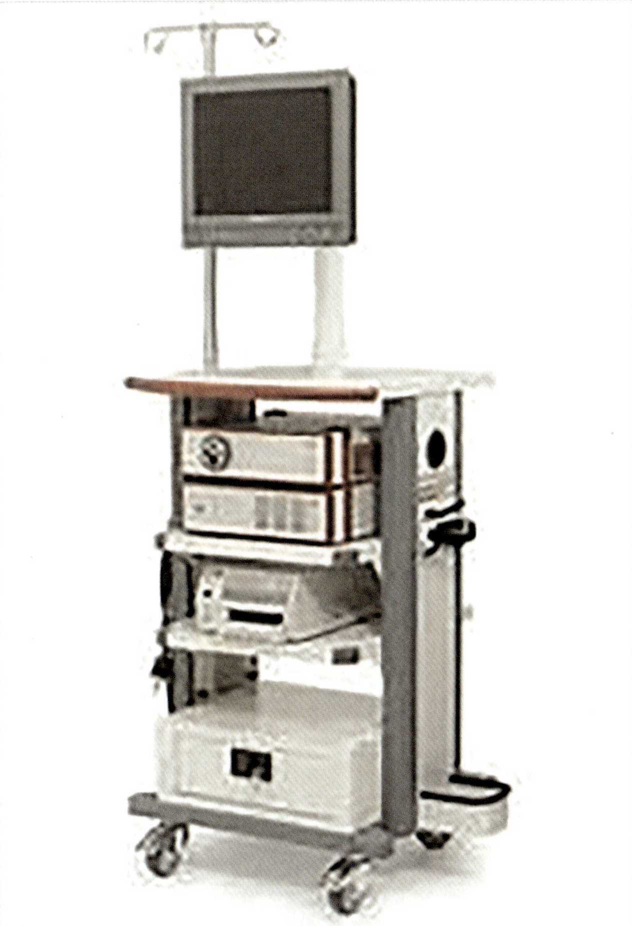

Fig. 6.9 RIWOmobil equipment cart, a video tower for spinal endoscopy. As a Class 1 medical device, the RIWOmobil, one of the few carts on the market, possesses a model verification for the warranty of operational safety. All swivel arm combinations are tested for load capacity and tilt safety. The position of the monitor can be adjusted to the requirements of the operating room team in all RIWOmobil versions. (Courtesy of Richard Wolf, Knittlingen, Germany.)

documentation should include the following: epiduroscopy type (purely diagnostic or therapeutic); subsequent treatments (laser, radio frequency (RF), or mechanical adhesiolysis); instrument type; volume of injected irrigation fluid; length of the procedure; side effects and complications.

Based on proprietary, prefabricated templates, a comprehensive operative report for the invasive-interventional procedure is generated in the shortest time. The AIDA system (Karl Storz) we use consists of database-supported application software and a corresponding hardware platform with a Windows operating system. The data for the epiduroscopy can be stored in by patient in an AIDA database, burned onto a CD/DVD, and/or transferred to another system (PACS [picture archiving and communication system], RIS [radiology information system], and KIS [KARE Information Services]) in DICOM3 [Digital Imaging and Communications in Medicine] and/or HL7 [Health Level 7 International] formats). With these systems, complete data of the diagnostic or therapeutic operation can be recorded directly from the sterile epiduroscopy area.

6.2.4 Equipment for Reprocessing of the Endoscope

The reusable epiduroscope undergoes preparation, manual cleaning and disinfection (or mechanical cleaning and disinfection), inspection and maintenance, packaging, sterilization, and storage before use.

One should check that their national health instructions for endoscope instrument outfitting and for organizing and performing epiduroscope reprocessing fulfill the current corresponding international capabilities and safety requirements.[10,15,21]

Endoscopes, which are inserted into tissue or into the neuraxial space, must be unconditionally sterile when they are used. To insure this, epiduroscopes must undergo a special sterilization procedure after cleaning and disinfection.[20] Endoscopes or endoscope parts or accessories that can withstand the corresponding temperatures should be sterilized in an autoclave. Thermolabile epiduroscopes or endoscope parts must at least undergo a plasma sterilization.

Visual Check

Flexible endoscopes, as encapsulated systems enclosed against the surroundings, are accessible only to a visual check. It is recommended to perform visual inspection of the epiduroscope for damage, particularly to the introducer sheath,, and the optical system (▸ Fig. 6.10).

Leak Test

The goal of the leak test is to identify and exclude damaged epiduroscopes from the preparation process. Before the endoscope is brought into the automated endoscope reprocessor, a leak test must be performed according to the manufacturer's instructions. The endoscope is placed in a water bath, and a slight excess pressure is introduced into the interior of the endoscope. In the case of damage, air bubbles escape past the point of damage.

Cleaning, Disinfection, Sterilization, and Repair of the Endoscope

Cleaning and disinfection

Epiduroscopes are very complex specialized instruments and are counted among the medical instruments most difficult to prepare.

For the processing of flexible reusable epiduroscopes, binding instructions exist, which are adapted to the respective current national hygiene prescriptions and recommendations (Medicine Product Law (MPG), BGBl. Nr. 657/1996 idgF). The essential hygiene instructions include:

- The epiduroscope is to be processed immediately after the endoscopy to prevent organic material from drying on it.
- The epiduroscope is to be cleaned of external soiling with a disposable cloth immediately after use.
- After spinal endoscopy, the epidurascope is to be mechanically cleaned (e.g., brush cleaning and flushing of the working channel) so that coarse uncleanliness is removed and the patency of all channels is ensured.

After the mechanical cleaning, a machine cleaning and chemothermal disinfection of the epiduroscope are to be carried out. Removable parts (e.g., valves and protective caps) are to be disinfected separately.

Disinfection tools in which precleaning and patency and leak tests are performed in the machine before disinfection are recommended. Single-chamber and multichamber automatic cleaning and disinfection machines are available. Machine

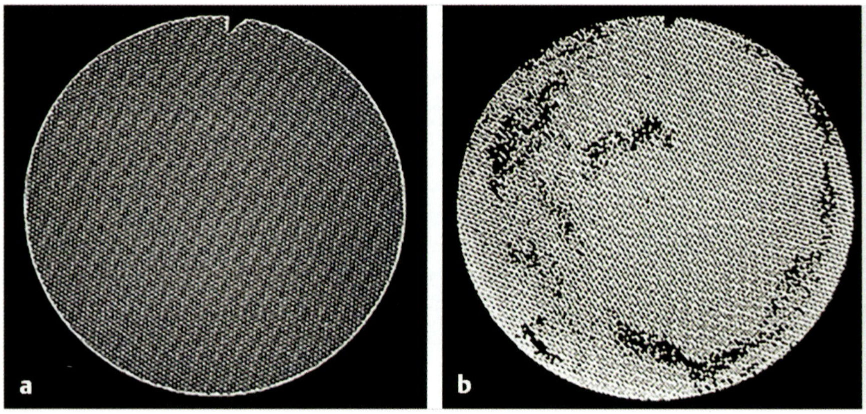

Fig. 6.10 Image conduit. (**a**) Undamaged condition of the fiber bundle. The image conduit is 100% intact. (**b**) Condition of the fiber bundle after multiple uses. The image conduit is damaged.

chemothermal disinfection is the state of the art. On the basis of being able to verify the time, temperature, chemistry, and mechanics of the disinfection process, machine cleaning and disinfection are preferred.

The disinfection and cleaning machine flushes the endoscope with an aldehyde-containing disinfection medium at about 60° C. This automated procedure can be used only with endoscopes that can be completely immersed in the disinfection solution and can also withstand the required temperatures. Currently only those procedures using special aldehyde-based instrument disinfection media qualify as disinfection procedures. But do note that these procedures are not sterilization.

With this type of processing it is important that after the flushing, the disinfection medium remains in the narrow channels without bubbles for the minimum contact time prescribed for disinfection.

Without a reliable and above all validated cleaning and disinfection of the epiduroscope, there can be no successful sterilization (▶ Fig. 6.11). In reprocessing of epiduroscopes the following problems can occur:

- Insufficient manual precleaning before the machine cleaning and disinfection.
- Incompatibility of ingredients of the cleaning medium (cleanser with and without antimicrobial effect) used for external and mechanical cleaning with the active substances of the final disinfection medium.
- Insufficient rinsing-out of the cleaning medium.

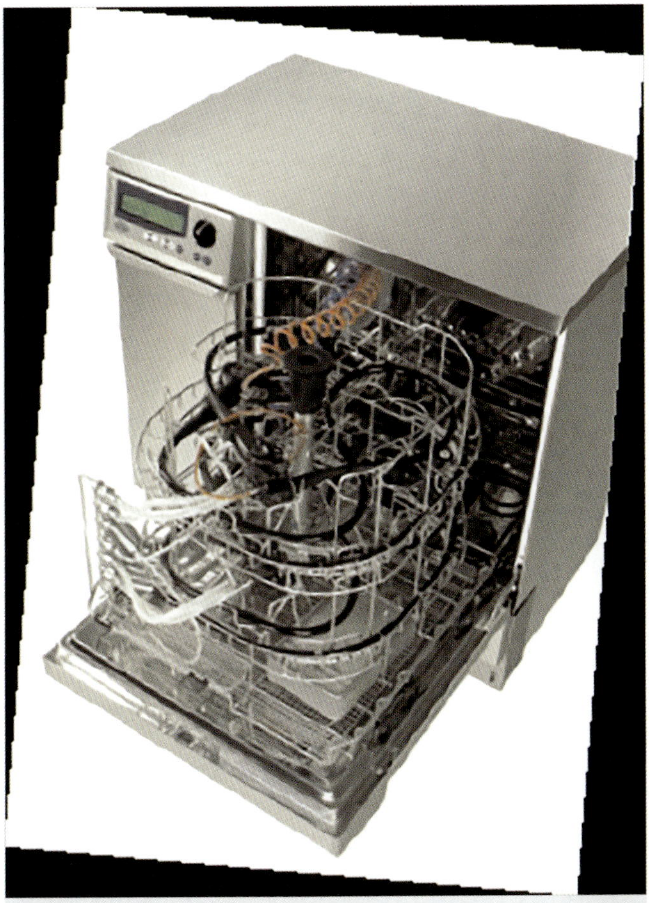

Fig. 6.11 ETD3 automatic washer/disinfector (Olympus, Tokyo, Japan). With the introduction of watertight endoscopes, automatic cleaning and disinfection of epiduroscopes could be performed. Beforehand, proper dismantling, as well as manual precleaning if necessary, of the epiduroscopic equipment is required, because without proper dismantling, processing is not possible.

Sterilization

In the ideal case of sterilization of an epiduroscope all microorganisms and their spores would be killed, and all viruses, plasmids, and DNA fragments would be destroyed.[22] In comparison with the technical limits for disinfection, sterilization requires that about a logarithmic step higher of germ-forming units are killed or inactivated; thus, in sterilization these units must be reduced to at least 10^{-6}, that is, one at maximum survives from a million germ-forming units. Hence, placing an epiduroscope in aldehyde-containing sporicidal solution is not considered a sterilization method.

Steam is not good for sterilization of all types of goods, above all, not for modern instruments that are made of combinations of materials.

According to the Swiss-NOSO CJD (Creutzfeldt-Jakob disease) Task Force the risk of a transmission of prions in the setting of endoscopic operations is neither approved nor quantified. Even if there has heretofore been no agreed on international sterilization procedure that is on the one hand is compatible with the materials of flexible, reusable epiduroscopes and on the other hand demonstrates a complete effectiveness against prions, according to Balmelli,[9] individual countries have published recommendations.

Also the packaging material must match the respective sterilization procedures and the corresponding strict prescriptions on the packaging of sterile goods are to be observed. In addition, the large number of European and U.S. prescriptions for sterilization are to be observed unconditionally.

Ethylene oxide sterilization

Ethylene oxide can be used for sterilization of medical instruments and materials. This process should, however, be used but only if classical thermal procedures with steam or hot air are not usable because of the heat sensitivity of the goods to be sterilized .

Ethylene oxide is a reactive, inflammable gas that forms an explosive mixture with air. Therefore, for use, it is mixed with inert gases. Ethylene oxide is a respiratory tract irritant, a poison to protoplasm, and acts as a carcinogen. It penetrates well and is absorbed by many materials; therefore outgassing times (to ≥ 1 ppm residual content) from 4 hours to 2 weeks are necessary. The sterilization temperature is 50 to 60°C. Ethylene oxide gas bonds to all types of albumin molecules, cells, viruses, and cell components and irreversibly inactivates them. Further, ethylene oxide at a sufficiently high concentration inactivates bacteria and viruses by itself.

The long outgassing times for sterilization with ethylene oxide, however, limit its practical use. The risks of handling ethylene oxide and the long outgassing times are problematic for epiduroscope sterilization.

In our own pain clinic, ethylene oxide sterilization is not done, but instead we exclusively use plasma sterilization.

Plasma sterilization

Plasma sterilization is useful for most medical thermolabile instruments and implements. Plasma sterilization is based on the high frequency excitation of hydrogen peroxide under low pressure from which a so-called peroxide plasma arises. It is a relatively low-temperature sterilization process that was developed principally for heat- and humidity-sensitive instruments and materials. Plasma sterilization has been for us as a crucial advance in the sterilization of highly sensitive epiduroscopes

Plasma sterilization is not effective against prions. As opposed to conventional sterilization processes, plasma sterilization has the essential advantage that it does not require the use of radioactivity, high temperatures, or toxic chemicals. Plasma sterilization leaves behind no residues dangerous to health; it is energy saving and environment friendly. Plasma sterilization is a dry sterilization at 45 to 50°C without toxic residue and allowing for quick reusability of the epiduroscope (▶ Fig. 6.12).

> Through the effect of hydrogen peroxide in combination with high frequency (plasma sterilization), the germ-free condition of the epiduroscope is achieved.

In plasma sterilization, the endoscopy goods to be sterilized, packaged in a container, are introduced into the sterilization chamber. The chamber is evacuated, and into this vacuum hydrogen peroxide at a concentration of up to 60% is evaporated.

All surfaces accessible to the H_2O_2 are sterilized. Then, the gas plasma chamber is evacuated to remove the majority of the H_2O_2. As a cleaning step, the rest of the gas is excited to a plasma by high-frequency voltage. In this plasma distinctive species of H_2O_2 arise, which finally recombine to form water (H_2O) and oxygen (O_2). In our pain clinic the advantages of the plasma sterilization process for flexible, reusable epiduroscopes have been definitive (▶ Fig. 6.13).

Transport and storage

For safe storage, transport, and sterilization of flexible epiduroscopes and their supporting instruments intelligent solutions are offered by many manufacturers. At the center stands the Sterisafe DURO A3 (Richard Wolf) or the Sterrad (Advanced Sterilization Products) plastic containers for sterilization with perforation and lids for use with flexible endoscopes (Sterrad-compatible 39405 AS, Karl Storz) especially for gas and hydrogen peroxide sterilization, or the plastic container for sterilization and storage of high definition (HD) camera heads (Sterrad and Sterrad-compatible 39301 HCTS, Karl Storz, respectively) also for steam, gas, and hydrogen peroxide sterilization.

To guarantee a good penetration of the packaging material with H_2O_2, cellulose-free packaging should be used. Paper and cardboard, unfortunately, cannot be used because of the interaction with H_2O_2 in the plasma sterilization process.

Endoscope repairs

Reusable epiduroscopes are subject, even with the most careful handling, to multiple changes and wear processes. Experience

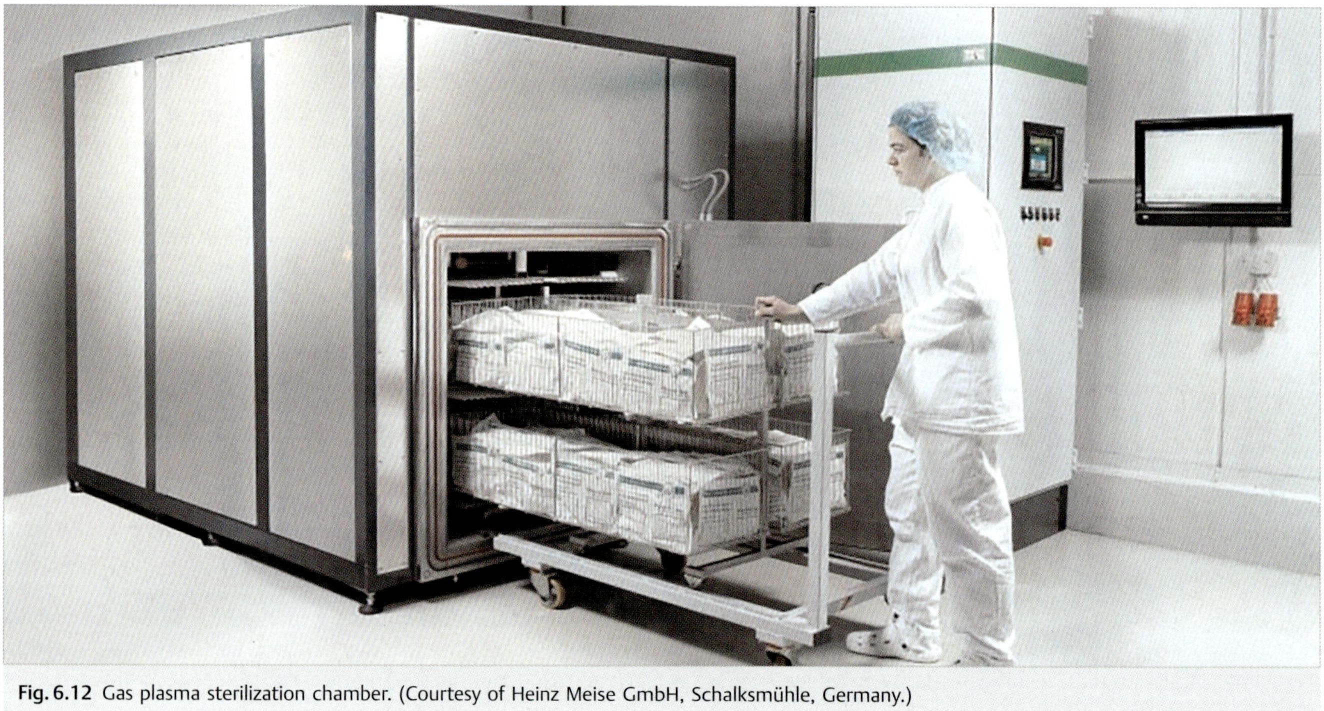

Fig. 6.12 Gas plasma sterilization chamber. (Courtesy of Heinz Meise GmbH, Schalksmühle, Germany.)

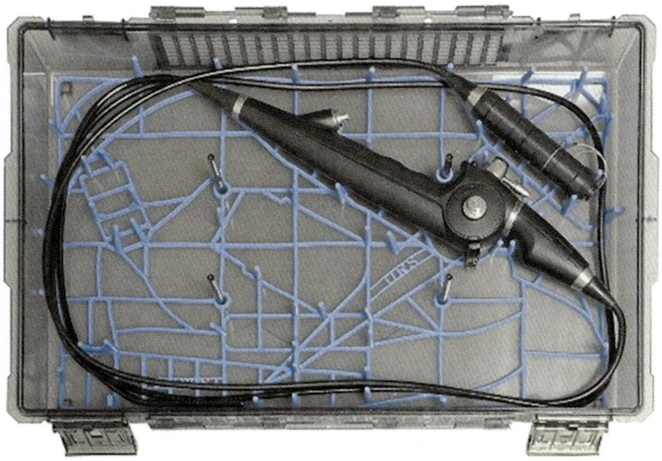

Fig. 6.13 Sterisafe DURO A3 container for introducing goods to be sterilized into the sterilization chamber. (Courtesy of Richard Wolf, Knittlingen, Germany.)

shows that the technical lifespan of the endoscope in hospital use frequently is 3 years.[23]

According to Bader et al,[12] potential interactions with cleaning and disinfection media during the reprocessing of epiduroscopes can trigger problems with the materials from which the epiduroscope is made, such as penetration into plastic, dissolution of adhesive connections, and corrosion of metal.

Following interaction of alcohols, amino acids, and amino acid derivatives with plastics and adhesive connections in the epiduroscope, penetration of these plastics and adhesive connections by disinfection media is also possible.

- Based on our own 2006 damage reports on defects of Karl Storz epiduroscopes, an epiduroscope could be used 16.4 times before it was sent to the manufacturer for repair because of leakage, damage to the fiber bundle or the jacket of the epiduroscope, or reduction in image sharpness.
- Due to optimized endoscopy techniques, including spinal navigation, and technical improvements in the endoscope, as of 2014, an epiduroscope could be used 21 times before being returned to the manufacturer for repair.

6.3 Equipment for Monitoring and Navigation

For the endoscopic inspection of the lumbar, thoracic, and cervical regions of the spinal canal, the use of a suitable flexible epiduroscope, including the microsurgical accessories for intervention, is an important precondition. To reach the corresponding neuraxial sections at all, flexible epiduroscopes in lengths of 40 cm, 70 cm, and 90 cm with different outside diameters (ODs) are available.

As a rough aid to orientation and to the location of the endoscope in the spinal canal for the investigator without a C-arm, external markings are located on the distal epiduroscope (Karl Storz) at intervals of 5 cm.

6.3.1 X-ray Image Intensifier (C-Arm), or Fluoroscopic Unit

For assessment of the location of the endoscope in the spinal canal and for endoscopic navigation, fluoroscopy in the anteroposterior beam path and in the lateral plane is recommended. The use of an X-ray image intensifier (C-arm) for endoscopic navigation is advantageous for an endoscopic procedure.

Clinical Note

Through the use of an X-ray image intensifier (C-arm) one can safely and without doubt position the endoscope, guidewire, catheter, electrodes, and microsurgical instruments in the sought-for ventral or dorsal epidural location.

In fluoroscopy the nonabsorbed X-radiation exiting the body is captured and presented on a monitor as an image or as a video with the help of an image-intensifier TV system. The diagnostic image quality depends above all on the efficiency of the X-ray equipment, the recording technique, and the properties of the image-intensifier TV system. The coordination of the entire image production system with the recording parameters and resolution possibilities of the image-intensifier TV system is the prerequisite for adequate image quality and acceptable radiation exposure (▶ Fig. 6.14).

Modern digital image intensifiers make possible the greatest possible reduction of the radiation dose used through an optimal intensification of the received X-rays. With the help of X-ray image conversion techniques, the investigating physician can observe functional sequences of events in the spinal canal continuously on the monitor. For image recording, various documentation possibilities are available. All patient-specific data are also documented. A magneto-optical drive allows problem-free, filmless archiving. The X-ray image intensifier system possesses, among other things, a video output and a CD drive on which the images can be output in tagged image format (TIF) specific to a personal computer.

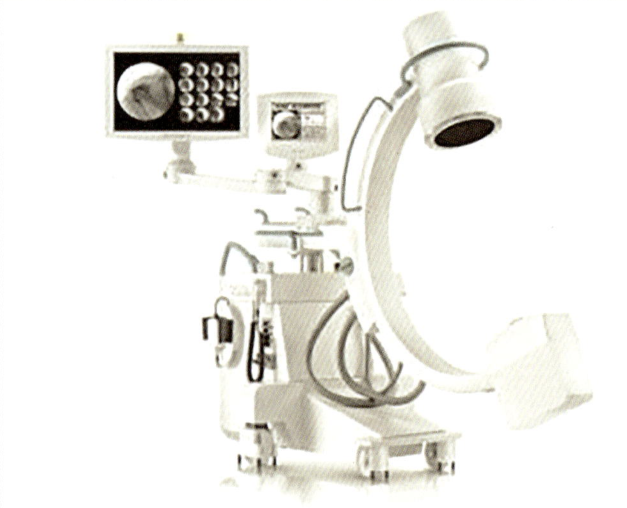

Fig. 6.14 C-arm image intensifier (Ziehm Exposcop CB7-D, mobile C-arm with MSA [maximum opacification sequence], RSA [roentgen stereophotogrammetric analysis], and DSA [digital subtraction angiography] functions with digital storage for 999 images with double monitor; resolution: 1.2 lines per mm; 23-cm HQ [high-quality]-BV [image intensifier]; Sony printer; small-format 23-cm HQ-BVb). (Courtesy of Ziehm Imaging GmbH, Nürnberg, Germany.)

Clinical Notes

Radiological support in the initial phase of spinal endoscopy

- C-arm technique, with endoscopy tip steering in a lateral X-ray beam path, provides the best help for success when anatomical–physiologic obstacles make endoscope insertion more difficult .
- C-arm technique can indicate required changes in direction of the endoscope or of the microsurgical instruments or show alternative pathways for successful movement.

C-Arm and Contrast Medium

Epidural pathologic structures are often difficult to visualize with conventional radiologic techniques alone. But the use of these conventional radiologic techniques with spinal endoscopy (epidurography or myelography) for has had some success.

In this technique an X-ray contrast medium is applied in the neuraxial space (epidural intrathecal space) and simultaneously X-ray images (C-arm image intensifier) and/or video sequences (epidurography or myelography) can be obtained. With the help of the epiduromyelogram, additional important information on the actual pathologic anatomical situation of the epidural space or neuraxial region of the vertebral column can be obtained.

Use of the radiologic technique (C-arm + contrast medium) with epiduroscopy allows for:

- Height localization and evidence of the ventral or dorsal location of the epiduroscope or of microsurgical instruments.
- Additional radiologic diagnostics (epidurogram or myelogram) before, during, or after an epiduroscopy.
- Checking for catheter placement and evidence of catheter misplacement.
- Checking the results of an epidural intervention.

Clinical Note

It is recommended that an epidurogram be obtained before every epiduroscopy after application of contrast medum, either over the cannula introduced sacrally or the sheath. Thereby the investigator has a current pathologic anatomical overview of the spinal canal. The anhydrous contrast medium iopamidol (Solutrast 200 M and 250M) contains 408.2 or 510.3 mg of iodine per mL. The iodine atoms absorb X-rays, on which the specific radiopaque effect rests.

Different volumes of the contrast medium, from 1 to 15 mL according to the situation, are injected in the cervical, thoracic, lumbar, or sacral region to be evaluated. No more than 3.75 g of iodine (15 mL of Solutrast 250) should be injected. In the literature a special complication with the occurrence of encephalopathy and rhabdomyolysis, induced by the application of iotrolan (a similar iodine-containing contrast medium to Solutrast), has been reported. The X-ray contrast medium entered the intrathecal space through an injury to the dura mater.[24]

Clinical Note

Due to the extraordinary breadth of anatomical variation in the area of endoscopic access to the sacral hiatus or sacral canal and the pathologic anatomical peculiarities of patients, even the technical feasibility of an epiduroscopic clinical intervention can be very well evaluated by the actual findings of epidurography.

6.3.2 Spinal Sonography and Epiduroscopy (SONO-EDS)

The use of ultrasound procedures today is part of the permanent medical repertoire. Ultrasound diagnostics have established themselves in anesthesiology and the subspecialties of intensive care and emergency medicine. Sonography is also steadily gaining significance in the realm of pain medicine.

The steerable epiduroscopes available at present indeed provide an endoscopic view of internal structures for orientation; however, size determination or area measurement of likely pathologic epidural and adjacent structures are difficult or impossible endoscopically.

To optimize endoscopic investigations with the epiduroscope, in 2000 Schütze[8] integrated microinvasive intraluminal ultrasound (intravascular ultrasound [IVUS]) into epiduroscopic technology creating SONO-EDS (▸ Fig. 6.15).

Schütze[8] first reported on clinical epiduroscopic-sonographic investigations (SONO-EDS) in patients with failed back surgery syndrome (FBSS, also called failed back syndrome, FBS).[3,8]

An ultrasound probe—a Visions Five-64 F/X catheter (IVUS Imaging Catheter)—placed over the working channel of the epiduroscope in the epidural space was used. The ultrasound probe produced over the ultrasound converter, ultrasound images with an angle of view of 360° vertical to the axis of the ultrasound probe, The emitted ultrasound waves enter the neuraxial space with varying effect on the

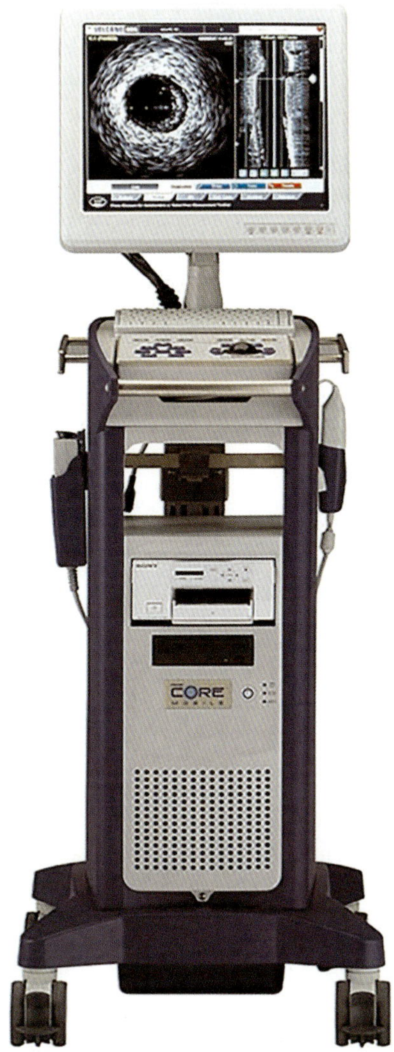

Fig. 6.15 CORE. (Courtesy of Volcano Corporation, San Diego, California.)

surrounding tissue. The tissue density and the water content in the epidural space affect this, as well as the beam angle, the form of the slice of the corresponding anatomical structure, or the pressure of the sound probe on the neuraxial tissue.[25]

Clinical Note

The ultrasound image of the spinal dura mater is particularly good and for that reason when consulted is outstandingly helpful for neuraxial navigation.

The advantages of sonographically supported epiduroscopy (SONO-EDS) are:

1. Broadening and optimizing of the endoscopic image information with information on anatomical structures outside of the live endoscopic image.
2. Improvement of spinal image production and navigation through the picture-in-picture (endoscopic- -sonographic) image presentation.
3. Provision of virtual histology information on pathologic structures from via color-coded endosonographic images.
4. Size determination and area measurement of structures.
5. Safeguarding against accidental injury to the spinal dura mater.

6.3.3 Monitoring Vital Functions, Brain Function, and Epidural Pressure and Volume

Vital Functions

Preparation of the patient that must be completed before the beginning of the spinal endoscopy includes placement of an IV access line, saline infusion, blood pressure measurement, EKG, and oxygen saturation measurement (SaO_2). For spinal endoscopy, analgesic sedation adequate to the patient's needs in the operative situation with an ultrashort awakening time following is advisable. With the use of anesthesia (analgesic sedation), equipment for monitoring the vital functions of the patient and for therapeutic intervention if complications arise must be provided.

Brain Function (EEG)

For optimizing of the operative patients' safety and their diagnostic and treatment quality during spinal endoscopy, monitoring of their brain activity is advantageous during analgesic sedation. We recommend the Danmeter Cerebral State Monitor (CSM, Danmeter A/S, Odense C, Denmark) The CSM is a noninvasive EEG measurement device, which measures variations in frequency of brain waves.

The EEG curve of the patient is derived from the signal recorded between the frontal and mastoid electrodes. EEG signal frequency changes that occur with changes in the patient's degree of awareness are continuously analyzed by the CSM, and it computes the Cerebral State Index (CSI) on the basis of these. Using the CSI, the patient's degree of awareness can be estimated. The CSM (Cerebral State Monitor) serves patient safety and optimizes the evaluation of an endoscopically supported pain provocation test (▸ Fig. 6.16).

- Through the continuous surveillance of brain activity during the invasive-interventional epiduroscopy via the CSM, an evaluation of an endoscopically supported pain provocation test can be carried out.
- The CSI is measured on a scale from 0 to 100. The value 0 is for a flat EEG, and the value 100 is for EEG activity that corresponds to an awake state; CSI values from 40 to 60 are found in cases of adequate narcosis.

Fig. 6.16 Cerebral State Monitor (CSM). (Courtesy of Danmeter A/S, Odense, Denmark.)

Epidural Pressure

For direct pressure measurements, epidural probes are available. It is also possible to install epidural pressure monitoring via an integrated transducer system for continuous surveillance of pressure.

A fundamental problem of every epidural pressure measurement is whether the place of measurement is also the place of highest pressure.

Based on our experience, with the use of an infusion pump for epidural infusions during an epiduroscopy, a pressure of no more than 60 mm Hg should be attained. Admittedly a pressure of 50 to 70 mm Hg can be reached with a cautious epidural injection; with a rapid bolus from a 10 mL syringe, reaching up to 300 mm Hg.[7]

Clinical Notes

- Depending on the epidural NaCl volume replacement and the neuraxial pathologic anatomical conditions in the epidural space, the epidural pressures when measured in our patients undergoing epiduroscopy varied between 22 mm Hg and 85 mm Hg.
- The impact of an epidural volume-controlled 0.9% NaCl infusion adequate to the situation on spinal pressure during an epiduroscopic investigation appears in our experience rather harmless.

Warming Epidural Saline Infusions

A prerequisite for epiduroscopy is a continuous or adapted-to-requirements epidural infusion with physiologic saline solution through the working channel of the epiduroscope. To avoid unnecessary spinal irritation by cold epidural NaCl infusions (at operating room [OR] temperatures), the epidural infusions should be warmed to body temperature.

Admittedly, with the use of conventional liquid warming systems, on the basis of the OR temperature, cooling of the infusion occurs very quickly between the warming device and the connection to the epiduroscope. This problem can be solved through active warming of the patient lines to the endoscope connection, for example, with the HOTLINE 2 warming system (Smith Medical International Ltd, Kent, UK) (▶ Fig. 6.17). Thereby, an infusion of NaCl solution with normothermic temperature (37° C) at pressureless flow rates between 50 and 4250 mL/h (or 0.8–71 mL/min) can be achieved.

The selection criteria for a suitable epidural infusion system, include not only the infusion rate and the infusion duration but also the dose accuracy.

According to the requirements, a system can be selected from one of the following types:

- *Gravity infusion:* These are manually directed epidural infusion systems. The sterile NaCl infusion occurs due to the drop in hydrostatic pressure drop between the infusion bottle or

Fig. 6.17 HOTLINE Blood and Fluid Warmers deliver normothermic fluid at routine flow rates. (Courtesy of Smiths Medical International Ltd., Kent, UK.)

bag and the patient. This type of system is recommended if low requirements are set for the infusion rate and the dose accuracy. Gravity infusion is of little help for epiduroscopy in our experience.

- *Pump-supported infusion:* In these systems, the infusion is performed by means of a pump, either volumetrically by use of an infusion system or by use of a syringe pump. This form of infusion ensures the highest dose accuracy as well as constant infusion rates. But because the infusion rate must constantly be adjusted to the actual pathologic anatomical situation in the neuraxial region during spinal endoscopy, its use for this is problematic at the least. This epidural infusion system is more suitablee for performing epidural analgesic therapy (EAT).[26]
- *Pressure infusion:* Pressure infusion guarantees rapid epidural volume replacement. By pressure, directed either manually or by a pressure cuff, the infusion rate can be adjusted to the changing requirements during epiduroscopy. The pressure can be held constant by means of a pressure infusion cuff and requires no afterpump as with manual systems. Reusable pressure infusion cuffs as well as cuffs for single-patient use are offered (▶ Fig. 6.18).
- *Pressure infusion cuffs:* Manual pressure infusion systems utilizing cuffs serve as rapidly manually regulatable controls of epidural NaCl infusions and are well suitable suited for epiduroscopy. Reusable cuffs as well as cuffs for single-patient use are available. Pressure infusion cuffs are handcrafted and stand for precision and quality (▶ Fig. 6.19).

Clinical Notes

For epiduroscopy, principally we use a sterile 0.9% NaCl irrigation solution To avoid spinal irritation of the patient, the solution is warmed to 37.5° C by a fluid warmer (Medi-Temp III, Stryker, Kalamazoo, MI).

Too rapid a drip sequence of the sterile physiologic NaCl irrigation solution can trigger pain in the patient, above all in the thoracic or cervical regions under investigation and in stenotic regions of the spinal canal. A reduction in the drip sequence will stop the acute pain.

We use on average 85 mL of sterile physiologic saline solution for an epiduroscopy. This is significantly below the maximal infusion volume of 200 mL for an epiduroscopic investigation recommended by the international consensus in Graz in 2006.[19] According to our clinical experience, only as much normothermic sterile 0.9% NaCl solution should be used as is necessary to create an open endoscopic field of view.

6.3.4 Equipment for Spinal Neuronavigation

In recent years, neuronavigation of the spine, visualization of the spinal anatomy without repeated X-rays, has become common during spinal operations. Spinal neuronavigation describes the computer-assisted transfer of patient image data to thc operative site of the spinal column.

Spinal neuronavigation is an indispensable prerequisite for the success of a spinal endoscopy for diagnosis and of neuraxial pain syndromes.

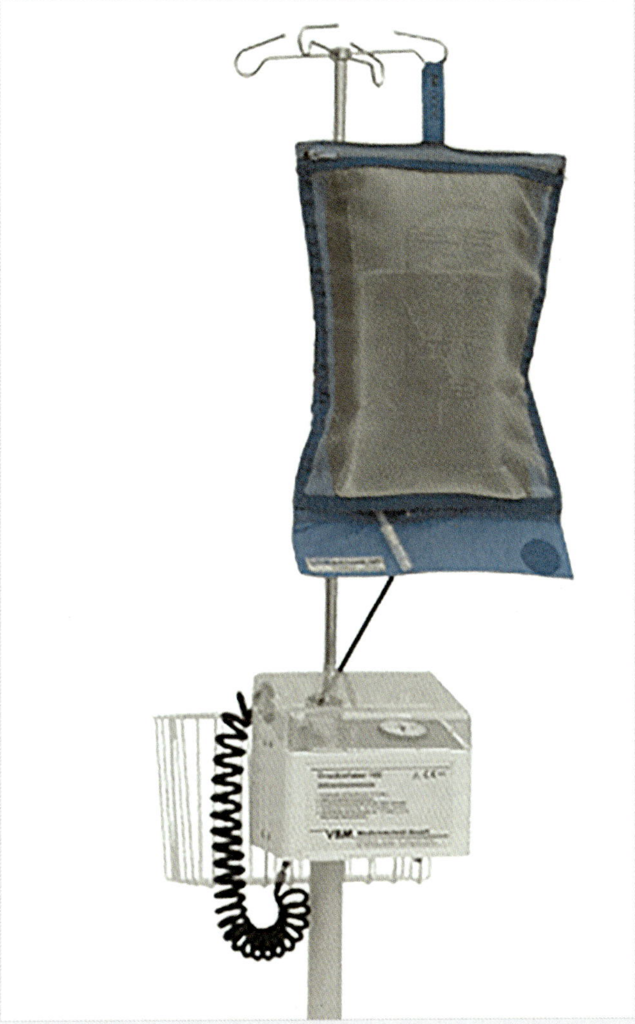

Fig. 6.18 Automatically regulated electrically powered pressure infusor. (Courtesy of VBM Medizintechnik GmbH, Sulz am Neckar, Germany.)

Epiduroscopic-Sonographic Radiologic-System

Through the use of this complex system, the highest technical prerequisites for a spinal endoscopy are met. The tight interweaving of the endoscopic, sonographic, and radiologic equipment (EDS-SONO-RO) essentially expands and optimizes the possibilities of spinal endoscopy for the diagnosis of and therapy for neuraxial pain syndromes.

One can become successful at spinal navigation through consistent, goal-directed use of this system. It requires, however, a high measure of effort, education, and experience.

If desired, X-ray contrast medium can be applied to the epidural space for epidurography or myelography,via a percutaneous injection, a sacral introduction aid, an epidurally placed catheter, or the epiduroscope itself, under targeted X-ray image converter technique.

The C-arm technique is outstanding for supporting spinal neuronavigation in the entire spinal column. Particularly

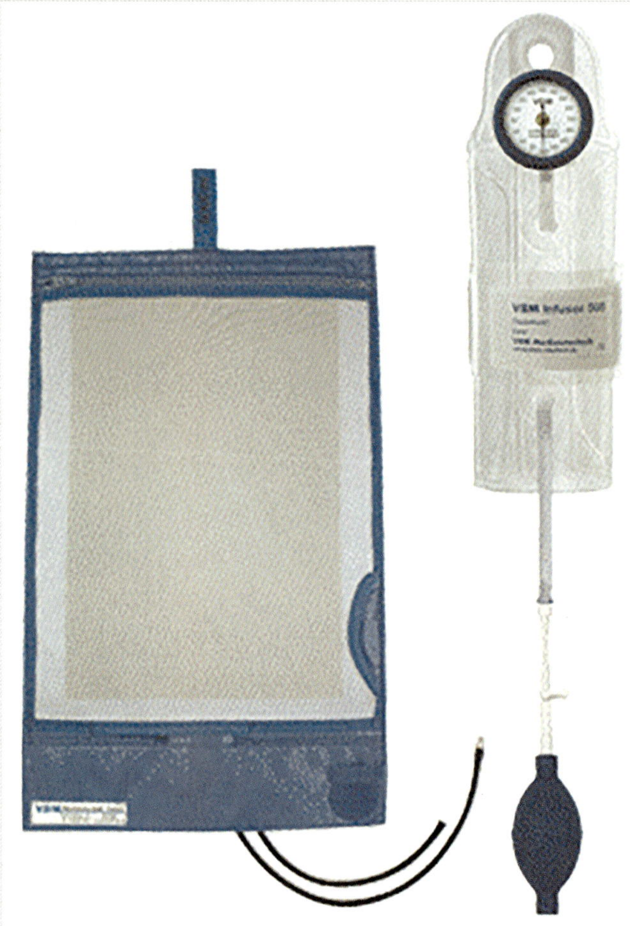

Fig. 6.19 VBM Infusor 500/1000, a transparent pressure infusion cuff for rapid administration of irrigation and infusion solutions. (Courtesy of VBM Medizintechnik GmbH, Sulz am Neckar, Germany.)

helpful is navigation and height determination of the endoscope in the thoracic and cervical regions.

In the overall endoscopic picture, neuraxial structures outside the endoscopic field of view (e.g., cerebrospinal fluid or spinal dura mater) that cannot be captured with the epiduroscope, can be brought in simultaneously by the X-ray image converter (C-arm).[24]

For that, the equipment required along with the epiduroscopic equipment is an ultrasound probe with the modern Cathscanner Imaging System (with its spectral analysis method), which can be connected with the EndoSonic or Volcano IVUS system without problems. This technology offers an automatic measurement tool to simplify the interpretation of the images and uses a predefined color coding for display of the tissue composition of, for example, a defined region of the dura mater or inside the epidural cavity.

With the use of established sonographic nomenclature, the additional simultaneous sonographic information from the epiduroscopy can be analyzed and described. An automatic tracing with very rapid computation of the sonographic findings makes solid diagnostic statements possible. The high-contrast resolution ensures reliable representation of the fine tissue structures in the sonographic sections imaged. The Cathscanner technology allows a realistic presentation of the neuraxial tissue localization and attains a sonographic section thickness as little as 1 millimeter.

The best preconditions for a useful and effective spinal navigation with a high patient safety factor is control of the epidural infusion with simultaneous use of a C-arm technique and endoscopic-sonographic image production in the so-called picture-in-picture format.

In addition to spinal navigation, the composite EDS-SONO-RO system is useful for:

- Support of the technique of sacral access to the epidural space.
- Additional imaging of pathologic anatomical neuraxial structures.
- Verification of the location of catheters, electrodes, and microsurgical instruments.

6.4 Equipment for Endoscopic Therapeutic Interventions

The indication for invasive–interventional epidural pain therapy for most patients is an unsolved pain problem. Neuraxial pain syndromes, even where radicular involvement is indicated, often are accompanied by pain that shows a nociceptive, or neuropathic, or mixed pain origin.

With the help of appropriate instruments, endoscopically supported interventions such as biopsies, adhesiolyses, resections of scar tissue, withdrawal of irrigation fluid or cerebrospinal fluid, removal of lipomas, hemostasis, extirpations of foreign bodies, and abscess drainages can be performed. The necessary instruments for these endoscopically supported therapeutic interventions will be presented in Chapter 8.

> We believe spinal endoscopy early for the diagnosis of and therapy for neuraxial pain syndromes is just as much a part of an intelligent, appropriate, interdisciplinary pain management plan as epiduroscopy along with CT and MRI is a part of the tools for diagnosis and monitoring of pain therapy.

6.4.1 Laser

In pain medicine, the possibilities for epiduroscopic operations are expanded via the use of laser techniques. The laser (meaning *l*ight *a*mplification by *s*timulated *e*mission of *r*adiation) is used in many ways in medicine, such as for hemostasis, recanalization of tumor stenoses, or destruction of deposits in blood vessel walls.

There are sufficient pilot studies, prospective studies, and prospective randomized studies of the use of 1064-nm neodymium:yttrium-aluminum-garnet (Nd:YAG), 1320-nm Nd:YAG, and 940-nm diode lasers in medicine. In 2002, Ruetten et al reported on 47 patients in whom the authors detected epidural adhesions epiduroscopically that they with a holmium:YAG laser.[17] In follow-up investigation, no patients exhibited signs of pathology at the site where the adhesions had been.. Nor did any edema or adhesions occur that could be traced to the epidurosopic laser treatment. According to the view of the authors, the appropriate use of a Ho;YAG laser (2100 nm) in the epidural space is safe.[27]

On the basis of its physical properties in water, the diode laser, along with the Ho:YAG laser, is suitable for use on highly sensitive neuraxial structures (▶ Fig. 6.20).

Diode Laser

The wavelength of the diode laser is 980 nm, the power range is selectable from 1 to 25 W, and the penetration depth into tissue is dependent on wavelength and, of prime importance in the infrared range, on the water content of the tissue. The penetration depth is specified as a few millimeters. The diode laser beam penetrates into the tissue and is absorbed neither by hemoglobin, by melanin, or by water to any extent.

A diode laser is constructed modularly. Its laser beam can in principle be used directly in medicine. The indications for the diode laser are the same as those for the Nd-YAG laser, primarily coagulation and vaporization of tissue.

Clinical Notes

For epidural adhesiolysis, scar tissue resection, or hemostasis, we use a diode laser. Through the working channel of the epiduroscope, the light guide, a 320-μm bare fiber, is introduced into the epidural space under epiduroscopic view. Under excellent endoscopic monitoring, the diode laser can be maneuvered and safely used. Fine-grade navigation of the relatively stiff laser fiber can succeed only through the flexible endoscope tip. With the help of the laser pilot beam, a pinpoint accurate ablation can be achieved without bleeding occurring. The diode laser is activated after careful optical control and under continuous 0.9% NaCl infusion.

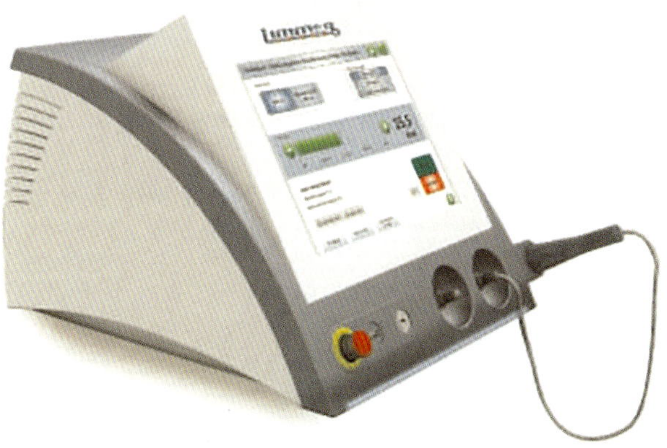

Fig. 6.20 Diode laser. (Courtesy of Limmer Laser GmbH, Berlin, Germany.)

Based on our experience a laser power range from 5.0 to 8.0 W is sufficient. The use of the diode laser for, for example, hemostasis, stenosis therapy, or flavoplasty is practicable, as is.endoscopically supported resection of epidural scar tissue relevant to pain within certain bounds depending on the pathologic anatomical findings and location. Endoscopically supported laser use also makes adhesiolysis possible in both the dorsal and ventral section of the epidural space.

Adhesions relevant to pain in the region of the nerve roots, such as those that cannot be mobilized with the endoscope tip by twisting and swinging motions, can be removed relatively easily with the laser. Our experiences with epidural laser therapy have been good, which are confirmed in the literature.[17,27]

Adhesions or fibroses in the immediate region of the nerve root should be avoided. If paresthesias occur, it is necessary to stop the laser process. Important for preventing laser complications are the knowledge and experience of the surgeon and excellent endoscopic visualization and management of the procedure. Complications of epidural laser therapy include accidental damage to neuraxial structures, bleeding, recurrence of stenosis through extended fibrin plaques, burns from the instruments and endoscopes through ignited gases, and toxic fumes. Accident prevention prescriptions and laser operation instructions in force nationally must be observed.

6.4.2 Radio-Frequency Therapy

Endoscopically supported radio-frequency therapy (RFT) offers wider possibilities for invasive-interventional methods to alleviate or eliminate chronic back pain.

In the original method, radio-frequency thermoablation, the tissue surrounding that which was vaporized was strongly heated during the continuous RFT application, which, along with pain reduction, could result in long-lasting damage to individual pain fibers. To avoid this additional unintended damage for the patient, at the end of the 1990s, pulsed RFT was developed, in which the surrounding tissue was heated only a little.

The indications for pulsed RFT (delivered by means of a multifunction electrode) are pain to any region of the spinal column of any origin, radicular pains, and other neuropathic pain syndromes.

The recommended standard access for a multifunction electrode for delivery of pulsed RFT for lumbar and sacral spinal column pain, as well as pains of the pelvic region and the legs, is percutaneous, paramedian access at the height of L3–L4 on the contralateral side.

For dorsolateral epidural electrode placement, the flexible multifunction PASHA-electrode (OMT GmbH, Frittlingen, Germany) is recommended. With this electrode, nerve stimulation with any desired frequency can be performed.

For the application of pulsed RFT to the nerve pathways in the framework of minimally invasive pain therapy, the PATRONA high-frequency generator (OMT GmbH, Frittlingen, Germany) has been developed.

Along with selecting an appropriate pulsed RFT program for a patient from those available, the highest possible patient safety is ensured through shutdown of the lesion program at

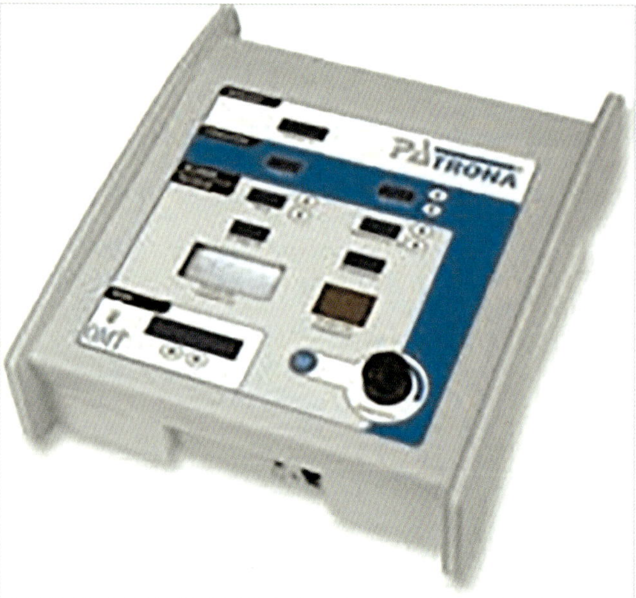

Fig. 6.21 PATRONA radio-frequency generator for interventional pain therapy with special features for the application of epidural pulsed radio frequency. (Courtesy of OMT GmbH, Frittlingen, Germany.)

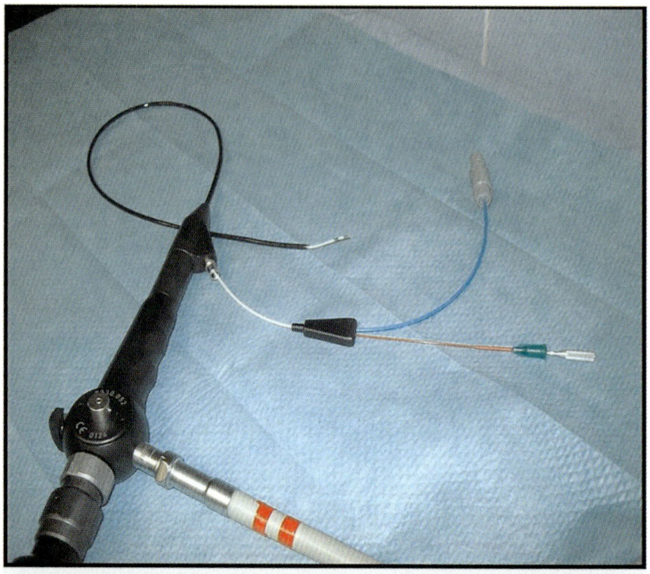

Fig. 6.22 PASHA multifunction electrode in epiduroscope.

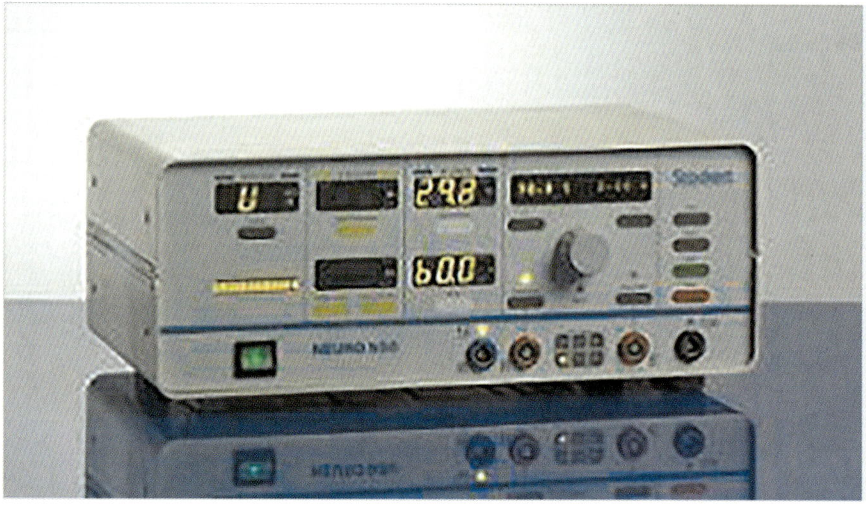

Fig. 6.23 Thermolesion generator Neuro N50. (Courtesy of Radimed Gesellschaft zur Medizintechnik mbH, Bochum, Germany.)

the time of attachment of the electrode. Documentation of the stimulation data depends on the storage capacity of the instrument.

The application of epidural pulsed RFT with an endoscopically placed PASHA multifunction electrode can be optimized through the program diversity built into the PATRONA generator. The combination of the PASHA multifunction electrode with the PATRONA generator offers a safe method of pain therapy (▶ Fig. 6.21).

Clinical Notes

- To be able to produce a corresponding electric field unerringly, we place the multifunction electrode under endoscopic control precisely epidurally over the sacral access desired to relieve the pain experienced by the patient.
- The electrode is inserted through the working channel of an epiduroscope with an OD of 3.5 mm to the epidural endpoint.
- Alternatively placement of the electrode parallel to the endoscope can also work (▶ Fig. 6.22).

In epidural pulsed RFT, a short (20 ms) active power pulse at 500 kHz is administered, followed by a pause of 480 ms. A maximum temperature of 42°C is all that is allowed by the computer-controlled instrument (Neuro N50, Radimed Gesellschaft zur Medizintechnik mbH, Bochum, Germany; 0.1–0.7 W). With voltages of under 0.8 V, stimulation perceptible to the patient can be produced (▶ Fig. 6.23).

The temperature-control program of the Neuro N50 also makes possible controlled lesioning techniques. All functional sequences are constantly supervised by the Neuro N50, providing maximal patient safety and highest operator confidence. Documentation of every operation, all settings and sequences, can be printed on an external printer or delivered to a PC.

R-ResAblator Epiduroscopy

Raffaelli and Righetti[28] reported on patients in whom they used an epiduroscopic high-frequency method, so-called R-ResAblator Epiduroscopy. This endoscopically supported RF therapy reduced pain by 90% in patients. The authors found that the RF epiduroscopy can show greater therapeutic benefit than traditional epiduroscopy or conventional surgery.

References

[1] Commission for Hospital Hygiene and Infection Prevention (KRINKO), Federal Institute for Drugs and Medical Devices (BfArM). Hygiene requirements for the reprocessing of medical devices. Recommendation of the Commission for Hospital Hygiene and Infection Prevention (KRINKO) at the Robert Koch Institute (RKI) and the Federal Institute for Drugs and Medical Devices (BfArM). [in German]. Bundesgesundheitsblatt Gesundheitsforschung Gesundheitsschutz. 2012; 55(10):1244–1310

[2] Rey JW, Kiesslich R, Hoffman A. New aspects of modern endoscopy. World J Gastrointest Endosc. 2014; 6(8):334–344

[3] Schütze G. Epiduroscopy—Spinal Endoscopy. New York, NY: Springer-Verlag; 2008

[4] Schütze G, Kurtze H. Percutaneous investigation of the epidural space using a flexible endoscope: A contribution to epiduroscopy. Reg Anesth Pain Med. 1993; 18:24

[5] Schütze G, Kurtze H. Direct observation of the epidural space with a flexible catheter-secured epiduroscopic unit. Reg Anesth Pain Med. 1994; 19 (2):85–89

[6] Shutse G, Kurtse G, Grol O, Enns E. Endoscopic method for the diagnosis and treatment of spinal pain syndromes. [in Russian]. Anesteziol Reanimatol. 1996; 4(4):62–64

[7] Saberski LR. (1996) Spinal endoscopy: current concepts. In: Winnie AP, Waldman SD, Dannemiller Memorial Education Foundation, eds. Interventional Pain Management. Philadelphia, PA: WB Saunders Company; 2006:137–149

[8] Schütze G. Epiduroskopie. Ein praxisorientierter Leitfaden zur epiduroskopischen Diagnostik und Therapie rückenmarksnaher Schmerzsyndrome. Lengerich, Germany: Pabst Science Publishers; 2006

[9] Balmelli C, Iffenecker A, Pittet D, Ruef C. Risk of prion transmission in epiduroscopy: current status of cleaning und disinfection of flexible endoscopes in Europe and recommendations of the Swiss-NOSO-CJD Task Force for Switzerland. [in German] Swiss-NOSO-CJD-Task Force 2003;10(4):25–32

[10] Kawanishi M, Kawase H, Kumagaya K. Equipment for epiduroscopy and its clinical applications. [in Japanese]. Masui. 2006; 55(9):1112–1117

[11] Becker V, Vercauteren T, von Weyhern CH, Prinz C, Schmid RM, Meining A. High-resolution miniprobe-based confocal microscopy in combination with video mosaicing (with video). Gastrointest Endosc. 2007; 66(5):1001–1007

[12] Bader L, Blumenstock G, Birkner B, et al. HYGEA (Hygiene in gastroenterology —endoscope reprocessing): Study on quality of reprocessing flexible endoscopes in hospitals and in the practice setting. [in German]. Z Gastroenterol. 2002; 40(3):157–170

[13] Hoffman A, Sar F, Goetz M, et al. High definition colonoscopy combined with i-Scan is superior in the detection of colorectal neoplasias compared with standard video colonoscopy: a prospective randomized controlled trial. Endoscopy. 2010; 42(10):827–833

[14] Kiesslich R, Burg J, Vieth M, et al. Confocal laser endoscopy for diagnosing intraepithelial neoplasias and colorectal cancer in vivo. Gastroenterology. 2004; 127(3):706–713

[15] Kiesslich R, Neurath MF, Jung M. Endomicroscopy: Is the endoscopist the pathologist of the future? [in German]. Gastroenterology. 2007; 2(2):87–92

[16] Kiesslich R, Götz M, Neurath MF, Galle PR. Endomicroscopy—technology with a future. [in German]. Internist (Berl). 2006; 47(1):8–17

[17] Ruetten S, Meyer O, Godolias G. Application of holmium:YAG laser in epiduroscopy: extended practicabilities in the treatment of chronic back pain syndrome. J Clin Laser Med Surg. 2002; 20(4):203–206

[18] Hancock S, Bowman E, Prabakaran J, et al. Use of i-Scan endoscopic image enhancement technology in clinical practice to assist in diagnostic and therapeutic endoscopy: a case series and review of the literature. Diagn Ther Endosc. 2012; 2012:193570

[19] Beltrutti D, Groen GJ, Saberski L, Sandner-Kiesling A, Schütze G, Weber G. Epiduroscopy: consensus decision March, 2006. Pain Clinic. 2007; 19 (2):47–50

[20] Kern-Wächter E. On the situation of continuing education for endoscopy service. [in German]. Endo-Praxis. 2009; 25(4):22–30

[21] Kovaleva J, Peters FT, van der Mei HC, Degener JE. Transmission of infection by flexible gastrointestinal endoscopy and bronchoscopy. Clin Microbiol Rev. 2013; 26(2):231–254

[22] Beekes M, Lemmer K, Thomzig A, Joncic M, Tintelnot K, Mielke M. Fast, broad-range disinfection of bacteria, fungi, viruses and prions. J Gen Virol. 2010; 91 (Pt 2):580–589

[23] Hilger C.. Flexible endoscopes: load characteristics and demand parameters [in German]. mt-Medizintechnik. 2003; 123(5):182–190

[24] Mizuno J, Gauss T, Suzuki M, Hayashida M, Arita H, Hanaoka K. Encephalopathy and rhabdomyolysis induced by iotrolan during epiduroscopy. Can J Anaesth. 2007; 54(1):49–53

[25] McLeod A, Roche A, Fennelly M. Case series: Ultrasonography may assist epidural insertion in scoliosis patients. Can J Anaesth. 2005; 52(7):717–720

[26] Schütze G. Interventionelles Schmerzmanagement—Bildgestützte Verfahren zur Diagnostik und Therapie rückenmarksnaher Schmerzsyndrome-Spinale Endoskopie, pharmakologische und elektrische Neuromodulation. Bremen, Germany: UNI-MED Verlag AG; 2011

[27] Ruetten S, Meyer O, Godolias G. Endoscopic surgery of the lumbar epidural space (epiduroscopy): results of therapeutic intervention in 93 patients. Minim Invasive Neurosurg. 2003; 46(1):1–4

[28] Raffaeli W, Righetti D. Surgical radio-frequency epiduroscopy technique (R-ResAblator) and FBSS treatment: preliminary evaluations. Acta Neurochir Suppl (Wien). 2005; 92:121–125

7 Epiduroscopic Surgical Tools

Günter Schütze

7.1 Introduction

The requirements for clinical use of spinal endoscopy medical products for diagnostics and therapy of neuraxial pain syndromes are regulated by the medical products statutes of the respective nations, for example, in the United States, by the Food and Drug Administration (FDA).[1] A large number of these endoscopic products, including the necessary tools, are licensed under the FDA 510(k) process or under country-specific or European Union (EU) medical product licenses.[2] In the EU the CE mark on a medical product is the manufacturer's declaration that the product meets all the legal requirements for that product for sale in the European Economic Area, and is recognized worldwide.

National and international medical guidelines exist for the performance of the most common neuraxial diagnostic and pain-therapeutic interventional procedures,, in particular, for the invasive operations for neuromodulation.[3] In this context, it is important to give the experienced physicians latitude for decisions and corridors for action that depart from the technical recommendations in well-founded individual cases.

This chapter contains systematically developed proposals for the solution of special neuraxial diagnostic and therapeutic problem situations from the realm of pain medicine (see also Chapter 6).

As a minimum standard, the following materials should be kept ready in the operating room (OR) for epiduroscopy (EDS) (▶ Fig. 7.1):

- Local anesthetic.
- Contrast medium.
- Saline ampoules.
- Syringes.
- Needles.
- Scalpel.
- Surgical sutures.
- An introducer set (introducer needle, safety J-guide, dilator, and sheath).
- Catheter (epidural catheter, Caud-A-Kath, balloon catheter, and ultrasound catheter).
- Electrodes (spinal cord stimulation electrodes [SCS] and multifunction electrodes).
- Laser fiber.
- Flexible microsurgical instruments.
- Facility for warming epidural infusion fluids.
- Transport vessels for tissue samples and swabs.
- Bandage materials.

Percutaneous spinal endoscopy is to be realized only through caudal access to the epidural space of the patient with special flexible endoscopic apparatus and supporting instruments.[4]

7.2 Instruments for Access to the Epidural Space

To gain percutaneous access to the epidural space via the sacrococcygeal ligament, a special set of instruments or a commercial introducer set are required. We recommend a set consisting of an introducer needle (also called a guide needle), a safety J-guide (Seldinger wire), a dilator, and a sheath. Percutaneous access is obtained with this set of instruments and utilizing the Seldinger technique, a method developed in 1953 by the Swedish radiologist Sven-Ivar Seldinger for venipuncture in order to introduce an angiography catheter. The Seldinger wire, the safety J-guide in this case, plays a special role in this process.

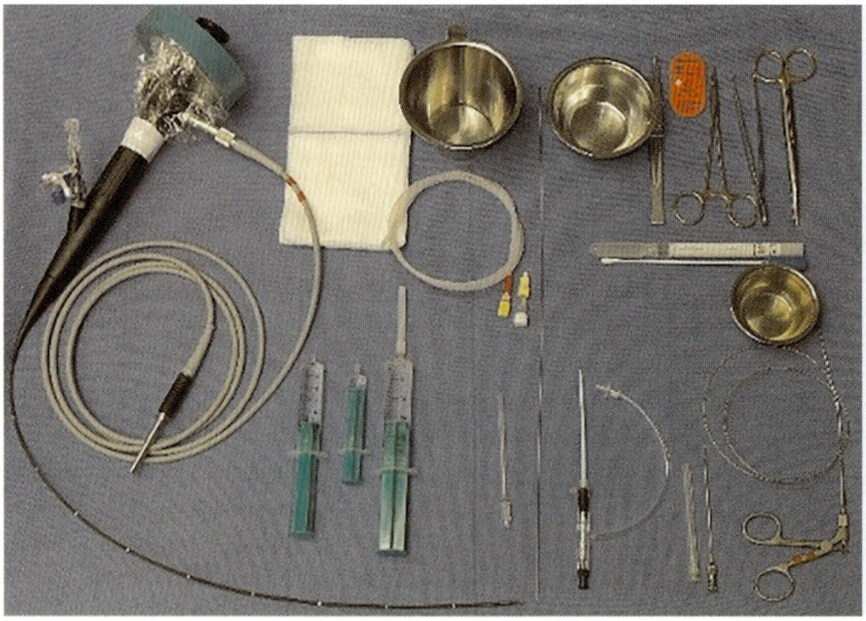

Fig. 7.1 Basic supplementary instruments for spinal endoscopy.

7.2.1 Scalpel

Prepare the area for the sacral incision, or puncture, following current clinical hygiene regulations. After local anesthesia for the percutaneous placement of the sacral sheath, with a common surgical scalpel, incise the skin in the designated region of the sacral hiatus from about 4 cm away from the anal opening, up to contact with bone (▶ Fig. 7.2).

7.2.2 Introducer Needle (Puncture Needle)

With a introducer needle provided with a precision triangular bevel for an easy and uniform percutaneous puncture in the region of the sacral hiatus, puncture the tectorial membrane at an angle of about 45°. After the introducer needle has perforated the sacrococcygeal ligament, and after a facultatively negative aspiration test, an introducer wire (Seldinger wire) can be placed through the funnel-shaped needle approach of the safety guide.

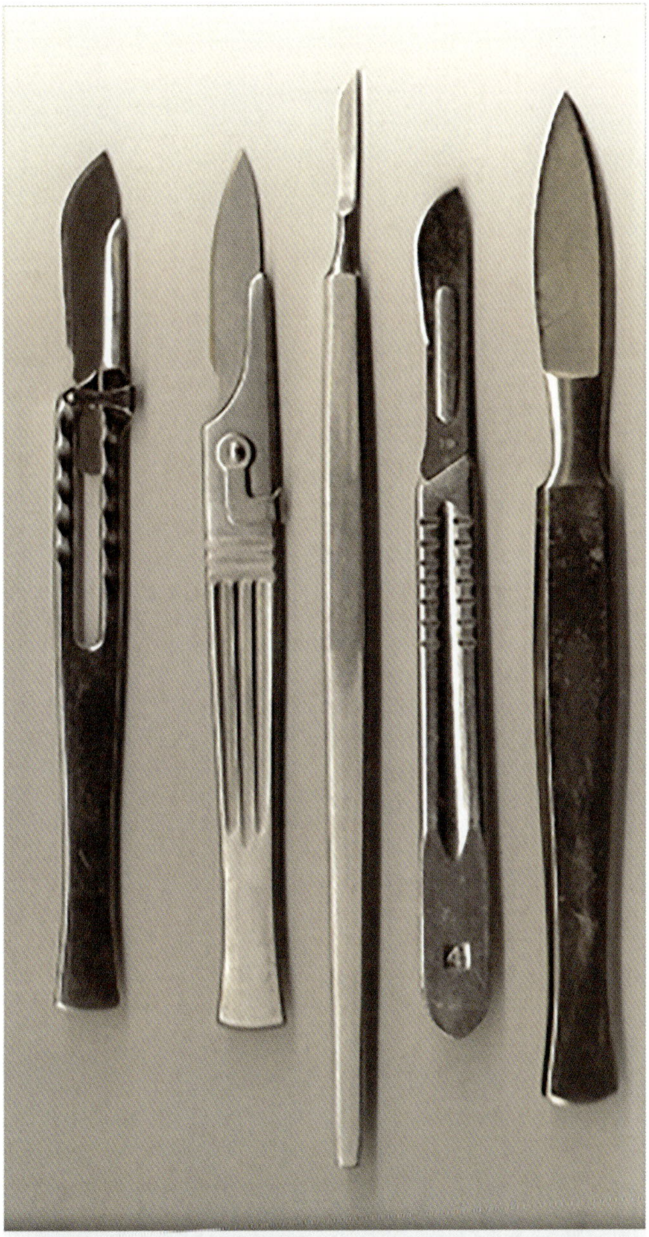

Fig. 7.2 Commercial scalpels from various manufacturers with adjustable and fixed blades. (Courtesy of Wikimedia.)

7.2.3 Tuohy Needle

In patients where anatomical access to the sacral canal is difficult to attain, the more robust 14-gauge Tuohy needle is recommended instead of the needle from the introducer set. The tip of the Tuohy needle is relatively blunt and formed at an angle of about 20°. Using this needle, an atraumatic positioning or guidance of the wire, catheter, or electrode in the sacral canal is made possible. A stylet attached to the Tuohy needle prevents tissue coring (biopsy effect) and increases the stability of the puncture.[5] The thin-wall needle technique ensures the penetration of guidewires and catheters of large diameter. Centimeter markings along the entire length of the needle make it possible to determine of the depth of penetration of the Tuohy needle. Tuohy 14- and 16-gauge introducer needles for sacral access are available in needle lengths of 9.2 cm, 11.4 cm, and 15.2 cm for SCS electrode placement. The Tuohy 15-gauge needle is available in needle lengths of 9.3 cm and 11.4 cm (▶ Fig. 7.3).

7.2.4 Guidewire

Through the introducer needle or Tuohy needle placed in the sacral canal, a guidewire (J-guide) can be introduced for a short length into the sacral canal. The guidewire (Seldinger wire) consists of a very tightly wound helical steel wire. The wire tip is bent back semicircularly in a J-shape to avoid perforation of surrounding tissue while within the sacral canal. For improved navigating ability, longer guidewires are ensheathed in a circularly bent plastic sleeve.

To identify and follow the J-guide (Seldinger wire), 50-cm long as a rule, in the sacral canal, fluoroscopic imaging (C-arm technique), primarily in anteroposterior (AP) but also in lateral projection, is very helpful.

7.2.5 Introducer Sheath

The sheath must be matched in size to the endoscope as well as adapted to the anatomical conditions of the patient (▶ Fig. 7.4).

A dilator with plastic sheath, for example, in sizes from 9.0 to 12.0 Fr and with or without a side port (PVB, Smiths Medical, Kirchseeon, Germany), can be placed over the recumbent guidewire (J-guide) percutaneously for a short distance in the sacral canal of the patient. The selection of the perfect dilator and sheath is a prerequisite for a smooth introduction of the sheath through the skin at the sacral hiatus and the tectorial membrane of the sacral canal (▶ Table 7.1).

After a check of the exact location of introducer, remove the dilator including the guidewire. The plastic sheath of the introduction system, which remains, ensures a relatively atraumatic, safe, and permanent access for the introduction of the epiduroscope into the epidural cavity and provides important protection against shear.

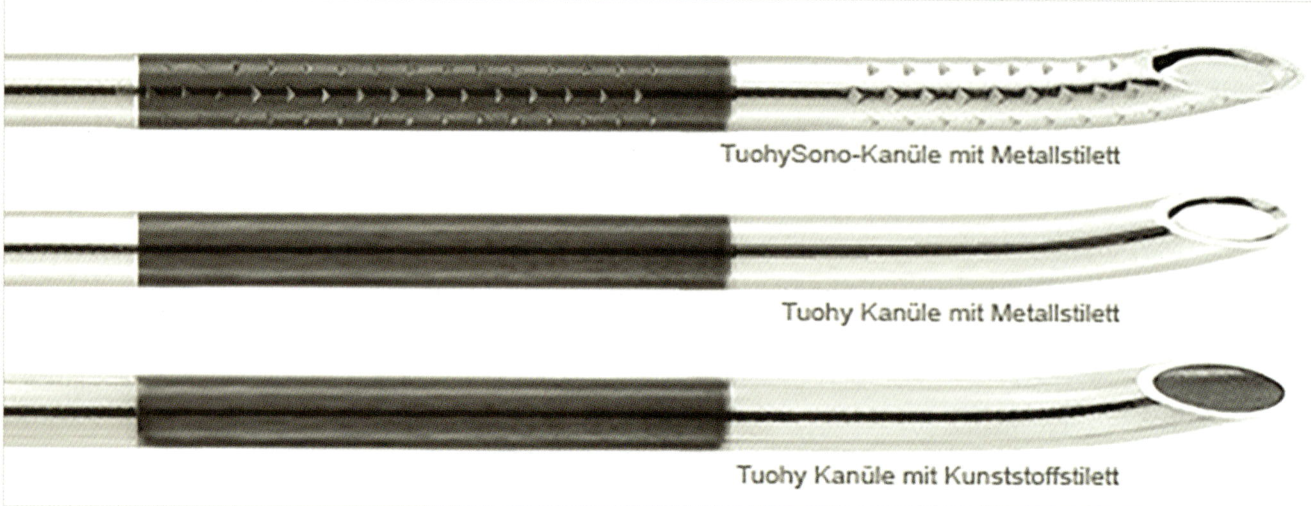

Fig. 7.3 Selection of Tuohy needles. (Courtesy of Pajunk GmbH Medizintechnologie, Geisingen, Germany.)

Table 7.1 Selected measurements of components of insertion set for technique of sacral access to the epidural space

Sheath		Endoscope		Guidewire		Needle		
ID (mm)	OD (mm)	Diameter (mm)	Diameter (Fr)	OD (mm)	Length (cm)	OD (mm)	Gauge	Length (mm)
2.4	3.1	2.3	7.0	0.96	50	1.4	17	70
2.8	3.4	2.7	8.0	0.96	50	1.4	17	70
3.1	3.7	3.0	9.0	0.96	50	1.4	17	70

Abbreviations: ID, inner diameter; OD, outer diameter.

Source: Data from Vygon GmbH & Co., KG, Medizintechnik, Aachen, Germany.

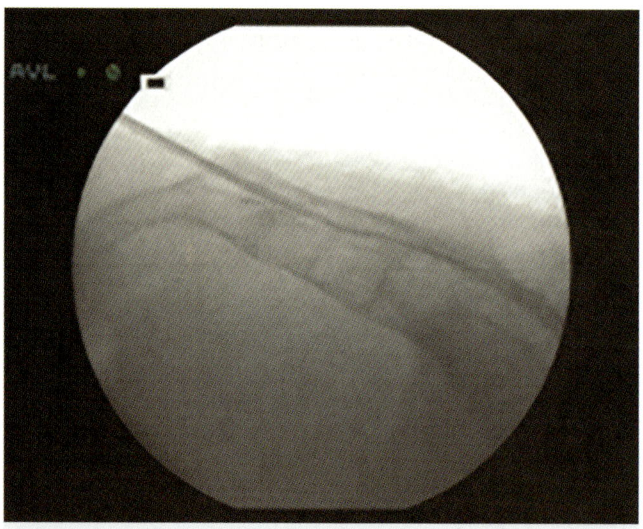

Fig. 7.4 Introducer sheath with Seldinger guidewire (J-guide) in the sacral canal.

The basic instrument set for percutaneous access to the sacral canal according to Seldinger technique includes (▶ Fig. 7.5):

- Puncture needle, diameter 1.3 × 63.5 mm (18 gauge).
- Guidewire, 70 cm, 0.035-in. diameter (0.89 mm).
- Flexible J-guidewire with introduction aid.
- Introducer kit 6–12 Fr with dilator and introduction lock.

7.2.6 Introducer Kits

The "Break-Away" Introducath Desilet (Vygon GmbH & Co. KG, Aachen, Germany) is available for percutaneous access to the sacral canal for placement of a flexible endoscope. This introducer kit contains an introducer needle, safety J-guide, dilator (20 cm), and a fully radiopaque splittable sheath (14 cm) as accessories. Through the correspondingly large, percutaneously placed shell, the sterile epiduroscope can be introduced into the epidural space carefully and easily under visual control via the monitor. After completion of the spinal endoscopy, the sheath can be split and removed without problems, by pulling apart the grip pieces.

The first introducer kit for percutaneous sacral access for epiduroscopy according to the Seldinger method (9045S, Karl Storz, Tuttlingen, Germany), the introducer kit according to Schütze, includes a single-use 9-Fr dilator and sheath and a puncture needle and guidewire. A separate vent as a portal for the epiduroscope with the possibility of irrigation is available.

Clinical Notes

- Gaining access to the epidural space through the sacral hiatus sometimes succeeds only with the help of special navigation maneuvers and, as necessary, also by the use of suitable size dilators and sheaths (9 Fr, 10 Fr, 11 Fr, or 12 Fr). For use and protection of instruments introduced epidurally parallel to the endoscope, it is desirable to use a sheath 1-Fr larger than the epiduroscope.

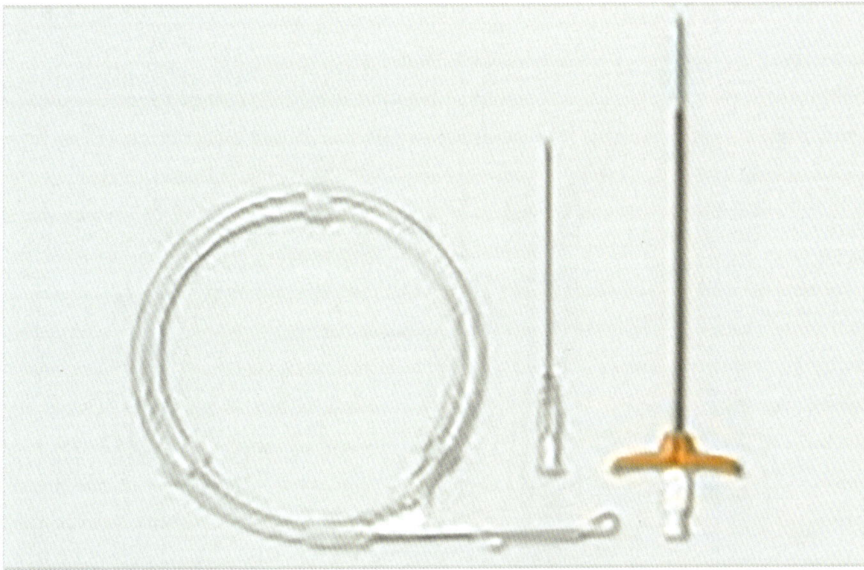

Fig. 7.5 Introducer kit. (Courtesy of PFM Medical AG, Cologne, Germany.)

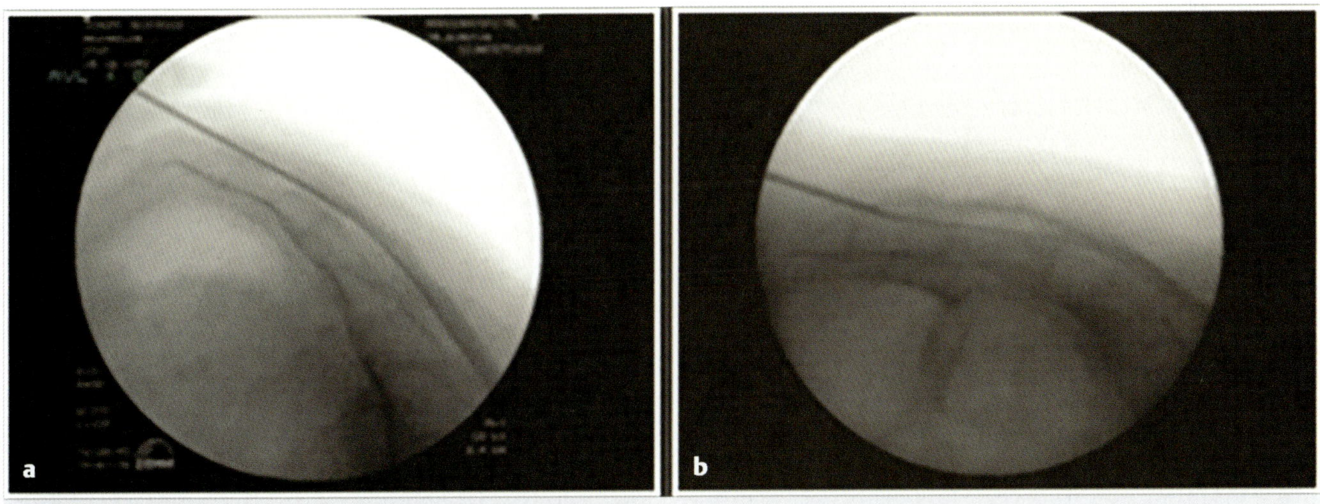

Fig. 7.6 (a) Guidewire (double flex, 0.038-in. diameter, 45-cm length) in the sacral canal. (b) Tuohy needle and guidewire in the sacral canal.

- With the percutaneous sacral access technique, the distance from the lower end of the dural sac to the sacral hiatus must absolutely be observed, as the distance is very individual. It can be as much as 16 mm (▶ Fig. 7.6).

7.3 Microsurgical Instruments

Along with the digital endocamera system and a fluoroscopic unit (C-arm), a corresponding sterile microsurgical instrument set is to be provided for the EDS and endoscopic operations. For a full list of EDS equipment to be kept in the OR, please refer to the list in 7.1 Introduction.

7.3.1 Forceps

There are a wide range of forceps available for taking targeted tissue samples under epiduroscopic view. In selecting the forceps to be used, you should be guided by the type and amount of tissue to be extracted and the neuraxial region where the operation will take place. Along with endoscopic tissue removal, mechanical epidural adhesiolysis or reduction or resection of epidural fibroses or scar tissue can be accomplished with the help of microsurgical forceps.[6]

For the extraction of neuraxial tissue for histologic investigation, the chosen microsurgical forceps are introduced through the working channel of the epiduroscope into the epidural space.

Flexible microsurgical forceps, or microforceps, are available in various lengths and diameters according to the epiduroscope used and to the tissue material to be removed. The following instruments, among others, are available:

- Oval flexible grasping forceps of various effective lengths up to 160 cm with an outside diameter (OD) at least 0.8 mm.
- Flexible grasping forceps, alligator jaws, of various lengths up to 160 cm with an OD of 1.0 mm.

For epiduroscopes with a shorter effective length, the selection of flexible microsurgical forceps include (▶ Fig. 7.7):

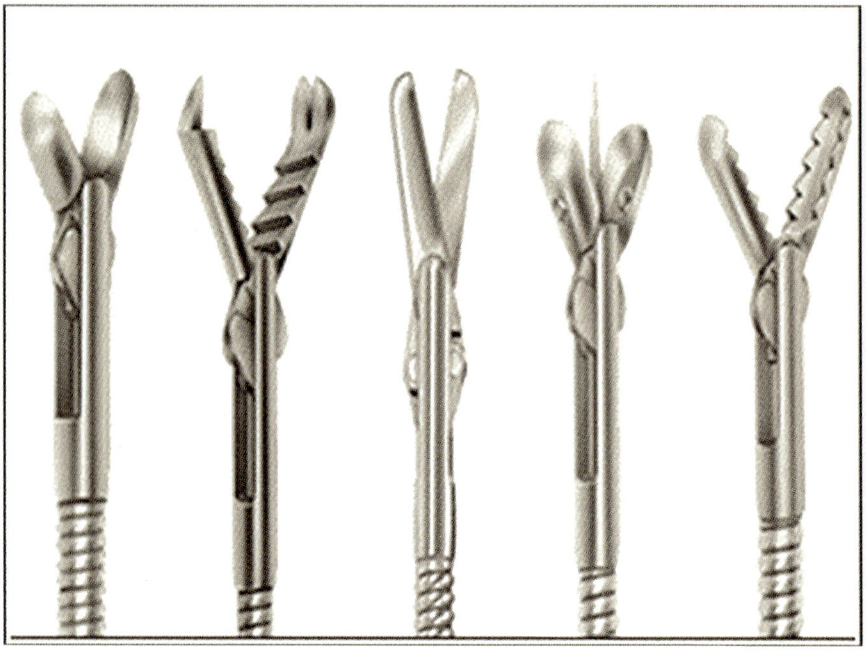

Fig. 7.7 Most common working ends for autoclavable flexible microsurgical instruments. (Courtesy of Günter Stoffel Medizintechnik GmbH, Wurmlingen, Germany.)

- Flexible grasping forceps with movable jaws (0.8-mm OD, length 60 cm); the same forceps with 100-cm length.
- Flexible grasping forceps with movable jaws (1.0-mm OD, length 60 cm); the same forceps with 100-cm length.

Basically, flexible forceps are opened and closed by a cable.

Because of the risk of damage to the working channel of the epiduroscope, for example, by a forceps thorn, forceps without thorns should absolutely be used (▸ Fig. 7.8; ▸ Fig. 7.9; ▸ Fig. 7.10).

Without sufficient endoscopic imaging, risky microsurgical procedures, above all in the thoracic and cervical regions of the spinal canal, should be forgone on the basis of patient safety. Special caution should be observed and attention given to biopsy and lysis procedures for inflammatory processes and in the region of the nerve roots! Additional safety can be ensured by epidural ultrasound navigation or endomicroscopy (see Chapter 8).

Clinical Notes

- Because of the vulnerability of neuraxial structures, it is essential to endoscopically observe the tissue removal with microsurgical forceps to help insure that no blood vessels, spinal dura mater, or nerve structures are injured.[6]
- If vessels are damaged by a biopsy, these can, as necessary, be coagulated with a laser fiber (diode laser) through the working channel of the endoscope.
- Epidural lysis or biopsy procedures with microsurgical instruments, above all in the cervical region of the spinal canal, must be carried out particularly carefully and scrupulously.[7]

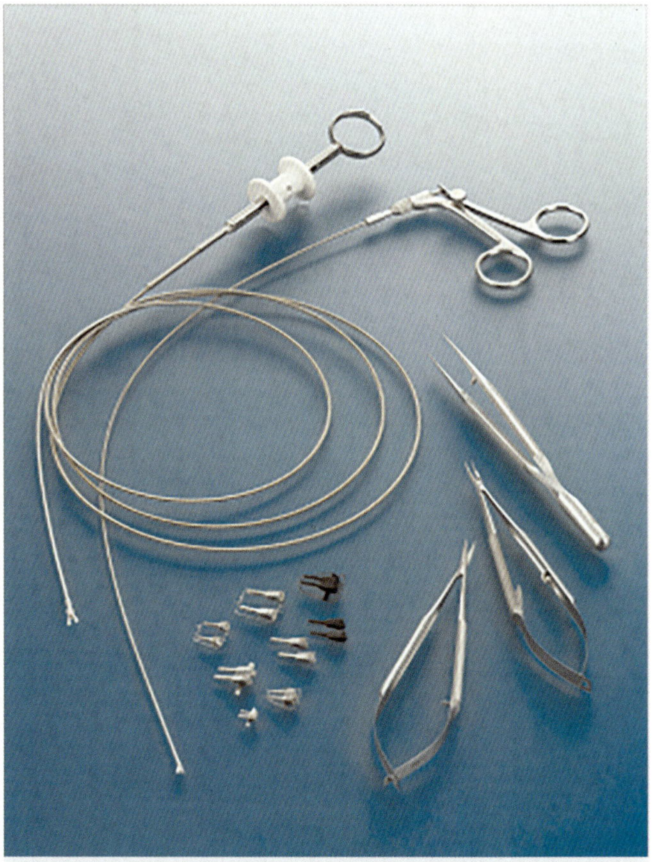

Fig. 7.8 Flexible autoclavable microsurgical grasping forceps of lengths 60 cm and 160 cm. (Courtesy of Karl Storz, Tuttlingen, Germany.)

7.3.2 Laser Fiber

Through the working channel of the epiduroscope, the light guide, the bare fiber, can be introduced under endoscopic view into the epidural space without problems.

The forward-emitting light guide is a standard accessory for minimally invasive endoscopic intervention. The bare fibers are available in diameters of 200 µm, 300 µm, 400 µm, and 600 µm. The 3-m long bare fiber provides the user with the highest

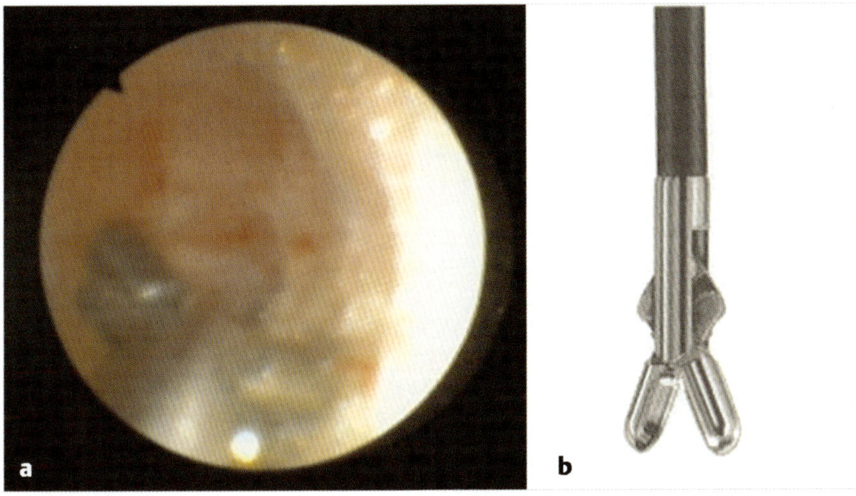

Fig. 7.9 Microsurgical instruments. (**a**) Open forceps in the epidural space. (**b**) Open forceps.

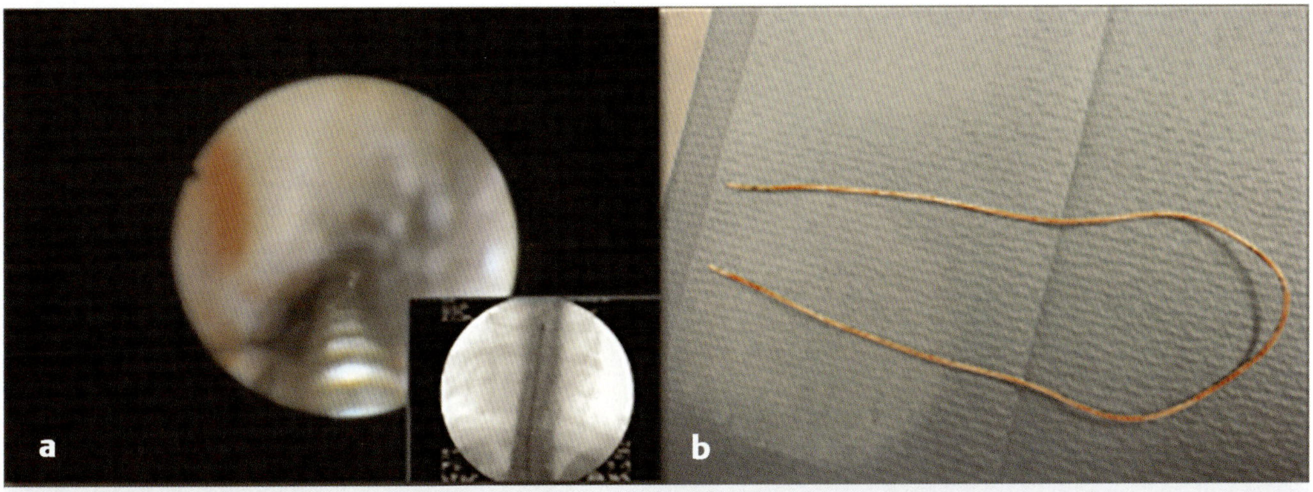

Fig. 7.10 (**a**) Spinal catheter endoscopically removed with microsurgical grasping forceps from the dorsal thoracic epidural space (picture-in-picture technique). (**b**) Endoscopically removed foreign body (a sheared-off epidural catheter).

possible flexibility and distinctive radii of curvature and expands the load-bearing capacity for light through an improved light guide core and the surrounding protective layer.

On the basis of its physical properties in water, the diode laser is also suitable, along with the long-established holmium; yttrium-aluminum-garnet (Ho:YAG) laser, for use on the highly sensitive neuraxial structures. The wavelength of the diode laser (DIOLAS, Limmer Laser GmbH, Berlin, Germany) is 980 nm.

The diode laser can also be used for all indications—primarily coagulation and vaporization of tissue—known for the neodymium;yttrium-aluminum-garnet (Nd:YAG) laser.[8] Rütten et al stress that the appropriate use of a Ho:YAG laser (wavelength 2100 nm) in the epidural space is safe.[9]

Only under excellent endoscopic monitoring is the bare fiber of the diode laser to be used for hemostasis, lysis, or even for scar tissue resection (▶ Fig. 7.11). Within the anatomically determined boundaries of the epidural space, navigation of the relatively stiff laser fiber can succeed only via the steerable endoscope tip. With the help of the laser pilot beam, a pinpoint ablation without bleeding can be achieved.

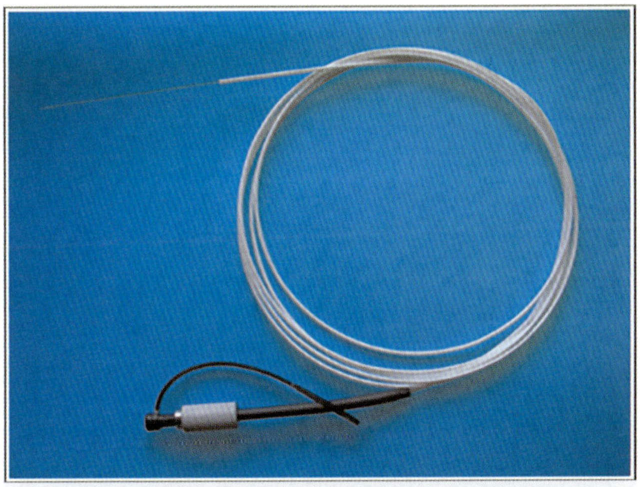

Fig. 7.11 Bare fiber. (Courtesy of Advanced Fiber Tools GmbH, Bonn, Germany.)

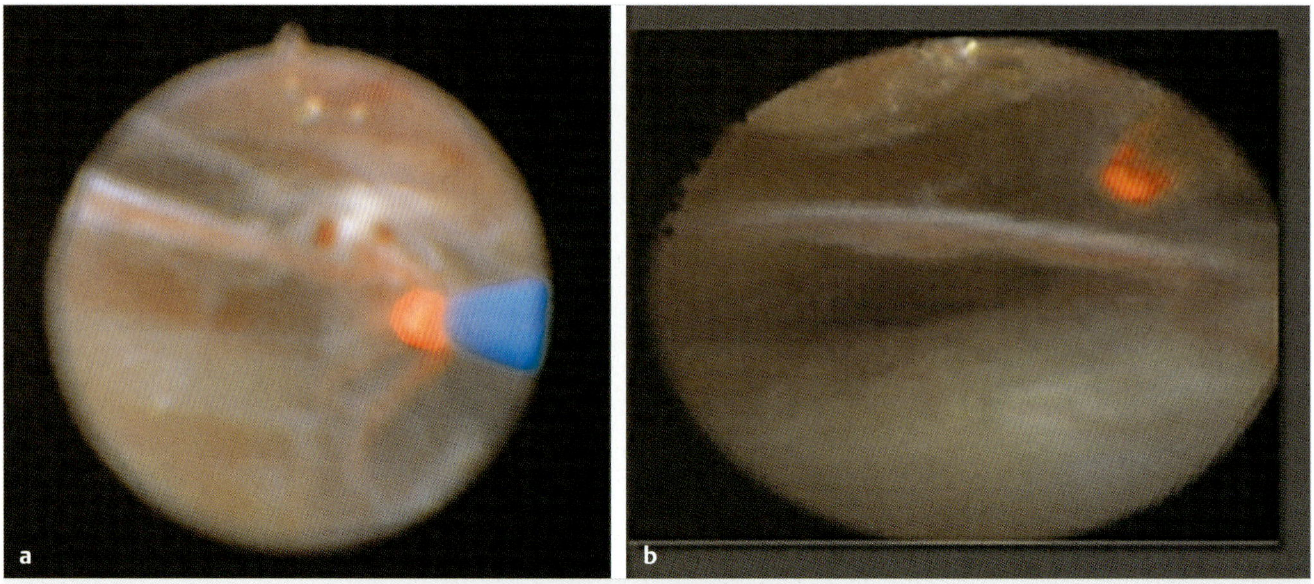

Fig. 7.12 (a) Laser fiber (*blue*) and pilot beam (*red*) after hemostasis. (b) Laser pilot beam (*red*) with blood vessel passing through the epidural cavity.

Clinical Notes

- The depth of penetration of the laser is reported as a few millimeters, primarily depending on the water content of the tissue.
- The laser should be activated after careful endoscopic surveillance and under controlled epidural saline infusion.
- Based on our experience, a laser power range of 5.0 to 8.0 W is more than sufficient.
- Only in the clear endoscopic target area should neuraxial structures relevant to pain or bleeds be treated with the laser. Nerve root regions should be spared from laser therapy as much as possible.
- At the slightest indication of paresthesias, it is urgent that oneterminate the laser procedure immediately.
- Valid national laser-operation instructions and accident-prevention prescriptions should be observed unconditionally (▶ Fig. 7.12).

7.4 Epiduroscopic Equipment

7.4.1 Catheters

A leading indication for endoscopically supported neuraxial management is catheter placement for epidural analgesic therapy (EAT), adhesiolysis, epidural removal of fluid, targeted application of medication, and obtaining smears.

Through the working channel of the epiduroscope, a catheter can be placed epidurally safely and accurately under endoscopic view dorally or ventrally in any region—sacral, lumbar, thoracic, or cervical.[6,10]

Effective EAT is possible only through endoscopically supported catheter placement, which makes possible targeting of the involved nerve roots (radiculitis, arachnoiditis) or the epidural region relevant to pain.

Epidural Catheter

Epidural catheters are designed in various shapes as single-hold or multi-hole catheters with or without stylet. The most commonly used catheters are manufactured in sizes that can pass through a 14-, 16-, or 18-gauge needle. Epidural catheters are made from plastics such as Teflon (PTFE, polytetrafluoroethylene), nylon (polyamide), polyurethane (PUR), or silicone (polyorganosiloxane) in various lengths and outside diameters.[5,11,12] It is recommended to select a catheter with a length of, for example, 900 to 1800 mm, depending on the neuraxial region to be investigated (lumbar to cervical).

The catheter should have an X-ray stripe, in case it is necessary to document its location in the spinal canal or to diagnose a loop formation or knotting in the catheter, for example, when a stylet is not used. Some catheters are additionally provided with markings at 5-cm intervals from the tip up (▶ Fig. 7.13).

To avoid irritation, epidural catheters are relatively soft and compatible with tissue, although they should possess a certain stiffness to facilitate epidural insertion if needed as far as the spinal cervical region. The catheter, with an opening placed at the end, is deemed atraumatic, that is, with low risk in insertion of inadvertently perforating the spinal dura mater. The tensile strength of the epidural catheter is also significant; it is approximately one-third less in a 19-gauge catheter than in an 18-gauge catheter.[11] Transparent catheters, in which a reflux of fluid or blood can be observed, are advantageous. Colored markings and a sure distal closure should also be available.

The integrated catheter and stylet makes possible the advance of the catheter through the working channel of the epiduroscope over numerous spinal segments from sacral to cervical.

Polyurethane catheter

The length-marked, radiopaque soft polyurethane (PUR) catheter, long clinically tested for EDS, is especially thermosensitive and, after placement, matches epidural temperature optimally. The PUR catheter is available in the distally open model with

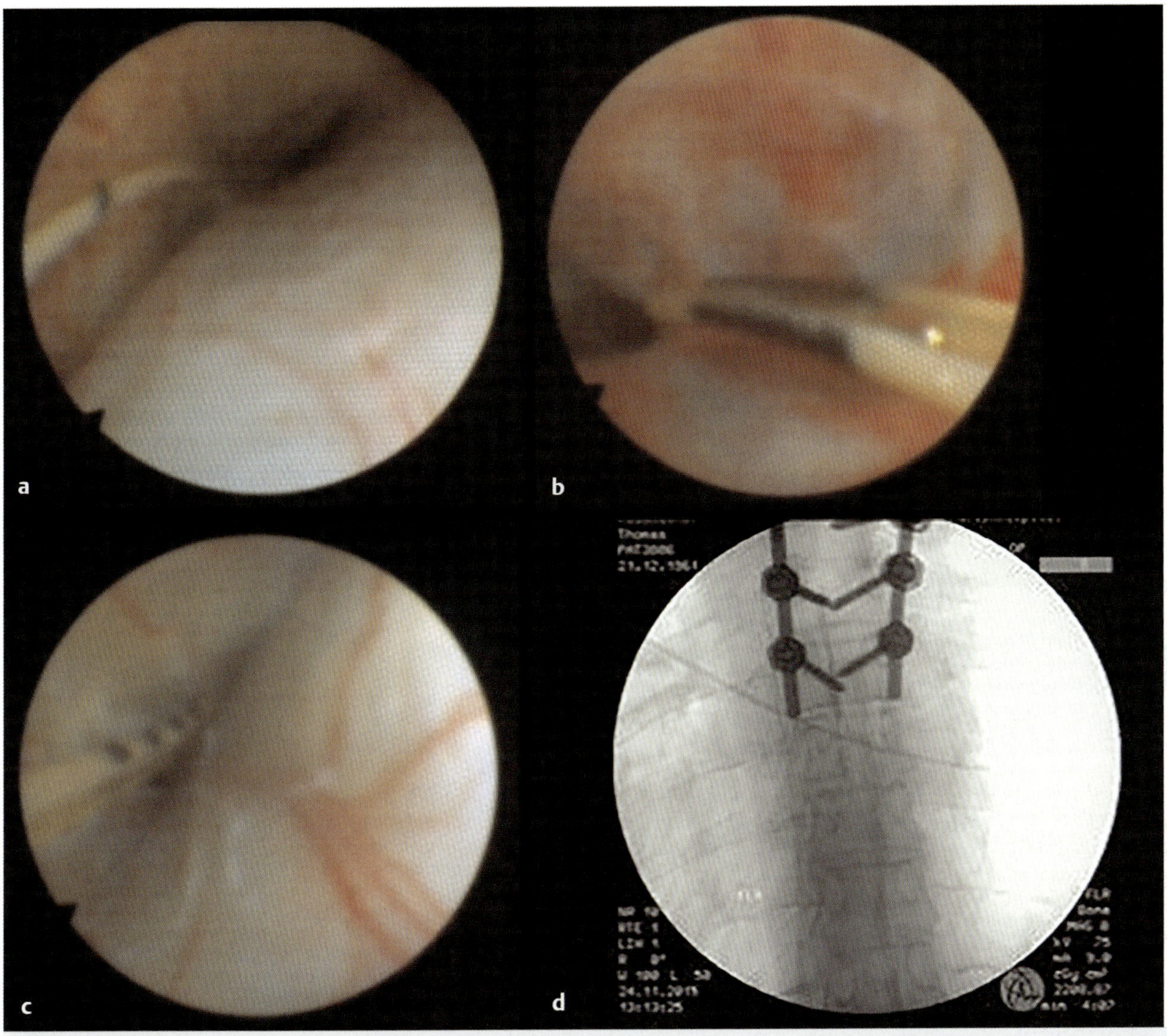

Fig. 7.13 **(a-d)** Endoscopically thoracically placed polyurethane epidural catheter with corresponding visible length markings and epidurographic position control.

stylet and has an outside diameter of 1.0 mm and an inside diameter of 0.6 mm in lengths of 900 mm and 1800 mm.

The distal catheter markings are visible at 1-cm intervals up to 10 cm, and additionally there is a marking at 20 cm. At the proximal end of the epidural catheter there is a supporting coil worked into the catheter material to avoid catheter occlusion. After placement in the central opening of the "easy-lock" proximal connector, the catheter end is inserted until it stops and is closed.

Multi-hole catheter

With the multi-hole epidural catheter, there are lateral holes and an opening in the head of the epidural catheter. Epidural catheters with multiple laterally placed openings have the advantage of enabling aspiration of blood in case of an inadvertent intravasal placement.[13] Partial epidural misapplications of medications are more frequently described with multi-hole catheters than with a single terminal opening.

Multiple-hole epidural catheter (U.S. Patent 5800407 A)

This special catheter is provided with an opening at the catheter tip as well as with three lateral openings some 1 mm apart and three additional openings axially spaced about 4 mm apart. All catheter openings are located within 1.5 cm of the catheter tip (▸ Fig. 7.14).

Stylet-guided single-hole catheter

Some epidural catheter models are offered with an extraordinarily prominent stylet.

Stylet-guided radiopaque catheters are commonly recommended for endoscopic epidural use. From Vygon (Aachen, Germany), an endoscopic epidural catheter even for a cervical EDS, (0.6 × 1.0 × 900 or 1800 mm) distally open with a stylet, is available. To prevent dura mater perforation, the catheter tip is soft and rounded. The catheter tip, marked black to make

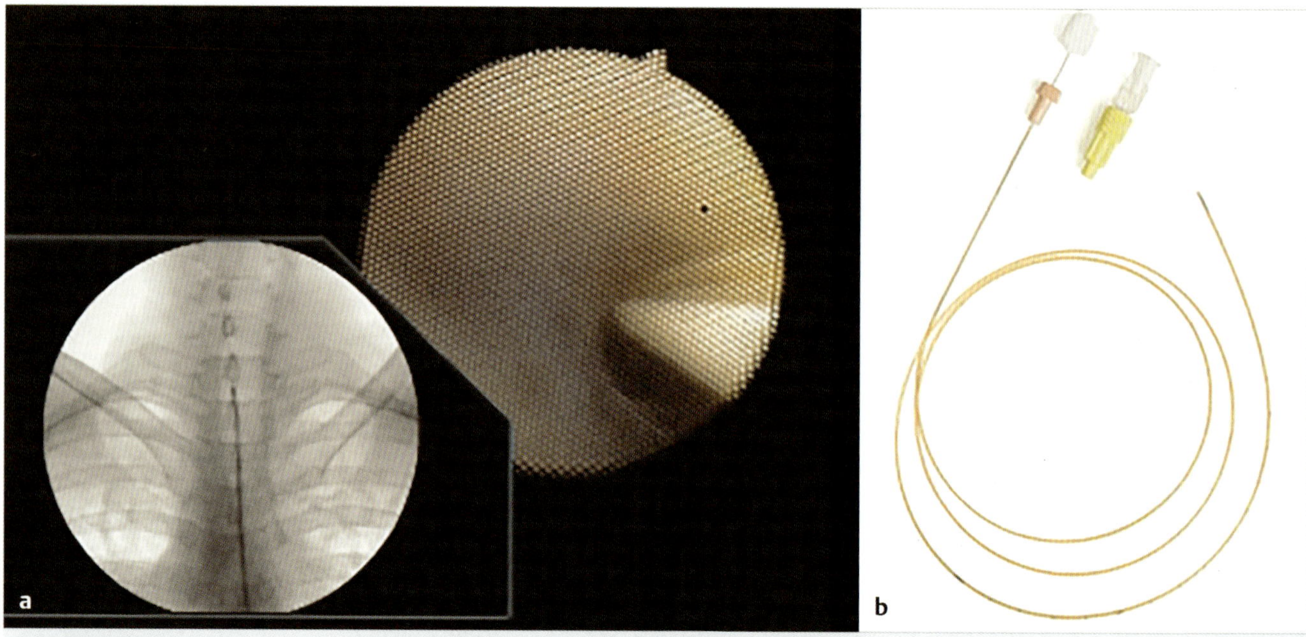

Fig. 7.14 (a) Catheter in the cervical epidural cavity, placed via the working channel of the epiduroscope (picture-in-picture technique). (b) Epidural catheter Vygon 185.101, an open tip, easy-lock endoscopic epidural catheter with specifications of 0.6 × 1.0 × 1800 mm. [(b) Courtesy of Vygon GmbH & Co. KG, Aachen, Germany.]

endoscopic control and navigation easier, is soft and rounded. Stability, absolutely necessary for endoscopic placement, is conferred to the polyurethane catheter by the use of a stylet as a guidewire. To prevent an accidental epidural injury or perforation of the spinal dura mater, the very important catheter stylet ends 2 cm before the catheter tip and is endoscopically visible within the catheter.

The soft, radiopaque, and length-marked polyurethane catheter is particularly thermosensitive and, after epidural placement, adapts itself optimally to the patient resulting in a high level of wearer comfort.

Catheter markings are visible over the distal 5 to 15 cm at 1-cm intervals and additionally there is a marking at the 20-cm length. To avoid catheter occlusion, a supporting coil is worked into the catheter material at the proximal end. Endoscopic recognition and endoscopic handling of the catheter are made easier by strong catheter coloring and corresponding marking aids (► Fig. 7.15; ► Fig. 7.16). These special catheters, similar to Racz catheters, are, based on our experience, are usable only with first-class flexible epiduroscopes, on the basis of their length from a minimum of 30.5 cm to a maximum of 84.5 cm.

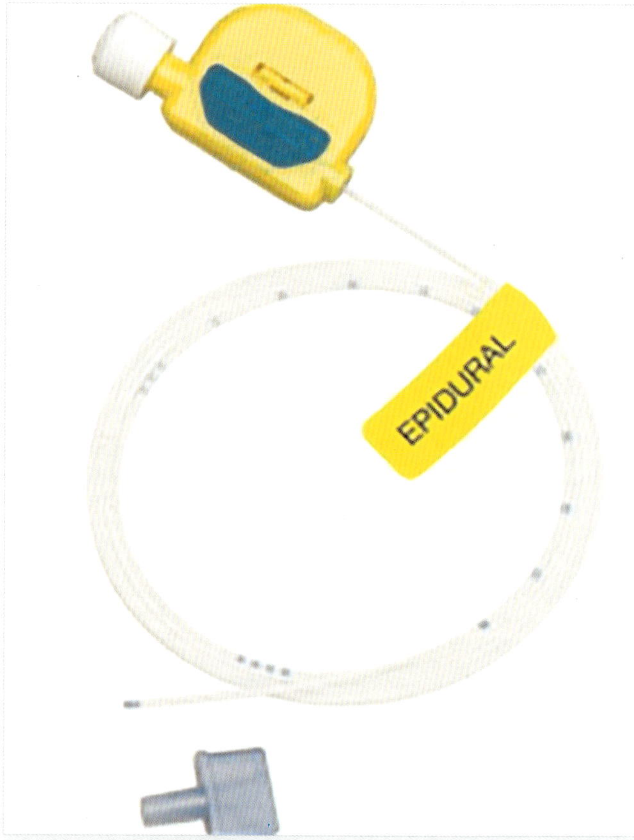

Fig. 7.15 Caud-A-Kath epidural catheter specially developed by Epimed International for percutaneous caudal catheter placement. (Courtesy of Epimed International, Gloversville, New York.)

Clinical Notes

- After the catheter arrives at the neuraxial target position, while the catheter is held in position, the stylet can be withdrawn from the catheter through the working channel of the endoscope without a problem. Even after initial correct catheter positioning, the catheter tip can become dislocated from epidural to subarachnoidal space or into a venous epidural blood vessel. After that, accidental intraspinal, subarachnoid, or

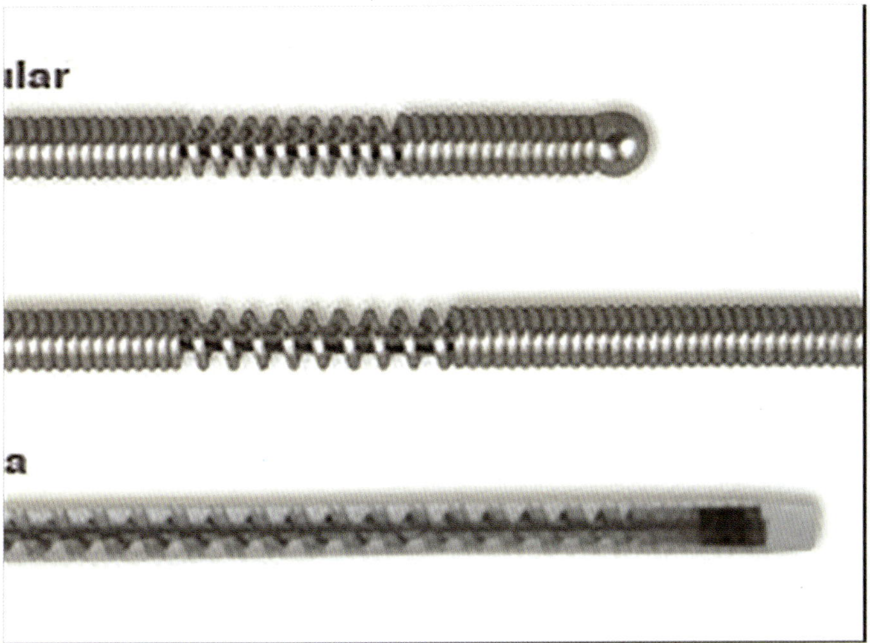

Fig. 7.16 Stainless steel spring–reinforced epidural catheters (Racz® Catheters) for strength in insertion, columnar strength, and lumen patency. The uncoated tips are flexible, smooth, and rounded. (Courtesy of Epimed International, Gloversville, New York.)

intravenous medication applications may occur. Through the use of a stylet-guided catheter under endoscopic view a rolling-up or a knot or loop formation in the catheter is practically impossible.

- With the use of multi-hole catheters, the likelihood exists that while some of the catheter openings will lie correctly in the epidural space, other catheter openings could lie in a vessel or even in the intrathecal space.
- For documentation, along with endoscopic tracking of the catheter, radiologic tracking with or without contrast (epidurogram) is useful.

Spinal Catheter

An endoscopically supported placement of a catheter for intrathecal pain therapy is possible. A prerequisite is a corresponding indication for an endoscopically supported catheter implantation.

For catheter selection, a large number are available from manufacturers, including spinal catheters such as the Spinal Catheter Set 4000 with introducer set (tricumed Medizintechnik GmbH, Kiel, Germany) and various models of intrathecal catheters (Medtronic, Minneapolis, Minnesota; Codman & Shurtleff, Inc., Raynham, Massachusetts; Arrow International, Inc., Reading, Pennsylvania) in various materials and sizes. The established catheter brands are mostly offered in radiopaque silicone rubber with an OD of 1.5 to 2.5 mm. The length of these silicone catheters range from 38 to 104 cm, and they have a longer shelf life and higher safety against kinking.

Intrathecal catheter (Ascenda Model 8780/8781)

The Ascenda Model 8780/8781 intrathecal catheter (Medtronic Inc., Minneapolis, Minnesota) is a two-part catheter. The catheter body consists of elastic, flexible, radiopaque silicone rubber and has an OD of 1.4 mm (4.2 Fr) and an ID of 0.53 mm. The distal end of the spinal segment is a radiopaque, closed catheter tip that has six lateral holes for medication delivery. To make the placement of the catheter easier, the spinal segment is provided from its start with markings at a 1-cm intervals. A guidewire in the lumen of the catheter causes additional stiffening and makes steering of the catheter tip easier during the implantation of the catheter (▶ Fig. 7.17).

Clinical Notes

- On the basis of the measurements of commercial catheters, placement of a spinal catheter through the standard working channel of commonly used epiduroscopes is occasionally problematic.
- Our experiences with endoscopically supported intrathecal catheter implantation are that only in absolutely select cases could the spinal catheter be guided to the intended intrathecal site. It succeeded only for those implantation sites that were through the sacral sheath and parallel to the epiduroscope.

Balloon Catheter

Epidural mechanical lysis, a supplementary endoscopically supported interventional therapy, can be carried out with the help of a balloon catheter. By methodical dilation of the catheter balloon the epidural adhesions or fibrotic tissue are broken down or reduced by mechanical lysis. The results of the invasive-interventional balloon catheter operation can be immediately assessed endoscopically and/or optically through the production of an epidurogram. The result of the catheter balloon dilation can be optimized through a simultaneous and targeted epidural application of medication (epidural analgesic therapy [EAT]) in the region relevant to the pain.

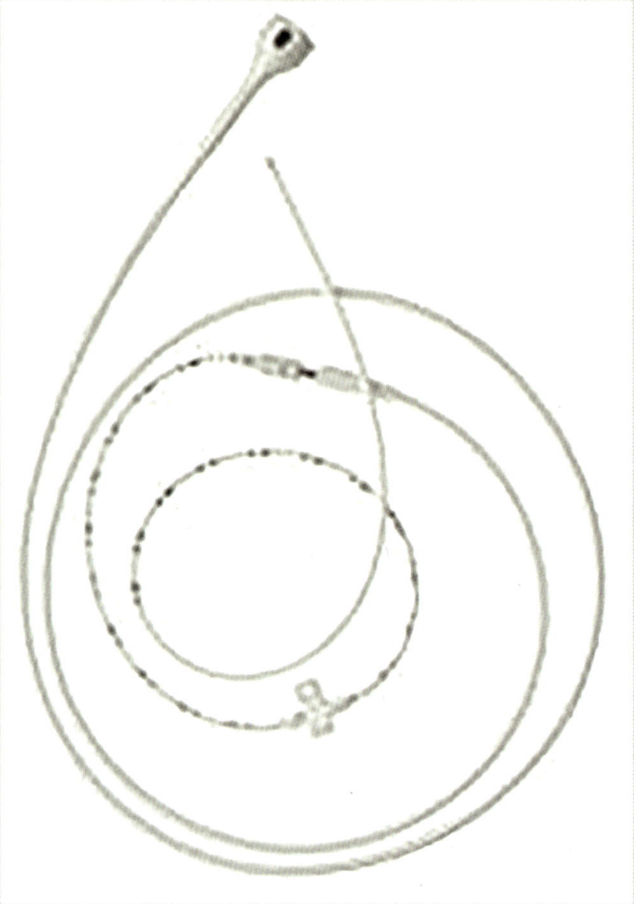

Fig. 7.17 Intrathecal catheter Model 8731SC with seamless pump connector. (Courtesy of Medtronic Inc., Minneapolis, Minnesota.)

Fogarty balloon catheter

The Fogarty catheter, named after Thomas J. Fogarty, is a 75-cm long plastic catheter with an inflatable balloon sleeve, available in different diameters, at the end, which was originally used for prograde or retrograde embolectomy.

For interventional endoscopically supported epidural lysis, the 2-Fr (0.67-mm) Fogarty balloon catheter (Edwards Lifesciences Corp., Irvine, California) can be filled with a volume of 0.67 mL. The outside diameter of the deflated balloon measures 1.13 mm; the diameter of the filled balloon is 4.0 mm. The maximum recommended pull force on the filled balloon of a 2-Fr Fogarty catheter is 0.23 kgf (0.5 lb.) according to the manufacturer.

If an endoscopically guided epidural balloon intervention is not planned, of course, a catheter of up to 7 Fr can be used. For better radiologic imaging of the balloon, filling the balloon with X-ray contrast medium is recommended in this special case (▶ Fig. 7.18; ▶ Fig. 7.19).

Latex-free balloon catheter

The latex-free catheter is equipped with a silicone balloon and stainless steel–reinforced walls, which provide firmness, flexibility, and radiopacity, with a smooth balloon top for easy insertion.

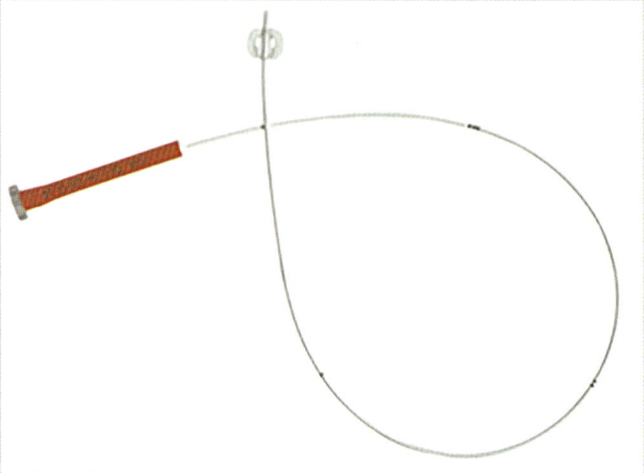

Fig. 7.18 Latex-free, stainless steel–reinforced Fogarty balloon catheter. (Courtesy of Edwards Lifesciences Corp., Irvine, California.)

Steerable fiber optic epidural balloon catheter (U.S. Patent 7273468 B2)

This balloon catheter is specially developed for neuroplasty procedures, mechanical lysis of adhesions, and decompression operations under direct endoscopic view in the epidural cavity.

Clinical Notes

- With saline balloon filling of the catheter, a very easy resistance is usually felt. If this resistance is lacking, one must assume a balloon rupture.
- Balloon filling or rupture can be observed endoscopically after the catheter tip becomes visible.
- In filling the balloon, absolutely the smallest possible respective volume must be injected, to reduce to a minimum the lateral wall pressure and shear forces on the internal epidural surfaces, such as the spinal dura mater and the posterior longitudinal ligament, or neural structures.
- The use of a strongly viscous or particle-containing X-ray contrast medium is problematic, because then the contrast medium injected into the catheter balloon can lead to plugging of the catheter lumen, in which case a complete discharge of the balloon volume can no longer occur.
- Under endoscopic view the yet-unfilled balloon catheter is brought into the target position of the epidural space and pushed out through the blocking tissue material. When the catheter is located in the desired position, the balloon can be filled with sterile saline solution or contrast medium (for C-arm control).
- For adhesiolysis or extirpation of blocking epidural tissue, the filled balloon catheter is drawn back carefully under constant endoscopic observation.
- The actual balloon volume is adapted to the respective degree of epidural stenosis observed by correspondingly increasing or decreasing the filling volume. After the balloon filling is stopped, the Fogarty catheter can be drawn back.

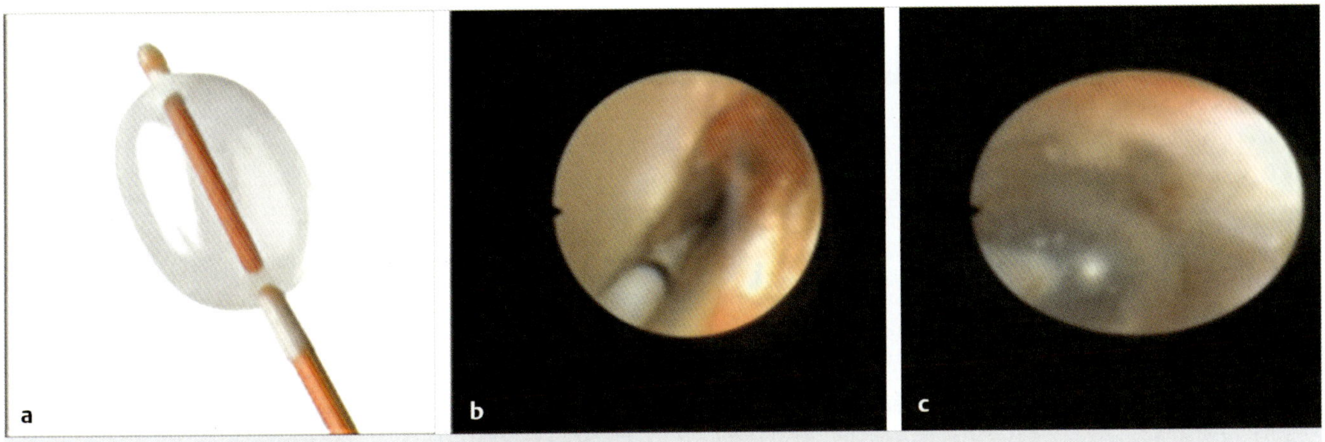

Fig. 7.19 (a) Fogarty catheter. (b) Fogarty catheter in the epidural cavity. (c) Fogarty catheter with filled balloon in the epidural cavity.

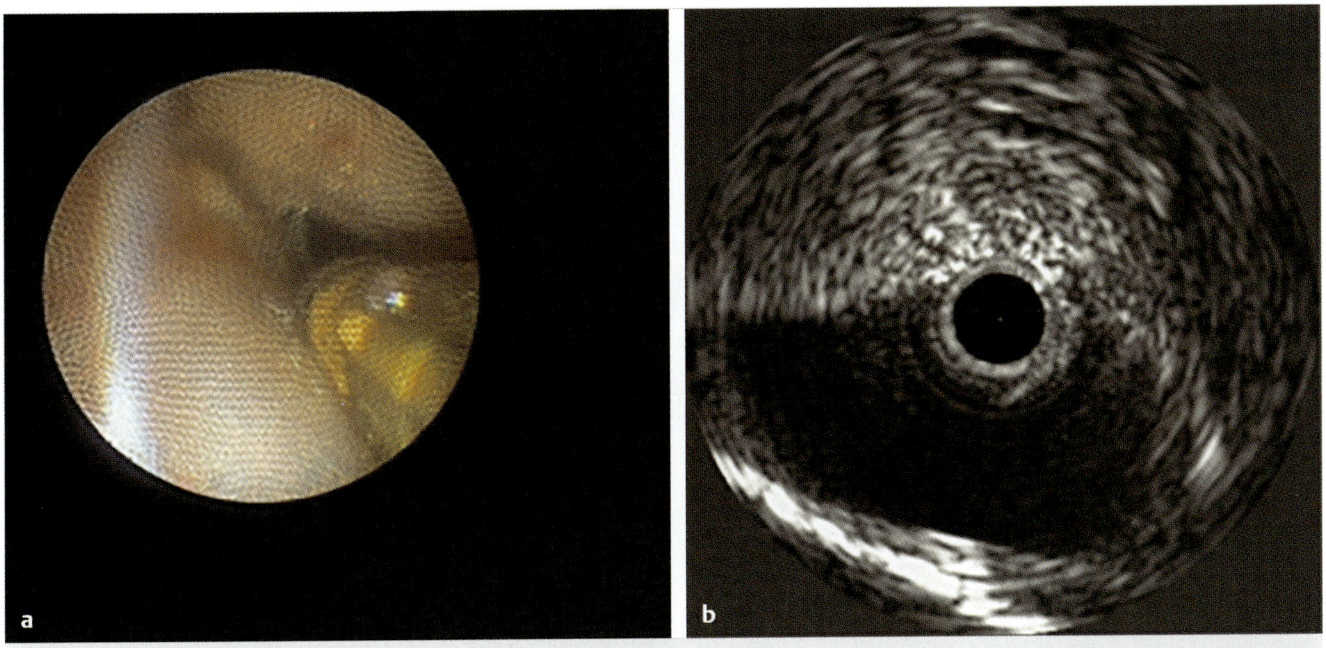

Fig. 7.20 (a) Sonographic–endoscopic (SONO-EDS) system in clinical use. (a) Epiduroscopic presentation of the tip of an intravascular ultrasound catheter. (b) Sonographic display of the epidural space.

- It is to be observed that an excessive pull in the removal of stuck epidural tissue material can lead to damage to the balloon.
- The maximum balloon-filling volumes and the applied pull force recommended by the manufacturer should not be exceeded. Thereby the risks of accidental trauma, balloon rupture, or detachment of the catheter tip can be minimized.

Ultrasound Catheter

To optimize microsurgical neuraxial interventions, we recommend the use of the neuraxial sonographic–endoscopic technique (SONO-EDS).[6,14]

The available flexible epiduroscopes do in fact provide endoscopic orientation of epidural and adjacent structures, but size and surface estimation, especially those with pathologic structural changes are difficult or impossible to make.

To solve this problem and optimize the endoscopic investigation technique, Schütze integrated a microinvasive intraluminal ultrasound system (intravascular ultrasound [IVUS]), with a 3.0/3.4-Fr monorail, 30-MHz phased-array ultrasound catheter, into epiduroscopic technology and introduced it in the clinic (▶ Fig. 7.20; ▶ Fig. 7.21).

Visions Five-64 Catheter F/X

We use a Visions Five-64 F/X ultrasound catheter (an IVUS imaging catheter from EndoSonics, Rancho Cordova, California), whose physical specifications include:

- Length: 135 cm.
- Length of the monorail: 25 cm.
- Shaft diameter: 3.0 Fr distal, 3.4 Fr proximal.

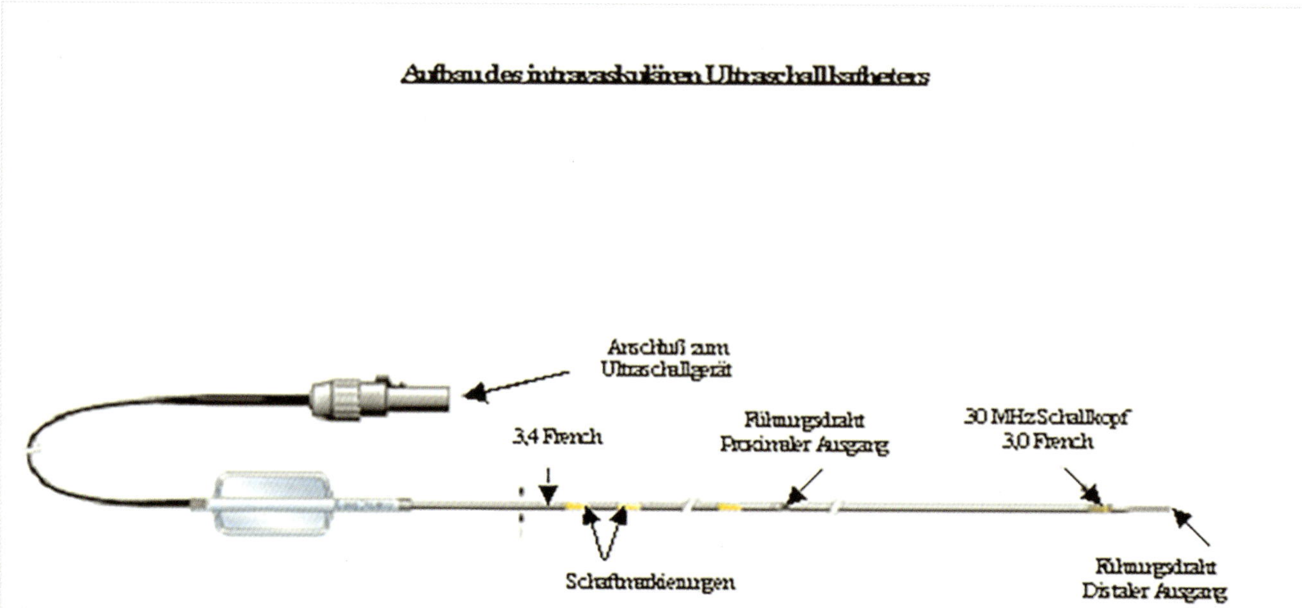

Fig. 7.21 Schematic diagram of the multi-element, phased-array intravascular ultrasound catheter of the type used for the sonographic–endoscopic (SONO-EDS) system (Visions Five-64, 30 MHz). (Courtesy of EndoSonics, Rancho Cordova, California.)

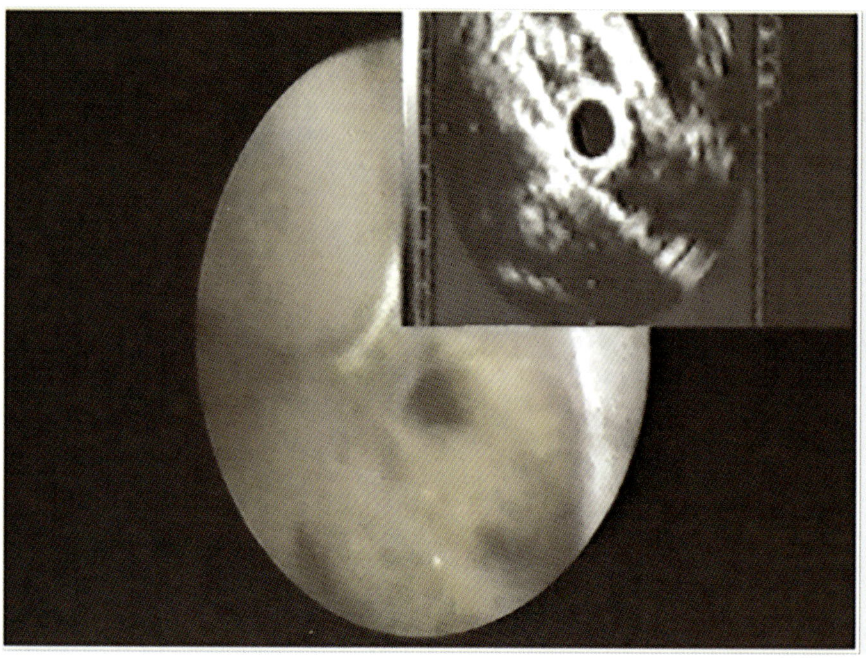

Fig. 7.22 Simultaneous sonographic and endoscopic representation (SONO-EDS) of the epidural space on one monitor (picture-in-picture technique). The multi-element, phased-array intravascular ultrasound catheter transducer is visible in the patient's epidural cavity.

- Floppy tip: 12 mm.
- Median frequency: 30 MHz.
- Bandwidth: 10 to 40 MHz.
- Imaging rate: 30 frames/s.
- Number of ultrasound elements: 64.
- Focusing: electronic via dynamic aperture array.
- Dynamic range: 68 dB.
- Penetration depth: 8 mm.
- Resolution: 150 to 300 µm lateral, 80 to 140 µm axial.

The 135-cm–long ultrasound probe has an outside diameter of 1.18 mm and is equipped with a transducer at its tip. The transducer module, with 1-mm diameter and a length of 3 mm, contains a cylindrical ultrasound converter. This miniaturized converter includes circular integrated chips. It has 64 piezoelectric crystals, arranged side by side, which are successively electronically driven to emit and receive ultrasound signals. Thereby a 360° two-dimensional image format is made possible. The axial resolution of the ultrasound catheter systems used lies between 0.08 and 0.14 mm, the lateral resolution between 0.15 and 0.3 mm. The resolution is dependent on the distance between the neuraxial structures and the transducer and decreases with increasing catheter–object distance.

The neuraxial placement of the ultrasound cathether is through the working channel of the epiduroscope. The ultrasound catheter is connected, along with the epiduroscopic equipment, with the Cathscanner Imaging System of EndoSonic or the Volcano Corporation ultrasound instrument (▸ Fig. 7.22).

IVUS technology offers the investigator detailed additional information about the composition of the patient's epidural anatomical structures. The virtual histology images from the IVUS technology are produced with the help of the Volcano image console and the Eagle Eye Gold IVUS catheter (Volcano Corporation, Rancho Cordova, California). The sonographic images can then be displayed in the EDS catheter laboratory in real time, so that the physician performing the operation and the OR team can evaluate them while the patient is still lying on the operating table (▸ Fig. 7.23).

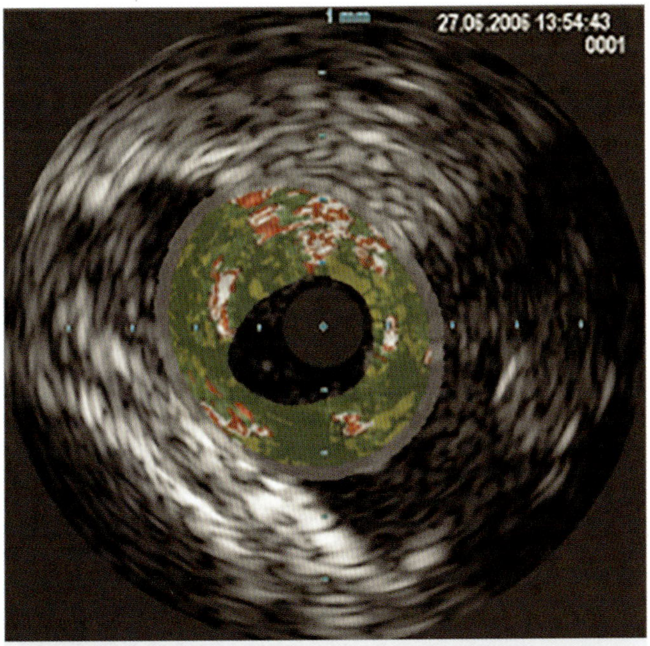

Fig. 7.23 Virtual histology via a sonographically color-coded spectrum analysis from inside the epidural space with the Visions Five-64 F/X catheter.

This technology offers an automatic measurement tool to simplify the interpretation of the images, and it uses a predefined color-coding for the display of the composition of a defined region of tissue of the spinal dura mater or within the epidural cavity.

7.4.2 Electrodes

For electrical epidural neuromodulation there is a diverse selection of products from each manufacturer of spinal cord stimulation electrodes. The basic distinction is between rod (percutaneous) electrodes and plate (paddle, surgical) electrodes. From the large number of 2- to 16-pole standard electrode models, the rod electrode has prevailed for endoscopically supported SCS electrode implantation.

Spinal Cord Stimulation Rod Electrodes

SCS rod electrodes are available in lengths of 60 cm, 75 cm, or 90 cm with a uniform diameter of 1.3 mm. The number of stimulation poles ranges from 4, 6, 8, to 16 poles. For fine tuning of the SCS stimulation array, the stimulation pole configurations are determined by length and distance. Through a single or paired epidural arrangement of the electrodes and programming of the SCS rod electrodes, simple or complex stimulation pole arrays are created.

On the basis of pole configuration, Vectris electrodes (Medtronic Inc., Minneapolis, Minnesota), for example, with epidural positioning reach over about two vertebral levels and thus contribute to improved coverage of the pain region of the patient. Changing the pole configurations of the SCS electrode make possible a so-called electronic repositioning, by which alterations of the stimulation region caused by vertical movement of the electrode can be compensated for (▸ Fig. 7.24).

The TargetStim programming tools (Medtronic, Inc., Minneapolis, Minnesota) make possible a smooth transition between different stimulation pole configurations and thus a rapid and individually adaptable repositioning of the stimulation region (▸ Fig. 7.25).

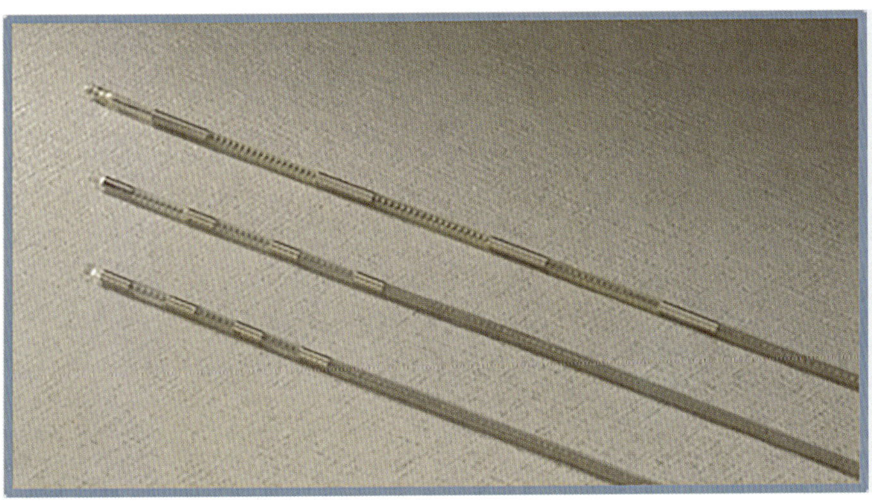

Fig. 7.24 Spinal cord stimulation rod electrodes allowed for whole-body MRI to be done while they are implanted and, as shown, they are lined up ready for use.

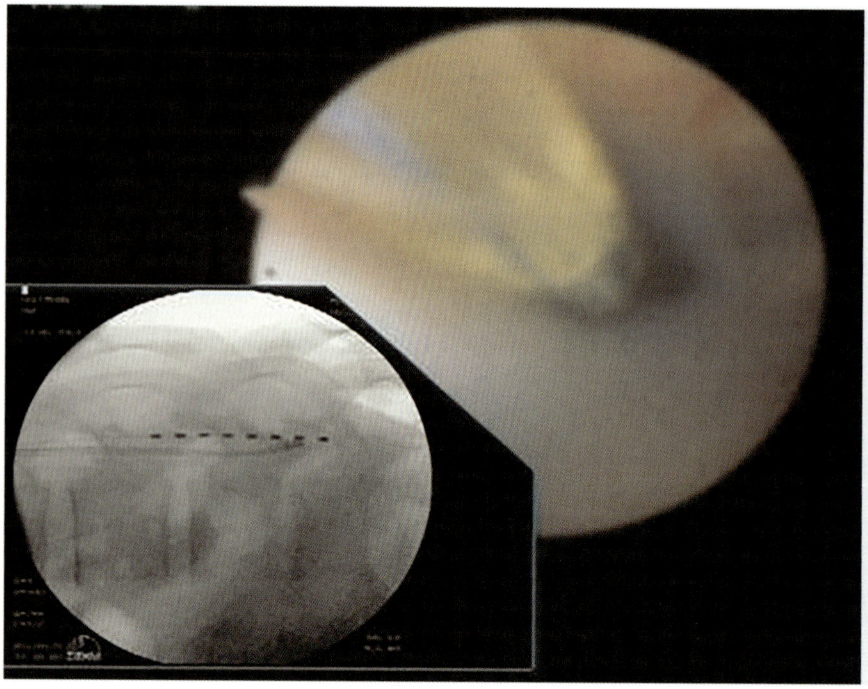

Fig. 7.25 Epiduroscopically supported spinal cord stimulation electrode implantation (8-pole) in the dorsal thoracic epidural space (picture-in-picture technique).

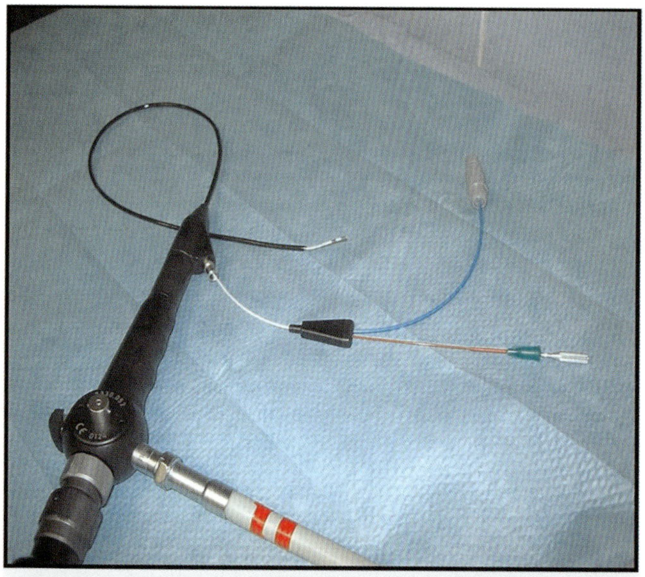

Fig. 7.26 PASHA multifunction electrode in the epiduroscope.

Multifunction Electrodes

With the help of a multifunction electrode (PASHA electrode, OMT GmbH, Frittlingen, Germany), a targeted epidural application of medication as a single shot or continuously, as well as electrical neuromodulation can take place, and nerve stimulation with any desired frequency can be performed. It is recommended for dorsolateral epidural placement.

Some anatomical structures, such as sacral ganglia or thoracic spinal roots, are difficult to access for conventional procedures. Endoscopically guided placement of a multifunction electrode lends itself to such procedures.[14] Endoscopically guided epidural placement of the electrode tip makes possible pinpoint-accurate neurostimulation of the posterior roots and posterior horns of the spinal cord (▶ Fig. 7.26).

The PASHA multifunction electrode with lumen and thermosensor has a diameter of 4 Fr (1.38 mm) and is available in two models, PC4–30 (30-cm length) and PC4–60 (60-cm length). It is equipped with a number of two bipolar electrodes with an electrode length of 3 mm and electrode spacing of 4 mm. The total lumen volume of the catheter is 112 or 225 µL. The straight stylet has a diameter of 0.34 mm and a length of either 427 or 727 mm.

The PASHA multifunction electrode offers:

- Pinpoint-accurate placement through sensory stimulation.
- Multiple targeted test blockades.
- Electrical neuromodulation through pulsed radiofrequency (RF) power.
- Epidural pharmacotherapy (EAT).

To be able to generate a corresponding electric field unerringly, the multifunction electrode is precisely placed epidurally under endoscopic conditions through the access technique already described. For that, the electrode is inserted through the working channel of an epiduroscope with an outside diameter of 3.5 mm to the epidural target point. With epidural pulsed radiofrequency therapy, short-duration active power pulses (0.1–0.7 W by Neuro N50, Radimed Gesellschaft fur Kommunikation dienstleistungen und Medizintechnik mbH, Bochum, Germany;) are administered (▶ Fig. 7.27).

R-ResAblator Epiduroscopy

Raffaelli and Righetti[15] reported on patients in whom they used an epiduroscopic high-frequency method, the so-called R-ResAblator epiduroscopy. The authors found that this RF epiduroscopy could show greater therapeutic benefit than traditional epiduroscopy or conventional surgery.

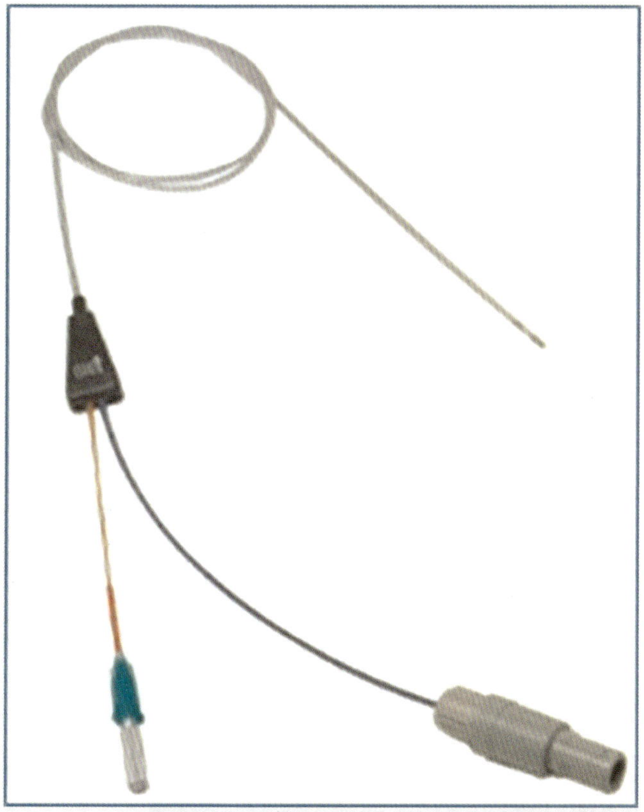

Fig. 7.27 PASHA multifunction electrode. (Courtesy of OMT GmbH, Frittlingen, Germany.)

7.5 Conclusion

The medical products described and shown are ones we use. Of course, comparable products not mentioned should be considered when selecting products for spinal endoscopy.

References

[1] Oxford Health Plans LLC. Epiduroscopy, Epidural Lysis of Adhesions and Functional Anesthetic Discography: Clinical Policy. Oxford Health Plans; 2015:9 https://www.oxhp.com/secure/policy/epiduroscopy_epidural_lysis_of_adhesions_and_functional_anesthetic_discography.pdf

[2] Manchikanti L, Falco FJ, Singh V, et al. An update of comprehensive evidence-based guidelines for interventional techniques in chronic spinal pain. Part I: introduction and general considerations. Pain Physician. 2013; 16(2) Suppl: S1–S48

[3] Boswell MV, Trescot AM, Datta S, et al. American Society of Interventional Pain Physicians. Interventional techniques: evidence-based practice guidelines in the management of chronic spinal pain. Pain Physician. 2007; 10 (1):7–111

[4] Schütze G. Epiduroscopy [multilingual DVD-ROM]. Tuttlingen, Germany: Karl Storz; 2006

[5] Wildsmith JAW, Armitage EN. Principles and Practice of Regional Anaesthesia. 2d ed. London, UK: Churchill Livingstone; 1987

[6] Schütze G. Interventionelles Schmerzmanagement; Bildgestützte Verfahren zur Diagnostik und Therapie rückenmarksnaher Schmerzsyndrome-Spinale Endoskopie, pharmakologische und elektrische Neuromodulation. Bremen, Germany: UNI-MED Verlag AG; 2011

[7] Schütze G. Epiduroskopie. Ein praxisorientierter Leitfaden zur epiduroskopischen Diagnostik und Therapie rückenmarksnaher Schmerzsyndrome. Lengerich, Germany: Pabst Science Publishers; 2006

[8] Ruetten S, Meyer O, Godolias G. Endoscopic surgery of the lumbar epidural space (epiduroscopy): results of therapeutic intervention in 93 patients. Minim Invasive Neurosurg. 2003; 46(1):1–4

[9] Ruetten S, Meyer O, Godolias G. Application of holmium:YAG laser in epiduroscopy: extended practicabilities in the treatment of chronic back pain syndrome. J Clin Laser Med Surg. 2002; 20(4):203–206

[10] Schütze G, Kurtze H. Percutaneous investigation of the epidural space using a flexible endoscope; A contribution to epiduroscopy. Reg Anesth Pain Med. 1993; 18:24

[11] Koch T, Hübler M. Thorakale epidurale Anästhesie und Analgesie. München, Germany: Arcis Verlag GmbH; 2003

[12] Prithvi Raj P. Interventional Pain Management: Image-Guided Procedure. Philadelphia, PA: Saunders Elsevier; 2008

[13] Jerosch J, Steinleitner W. Minimal-invasive Wirbelsäulen-Intervention. Köln, Germany: Deutscher Ärzte-Verlag Köln; 2005

[14] Schütze G. .Epiduroscopy—Spinal Endoscopy. New York, NY: Springer-Verlag; 2008

[15] Raffaeli W, Righetti D. Surgical radio-frequency epiduroscopy technique (R-ResAblator) and FBSS treatment: preliminary evaluations. Acta Neurochir Suppl (Wien). 2005; 92:121–125

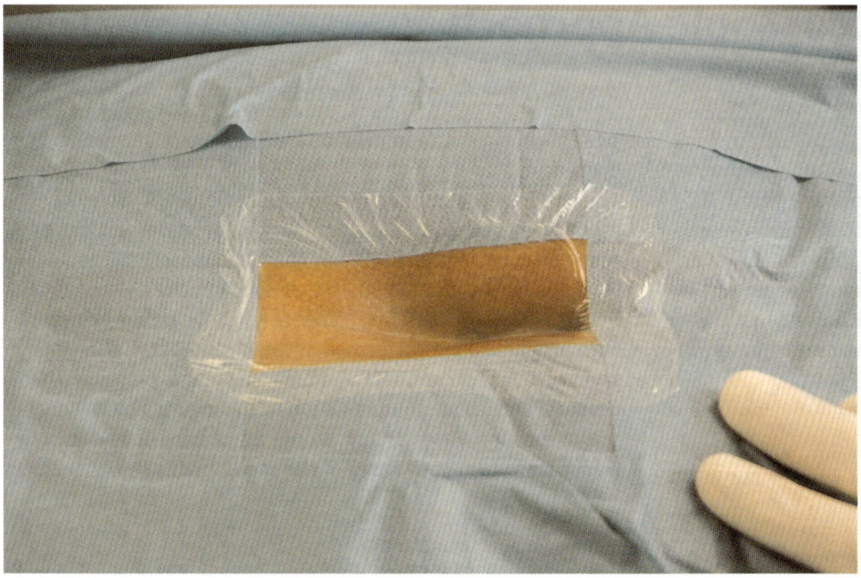

Fig. 9.2 Surgical field after draping.

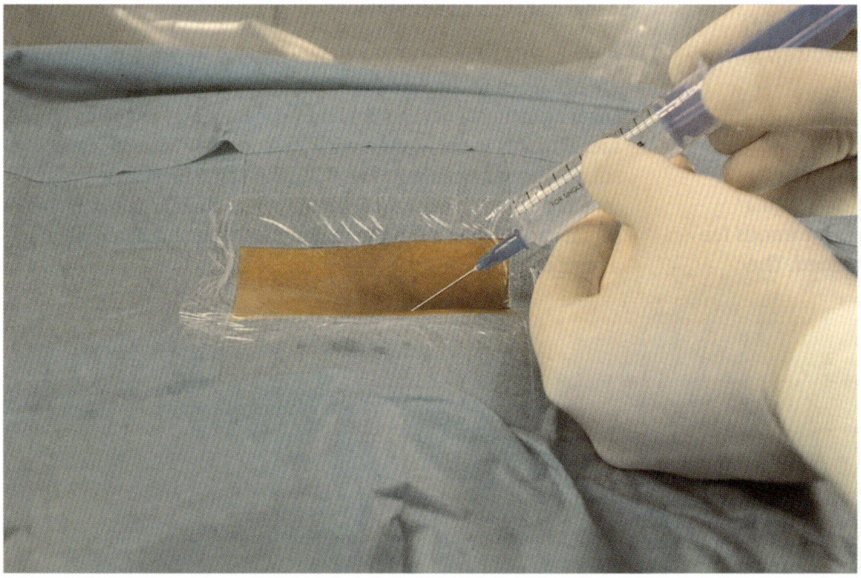

Fig. 9.3 Local anesthesia of the skin and subcutaneous tissue of the region over the sacral hiatus.

9.4 Step 3: Needle Insertion Technique and Epidurography

- An 18-gauge Tuohy needle is inserted into the sacral canal through the sacral hiatus under fluoroscopic control in lateral view (▶ Fig. 9.5). The skin entry point is chosen, assuring that a shallow angle is achieved between the needle and the sacral canal (typically 2–4 cm caudal to the sacral hiatus; ▶ Fig. 9.6).[1]
- The needle position may be confirmed by injecting nonionic contrast (e.g., Ultravist 240, Bayer Schering Pharma A.G.).

9.5 Step 4: Setting the Approach for the Epiduroscope

- The stylet of the Tuohy needle is withdrawn and a "J-tip" flexible guidewire is introduced through the lumen into the sacral canal up to the level of S3 (▶ Fig. 9.7).[6]
- A 4-mm-length skin incision is made with a no. 11 scalpel blade, centered on the entry point, and the Tuohy needle is gently removed, ensuring that the guidewire is kept in place (▶ Fig. 9.8 and ▶ Fig. 9.9).

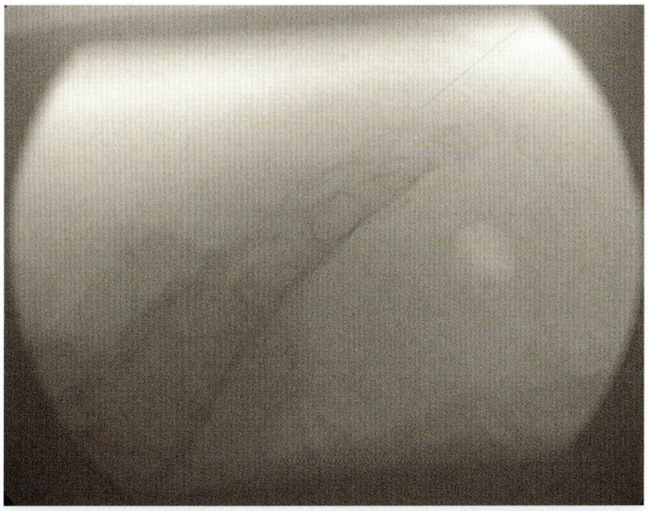

Fig. 9.4 Local anesthesia of the caudal end of the sacral canal using a spinal needle inserted through the sacral hiatus.

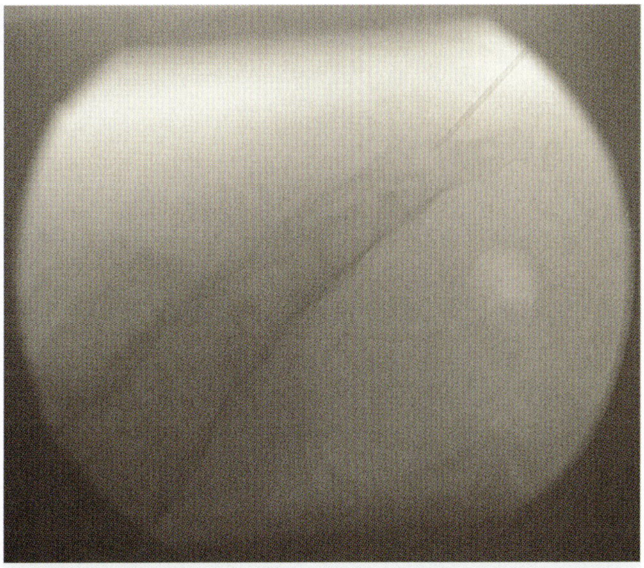

Fig. 9.5 A Tuohy needle being inserted through the sacral hiatus.

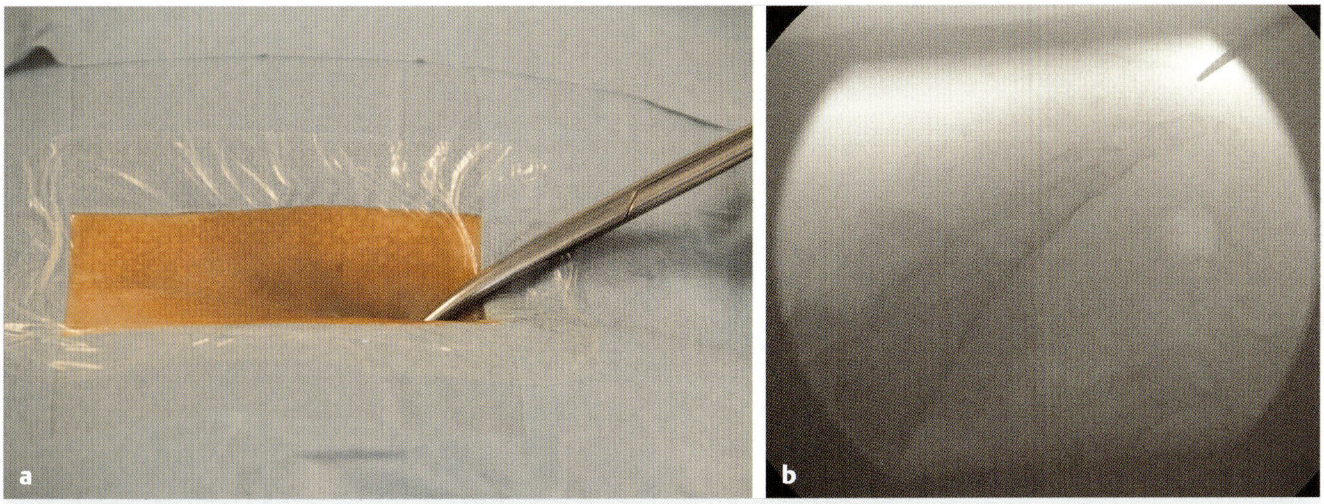

Fig. 9.6 (a,b) Localizing the skin entry point with lateral fluoroscopy.

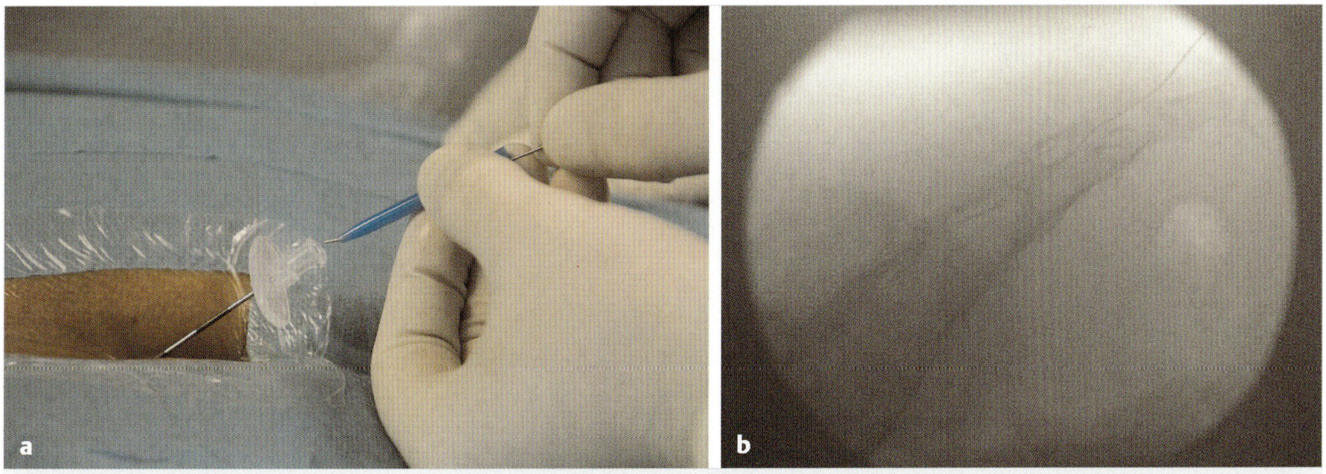

Fig. 9.7 (a,b) Insertion of the guidewire through the Tuohy needle.

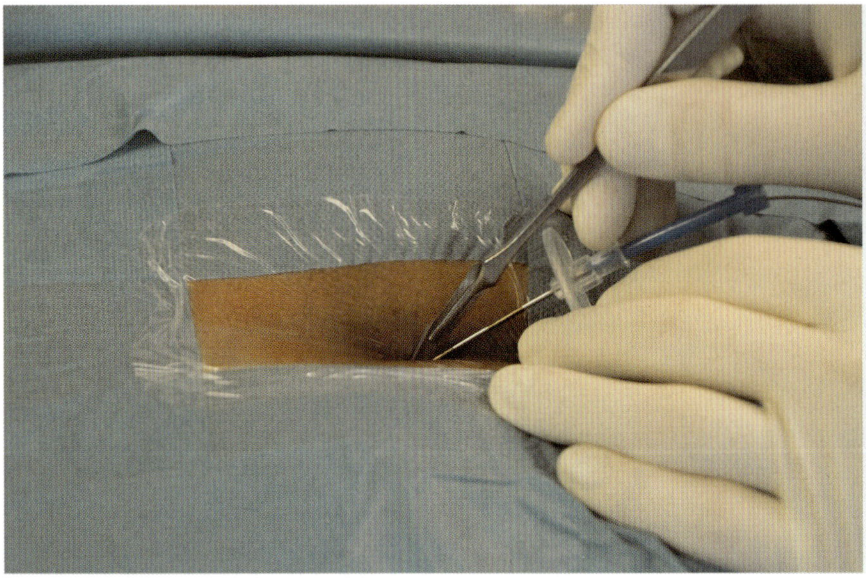

Fig. 9.8 Enlarging the skin incision.

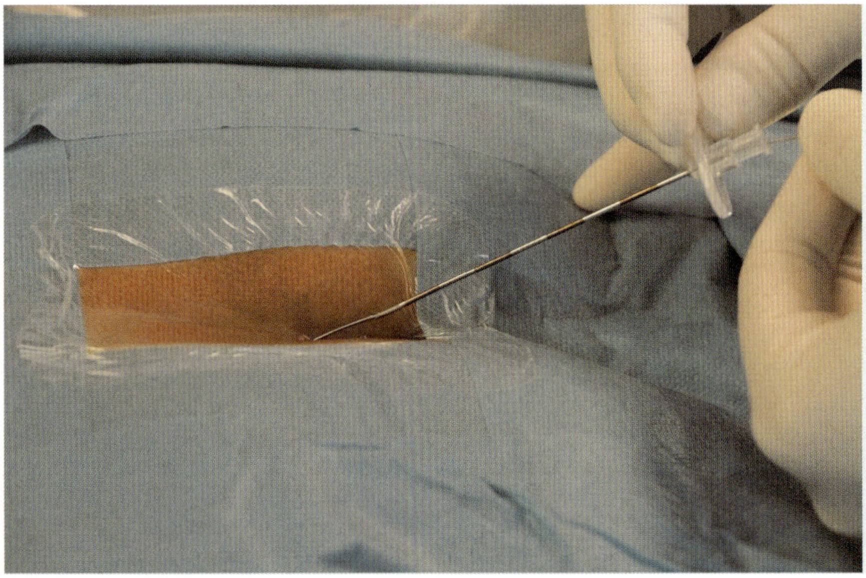

Fig. 9.9 Removal of the Tuohy needle with care so as not to pull the guidewire.

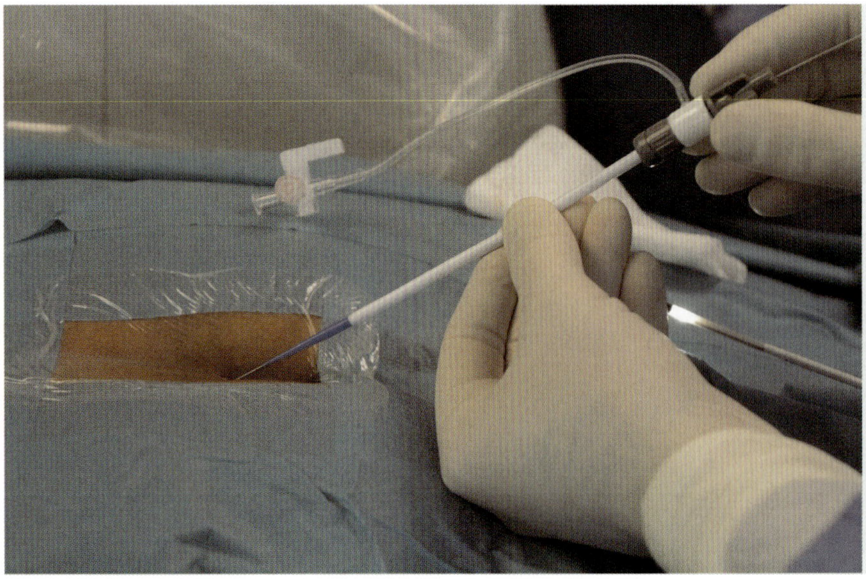

Fig. 9.10 Inserting the dilator surrounded by a plastic sleeve around the guidewire into the sacral canal.

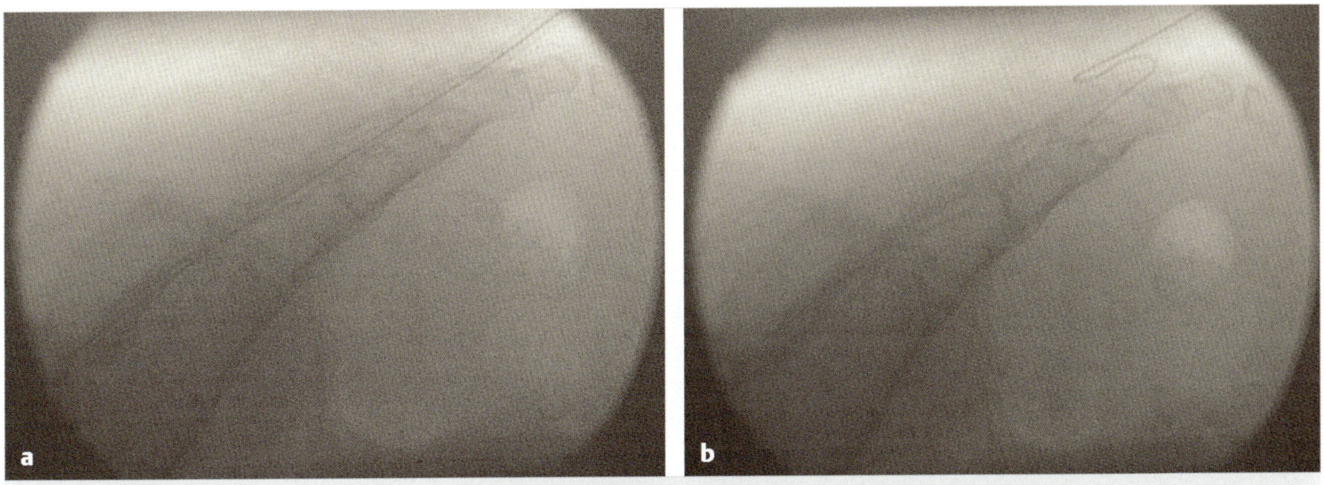

Fig. 9.11 **(a)** Dilator, plastic sleeve, and guidewire in the sacral canal. **(b)** Kinking of the guidewire at the sacral hiatus.

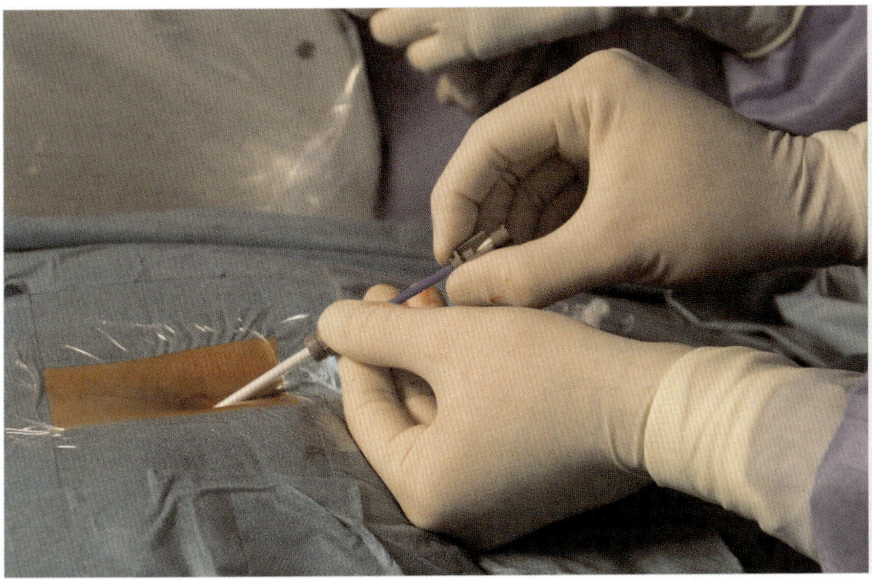

Fig. 9.12 Removal of the plastic dilator and the guidewire.

- A rigid plastic dilator surrounded by a plastic sleeve is passed around the guidewire taking care to avoid its kinking (▶ Fig. 9.10 and ▶ Fig. 9.11). Finally, the dilator and the guidewire are removed (▶ Fig. 9.12).

9.6 Step 5: Epiduroscopic Approach

- An appropriate epiduroscope should be chosen from the several models commercially available. The device should be sterile, flexible, and steerable and contain at least two internal operating channels (one for the flexible optics and irrigation and the other for working tools).[7]
- After adjusting the focus, brightness, and white balance, the epiduroscope is introduced into the plastic sleeve and slowly advanced in the epidural space using small *boluses* of physiologic saline solution to distend and allow visualization of the epidural space and to clean the visual field (▶ Fig. 9.13).
- During the initial progression of the epiduroscope in the sacral canal, an endoscopic view of epidural fatty tissue is achieved, usually with the characteristic scarce vascularization (▶ Fig. 9.14).

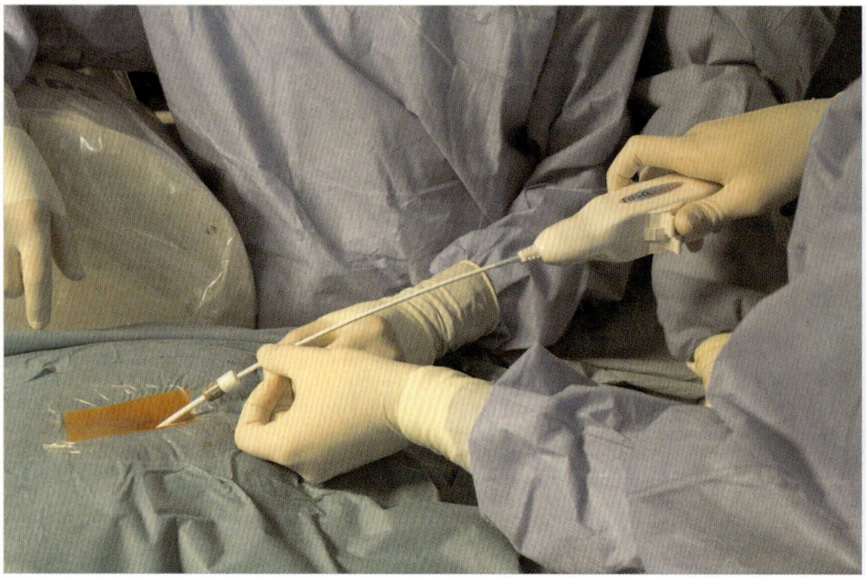

Fig. 9.13 Epiduroscope being introduced in the plastic sleeve into the spinal canal.

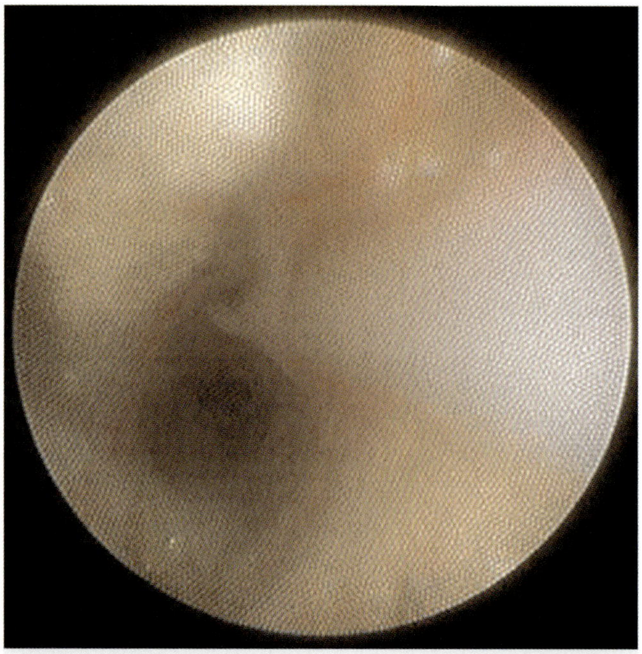

Fig. 9.14 Epiduroscopic view of the sacral canal with the characteristic identification of epidural fat, few small arterial vessels, and the distal filum terminale on the right side.

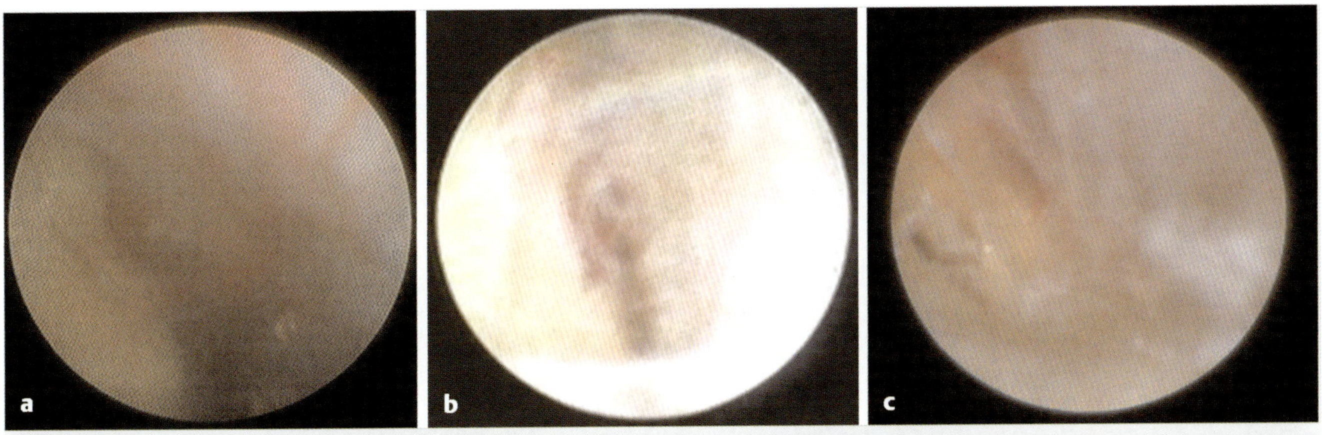

Fig. 9.15 Epiduroscopic views. **(a)** Dura mater on the superior and right side, loose fibrous tissue in the center, and epidural fat on the left. Small arteries can be seen at the surface of the dura mater and on the epidural space. **(b,c)** Epidural fibrosis.

- As the endoscope is progressed more cranially, it becomes possible to visualize the *dura mater*, nerve roots, epidural blood vessels, and, particularly in postoperative patients, epidural fibrosis and inflammation tissue (▶ Fig. 9.15).
- The position of the endoscope inside the spinal canal is guided and confirmed by fluoroscopic imaging, usually in anteroposterior (AP) view (▶ Fig. 9.16).
- Cadaveric dissection demonstrates the anatomic location of the epiduroscope in the dorsal epidural space (▶ Fig. 9.17a). Fluoroscopy confirms the position of the epiduroscope (▶ Fig. 9.17b). The epiduroscope can be in the dorsal or ventral epidural space (▶ Fig. 9.18 and ▶ Fig. 9.19).

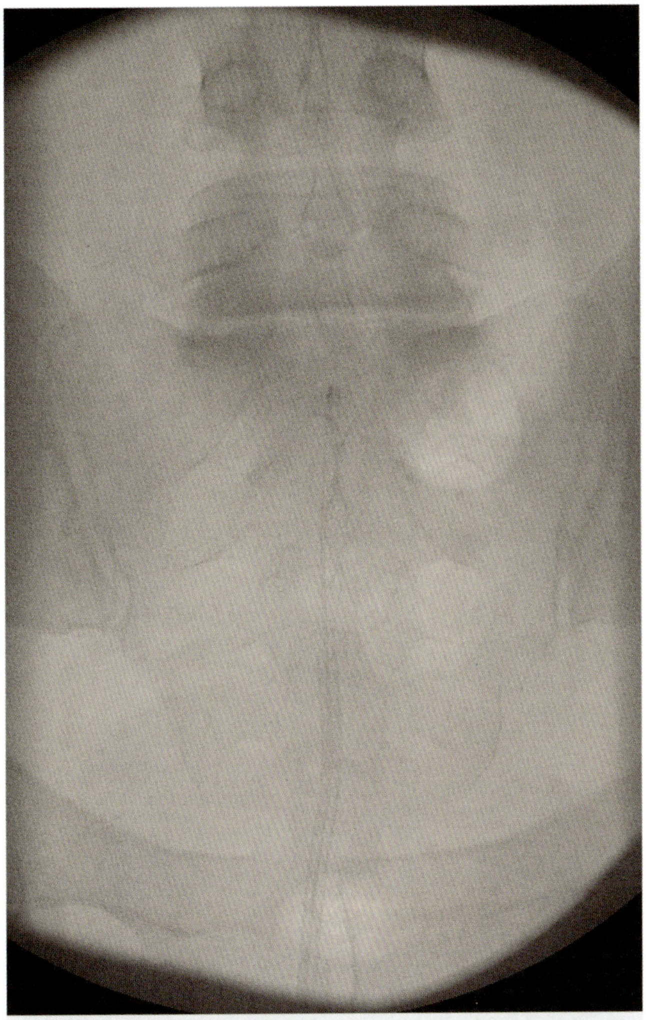

Fig. 9.16 Fluoroscopic anteroposterior view with visualization of the tip of the epiduroscope at S1 level.

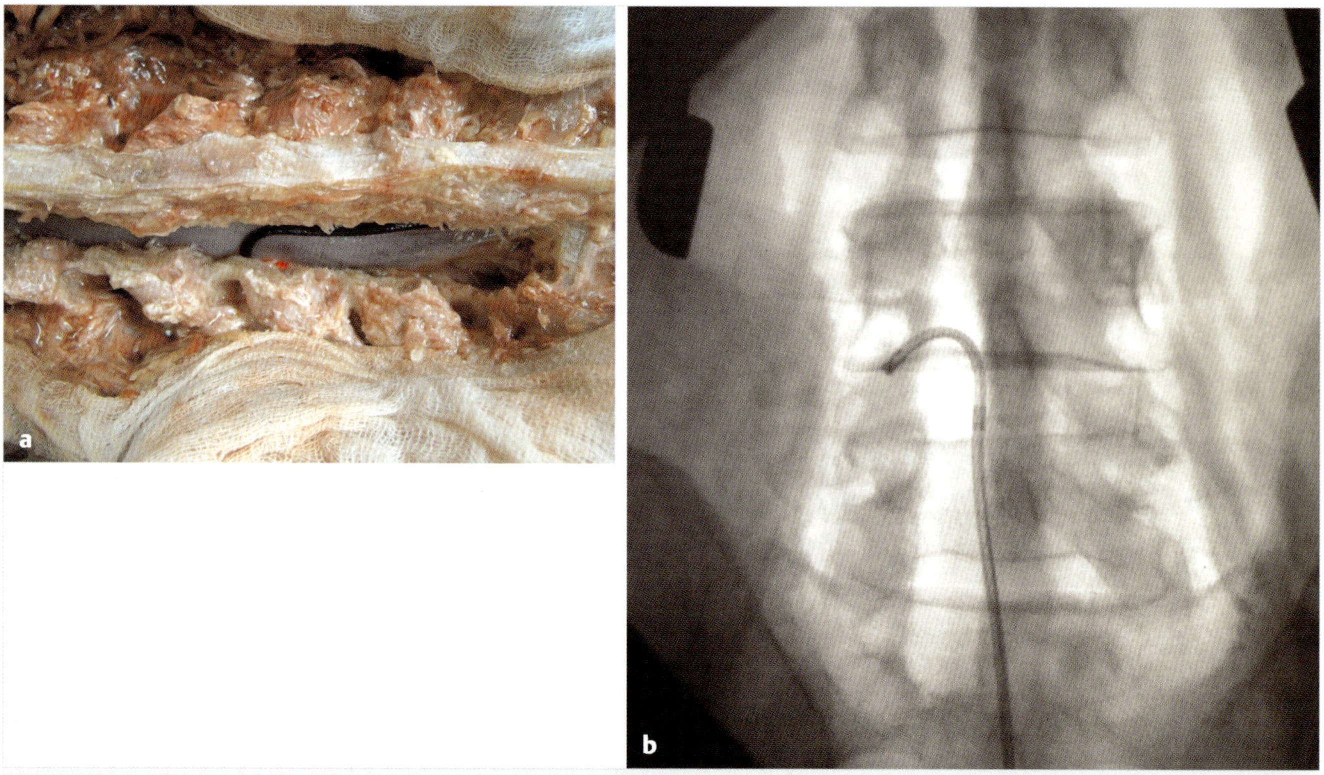

Fig. 9.17 (a) Cadaveric dissection of the lumbar spine with a left L3–L5 hemilaminectomy. The tip of the epiduroscope can be seen in the dorsal epidural space at the left L4–L5 neural foramen. **(b)** Anteroposterior fluoroscopy of the cadaveric dissection of the lumbar spine with a left L3–L5 hemilaminectomy. The tip of the epiduroscope can be seen in the epidural space at the left L4–L5 neural foramen.

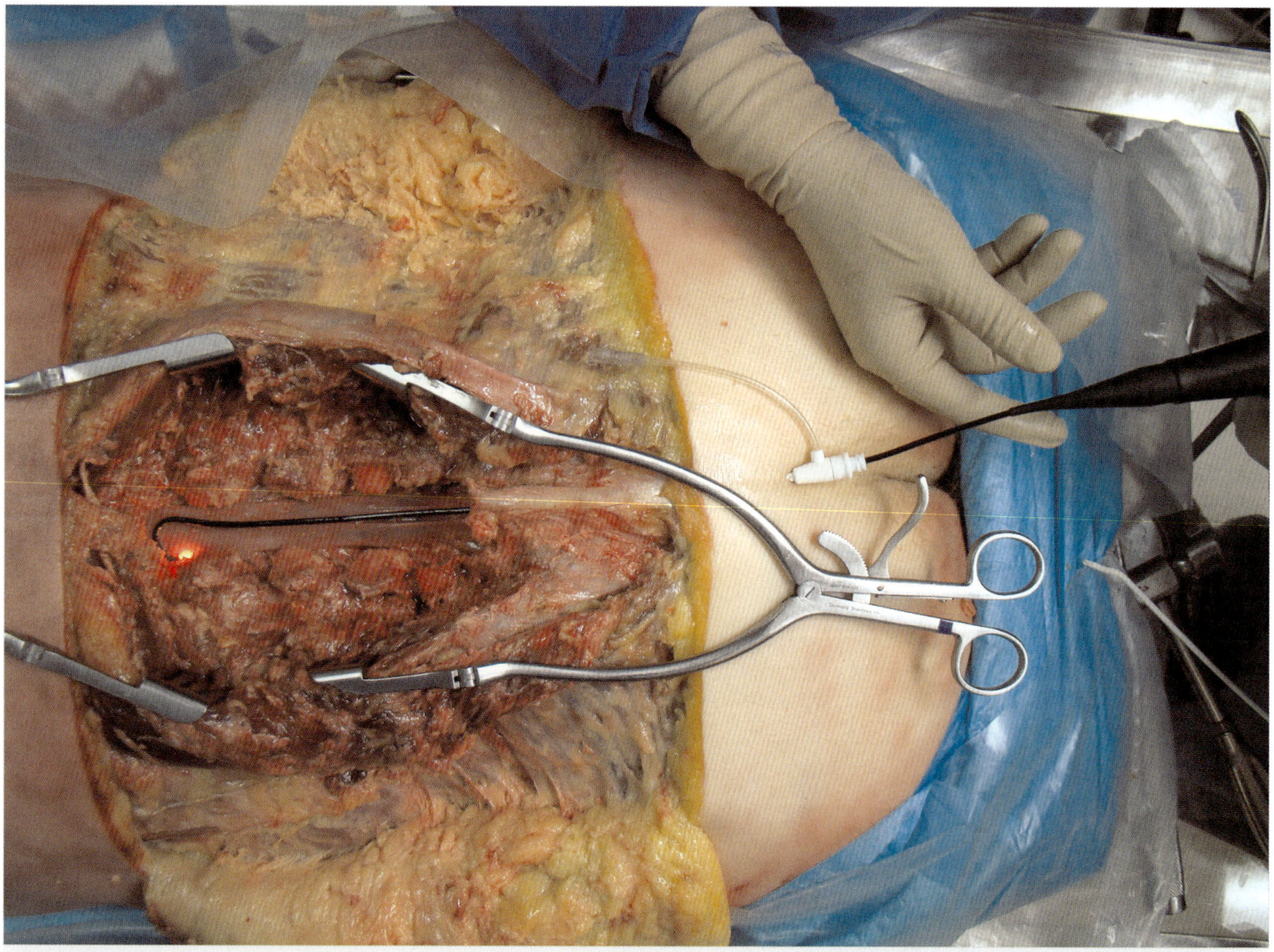

Fig. 9.18 Cadaveric dissection of the lumbar spine with an L1–L5 laminectomy and proximal sacrum unroofed. The tip of the epiduroscope can be seen in the dorsal epidural space at the left L1–L2 neural foramen. Percutaneous entry site of the epiduroscope can be seen on the right side of the image.

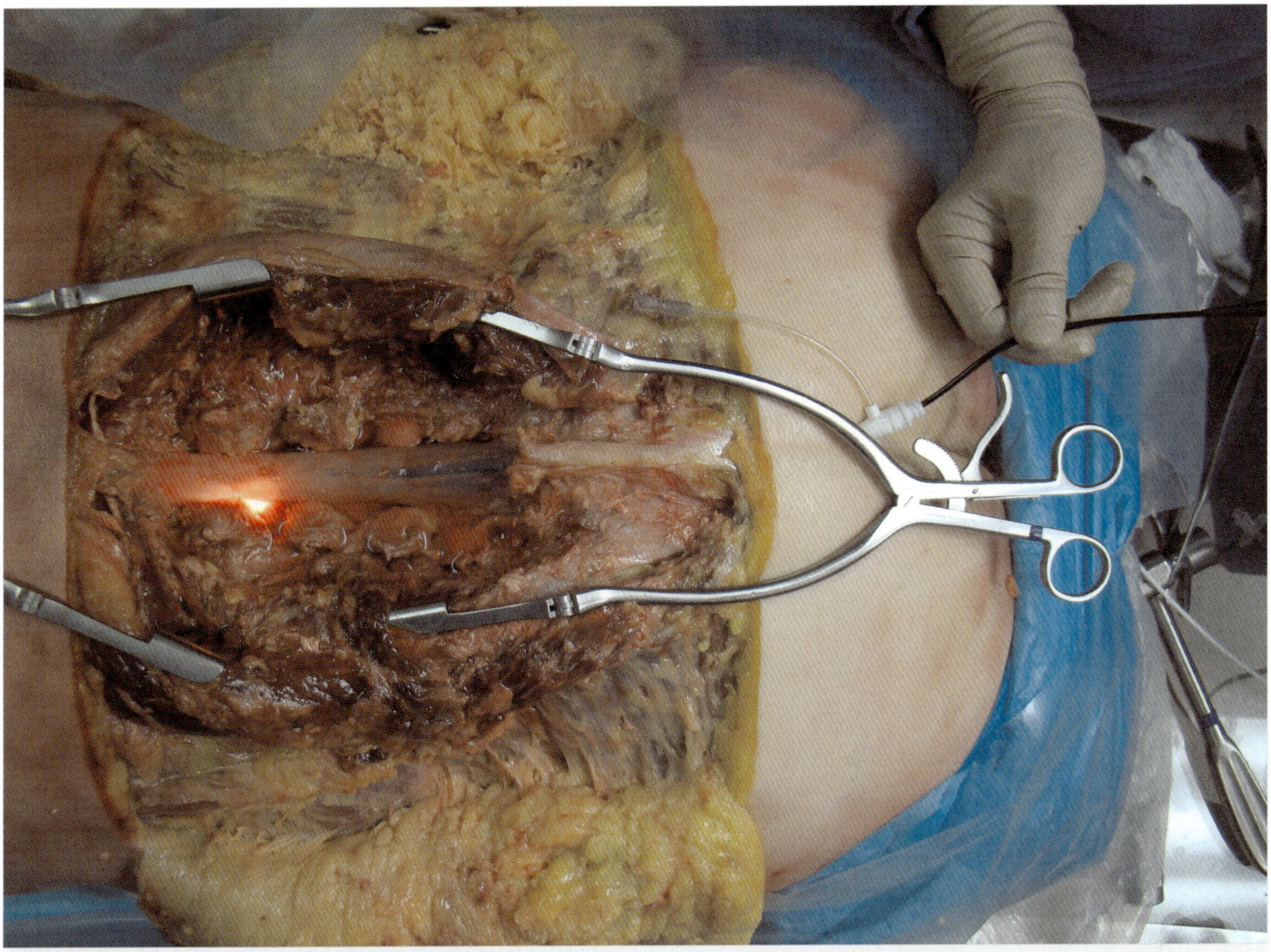

Fig. 9.19 Cadaveric dissection of the lumbar spine with an L1–L5 laminectomy and proximal sacrum unroofed. The tip of the epiduroscope can be seen in the ventral epidural space at the left L2–L3 neural foramen.

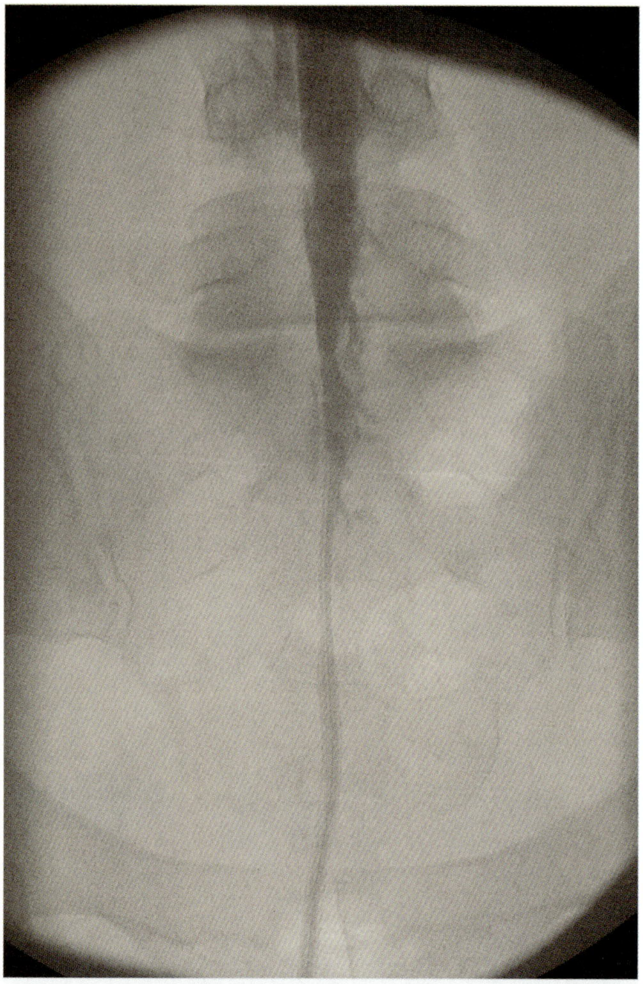

Fig. 9.20 Epidurogram with dye injection at S1–S2 level, showing a good upward progression of the contrast in the spinal canal.

9.7 Step 6: Epidurography

- When the tip of the epiduroscope reaches the S1–S2 level, an epidurogram is performed, injecting 5 to 10 mL of nonionic contrast (e.g., Ultravist 240, Bayer Schering Pharma A.G.; ▸ Fig. 9.20).[4]
- In a normal epidurogram, the dye is seen to progress through the spinal canal and exiting through the intervertebral *foramina* around the nerve roots, forming the typical "Christmas tree" pattern (▸ Fig. 9.21).
- When extensive epidural fibrosis is present inside the vertebral canal, it can hinder the progression of the contrast dye beyond the affected level and a "stop sign" can arise on the epidurogram (▸ Fig. 9.21). The "stop sign" can also be seen on periradicular spread of contrast if the fibrosis blocks the epidural space on the intervertebral *foramina*.
- The epidurogram provides information during the procedure to be integrated with the patient complaints, previous magnetic resonance images, and endoscopic findings.
- Epidurography may be repeated at the end of the procedure for comparison and assessment of the adhesiolysis.

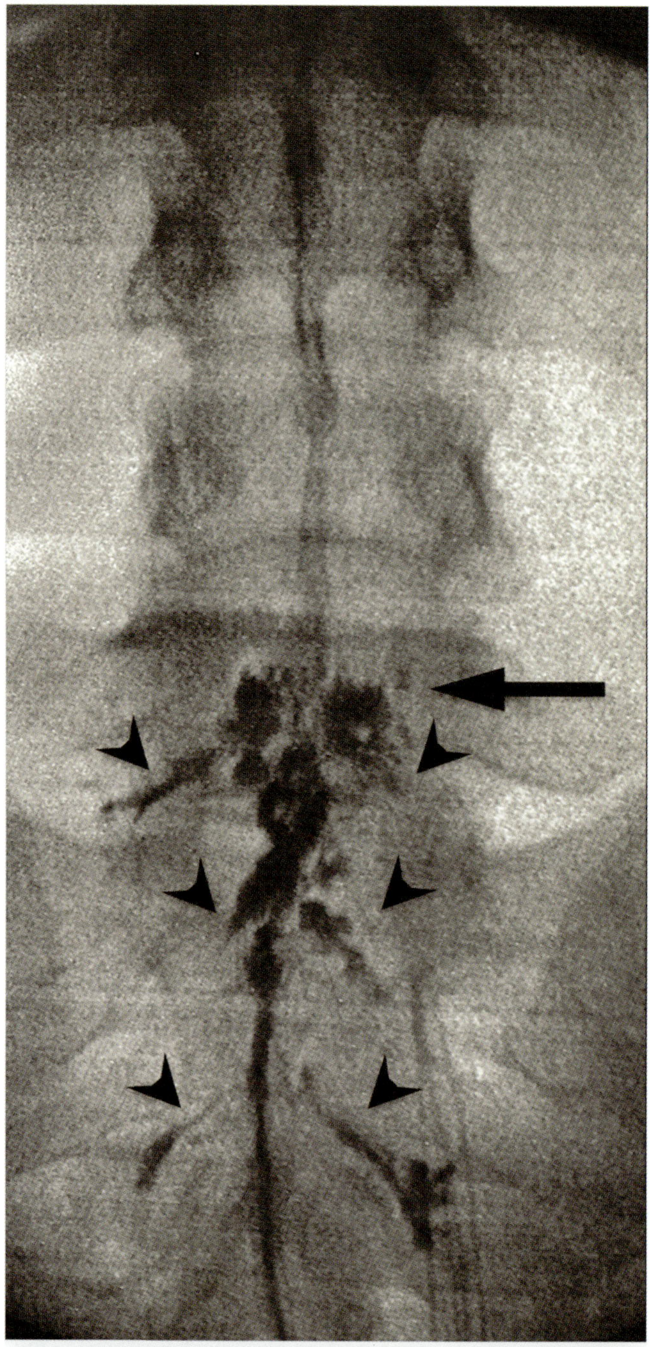

Fig. 9.21 Epidurogram showing a "stop sign" at the level of L5 pedicles with no upward progression of the contrast dye (*black arrow*). Contrast progression around the dural sleeves of the L5, S1, and S2 nerve roots (*arrowheads*) can be seen, forming the "Christmas tree" pattern.

9.8 Step 7: Epiduroscopy

- During epiduroscopy, several diagnostic and therapeutic procedures may be undertaken depending on the patient's clinical picture, preoperative imaging, intraoperative findings, purpose of the intervention, and physician's option.[2]
- From a diagnostic perspective, besides the endoscopic visualization of the epidural structures and related pathology, the tip of the epiduroscope or a Fogarty catheter may be used to probe different anatomic structures, looking for concordant pain with the patient's usual one (epidural pain provocation test; ▸ Fig. 9.22). Moreover, an endoscopic biopsy forceps may be used to obtain samples for a number of studies (▸ Fig. 9.23).
- As a therapeutic tool, epiduroscopy allows the proinflammatory cytokines present in the epidural space to be diluted and washed out by copious saline irrigation. Adhesiolysis may be performed and therapeutic drugs may be administered under endoscopic view at the site of the identified pathology and pain provocation.

9.9 Step 8: Finishing the Procedure and Closing

- After performing the intended epidural procedures, the epiduroscope and the plastic sleeve are removed and the skin entry point is closed with a stitch using a resorbable surgical thread and appropriate sterile dressing (▸ Fig. 9.24).

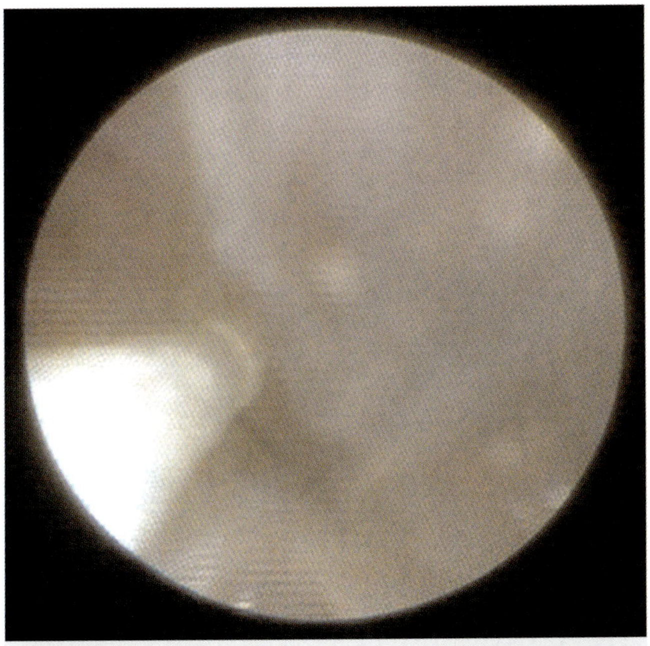

Fig. 9.22 Epidural pain provocation test.

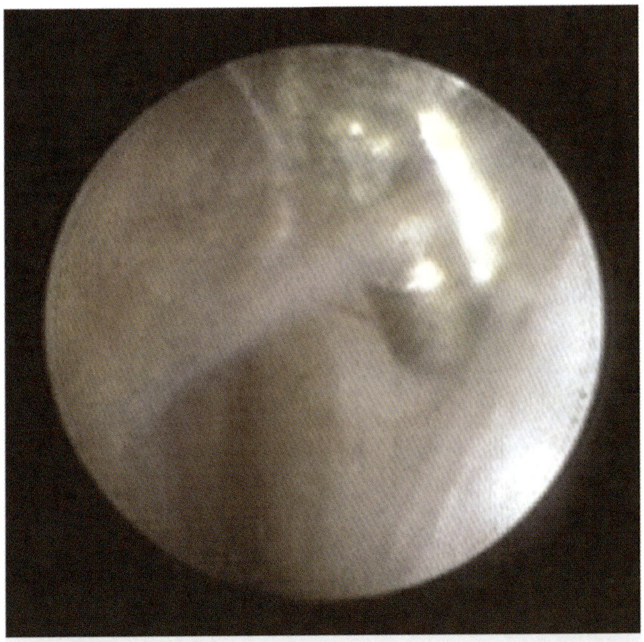

Fig. 9.23 Endoscopic biopsy forceps.

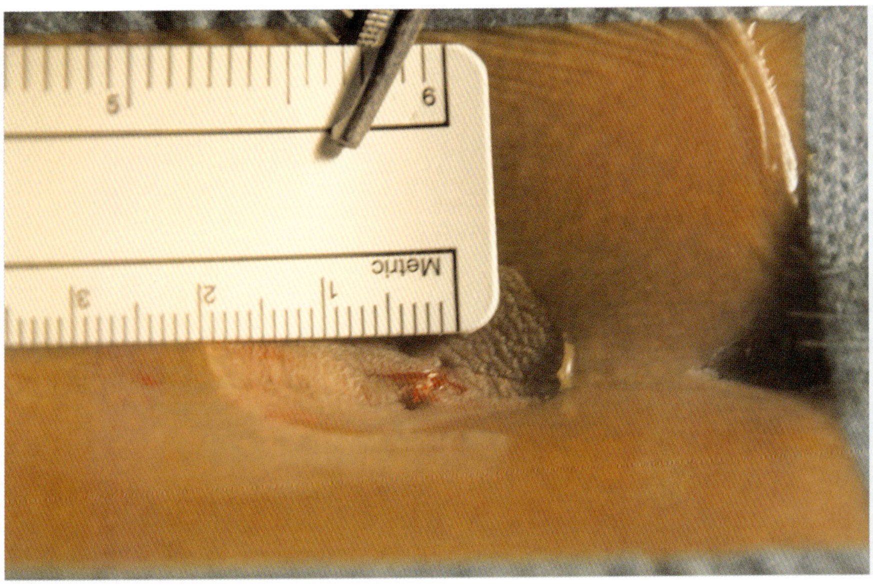

Fig. 9.24 Skin closure.

9.10 Relevant Sacral Anatomy

- Sacrum:
  - Five fused sacral vertebrae, dorsally convex, articulating laterally with the two iliac bones, cranially with the fifth lumbar vertebra through the intervertebral disk and the zygapophyseal joints, and caudally with the coccyx.
  - There are four pairs of *foramina* that are actually channels through which the sacral nerves exit the sacral canal. The posterior *foramina*, which are smaller than the anterior ones, contain the dorsal primary rami of the sacral nerves and the anterior foramina give passage to the ventral primary *rami* of the S1–S4 nerves.
- Sacral hiatus:
  - Inverted U-shaped opening in the midline, in the posterior wall of the sacral canal, at S4–S5 level, limited laterally by the sacral *cornua* (inferior articular processes remnants that project downward).
  - Covered by the posterior sacrococcygeal ligament.
- Sacral canal:
  - The caudal extension of the lumbar spinal canal until the sacral hiatus.
  - Contains the caudal end of the dural sac, usually between S1 and S3.
  - The S1–S4 nerve roots exit through their foramina while S5 and the coccygeal nerve exit through the sacral hiatus.
  - There is an epidural venous plexus, mostly anterior, which generally ends at S4.
  - Fat tissue occupies most of the remaining space in the sacral canal.

References

[1] Saberski LR, Kitahata LM. Direct visualization of the lumbosacral epidural space through the sacral hiatus. Anesth Analg. 1995; 80(4):839–840

[2] Kallewaard JW, Vanelderen P, Richardson J, Van Zundert J, Heavner J, Groen GJ. Epiduroscopy for patients with lumbosacral radicular pain. Pain Pract. 2014; 14(4):365–377

[3] Pereira P, Severo M, Monteiro P, et al. Results of lumbar endoscopic adhesiolysis using a radiofrequency catheter in patients with postoperative fibrosis and persistent or recurrent symptoms after discectomy. Pain Pract. 2016; 16 (1):67–79

[4] Schuetze G. Epiduroscopy for diagnosis and treatment of spinal pain syndromes. In: Medimond International Proceedings. 12th International Pain Clinic Congress of the World Society of Pain Clinicians. Turin, Italy; 2006:215–221

[5] Beltrutti D, Groen GJ, Saberski L, et al. Epiduroscopy – Consensus Decision. World Initiative on Spinal Endoscopy (WISE). Graz, Austria; 2006. Available online at http://www.epiduroscopia.com/consensus2006.pdf

[6] Bosscher HA, Heavner JE. Incidence and severity of epidural fibrosis after back surgery: an endoscopic study. Pain Pract. 2010; 10(1):18–24

[7] Bosscher HA, Heavner JE. Lumbosacral epiduroscopy findings predict treatment outcome. Pain Pract. 2014; 14(6):506–514

10 Transforaminal and Interlaminar Approaches in Epiduroscopy

Alan David Kaye, Diana Mekler, Adam Shomstein, and Richard D. Urman

10.1 Introduction

An important indication for epiduroscopy has been to further address chronic back pain following failed back surgery. Epiduroscopy in chronic radicular pain demonstrated fibrosis in almost 100% of patients.[1] The high prevalence of scar tissue in these patients has created interest in using epiduroscopy for performing adhesiolysis to remove the fibrosis and possibly alleviate pain associated with irritation of affected nerves.[2] It can also be used to deliver corticosteroids to more specific sites. In recent years, the interlaminar and transforaminal approaches have been developed and refined (▶ Fig. 10.1). However, they are not widely used, especially in the United States, and there is limited literature about the technique and benefits of one approach versus another. Epidural lysis and diskectomy appear to be more common uses of epiduroscopy at this time. Among surgeons, the significance of adhesions causing pain is not uniformly accepted. Adhesions are part of the healing process and occur in every surgery, but not every surgery or even a significant number of spine surgeries have pain from adhesions. In addition, it should also be noted that there may be a lack of correlation between the extent of radicular symptoms and the amount of displacement of a herniated disk.[3]

There is a significant body of evidence over the past decade involving epiduroscopy utilizing caudal insertion. Recently, techniques have involved utilizing smaller instruments in an attempt to employ epiduroscopy via transforaminal and interlaminar techniques. Due, in part, to the large needles required, the interlaminar approach is not commonly practiced. Interventional pain specialists not trained in surgery will seldom use these techniques secondary to the innate complexity requiring significantly more training with exponentially more risk for medical errors. This chapter focuses on these newer techniques and provides a very general introduction for people interested in these approaches to pain management. Adhesions can be observed in every epidural space, but do not correlate with these flimsy adhesions. Just irrigating the epidural space can bring temporary relief from pain caused by inflammation. Furthermore, relief can be obtained with caudal epidural blocks without the use of an epiduroscope.

10.2 Transforaminal Approach

The transforaminal approach can be advantageous to gain access to the epidural space and facets without disrupting or

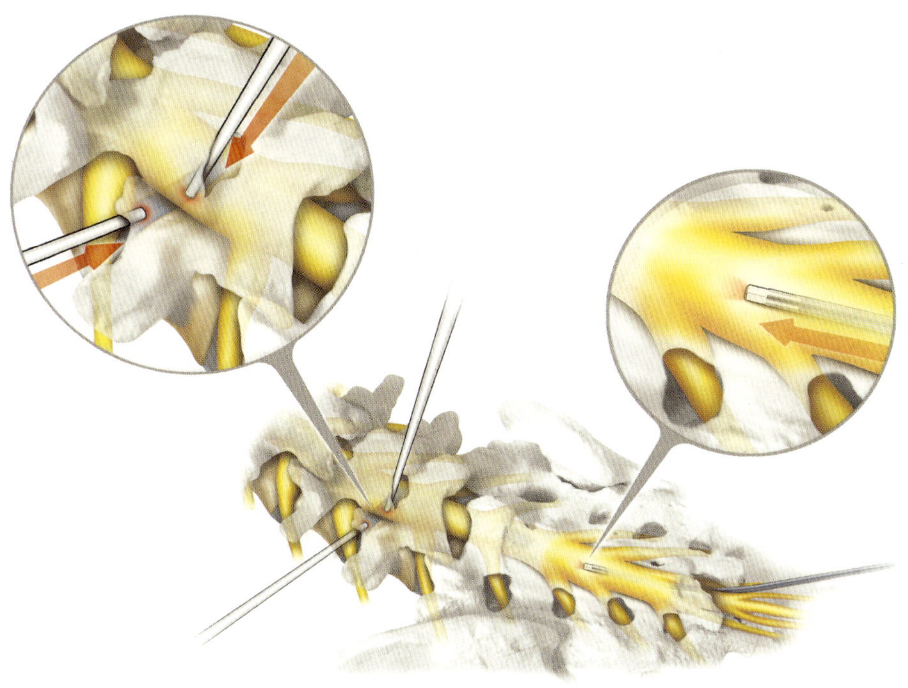

Fig. 10.1 The various approaches to the epidural space include the transsacral hiatus, transforaminal, and interlaminar approaches.

denervation the multifidus and sacral spinalis muscles.[4] Transforaminal endoscopic approach can also provide the least invasive basic access to the disk and foramen to cover a large spectrum of painful pathologies. Indications for using the transforaminal approach as stated by Gore and Yeung[3] include removal of foraminal fibrosis from failed back surgery, foraminal and extraforaminal stenosis, foraminal osteophytosis, and annular tears with diskogenic lumbar pain. Yeung and Gore also note that the absolute ideal indication of the transforaminal approach is for foraminal and extraforaminal disk herniations and diskitis.

10.2.1 Preoperative

Preoperative administration of moderate sedation to the patient is dependent on surgeon preference, but is largely a requirement. Experienced endoscopic surgeons use less and less sedation in order to obtain greater feedback from the patient during surgery—a safety factor and a condition that produces the best results while monitoring the procedure during surgery. In reviewing the literature, either a 14- or 18-gauge needle is employed, allowing for appropriate delivery of local anesthetic imperative for a smooth procedure and comfort of the patient. In this regard, the patient must continually be conscious in order to provide an indicator to whether there the structure visualized is a pain generator or if there is any nerve compromise,[2] as this could result in a catastrophic outcome. Generally, midazolam, 1 to 2 mg intravenously, is administered half an hour before the surgery. Additionally, fentanyl, 25 to 50 µg intravenously, is typically administered 15 minutes prior to the procedure and continued intermittently intraoperatively.[5] In very rare patients, propofol can be administered in a small dose, 25 to 50 µg/kg/min intravenously, to provide a uniform plane of anesthesia to accomplish intraoperative moderate sedation.

10.2.2 Intraoperative

Step 1: Positioning

- The patient is faced prone on the operating table to ensure the patient is comfortable for the duration of the procedure (▶ Fig. 10.2). The trajectory to the neural foramen is determined on preoperative computed tomography scan or magnetic resonance imaging and a skin entry site marked (▶ Fig. 10.3).

Step 2: Local Anesthesia

- A wheel is formed using 2 mL of lidocaine 1%. Under fluoroscopic guidance and using the posterolateral approach, an 18-gauge spinal needle approximately 7 inches in length is used to create a track along the anticipated course, while evenly administering 10 mL lidocaine 1% (▶ Fig. 10.4).[5]

Step 3: Transforaminal Epidural Approach

- Related to the pain of this procedure, a concomitant epidural can be preliminarily performed at this step to further assist with pain reduction. First, a 20-gauge spinal needle is inserted into the epidural space.[5] Fluoroscopy confirms the trajectory to the neural foramen (▶ Fig. 10.5). The needle is directed toward the foramen (▶ Fig. 10.6) and a fluoroscopic picture taken from an anteroposterior (AP) view should show the tip of the needle located at the corresponding midpedicular line within the epidural space (▶ Fig. 10.7a) and inferior aspect of the foramen on lateral view (▶ Fig. 10.7b).[4] The stylet is removed and the guidewire is passed through the spinal needle (▶ Fig. 10.8). The position of the guidewire is confirmed by fluoroscopy (▶ Fig. 10.9). The spinal needle is removed over the guidewire (▶ Fig. 10.10) and the 14-gauge Tuohy needle is passed over the guidewire (▶ Fig. 10.11).

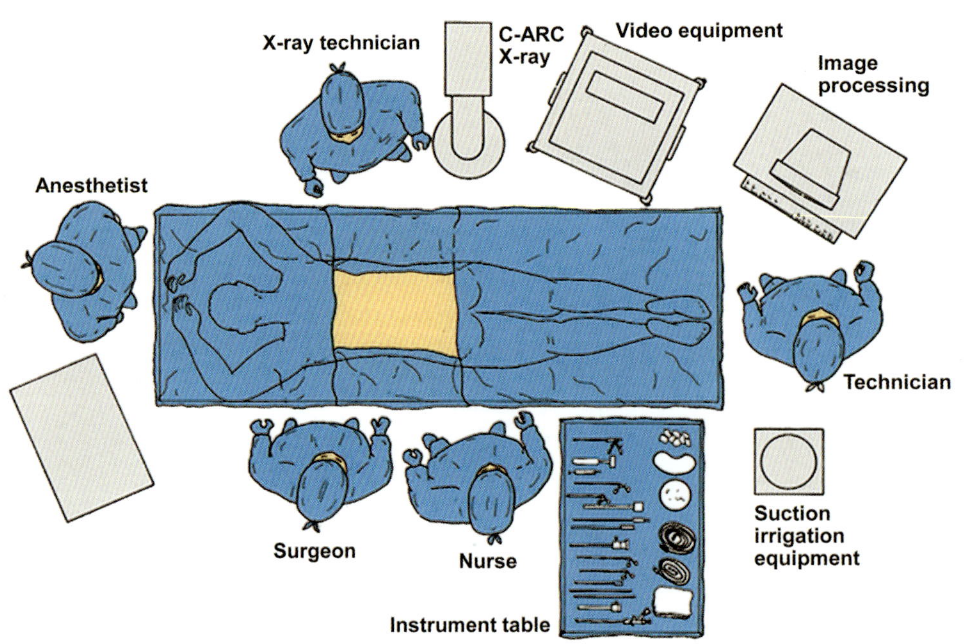

Fig. 10.2 Operating room setup.

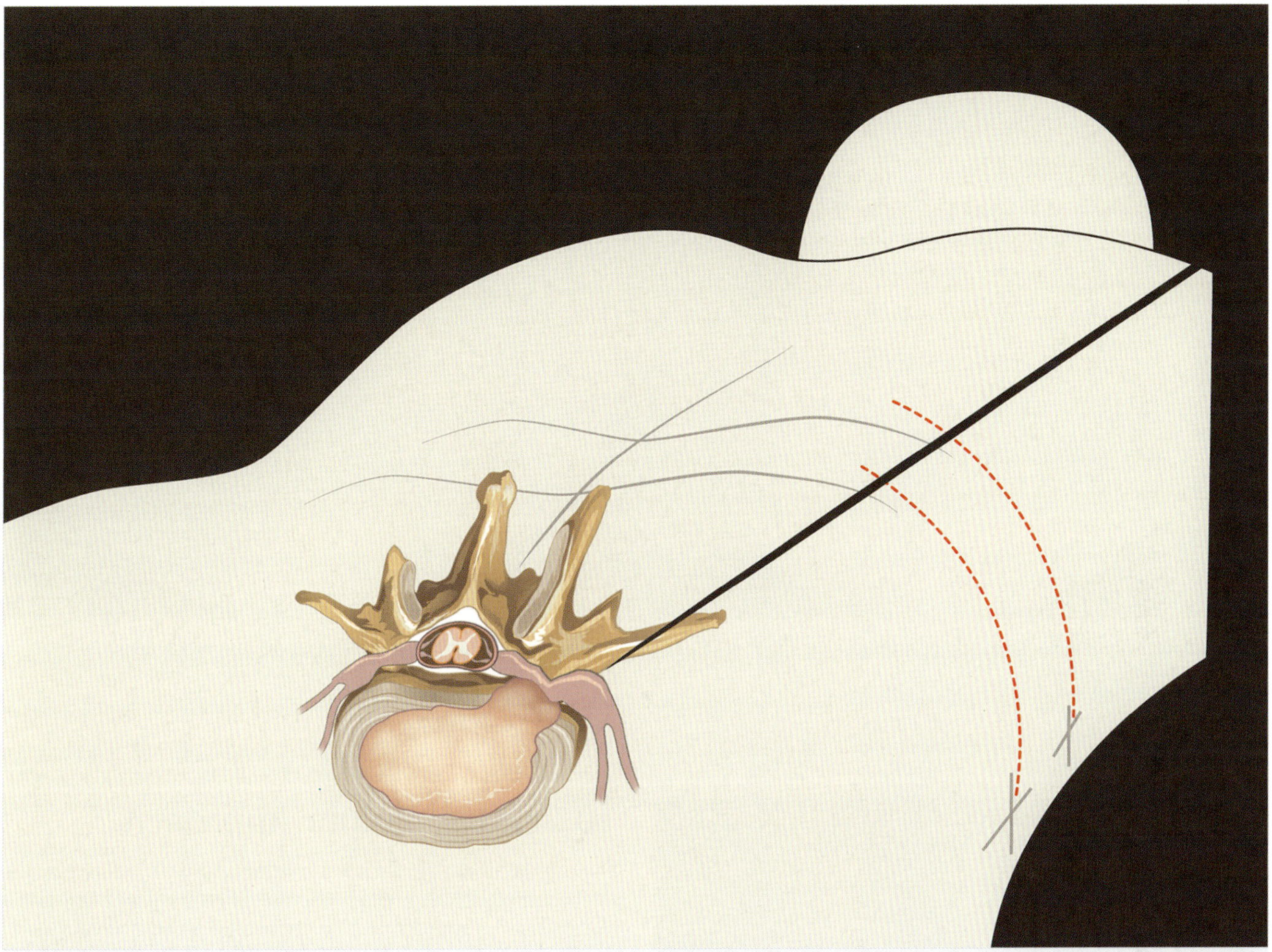

Fig. 10.3 Estimation of the trajectory angle from the skin entry to the neural foramen.

Fluoroscopy is used to confirm the position of the Tuohy needle (▶ Fig. 10.12). The Tuohy needle is removed and the skin entry site is enlarged with a 15 blade (▶ Fig. 10.13). The dilators is then passed over the guidewire (▶ Fig. 10.14) and confirmed on fluoroscopy (▶ Fig. 10.15). The guidewire is then removed and radiopaque dye is injected, confirming proper placement (▶ Fig. 10.16 and ▶ Fig. 10.17).[4]

Step 4: Dilation and Sheath Insertion

- The guidewire is reinserted through the dilator (▶ Fig. 10.18). With the guidewire in position, the dilator is removed (▶ Fig. 10.19) to allow passage of the dilator with sheath (▶ Fig. 10.20). The position of the dilator and sheath is confirmed on fluoroscopy (▶ Fig. 10.21). The dilator and guidewire are removed, leaving the sheath in position (▶ Fig. 10.22).

Step 5: Endoscope Insertion

- The endoscope is brought in and inserted through the sheath (▶ Fig. 10.23). Should the patient report symptoms of nerve

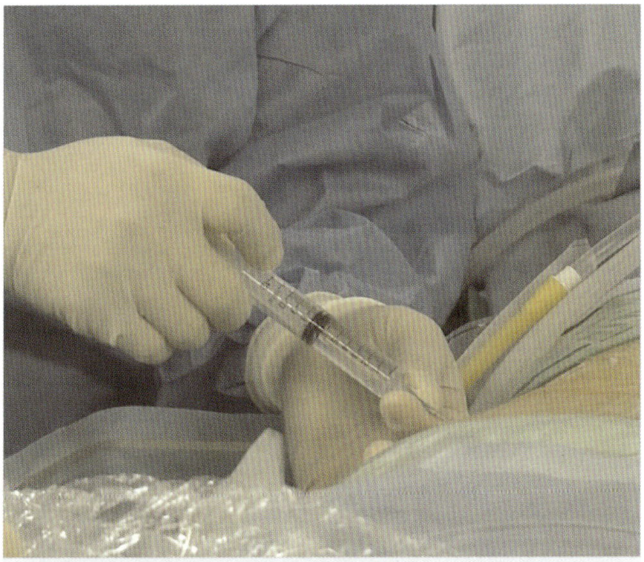

Fig. 10.4 Determination of skin entry site. Local anesthesia is injected along the trajectory of the spinal needle.

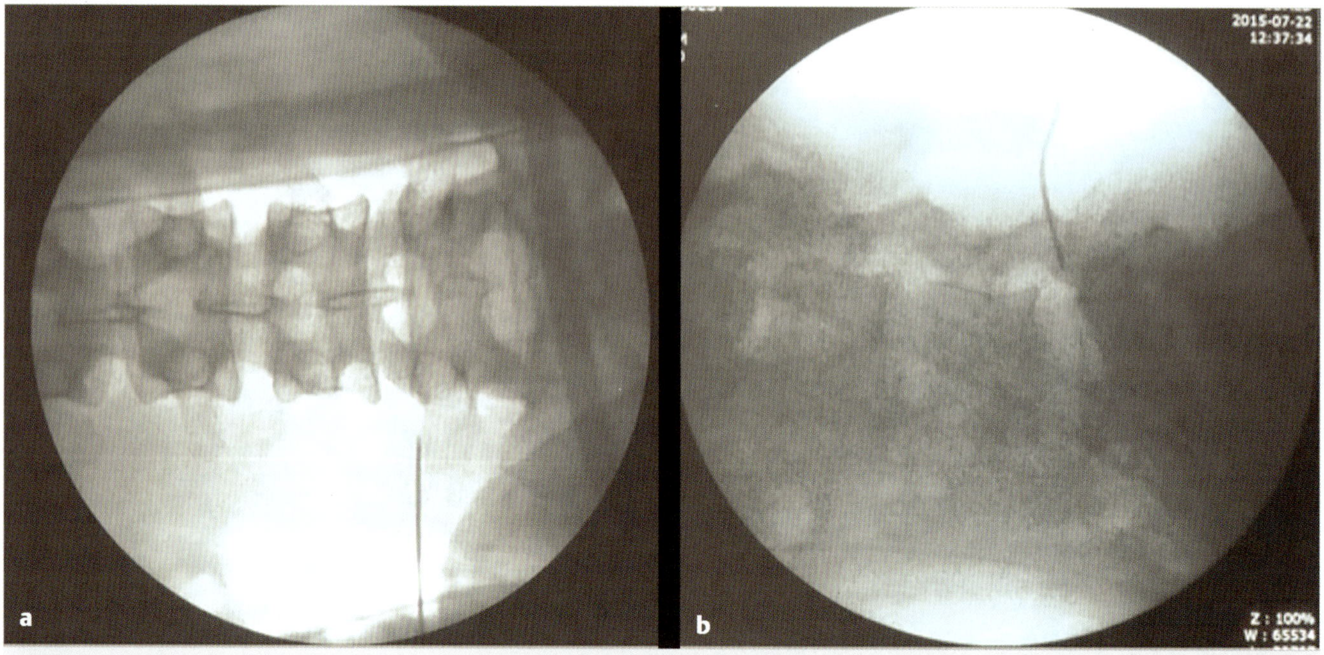

Fig. 10.5 (a) Anteroposterior fluoroscopic view of the spinal needle approaching the foramen. (b) Lateral fluoroscopic view of the spinal needle approaching the foramen.

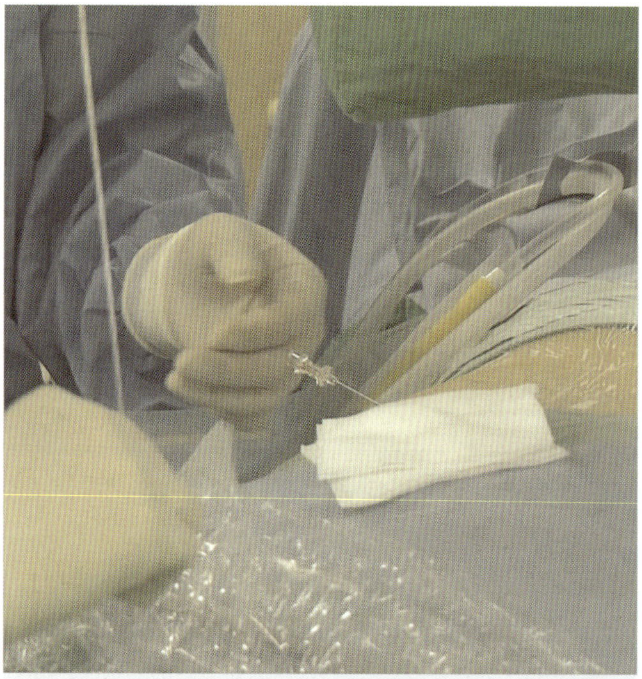

Fig. 10.6 Spinal needle in position at the posterolateral aspect of the annulus.

irritation, the sheath can be maneuvered to avoid the exiting nerve root.[4] The endoscope is now in place for viewing (▸ Fig. 10.24). After the procedure is completed, the cannula and endoscope are slowly withdrawn, taking caution not to damage any of the nerve roots.

10.3 Interlaminar Approach

The interlaminar approach is a routine approach utilized by the interventionalist using open nonmagnified or magnified with a microscope or endoscope. The endoscope has the advantage of smaller cannular, which are mobile rather than fixed. This is considered the "gold standard" and is a well-known surgical approach to the lumbar spine. Epiduroscopy is not needed for the translaminar approach. Epiduroscopy has commonly been done using the caudal approach but is widely known to be painful for many patients[4]; as a result, some effort has been made to use an interlaminar approach. However, due, in part, to the large needles required, the interlaminar approach is not as commonly practiced.

10.3.1 Preoperative

General anesthesia is the routine form of anesthesia for the translaminar approach. In this case, the anesthesiologist has total control of the airway with the patient in the prone position. Moderate sedation is the alternative method for the interlaminar approach. Midazolam, 1 to 2 mg intravenously, is administered half an hour before the surgery. Additionally, fentanyl, 25 to 50 μg intravenously, is typically administered 15 minutes prior to the procedure and continued intermittently intraoperatively.[5] Midazolam can be repeated in the operating room at approximately half the dose that was given initially to facilitate anxiolysis. Fentanyl is also given at a bolus dose of 50 μg before the procedure begins. The Fentanyl can be given every 3 to 5 minutes, with a maximum dose of 150 μg.[5] In certain patients, propofol can be administered in a small dose, 25 to 50 μg/kg/min intravenously, to provide a uniform

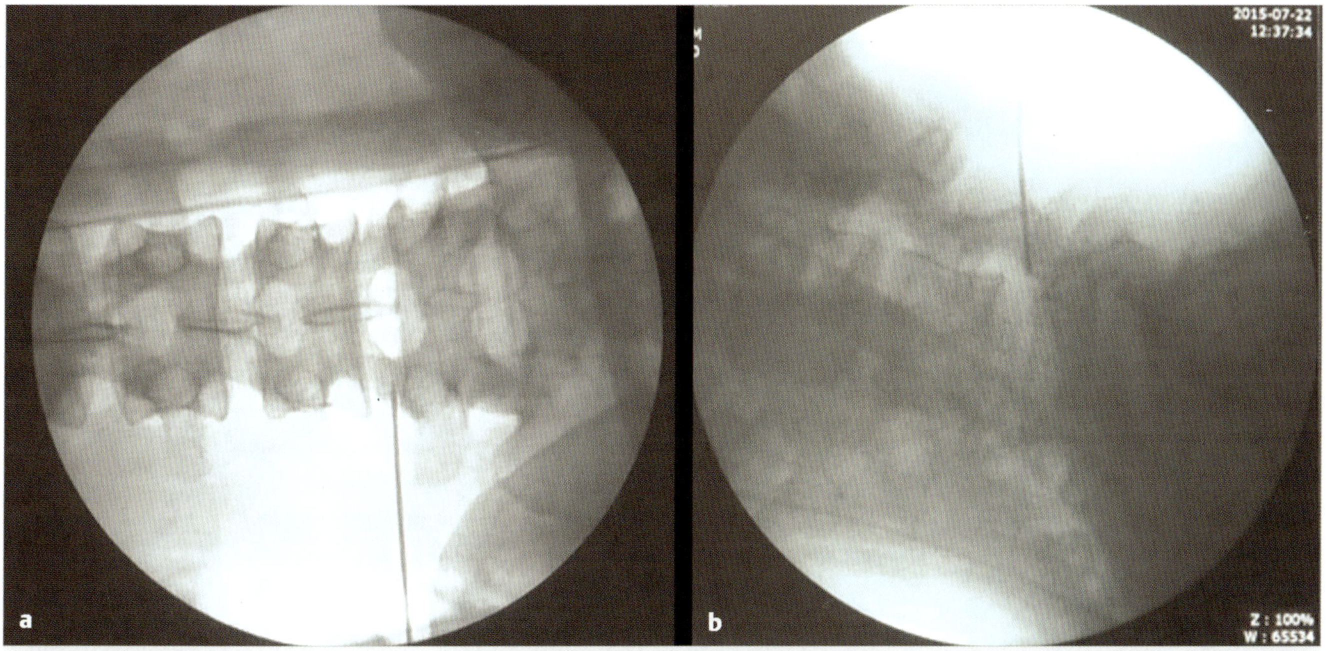

Fig. 10.7 **(a)** Anteroposterior fluoroscopic view of the spinal needle at the foramen. **(b)** Lateral fluoroscopic view of the spinal needle at the foramen.

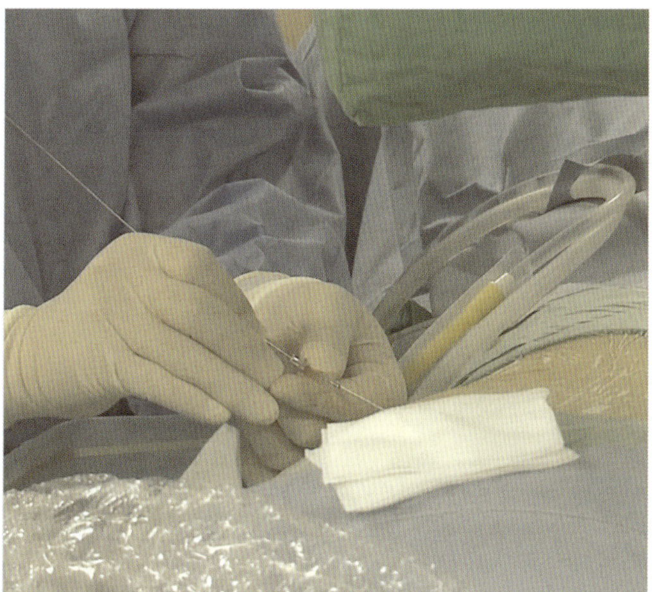

Fig. 10.8 The stylet is removed and the guidewire is passed through the spinal needle.

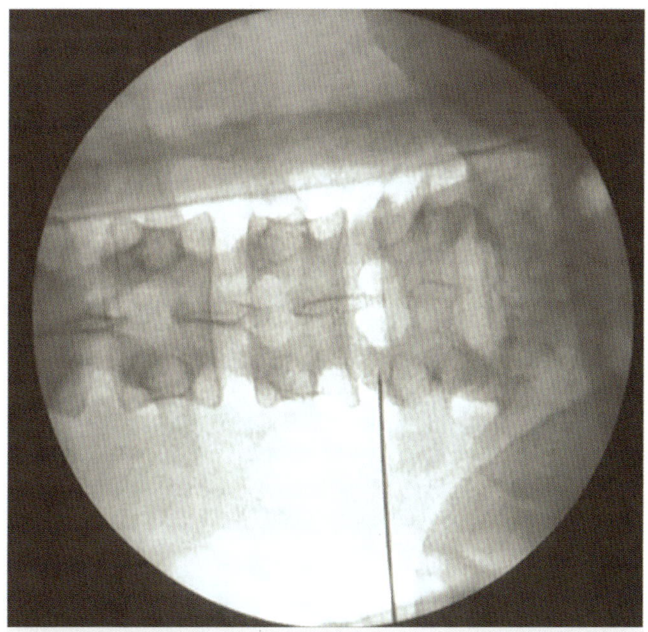

Fig. 10.9 Anteroposterior fluoroscopic view confirms position of the guidewire.

plane of anesthesia to accomplish intraoperative moderate sedation.

10.3.2 Intraoperative

Step 1: Positioning

- As previously described, the patient is most commonly placed in the prone position for optimum comfort, but sometimes is placed in the lateral position.[6]

Step 2: Local Anesthesia

- A caudal epidural block is recommended before the start of the procedure in order to ensure that adequate pain control can be provided during the surgery.[5] This is an important consideration since the interlaminar approach typically uses a 14-gauge needle, which can be significantly painful; therefore, a caudal epidural block should be considered.[4]

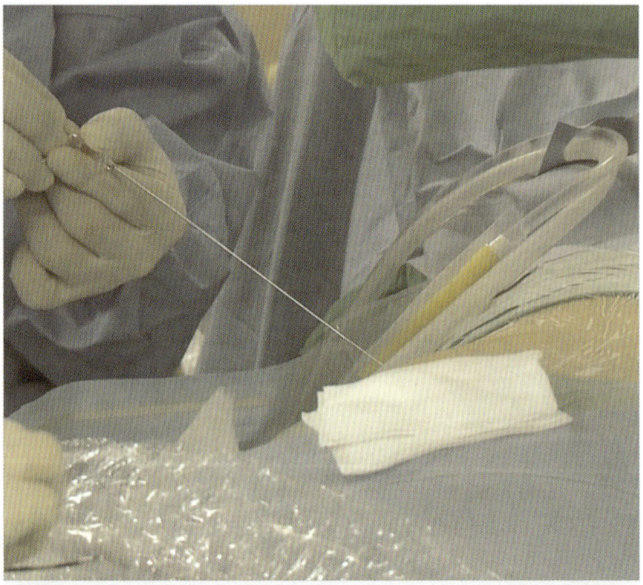

Fig. 10.10 The spinal needle is removed over the guidewire.

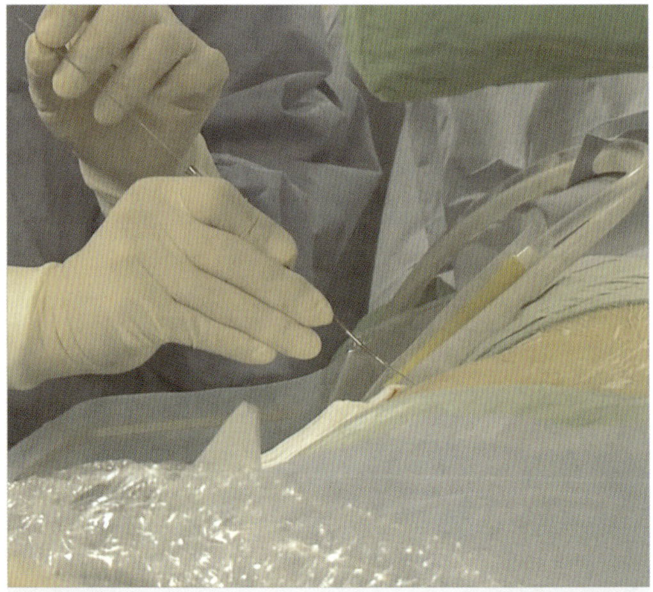

Fig. 10.11 The 14-gauge Tuohy needle is passed over the guidewire into the epidural space.

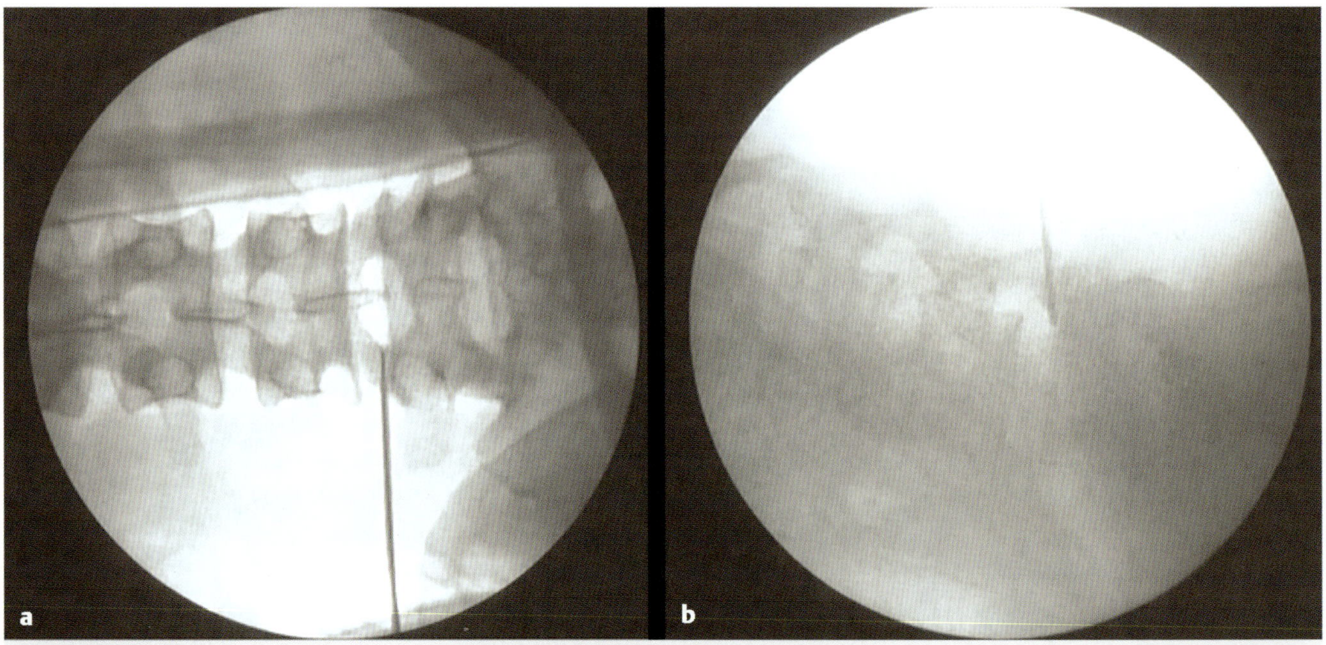

Fig. 10.12 (a) Anteroposterior fluoroscopy confirms the position of the 14-gauge Tuohy needle. (b) Lateral fluoroscopy confirms the position of the 14-gauge Tuohy needle.

- Local anesthetic of lidocaine 1% is recommended. Using 3 to 5 mL of lidocaine, a wheel is formed on the skin and directed into the underlying tissue.[7]

Step 3: Needle Insertion

- Following adequate anesthesia, an epidural needle is slowly inserted at an angle of 25 to 30 degrees from the surface of the skin.[7] The needle is advanced under fluoroscopic guidance until loss of resistance is identified using saline. Then, contrast dye is injected in order to confirm placement of the needle in the epidural space.

Step 4: Diskography

- Indigo carmine is injected to help identify a herniated disk. As the cannula is inserted further, blue-stained herniated disk tissue should become visible along with possibly the posterior longitudinal ligament.[5]

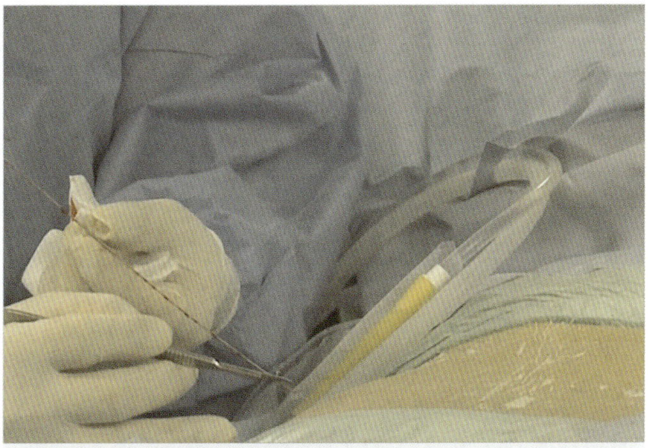

Fig. 10.13 The Tuohy needle is removed over the guidewire. A no. 15 blade is used to increase the skin entry site.

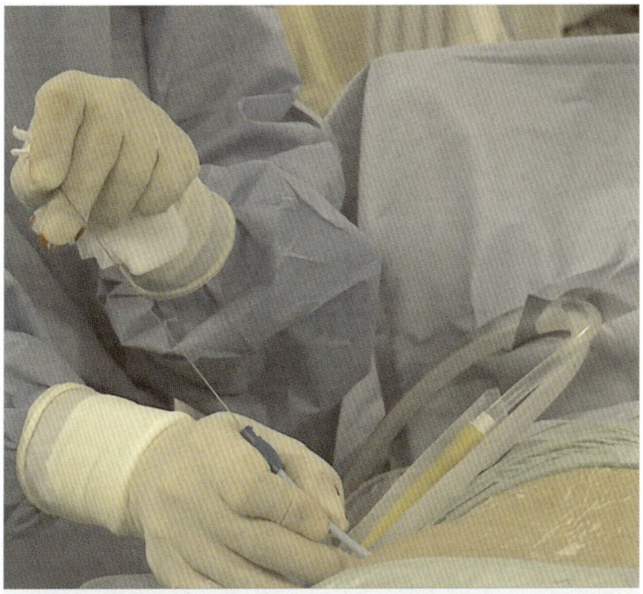

Fig. 10.14 The dilator is passed over the guidewire.

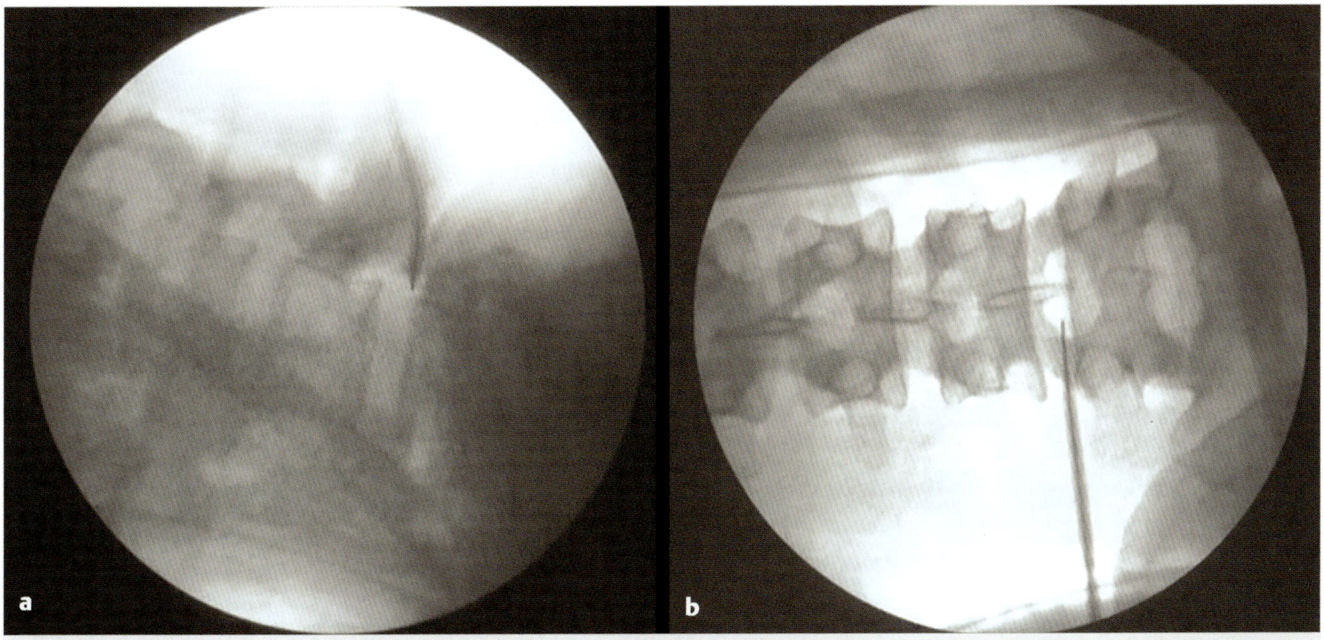

Fig. 10.15 **(a)** Lateral fluoroscopy demonstrates the location of the dilator. **(b)** Anteroposterior fluoroscopy demonstrates the location of the dilator.

Step 5: Instrument Placement

- A guidewire is then placed through the needle followed by serial dilators. When an adequate space is created by the dilators, a working cannula can be inserted. The tip of the cannula should be located slightly to the side of the spinous process and can be confirmed with a lateral fluoroscopic view.[5]
- When the cannula's position is confirmed, the guidewire and the dilators can be removed. Then, the endoscope can be used to visualize surrounding structures in the epidural space. Intermittent saline flushes are done to provide better visualization of structures.
- The guidewire may need to be placed again under fluoroscopic guidance if visual space is not sufficient. When the guidewire is replaced, the endoscope is withdrawn so that additional dilators can be inserted using the guidewire. Then, the cannula can be inserted again over the guidewire with subsequent removal of the dilators and the guidewire.[5]
- After the procedure is completed, the cannula and endoscope are slowly withdrawn, taking caution not to damage any of the nerve roots.

10.4 Clinical Outcomes

There were 15 studies involving the transforaminal approach during the period from 1995 to 2005.[8,9,10,11,12,13,14,15,16,17,18,19,20] There were nine significant studies after 2005. Yeung and Tsou[8] reported at 1-year follow-up that an average visual analog score decreased from 6.6 to 2.5 and the Oswestry Disability Index decreased from 46 to 32%. Thirty-seven patients (15%) developed temporary dysesthesia in the 2-week postoperative period. Even with dysesthesia, improvement in the Visual Analog Pain Scale and Oswestry Disability Index was comparable to microlumbar diskectomy.[4] Approximately 91% of the patients were satisfied with their results and would opt to undergo the procedure again if they had the same diagnosis and symptoms. The overall complication rate was reported to be 4%.[4]

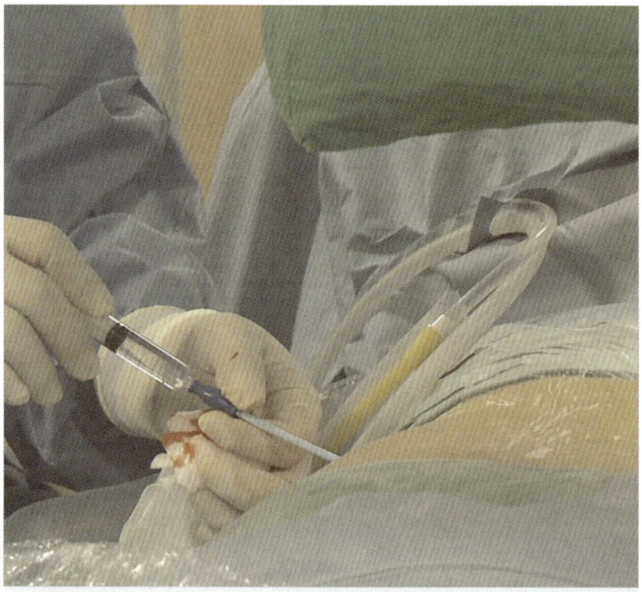

Fig. 10.16 The guidewire is removed from the dilator. Contrast is injected to confirm epidural location.

Yeung and Tsou[8] separated out a subgroup of 219 patients with noncontained herniations and reported the results at 1-year follow-up. Patient satisfaction was 91%. Transforaminal epiduroscopy provided equivalent results to reported results of open microdiskectomy, even with noncontained, extruded herniations.

In 2013, Yeung reported 50 cases of microlumbar diskectomy compared with transforaminal epiduroscopy: microlumbar diskectomy L4–L5 = 15, L5–S1 = 35.[4] The average Visual Analog Pain Score decreased from 6.5 to 1.7 and the average Oswestry Disability Index decreased from 44 to 30%. Complications included one case of seroma and one case of durotomy. Patients receiving microdiskectomy were usually for extruded, migrated, or sequestered herniated nucleus propulsus, believed better suited for microlumbar diskectomy and more difficult for epiduroscopy. Patient receiving microlumbar diskectomy was due to surgeon advice and preference. Patient satisfaction was 92%. The transforaminal epiduroscopy group numbered 137, with 209 total levels: L1–L2=1, L2–L3 = 3, L3–L4 = 31, L4–L5 = 94, L5–S1 = 80. The average Visual Analog Pain Score was 6.6 and Oswestry Disability Index was 46%. In the transforaminal epiduroscopy group, improvement in Visual Analog Pain Score and Oswestry Disability Index was 4.1 and 32%, respectively. Endoscopic decompression included foraminoplasty for lateral stenosis. In the transforaminal epiduroscopy group, 20 patients (14.7%) developed dysesthesia in the 2-week postoperative period. In spite of the inclusion of more complex degenerative spine problems in patients who were also candidates for decompression and fusion, improvement in Visual Analog Pain Score and Oswestry Disability Index was comparable to microlumbar diskectomy at 4.1 and 32%, respectively. Patient satisfaction was more than 90% as many chose the transforaminal epiduroscopy over microlumbar diskectomy, even in the face of disk extrusion at L5–S1.

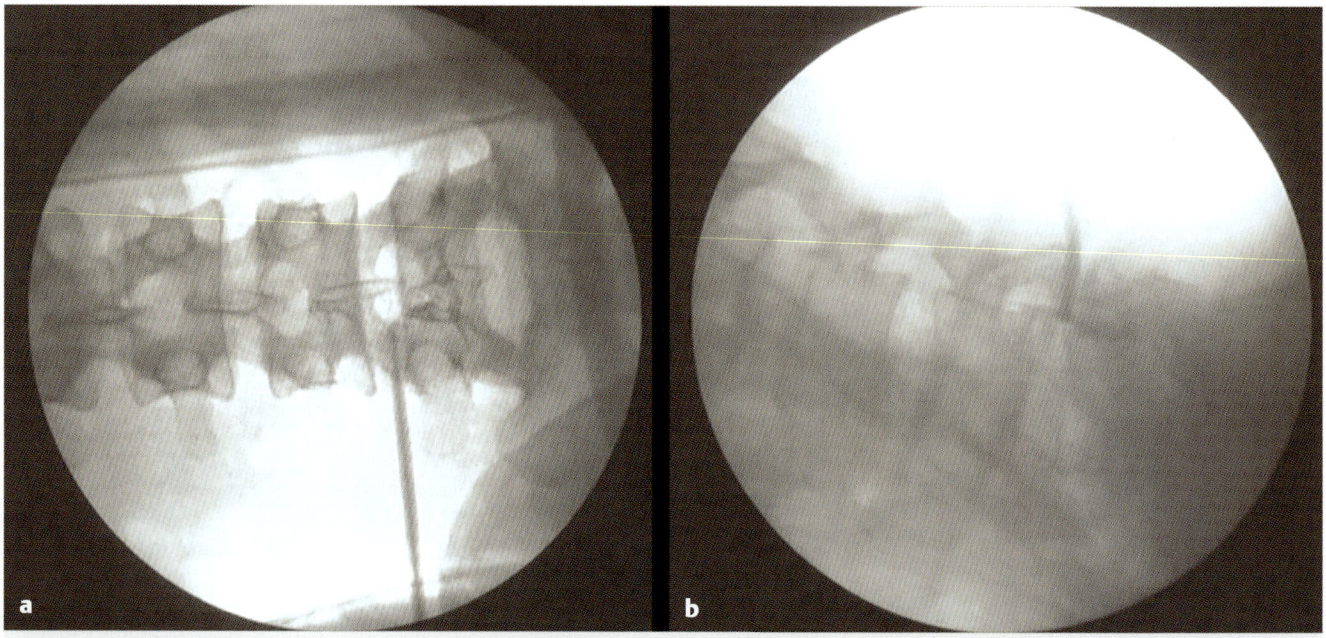

Fig. 10.17 (a) Anteroposterior fluoroscopy demonstrating contrast in the epidural space at the foramen. (b) Lateral fluoroscopy demonstrating contrast in the epidural space at the foramen.

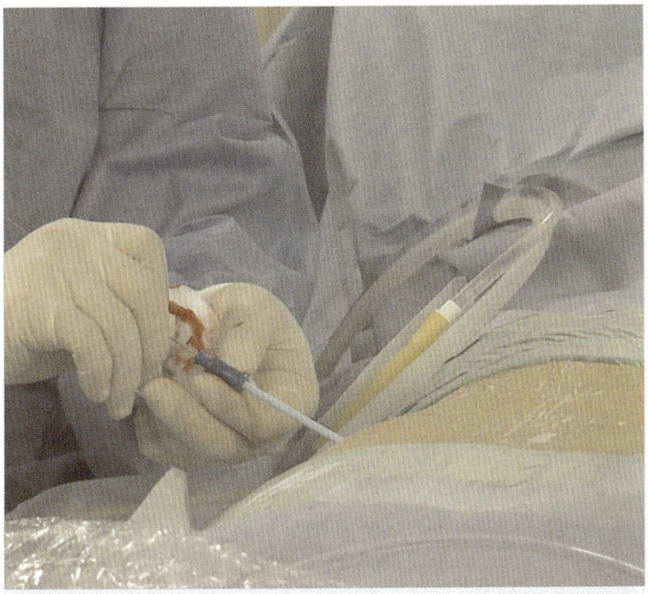

Fig. 10.18 The guidewire is reinserted through the dilator.

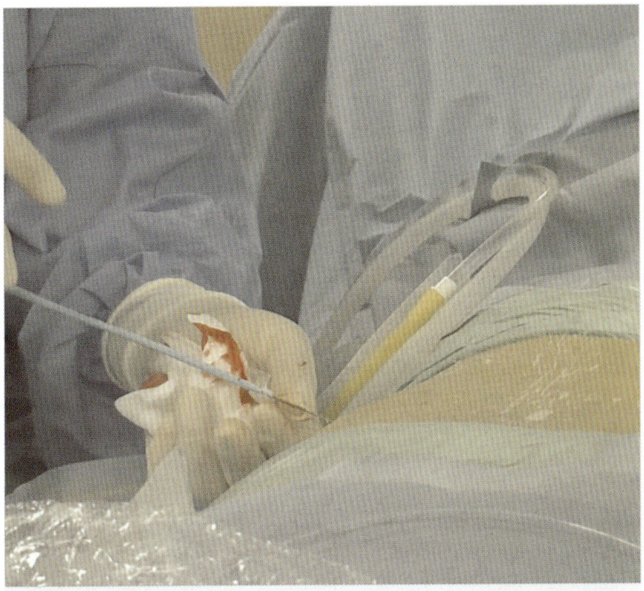

Fig. 10.19 The dilator is removed over the guidewire.

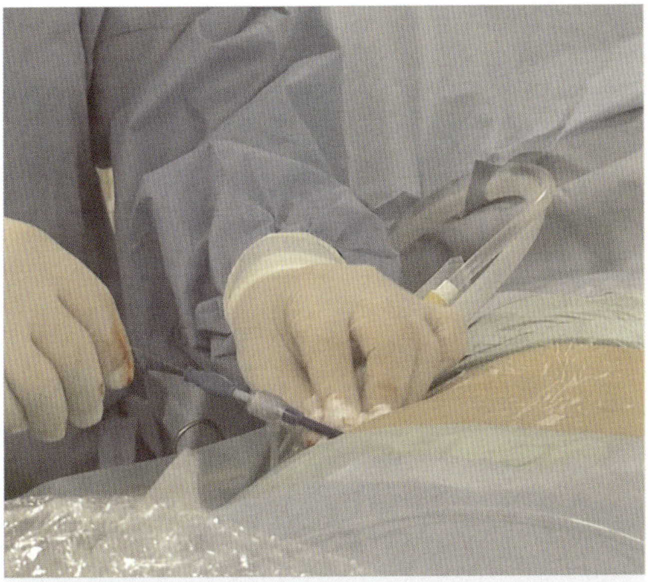

Fig. 10.20 The dilator and sheath are passed over the guidewire.

There are limited studies involving the interlaminar endoscopic approach compared with the transforaminal approach. In one study by Avellanal and Diaz-Reganon,[5] 19 patients were included. The mean number of operations at the lumbar level was 2.26. The Visual Analog Pain Score was 7.89 at baseline, 5.95 ($p = 0.001$) 3 months later, and 6.05 ($p = 0.001$) 6 months later. Six patients (31.6%) did not show any improvement, and six other patients showed a very significant improvement (at least three-point reduction in the Visual Analog Pain Scale) 3 months later. Six patients (31.6%) did not show any improvement 3 months later, and six other patients showed a very significant improvement (at least three-point reduction in the Visual Analog Pain Scale) at 3 and 6 months. In the study, seven patients (36.8%) experienced a mild-to-moderate improvement (one- or two-point reduction in the Visual Analog Pain Scale). Transient neurologic symptoms (headache and hypoacusia) were recorded in four patients, all of them directly related to the injection of saline boluses. There were four cases of dural puncture, with one patient requiring hospital admission.[7]

In another study, 80 patients each underwent conventional microsurgical or full-endoscopic interlaminar decompression. All operations were performed by two surgeons with many years of experience in both techniques. Eleven interventions were performed at the L5–S1 level (4 × microsurgical, 7 × full-endoscopic interlaminar), 80 at L4–L5 (42 × microsurgical, 38 × full-endoscopic interlaminar), 55 at L3–L4 (25 × microsurgical, 30 × full-endoscopic interlaminar), and 14 at L2–L3 (9 × microsurgical, 5 × full-endoscopic interlaminar). In terms of efficacy, excluding the patients revised by fusion, 132 patients remained after 2 years. Constant and significant ($p < 0.001$) improvement in leg pain and daily activities was seen in both groups. Approximately 95 patients (72%) no longer had leg pain or the pain was almost completely eliminated (42 × microsurgical = 67.7%, 53 × full-endoscopic interlaminar = 75.7%), 28 (21.2%) had pain occasionally or the pain was greatly reduced (16 × microsurgical = 25.8%, 12 × full-endoscopic interlaminar = 17.2%), and 9 (6.8%) experienced no essential improvement (4 × microsurgical = 6.5%, 5 × full-endoscopic interlaminar = 7.1%). Walking time improved from an average of less than 20 minutes to more than 50 minutes. The differences in the clinical results between the groups were not significant. Approximately 90 patients suffered increasing back pain (7 × microsurgical, 2 × full-endoscopic interlaminar, $p < 0.01$). Overall, 13 patients (9.6%) had a poor result in terms of no leg pain reduction (9 patients) or had to undergo conventional revision surgery later for persistent pain (4 patients). Approximately 119 (88.1%) patients reported subjective satisfaction and would undergo the operation again (56 × microsurgical = 87.5%, 63 × full endoscopic interlaminar – 88.7%). Neurologic deficits were significantly ($p < 0.001$) reduced when the patient's history of weakness was less than 6 weeks. Postoperative pain and pain medication were significantly reduced in the full endoscopic interlaminar group ($p < 0.01$).

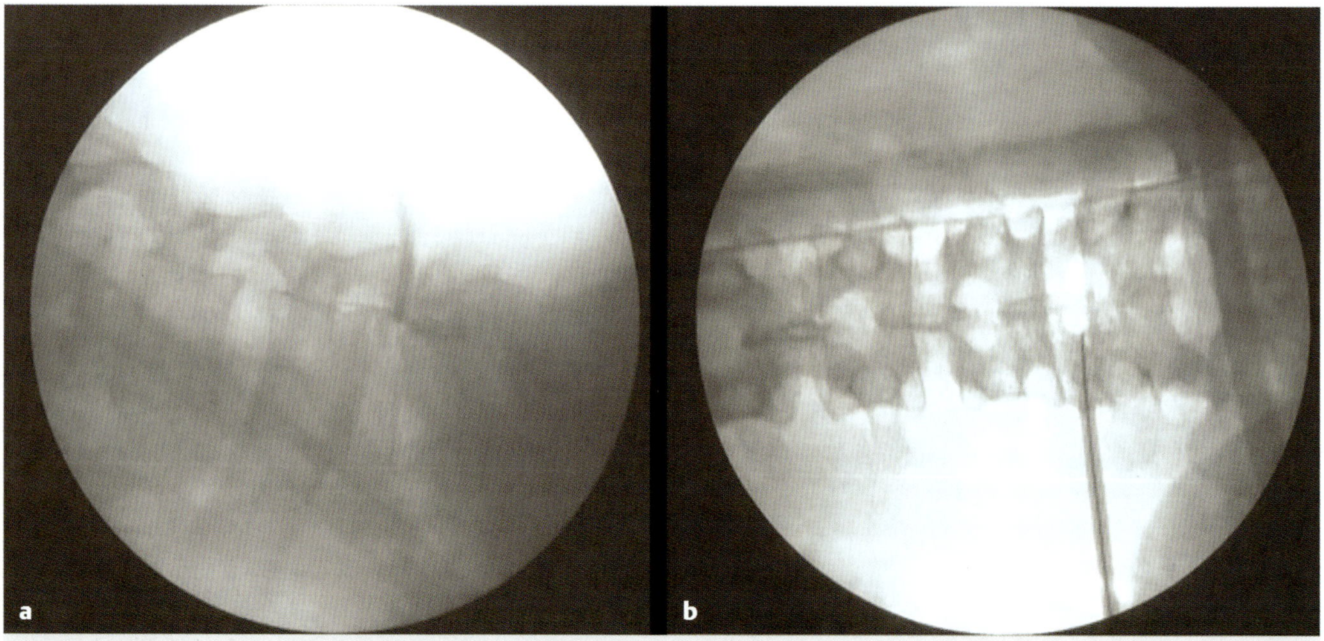

Fig. 10.21 **(a)** Lateral fluoroscopy confirms location of the dilator and sheath. **(b)** Anteroposterior fluoroscopy confirms location of the dilator and sheath.

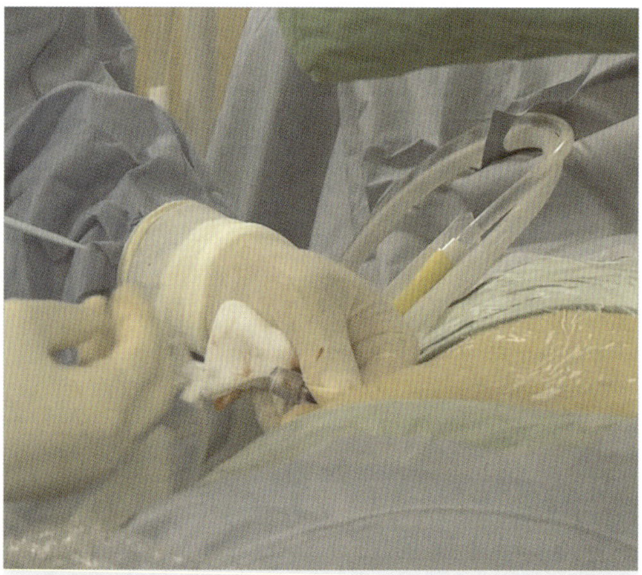

Fig. 10.22 The dilator and guidewire are removed, leaving the sheath is position.

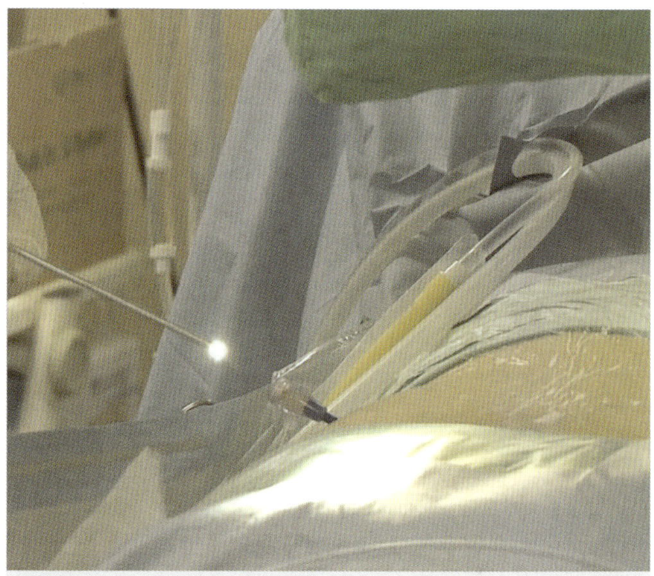

Fig. 10.23 The endoscope is brought in and inserted into the sheath.

On the second postoperative day, 48% of the patients in the microsurgical group and 11% of the patients in the full-endoscopic interlaminar group requested analgesics. The maximum time in hospital was 8 days in the microsurgical group and 3 days in the full-endoscopic interlaminar group. With regard to side effects, 11 patients had transient postoperative dysesthesia (7 × microsurgical, 4 × or full-endoscopic interlaminar), 4 had transient urinary retention (3 × microsurgical, 1 × or full-endoscopic interlaminar), and 5 had dura injuries (3 × microsurgical = 3.8%, 2 × or full-endoscopic interlaminar = 2.5%). The authors concluded that the full-endoscopic technique offers ease of operation for the surgeon due to excellent visualization, good illumination, and expanded field of vision with 25 degree scope, cost-effective procedure due to short operating time, rapid rehabilitation, and low postoperative costs of care, and, finally, reduced anatomic trauma.[21]

10.5 General Considerations and Conclusion

Endoscopic removal of disk herniation is only limited by the accessibility of endoscopic instruments to the herniation site. The location of the disk herniation, the extent of the extrusion

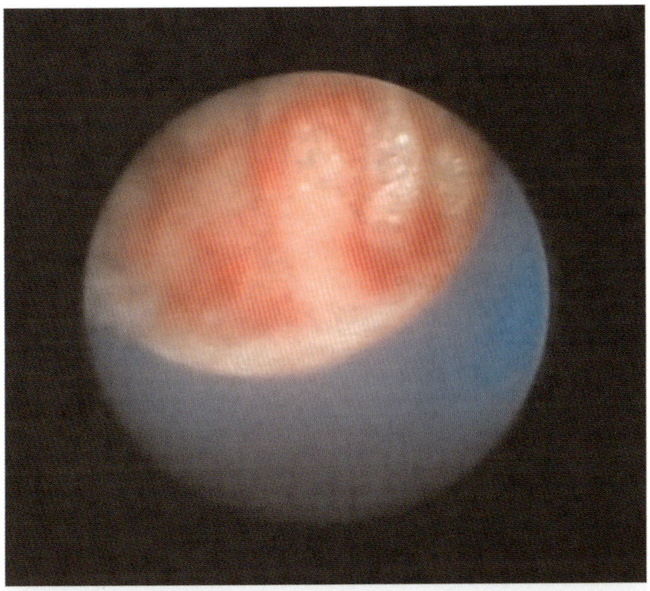

Fig. 10.24 Epiduroscopic view on entering the neural foramen. Epidural fat is visualized.

and sequestration, and the experience level of the surgeon are vital considerations for patient selection. Some patients with high iliac crests, horizontal L5–S1 disk spaces, or degenerative scoliosis make surgical access through the foramen difficult. Contraindications thus are relative and depend on the anatomic factors involved.

The risks of serious complications or injury are relatively low and approach 1% or less. Surgeon experience is a significant factor in looking at side effects and complications in published literature. Also, evolving technology reflects improvement in many aspects related to outcomes and to complications. As with any surgical intervention, risks include bleeding, development of scarring, infection, nerve injury, and dural tears. The most common postoperative complaint is transient dysesthesia, often reflecting nerve recovery and seen even without any demonstrated nerve irritation utilizing electromyography and somatosensory evoked potentials, is seen in approximately 5 to 15% of cases.[4] Postoperative dysesthesia can be treated with transforaminal epidurals, sympathetic blocks, and off-label use of gabapentinoid preparations.

References

[1] Van Boxem K, Cheng J, Patijn J, et al. 11. Lumbosacral radicular pain. Pain Pract. 2010; 10(4):339–358

[2] Kallewaard JW, Vanelderen P, Richardson J, Van Zundert J, Heavner J, Groen GJ. Epiduroscopy for patients with lumbosacral radicular pain. Pain Pract. 2014; 14(4):365–377

[3] Gore S, Yeung A. The "inside out" transforaminal technique to treat lumbar spinal pain in an awake and aware patient under local anesthesia: results and a review of the literature. Int J Spine Surg. 2014; 8:28

[4] Blomberg RG, Olsson SS. The lumbar epidural space in patients examined with epiduroscopy. Anesth Analg. 1989; 68(2):157–160

[5] Avellanal M, Diaz-Reganon G. Interlaminar approach for epiduroscopy in patients with failed back surgery syndrome. Br J Anaesth. 2008; 101(2):244–249

[6] Racz GB, Day MR, Heavner JE,et al. Epidural lysis of adhesions and percutaneous neuroplasty. In: Racz GB, ed. Pain Management – Current Issues and Opinions. InTech; 2012:338–367

[7] Kim DH, Choi G, Lee SH. Percutaneous endoscopic lumbar diskectomy. In: Conerly K, ed. Endoscopic Spine Procedures. New York, NY: Thieme; 2011:108–118,135–142

[8] Yeung AT, Tsou PM. Posterolateral endoscopic excision for lumbar disc herniation: surgical technique, outcome, and complications in 307 consecutive cases. Spine. 2002; 27(7):722–731

[9] Wójcik AS. Endoscopically assisted percutaneous lumbar discectomy. Ortop Traumatol Rehabil. 2004; 6(3):259–263

[10] Lew SM, Mehalic TF, Fagone KL. Transforaminal percutaneous endoscopic discectomy in the treatment of far-lateral and foraminal lumbar disc herniations. J Neurosurg. 2001; 94(2) Suppl:216–220

[11] Hoogland T. Transforaminal endoscopic discectomy with forminoplasty for lumbar disc herniation. Surg Tech Orthop Traumato. 2003; 40:55–120

[12] Eustacchio S, Flaschka G, Trummer M, Fuchs I, Unger F. Endoscopic percutaneous transforaminal treatment for herniated lumbar discs. Acta Neurochir (Wien). 2002; 144(10):997–1004, discussion 1003–1004

[13] Chiu JC. Evolving transforaminal endoscopic microdecompression for herniated lumbar discs and spinal stenosis. Surg Technol Int. 2004; 13:276–286

[14] Ahn Y, Lee SH, Park WM, Lee HY, Shin SW, Kang HY. Percutaneous endoscopic lumbar discectomy for recurrent disc herniation: surgical technique, outcome, and prognostic factors of 43 consecutive cases. Spine. 2004; 29(16): E326–E332

[15] Suess O, Brock M, Kombos T. Motor nerve root monitoring during percutaneous transforaminal endoscopic sequestrectomy under general anesthesia for intra- and extraforaminal lumbar disc herniation. Zentralbl Neurochir. 2005; 66(4):190–201

[16] Schubert M, Hoogland T. Endoscopic transforaminal nucleotomy with foraminoplasty for lumbar disk herniation. Oper Orthop Traumatol. 2005; 17 (6):641–661

[17] Ruetten S, Komp M, Godolias G. An extreme lateral access for the surgery of lumbar disc herniations inside the spinal canal using the full-endoscopic uniportal transforaminal approach-technique and prospective results of 463 patients. Spine. 2005; 30(22):2570–2578

[18] Ramsbacher J, Kern BC, Kombos T, et al. Transforaminal endoscopic sequestrectomy: indications, operative technique, and first clinical experience. Neurosurg Q. 2000; 10:224–227

[19] Scheckenbach C, Hoogland T. Endoskopische transforaminale diskektomie (ETD)—Ergebnisse nach 2 jahren. Orthop Prax. 1999; 35:104–105

[20] Morgenstern R, Morgenstern C, Abelleo A, et al. Eine studie von 144 fallen nach unterzogener endoskopischer endenwirbelsaulenchirurgie—klassische rehabilitation im vergleich zur FPZ methode [A study on 144 cases after lumbar spine endoscopic surgery. Classical rehabilitation vs. FPZ machines system]. Orthop Prax. 2005; 41:674–681

[21] Komp M, Hahn P, Oezdemir S, et al. Bilateral spinal decompression of lumbar central stenosis with the full-endoscopic interlaminar versus microsurgical laminotomy technique: a prospective, randomized, controlled study. Pain Physician. 2015; 18(1):61–70

11 Epiduroscopic Images of Spinal Anatomy

Dae Hyun Jo and Jee Youn Moon

11.1 Introduction

Epiduroscopy has become a popular procedure for patients with low back pain and/or leg pain. As a direct three-dimensional imaging procedure, epiduroscopy allows the physician to explore the anatomic structures in the epidural space, such as spinal dura mater, epidural fat tissue, epidural blood vessels, spinal nerve roots, and posterior longitudinal ligament (PLL). The physician can diagnose the epidural pathology by an epiduroscope and treat it under direct vision at the same time. An understanding about static and functional anatomy of the epidural space is imperative to physicians who perform an epiduroscopy. However, the learning curve for epiduroscopy is very slow because the epidural space is narrow, especially in cases of spinal stenosis, and the epiduroscope is too small to observe the whole epidural space. During epiduroscopy, the structural view of the epidural space is changing because the dura is sloping around in the epidural space. To obtain a good view of the epidural space using the epiduroscope, the whole procedure should be taken with prudence to avoid injury of the blood vessels in the epidural space. Bleeding can occur easily, which can obscure the field of vision, making it difficult to differentiate the epidural structures. If there is severe adhesion in the epidural space, it may not be possible to distinguish each structure, and the scope cannot be advanced further.

Understanding the anatomy of epidural space of the extradural space in the spinal canal is essential for understanding the epiduroscopic views of the epidural space. Epiduroscope can be advanced into the anterior epidural space as well as the posterior epidural space. Epiduroscopic view of the epidural space varies according to the angle from the anterior and posterior epidural space. Physicians should also know about the limitation of the epiduroscope, which can angulate only left and right, not up and down. The more we view the epidural space directly, the more we can understand the epidural pathology.

This chapter discusses the normal epiduroscopic and pathologic images, and also mentions about instrumentation images under epiduroscopy.

11.2 Boundaries of the Epidural Space

The boundaries of the epidural space are foramen magnum to the superior; sacrococcygeal membrane to the inferior; PLL, vertebral bodies, and disks to the anterior; pedicles and intervertebral foramen to the lateral; and ligamentum flavum, facet joint capsules, and lamina to the posterior (▶ Fig. 11.1). The epidural space contains fatty connective tissue, the dural sac, spinal nerves, epidural blood vessels, and other types of connective tissue. Since the dural sac ends at the level of the S2 vertebral body, the tissues from S2 to sacrococcygeal membrane are epidural fatty tissue, filum terminale externum, and the proximal parts of the nerve roots of S2 to Cocc1.

11.3 Normal Epiduroscopic Anatomy

Characteristics of the epiduroscopic images of epidural structures are summarized in ▶ Table 11.1.

11.3.1 Fatty Connective Tissues

The epidural fatty connective tissue is generally the first structure to be identified during the epiduroscopic procedure. The epidural fatty connective tissue appears as a greasy, yellow-colored bubble (▶ Fig. 11.2).

- The fatty tissue is often traversed by small blood vessels.
- It is mainly located in the anterolateral and dorsomedial parts of the epidural space and acts as a lubricant for the movements of the nerve roots in the spinal canal.
- The epidural fatty tissue is important for the nerves to be used up after chronic inflammatory processes; this can lead to involvement of the neural structures in the scar tissue being formed.

11.3.2 Dura

- Spinal dura mater is important as an optical guide rail when navigating the epidural space.
- In the epiduroscopic image, the dura mater appears as a blue-gray or gray-white connective tissue structure (▶ Fig. 11.3) with small blood vessels traversing it.
- The elasticity of the spinal dura mater is easy to determine by triggering "the tenting phenomenon," which is only positive for the spinal dura mater.
- The spinal dura mater has an average thickness of 0.321 (SD 0.127) mm with significant variations depending on the particular segment.

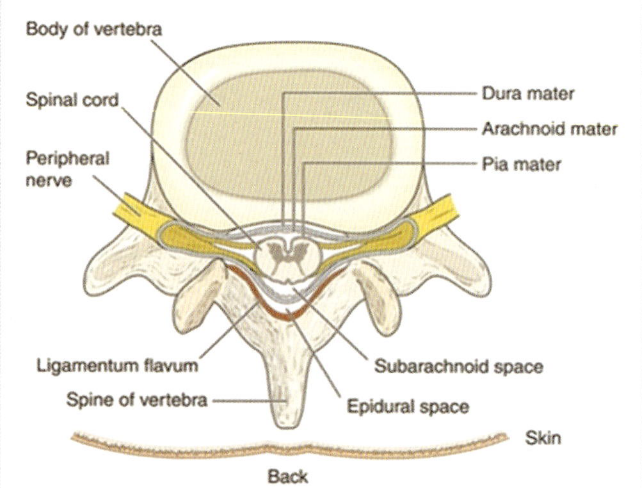

Fig. 11.1 Boundaries of the epidural space.

Table 11.1 Characteristics of structures observed in epiduroscopic images

Structures	Characteristics
Epidural fat	Glistening, white to yellowish, and often traversed by small blood vessels
Dura mater	Blue-gray or gray-white, and traversed with small blood vessels Elasticity triggered with "the tenting phenomenon"
Nerve root	White with a pin-yellow tinge containing of blood vessels Easily mistaken for fibrotic tracts
Posterior longitudinal ligament	White with a fibrous structure without visible blood vessels Distinguished with the spinal dura mater by no vascularization and rarely elastic without "tenting phenomenon"
Ligament flavum	White to yellowish concave surface without visible vessels
Blood vessels	
Adhesions	As tissue organized in strings and sheets of white fibers or the presence of impenetrable dense white tissue Generally avascular regardless of their severity
Inflammation	Hyperemia observed as a discrete area of dura root sleeve, peridural membrane, or other epidural structure, as compared to normal appearing areas of the epidural space Changes in vascularity is identified in chronic inflammation Veins appear bright and dark red, curved, balloon shaped, or tortuous and arteries appear as pulsating, enlarged bright blood vessels Thickened dura mater and increased vascularization in the case of arachnoiditis and perineural nerve sheath fibrosis

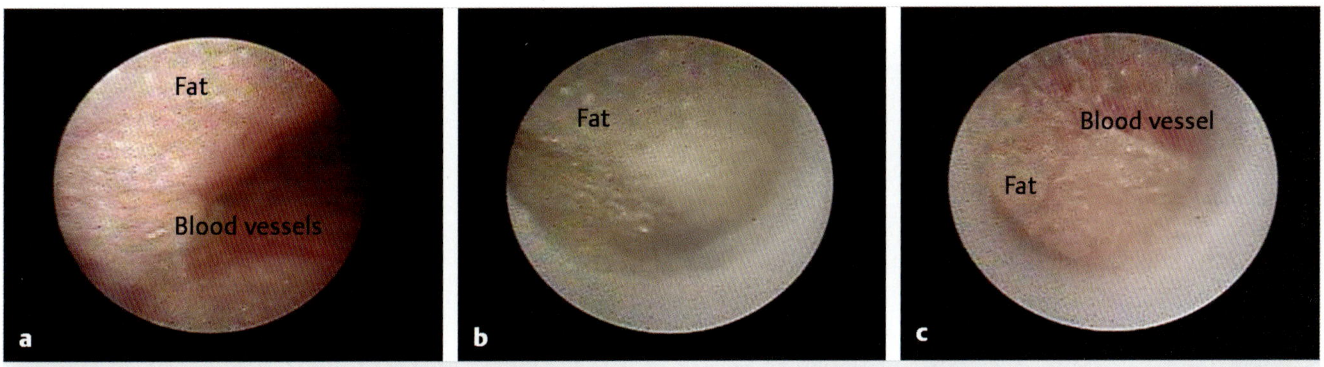

Fig. 11.2 (a–c) Fatty connective tissues.

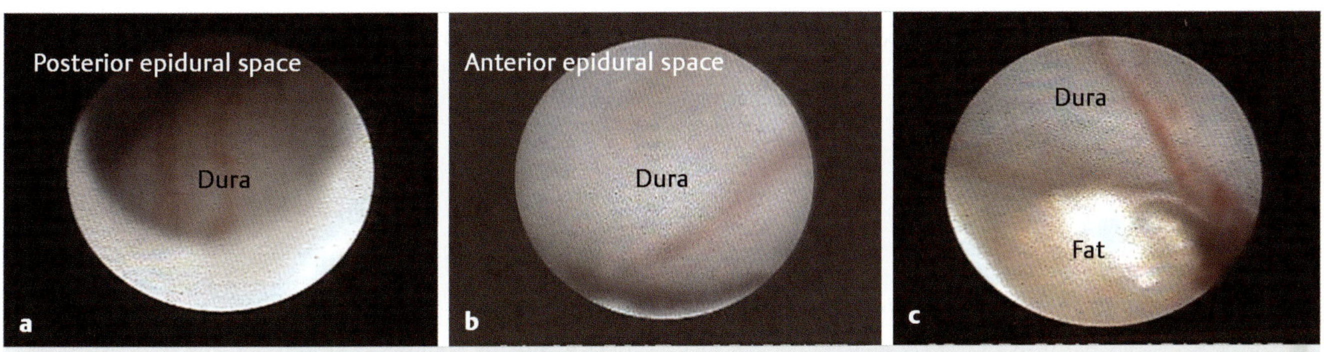

Fig. 11.3 (a–c) Dura mater in the anterior or posterior epidural space, with small blood vessels traversing it.

11.3.3 Spinal Nerves

- On a transverse section, the dural nerve root sleeves are located in the anterolateral quadrant of the spinal canal at 10 o'clock and 2 o'clock positions (▶ Fig. 11.4a,b), which contains the anterior and posterior nerve roots as well as the dorsal root ganglia.
- Epiduroscopic assessment in the area of the nerve root is sometimes difficult. However, it generally allows an inspection of the nerve root. In the epiduroscopic image, nerve roots within the dural sleeves are generally depicted as white with a pink-yellow tinge containing the blood vessels that can extend across the surface of the nerve root (▶ Fig. 11.4c).

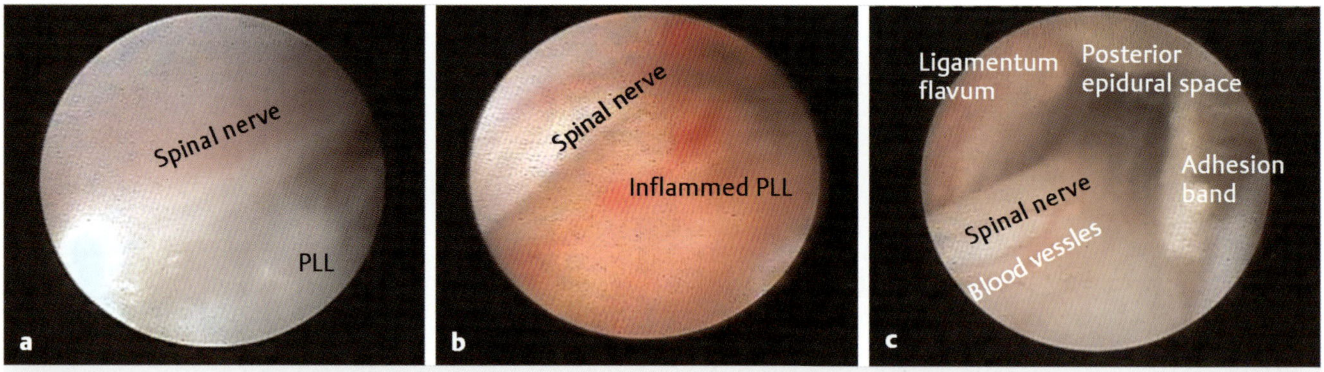

Fig. 11.4 Spinal nerves. **(a,b)** The dural nerve root sleeves are located in the anterolateral quadrant of the spinal canal. **(c)** Nerve roots within the dural sleeves generally contain blood vessels that can extend across the surface of the nerve root.

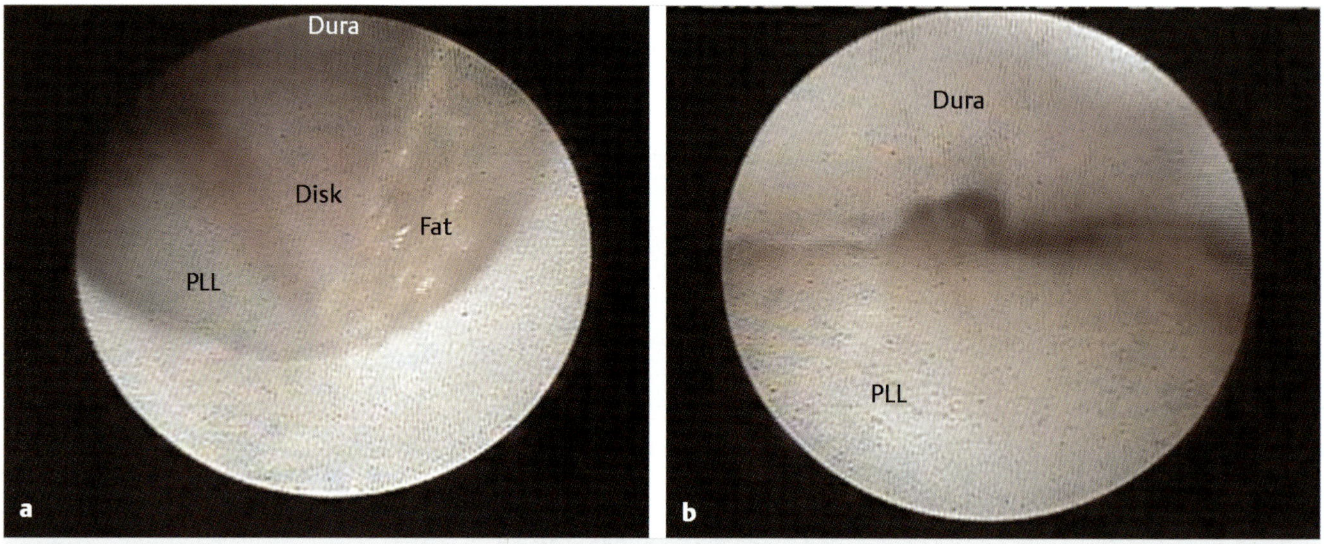

Fig. 11.5 (a,b) Posterior longitudinal ligament (PLL) without visible blood vessels in the anterior epidural space.

- The spinal nerve roots can be identified relatively more easily because of their white color. However, they can also be readily mistaken for fibrotic tracts.

11.3.4 Posterior Longitudinal Ligament

- The PLL is easily identified in the epiduroscopic image of the ventral epidural space.
- In the epiduroscopic image, the PLL appears as a white fibrous structure without visible blood vessels (▶ Fig. 11.5).
- Distinguishing the spinal dura mater and the PLL is initially possible on the basis of the differences in vascularization. The other feature to distinguish them from each other is the elastic "tenting phenomenon," which is only positive for the spinal dura mater.[1]
- The differences in the structure and color of the ventral epidural space can be used to assess which region of the epidural space the epiduroscopy is taking place.

11.3.5 Ligamentum Flavum

- The yellowish color of the ligamentum flavum is caused by the elastic fibers arranged like a folding grille with a parallel course.
- In the epiduroscopic image, the ligamentum flavum can be identified by its white to yellowish concave surface without visible vessels (▶ Fig. 11.6).
- Thickening of the ligamentum flavum can cause spinal canal stenosis.

11.3.6 Blood Vessels

- The epidural space contains arteries and veins supplying the spinal column; the arteries branch off from the segmental arteries and the veins interconnect, forming a venous plexus that is valveless, comprising an anterior part and a posterior venous part. In the lumbosacral part of the vertebral column, the ventral venous plexus is generally larger than the dorsal plexus.
- In the epiduroscopic image, both arterial and venous valveless epidural vessels pervade the epidural space (▶ Fig. 11.7), which can be easily distinguished from the pulsations.
- In the region of the lateral ventral epidural space or in the lateral dorsal epidural space, small artery can be observed.
- Obstructed epidural veins are primarily found in segments of the spinal canal that are growths and usually fibrotic.

11.4 Pathologic Epiduroscopic Findings

11.4.1 Adhesions

- Diagnosis of epidural adhesions with fibrosis associated with stenosis, instabilities, and herniated disks are frequently made by epiduroscopy.[1,2]
- In the epiduroscopic image, adhesion is observed as tissue organized in strings and sheets of white fibers or the presence of impenetrable dense white tissue (▶ Fig. 11.8), which appears generally avascular, regardless of their severity.
- During epiduroscopy, pathologically changed nerve roots can appear to be avascular, most likely due to the epidural adhesions surrounding them. The further process ultimately leads to an impairment of the epidural blood flow.
- In mild epidural adhesion with fibrosis, the scope can easily be advanced. In severe cases, however, it is difficult or impossible to advance the scope.
- In some patients, adhesions are too strong to perform adhesiolysis. In this case, the chances of the dural puncture will be increased if performed by force. To avoid the dural puncture, the physician should not perform adhesiolysis by force.

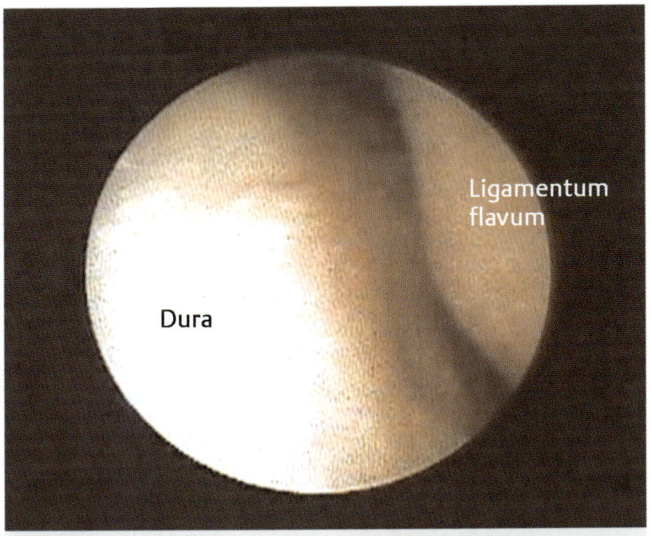

Fig. 11.6 Ligamentum flavum in the posterior epidural space.

11.4.2 Chronic Inflammation

- Chronic inflammatory processes in the epidural space, such as epiduritis and radiculitis, are depicted under epiduroscopy as clear, edematously enlarged tissue structures. These inflammatory processes are the source of pain (▶ Fig. 11.9).
- The inflammatory responses in the epidural space occur in stages.[1] The early stage involves an inflammatory response with hyperemia and swelling around the nerve root. In the next stage, a proliferation of fibroblasts causes adhesions with tortuous blood vessels that causes atrophy of the nerve root (▶ Fig. 11.10).
- In the epiduroscopic image, hyperemia can be observed as a discrete area of dura root sleeve, peridural membrane, or other epidural structure, as compared to normal appearing

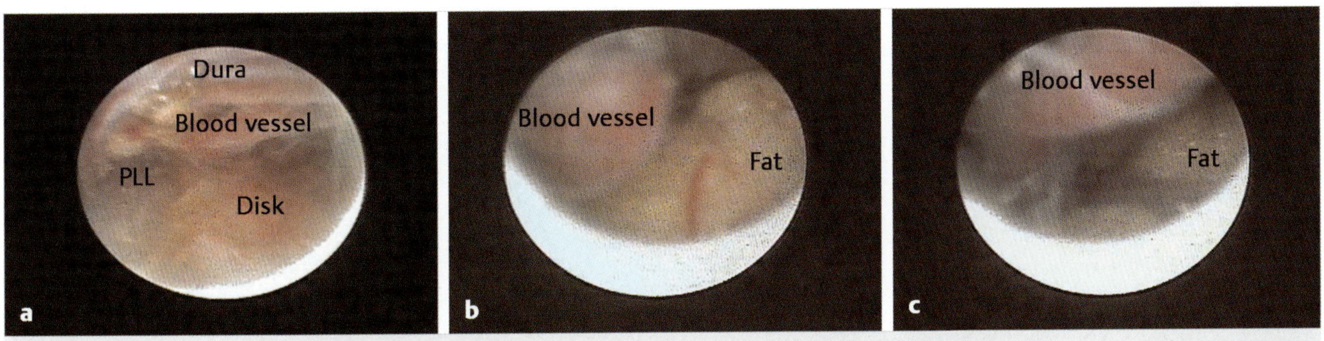

Fig. 11.7 (a–c) Tortuous blood vessels in the epidural space. PLL, posterior longitudinal ligament.

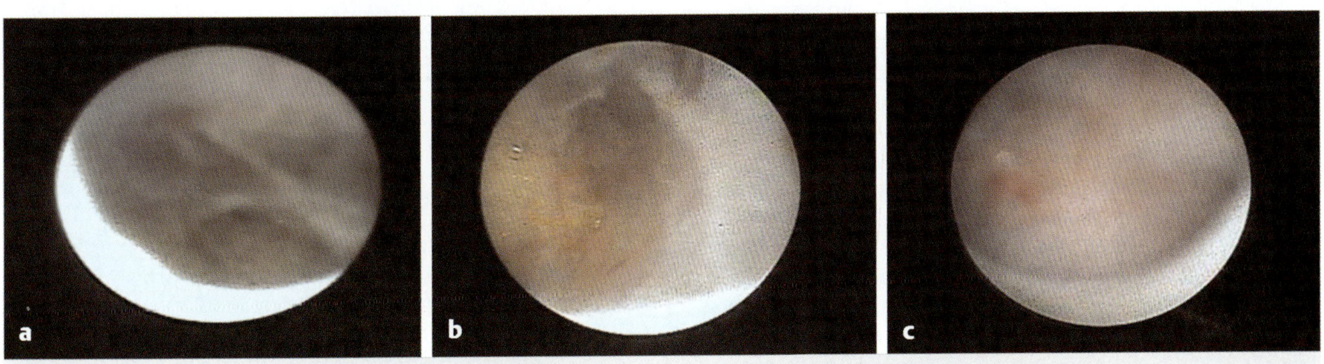

Fig. 11.8 Epidural adhesions. (a,b) Distinguishable solid fibrous bands in the epidural space are observed. (c) The epidural space is completely obliterated by fibrous tissue which does not allow a scope to penetrate.

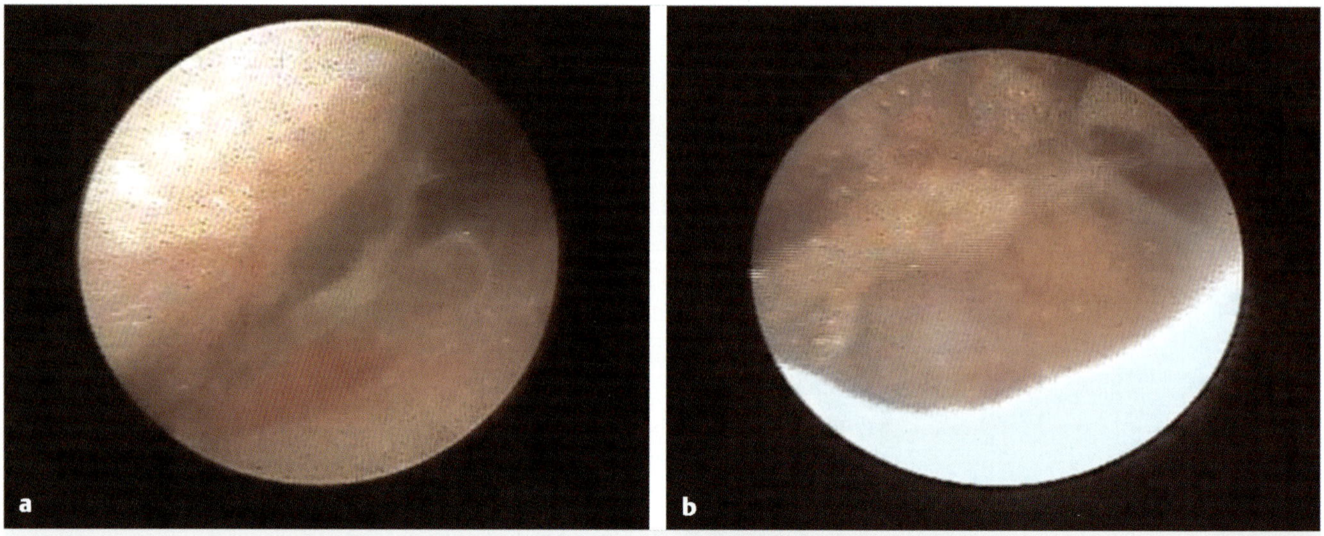

Fig. 11.9 (a,b) Inflammation of fatty tissue in the epidural space.

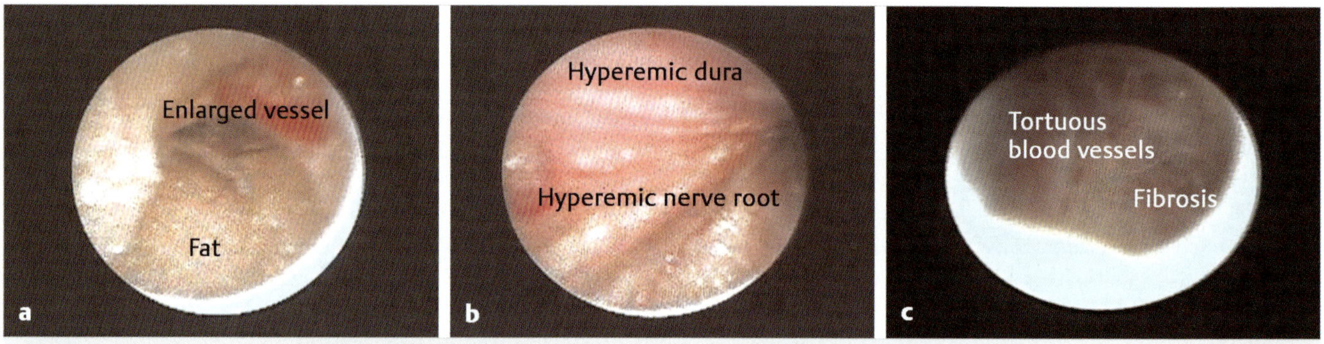

Fig. 11.10 Inflammation in stages. **(a,b)** The early stage involves an inflammatory response with hyperemia and swelling around the nerve root. **(c)** In the next stage, proliferation of the fibroblasts causes adhesions with tortuous blood vessels.

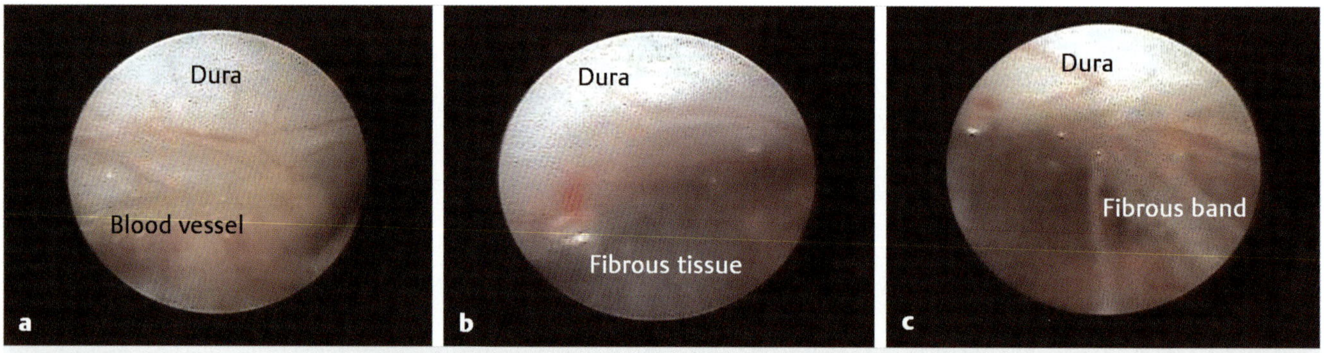

Fig. 11.11 (a–c) Dura mater with inflammation with or without fibrous band.

areas of the epidural space (▶ Fig. 11.11). In chronic inflammation, changes in vascularity is also identified (▶ Fig. 11.12). Veins appear bright and dark red, curved, balloon shaped, or tortuous. Arteries appear pulsating enlarged bright blood vessels.

- In the case of arachnoiditis and perineural nerve sheath fibrosis, the spinal dura mater appears thickened and the tissue appears to have increased vascularization.

11.4.3 Disk Herniation

- When the epiduroscope is administered into the anterior epidural space, disk can be visualized (▶ Fig. 11.13).
- If the disk bulging is large without an annular tear, then the disk can be pushing the dura upward. The dura is observed as white-yellowish structures with traverse blood vessels. Sometimes, it looks gray-white.

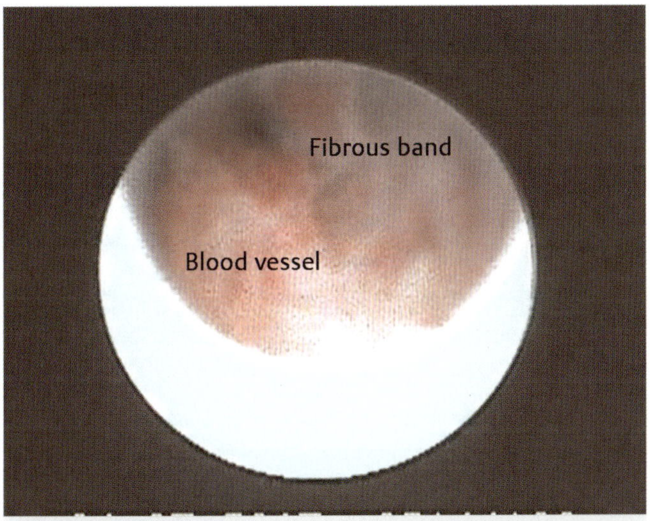

Fig. 11.12 Inflamed blood vessels with fibrous tissue.

- If there is an annular tear, yellowish inflammatory soup and whitish nuclear pulposus can be observed. If disk is migrated downward, the migrated disk is shown as mass-like shape (see Chapter 13 of this book).

11.4.4 Other Findings

- Fibrosis and scar tissue formation is responsible for hypertrophy of the ligamentum flavum.
- Cysts of the ligamentum flavum or the facet joint capsule are rarely observed. These cysts can lead to nerve root compression (see Chapter 12).
- During the epiduroscopic procedure, an accidental dural perforation can occur with a view of the intrathecal space (▶ Fig. 11.14 and ▶ Fig. 11.15).
- Other findings such as a broken catheter sheath can be seen during the epiduroscopy (▶ Fig. 11.16).
- It is possible to remove the protruded disk using laser during the epiduroscopic procedure (▶ Fig. 11.17).

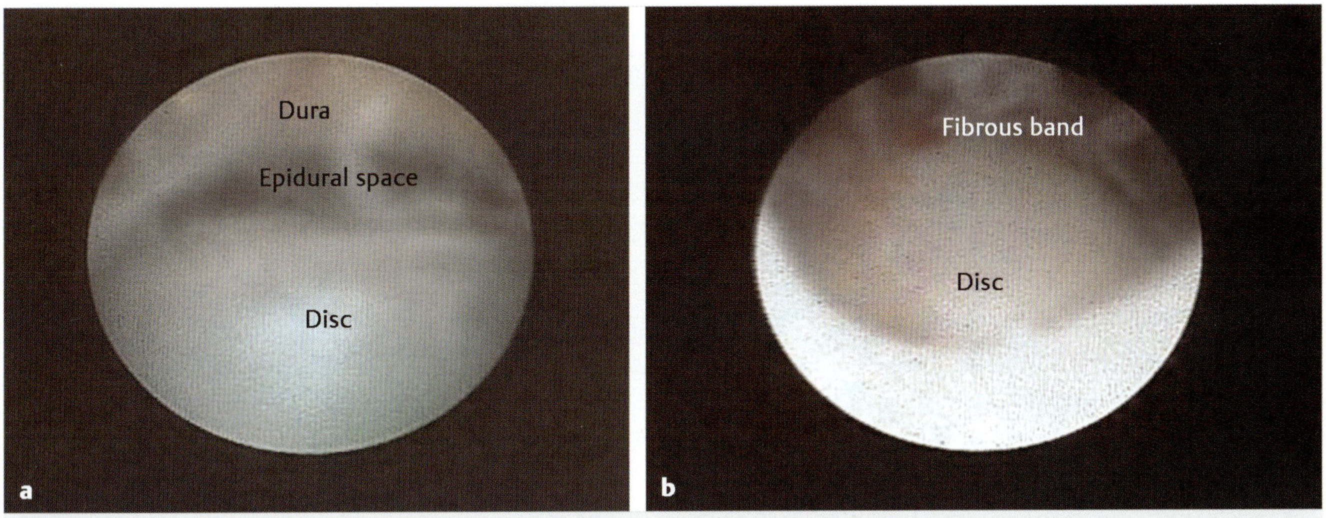

Fig. 11.13 Disks. (a) Disk stained with methylene blue via diskography. (b) Protruded disk with fibrous bands.

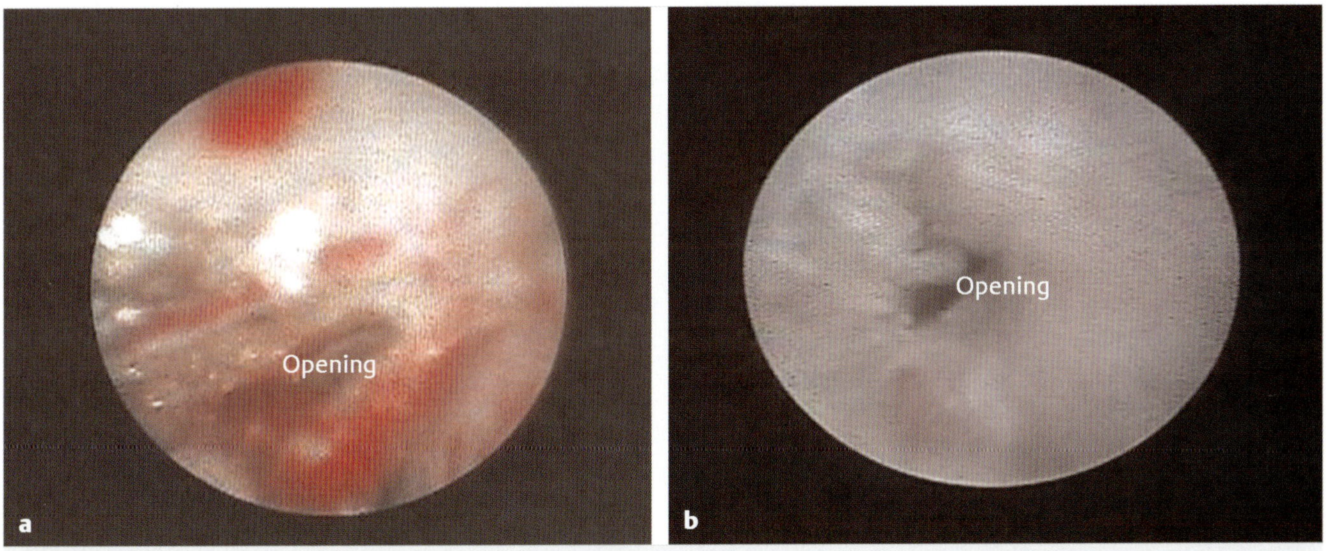

Fig. 11.14 (a,b) Accidental dural perforation. The opening of the subarachnoid space is seen.

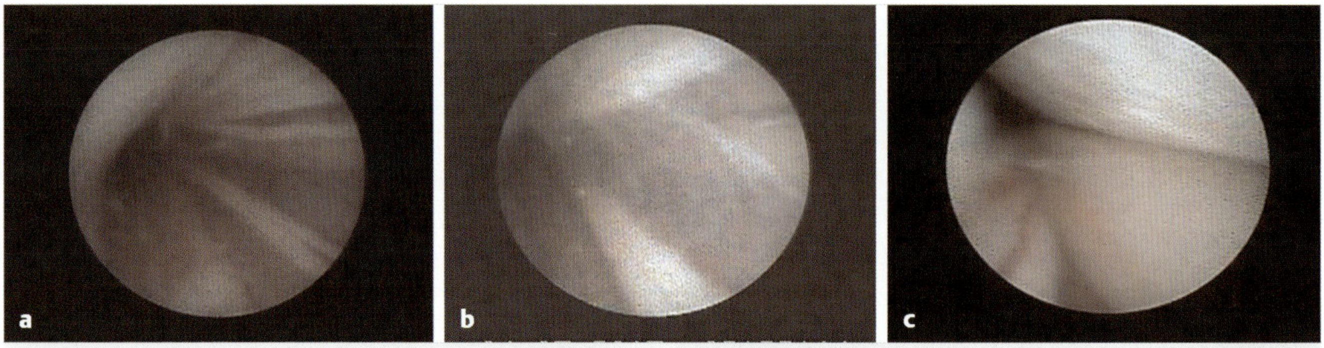

Fig. 11.15 (a–c) Dural perforation with a view of the intrathecal space.

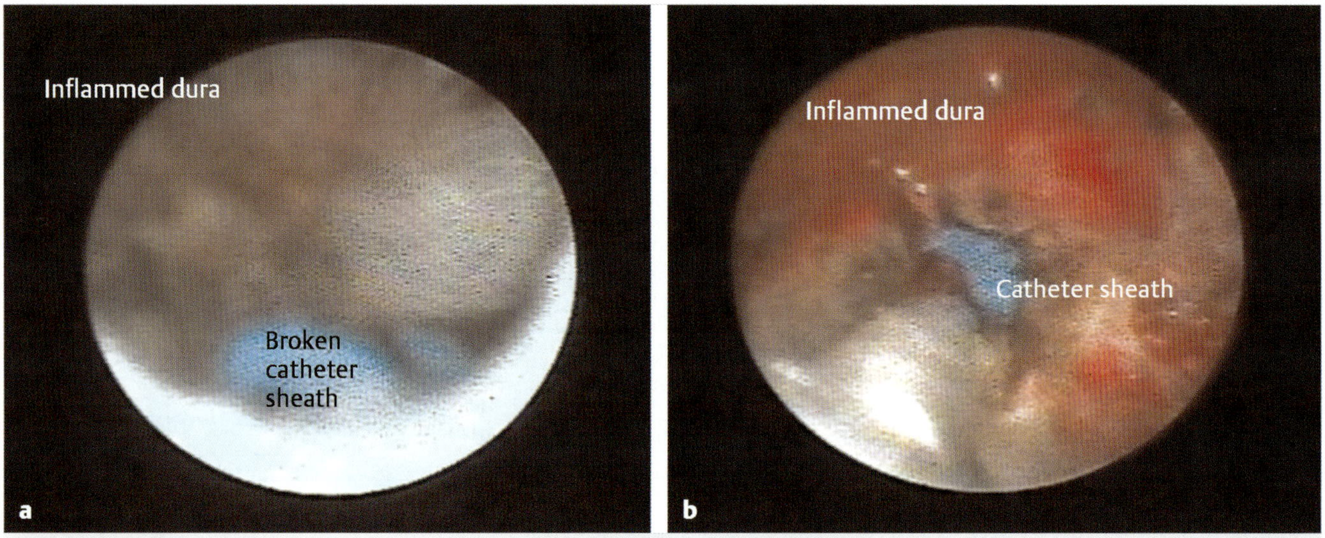

Fig. 11.16 (a,b) Broken catheter sheath.

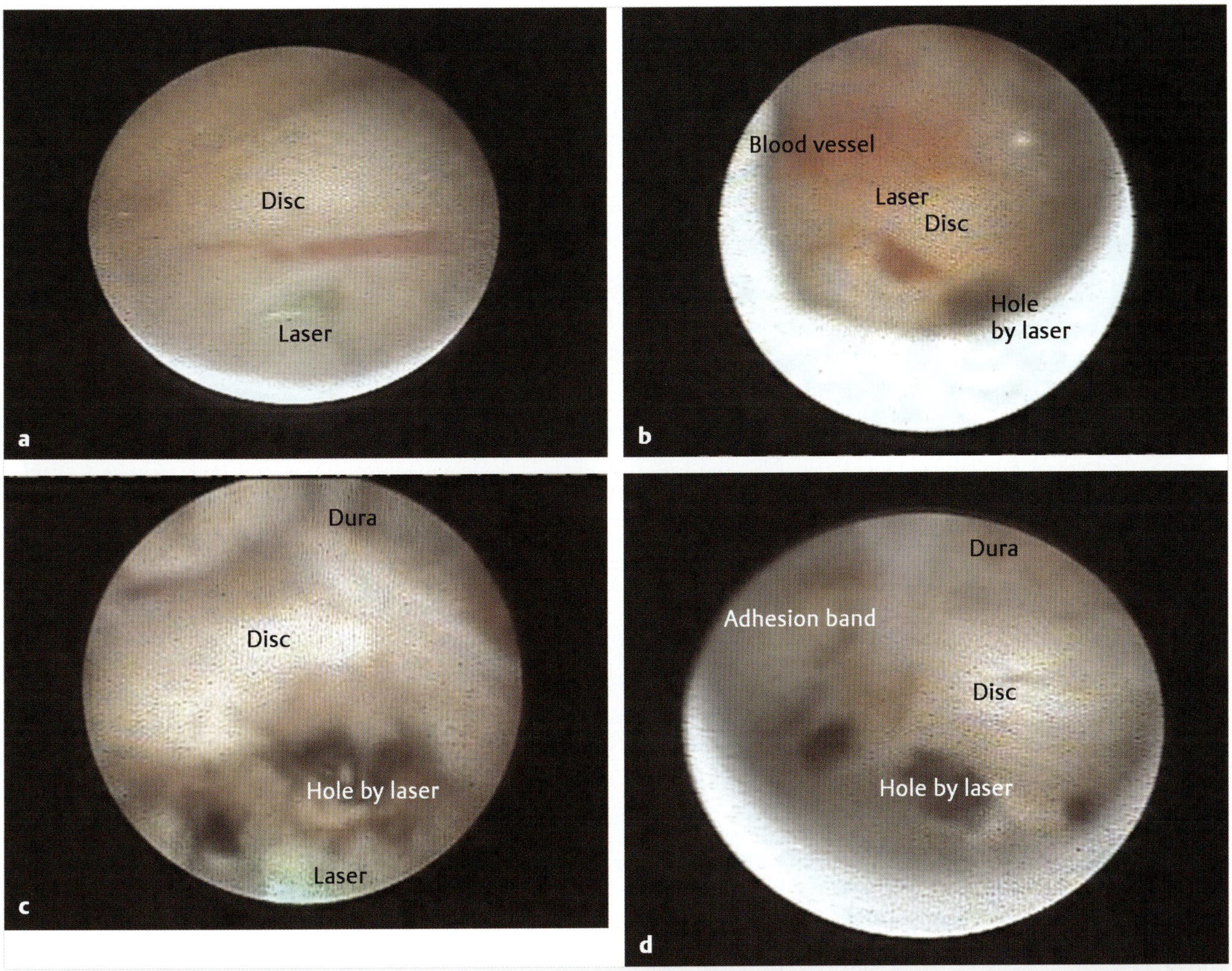

Fig. 11.17 (a,b) Lasering the protruded disk. **(c,d)** Disk removal using laser.

References

[1] Schutze G. Epiduroscopy – Spinal Endoscopy. Heidelberg: Springer; 2008

[2] Kallewaard JW, Vanelderen P, Richardson J, Van Zundert J, Heavner J, Groen GJ. Epiduroscopy for patients with lumbosacral radicular pain. Pain Pract. 2014; 14(4):365–377

12 Evaluation of Epidural Pathology and Epiduroscopic Images

Günter Schütze

12.1 Introduction

Among the most exacting and difficult tasks for any pain therapist are the diagnosis of and treatment for neuraxial pain syndromes For carrying out invasive–interventional pain medicine procedures in the vicinity of the spinal column, knowledge of the topographic anatomy of the neuraxial region is an absolutely indispensable prerequisite (▶ Fig. 12.1). This is especially true for spinal endoscopy. Knowledge of endoscopic anatomy is the basis for a proper professional evaluation of the endoscopic neuraxial anatomical findings. Epiduroscopy (EDS) is carried out percutaneously through the sacral hiatus or sacrococcygeal ligament anywhere from the sacral to the cervical region of the spinal column of the patient depending on the location of the pathologic anatomical findings and utilizing a professional investigative technique.[1,2]

A review of the literature for endoscopic investigations of the spinal column, makes clear that these are primarily investigations of the epidural space of the dorsal sacral or dorsal lumbar regions of the spinal canal. The cause for this, among other things, is that a majority of pathologic processes occur in the lumbar section of the spinal column. Some epiduroscopes that are used, on the basis of their technical construction and outfitting, allow only very limited endoscopic investigations, maximally in the lumber spinal column region and often with restricted image quality.

In most published clinical reports of the findings and therapeutic results of epiduroscopies with flexible endoscopes,, the images that identify, verify, or clarify the normal or pathologic anatomical findings are not included. Only isolated reports and very few useful endoscopic images exist about normal endoscopic anatomy and its pathologic alterations. The sporadically published endoscopic images did not include comparative images of similar pathologic findings from different patients either. The Epiduroscopy DVD (2006) by Schütze[3] was an attempt to make the first visual epiduroscopic atlas and to promote a unified nomenclature for pathologic anatomical findings in the neuraxial region.

In this chapter the pathologic anatomical findings from patients with pain syndromes, observed exclusively via a percutaneous flexible endoscopic investigation technique (spinal endoscopy [epiduroscopy]) are presented from a systematic, topographic, clinical, and practical point of view.

Neuraxial endoscopic investigation technique makes it possible to visualize the pathologic anatomical structures accessible to the epiduroscope in the sacral, lumbar, thoracic, and cervical regions of the spinal column of a patient with a pain syndrome. Epiduroscopy also conveys the dynamics of the epidural space as well as a visual understanding for this topographic region so important for pain medicine. For patients with neuraxial pain syndromes, spinal endoscopy is an excellent minimally invasive endoscopic procedure for diagnosis of pathologic structural changes.

12.2 Pathogenesis of Neuraxial Pathologic Findings

For an understanding of a neuraxial pathologic anatomical change, knowing the extent of the change, the height

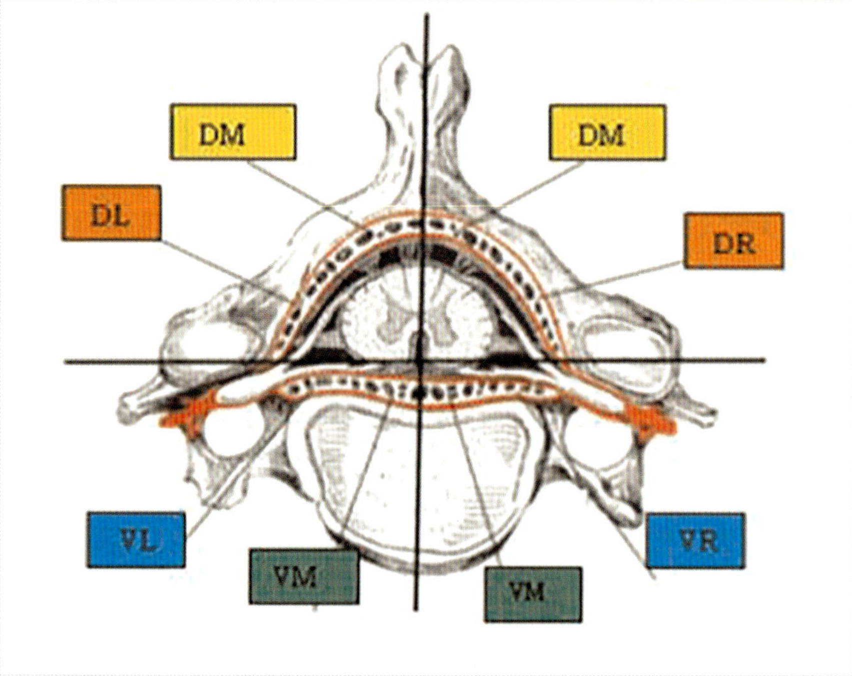

Fig. 12.1 Illustration of simplified orientation and documentation of spinal topography and endoscopic navigation. The epidural space is shown in red, and the intracanthal space in black. DM, dorsal medial epidural space; DL, dorsal left lateral epidural space; DR, dorsal right lateral epidural space; VR, ventral right lateral epidural space; VL, ventral left lateral epidural space; VM, ventral medial epidural space. (Reproduced with permission of Schütze 2008.[6])

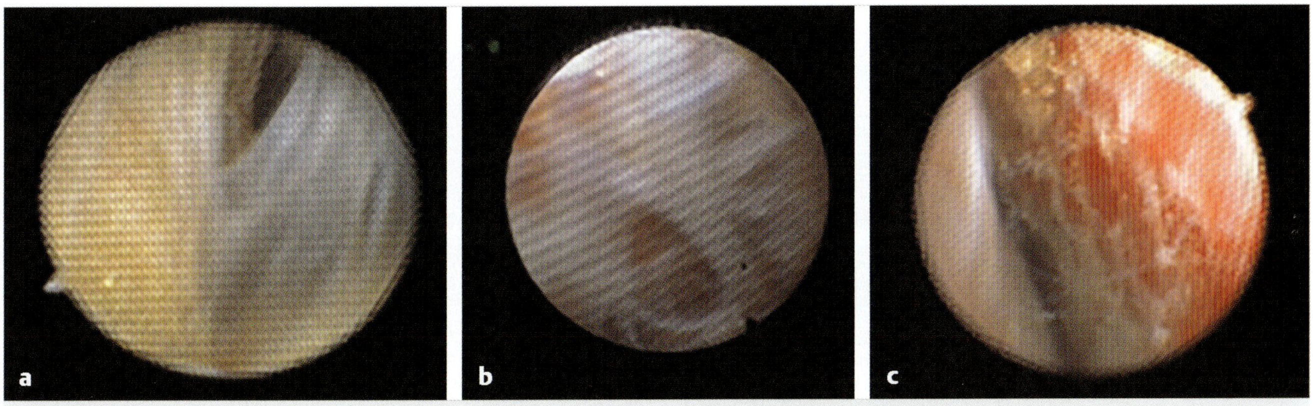

Fig. 12.2 Common pathologic changes in the epidural space in patients with neuraxial pain syndromes: (a) adhesions, (b) fibroses, and (c) chronic inflammation.

localization of the change, and the time course of the change are important. The clinical image varies depending on these factors. On that account, knowledge of the origin of the pathologic change aids in interpretation of the clinical findings and is indispensable with regard to the predicting locale and dimensions of the pathologic process near the spinal cord as well as formulating a diagnosis.

Neuraxial pain syndromes are characterized by pathomorphological changes in the spinal canal both for patients with pain syndromes who have been operatively treated and those who have been nonoperatively treated (▶ Fig. 12.2).

Via endoscopic investigation, epidural pathologic anatomical structures such as adhesions, fibroses, radiculopathies, ischemias, perineural edemas, congested veins, chronic inflammatory processes (arachnoiditis, epiduritis, and radiculitis), sequestration and nerve root compressions, or cysts and tumors can be diagnosed.[4,5,6] Pathologies such as ependymomas, schwannomas, or chordomas of the sacrum, and tethered cord syndrome in the neuraxial section of the cauda equina, filum terminale, and conus medullaris are difficult to diagnose, above all in the initial stage these diseases, but can be identified early endoscopically. Also, through spinal endoscopy the possibility exists of taking out adequate tissue samples for fine tissue pathologic diagnostics on site.[6]

In the future, technological advances such as virtual chromoendoscopy, an endomicroscopic technique, will also be available with spinal endoscopy. Endomicroscopy can endoscopically visualize and distinguish cellular, vascular, and connective tissue structures, for example through laser-supported and confocal fluorescence technology or special visualization tools. Thereby, histologic evaluation of pathologic anatomical structures becomes possible in real time during the endoscopy (see Chapter 14).

It is known that no single mechanism underlies the origin and maintenance of pain. Not only pathologic structural changes but also biochemical processes in the neuraxial region are responsible for the generation of pain. Inflammatory and immunologic processes exist, with activation of nociceptors, in which mechanical influences are involved more or less predominantly in the generation of pain.[7,8,9]

With the help of immunohistochemical investigation techniques it has been determined that, for example, autologous intervertebral disk material, if it comes in contact locally or diffusely epidurally with nerve tissue, provokes immunologic inflammatory processes in the adjacent nerve root.[10] Tissue of the nucleus pulposus, when it protrudes into the epidural space from a herniated intervertebral disk, can trigger chemically induced inflammatory processes. These processes are immunocompetent cellular reactions against expressed intervertebral disk material. The evidence of monoclonal antibodies against T-lymphocytes and macrophages supports this interpretation.[11] The results confirm the conjecture that even with radicular symptoms of pain, the pain is not due purely to mechanical changes (such as pressure) on the nerve root.[12] Also in clinical studies it has been shown that only with irritated or inflammatorially changed nerve roots is compression or traction perceived as painful.

These pathologic changes, combined with impeded cerebrospinal fluid (CSF) circulation and insufficient fibrinolysis, lead to, for example, impaired nutrition of the affected nerve root, which subsequently leads to changes in the CSF and cell body. The nerve root can become compromised, deformed, or stretched through pathologic changes in the intervertebral disk, facet joints, or intervertebral foramina.[13] These deformations of the nerve root or the dorsal spinal ganglion lead to impairment of the local microcirculation, with wide-ranging clinical consequences.[14,15,16]

12.3 Pathologic Findings in the Epidural Space: Adhesions, Fibroses, and Chronic Inflammatory Processes

To be able to interpret pathologic situations found near the spinal cord also requires experience obtaining and interpreting epiduroscopic images.

On the basis of our clinic's long-running experiences with more than 2,000 spinal epiduroscopic investigations in patients suffering from pain, I will attempt, in what follows, to classify and evaluate the common pathologic anatomical findings (▶ Fig. 12.3).

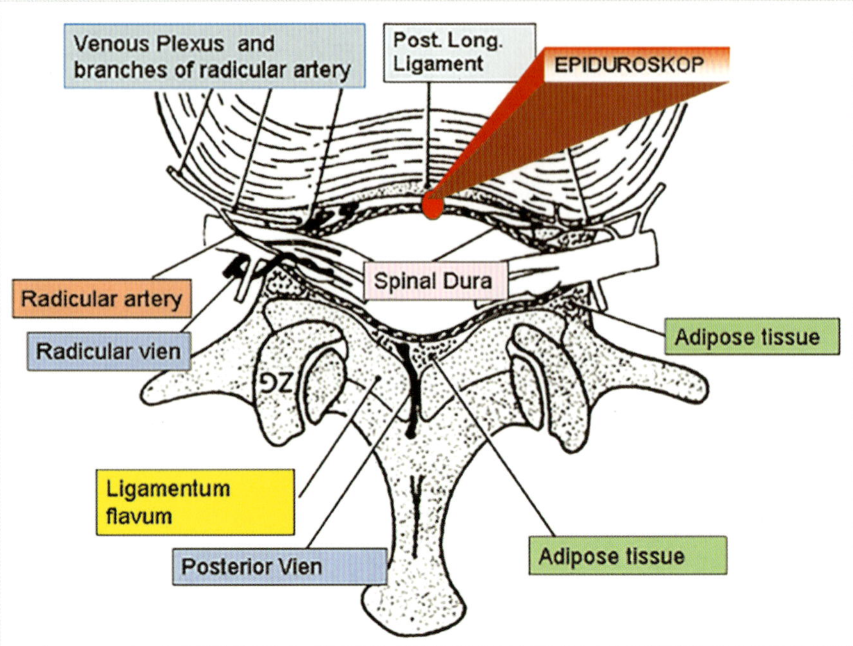

Fig. 12.3 Schematic transverse section through the spinal canal with ventral positioning of the epiduroscope for orientation. (Reproduced with permission of Schütze 2008.[6])

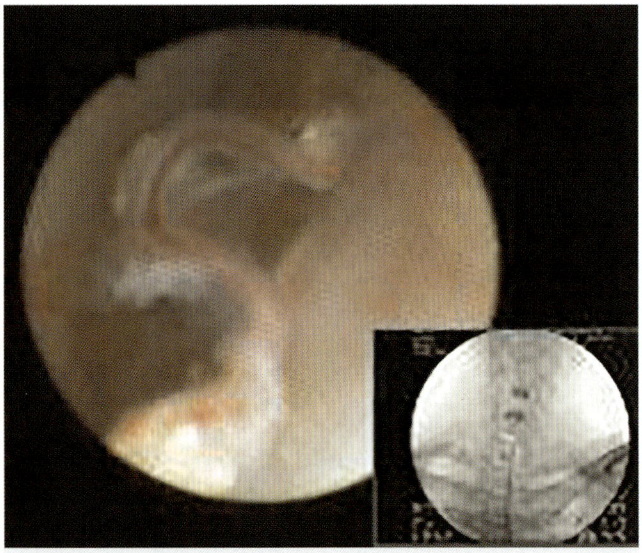

Fig. 12.4 Blood vessel in the postoperative cervical epidural space with beginning adhesive encasement (picture-in-picture technique).

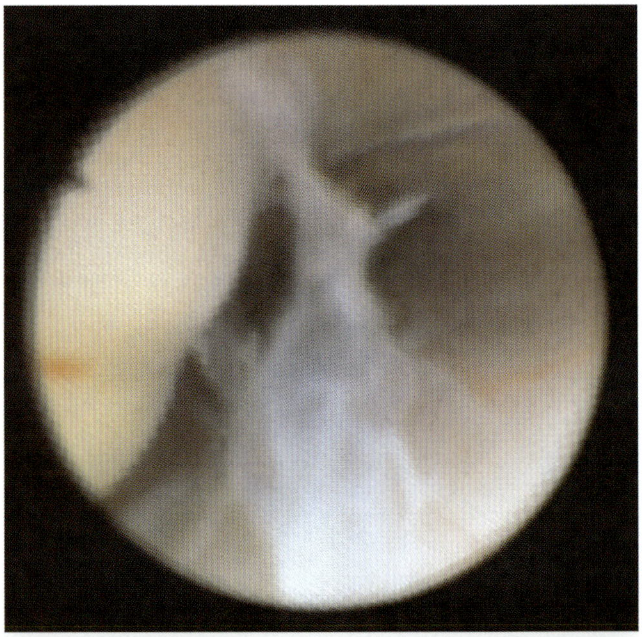

Fig. 12.5 Epiduroscopic view of an intervertebral disk herniation (*left*) in the cervical epidural section with formation of an adhesion between the spinal dura mater (*right*) and the posterior longitudinal ligament (*left*).

12.3.1 Adhesions

For a long time, mechanical irritation was favored as the cause for neuraxial pain syndromes. Now increasingly, more significance is being attached to secondary inflammatory processes.

The formation of adhesions can be explained pathophysiologically on the basis of special healing mechanisms in the neuraxial space and on an imbalance, arising from injury, between fibrinogenesis and fibrinolysis (▸ Fig. 12.4).[11,17]

Epidural adhesions or adhesive regions occur very frequently in our patient population. Richardson et al diagnosed epidural adhesions as the cause of pain in all endoscopically investigated patients with chronic radicular back pain.[18]

Special Pathogenesis

As an example of the special pathogenesis of epidural adhesions, material from a herniated intervertebral disk can escape into the epidural space through a ruptured anulus fibrosus. In response to the proteoglycans from nucleus pulposus in the epidural cavity, serotonin, histamine, bradykinin, prostaglandin, and phospholipase A2 are released. This initiates inflammatory reactions, which in turn lead to epidural adhesions (▸ Fig. 12.5). This example demonstrates that epidural adhesions can arise

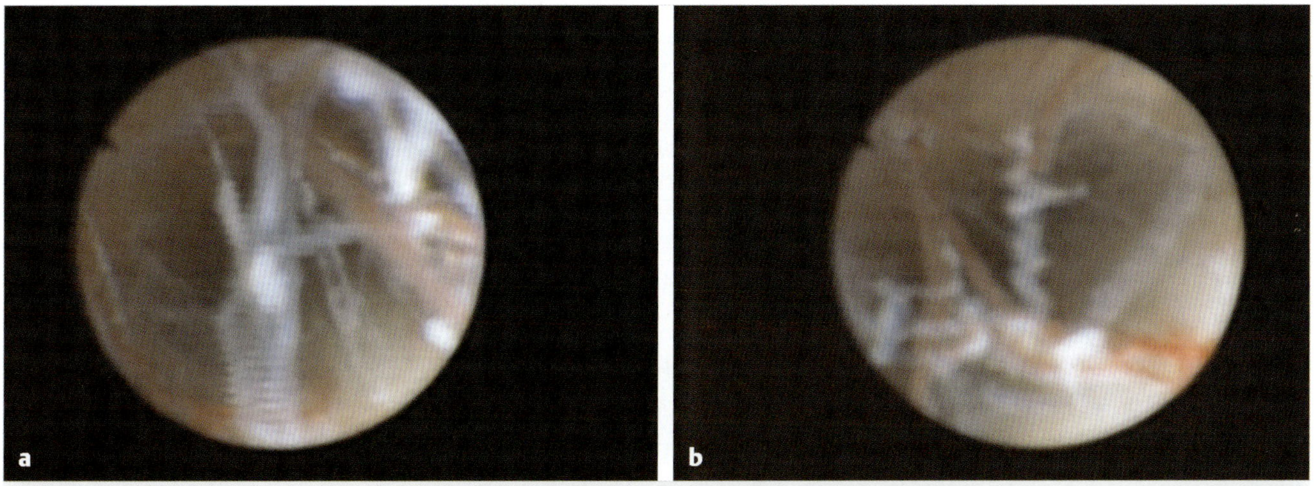

Fig. 12.6 (a,b) Inclusion of epidural vessels in developing fine connective tissue adhesion strands.

even without a preceding operative neuraxial intervention or inflammatory process.

The spinal dura mater in the ventral region, the ventral leaf of the dura, is richly innervated with silent nociceptors with high activation thresholds. After these receptors are sensitized by substances such as serotonin, histamine, bradykinin, prostaglandin, and phospholipase A2, they behave as mechanosensitive, that is, sensitive to mechanical stresses like compression.

The extent of epidural adhesion formation depends decisively on the presence of local ischemia, inflammation-reaction hematomas, foreign body particles, mechanical and thermal irritation, dehydration, and allergic reactions (▸ Fig. 12.6).[19,20,21,22]

On that basis, marked to excessive epidural adhesions, which are bound fast to the spinal dura mater, can provoke quite considerable pain in response to mechanical stress even after proper endoscopic observation that does not disturb the structures in any way to cause pain.

Endoscopic Identification

In the endoscopic image, easily identifiable, diversely structured epidural adhesions appear as irregular, often bizarre, delicate to strong, white or white-gray connective tissue structures.

Roughly simplified, soft adhesions can be distinguished from stiff ones by contact with the endoscope itself or microsurgical instruments.

Epidural adhesions can be observed to cover up blood vessels, nervous structures, and foreign bodies (sequestration, implanted catheters, electrodes) with thick connective tissue structures or else to occupy space, filling up the epidural cavity (▸ Fig. 12.7).

Nervous structures (nerves and nerve roots) completely enclosed by adhesions react, often with a pain reaction, on contact.

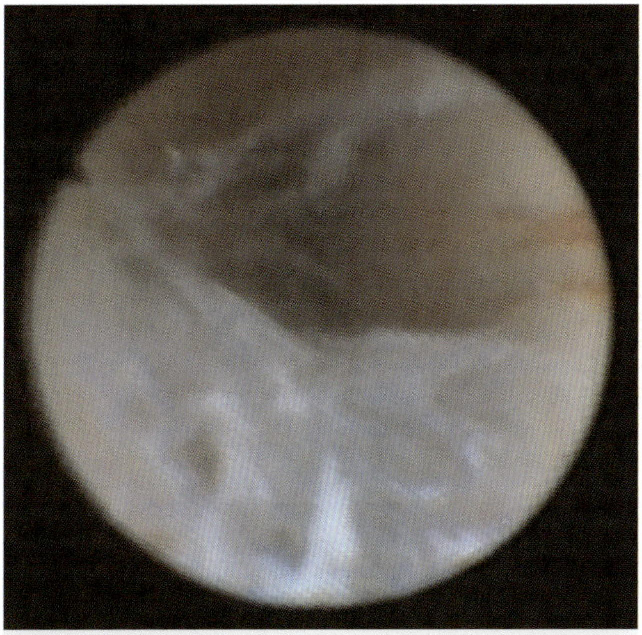

Fig. 12.7 Space-occupying epidural adhesions (*left*) in contact with spinal dura mater (*right*).

Topographic Aspects

Sacral and lumbar regions

In most endoscopic investigations on frequent failed back surgery syndrome (FBSS) cases, postoperative adhesions of varying degrees occur mainly in the lumbar region operated on. The adhesions are frequently associated with marked fibrosis. Often they can be overcome to expand the endoscopic region of investigation only with the use of microsurgical instruments or modified access techniques.

Thoracic and cervical regions

Adhesions and areas of adhesions in the thoracic and cervical regions of the spinal column have special clinical significance (▸ Fig. 12.8).

Especially in topographic spinal column regions of intervertebral disk prolapse and herniation, after standard interventional microsurgical operative procedures marked cervical and thoracic epidural adhesions are frequently diagnosed endoscopically.

Our experience verifies that increased adhesions are endoscopically encountered especially in the ventral thoracic and cervical pain regions. The adhesions are frequently in loose connection with the posterior longitudinal ligament. Highly marked or space-filling adhesions occurred in our investigations in the ventral T4 and T10 segments. The cervical adhesions were diagnosed endoscopically in our patient population primarily in the cervical spine segments with vertebral bodies C5–C6 and C6–C7, and less often with C7–T1 and C4–C5.

12.3.2 Fibroses

Endoscopically visible epidural fibroses are very often diagnosed in connection with spinal stenoses, spinal column instabilities, and intervertebral disk prolapse and herniation (▶ Fig. 12.9). According to Laus et al, excessive fibrosis is also frequently a consequence of an insufficient intraoperative hemostasis.[23]

A diffuse increase in connective tissue in the epidural space leads to fibrosis (wound healing), which is also easily identified

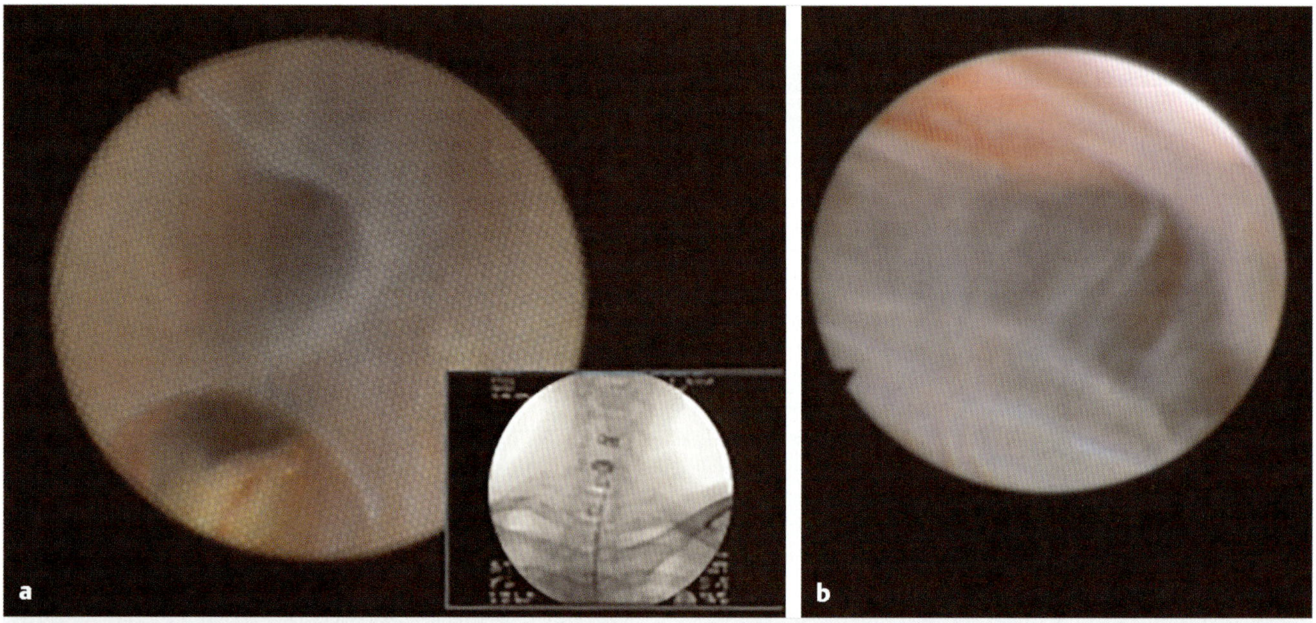

Fig. 12.8 (a,b) Cervical space-occupying epidural adhesions (picture-in-picture technique).

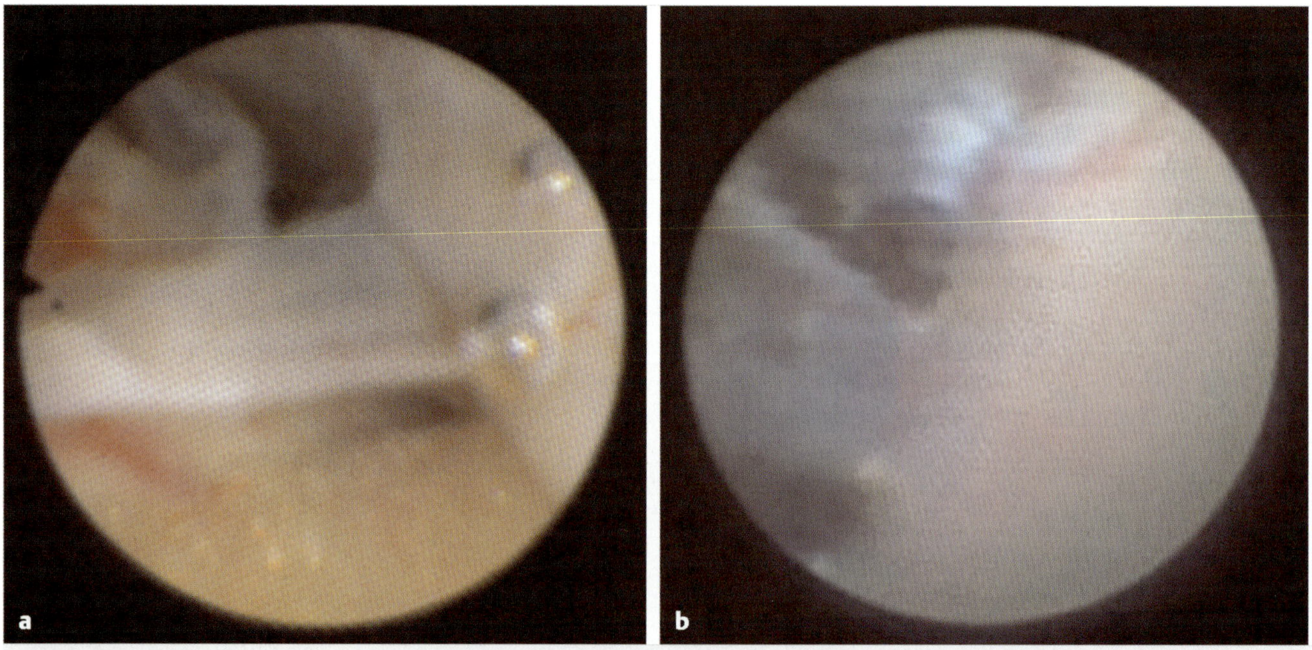

Fig. 12.9 (a,b) Pain-relevant, ribbon-shaped lumbar epidural fibrosis that is connected with the spinal dura mater and the ligamentum flavum.

endoscopically. If the increased collagen production leads to a pathologic hardening (induration), an epidural sclerosis occurs (▶ Fig. 12.10).[11]

The clinical phenomena of an atypical spread of regional anesthesia (missed segments) in classical epidural anesthesia may also be explained by epiduroscopically visible pathologic structures such as marked adhesions and fibroses. The unfavorable result in patients can be explained by the neuraxial pathologic anatomical changes such as inflammatory processes, adhesions, and/or fibroses that alter resorption relationships.[24]

Special Pathogenesis

Epidural fibrosis possesses central significance in pain medicine (▶ Fig. 12.11).

Fibroses observed endoscopically along the dural sac and the nerve roots are often a cause for postoperative pain syndromes such as FBSS. A regional inflammatory reaction with edema formation in neurogenous structures, local circulatory disturbances, as well as adhesions can explain the emergence of an epidural fibrosis.

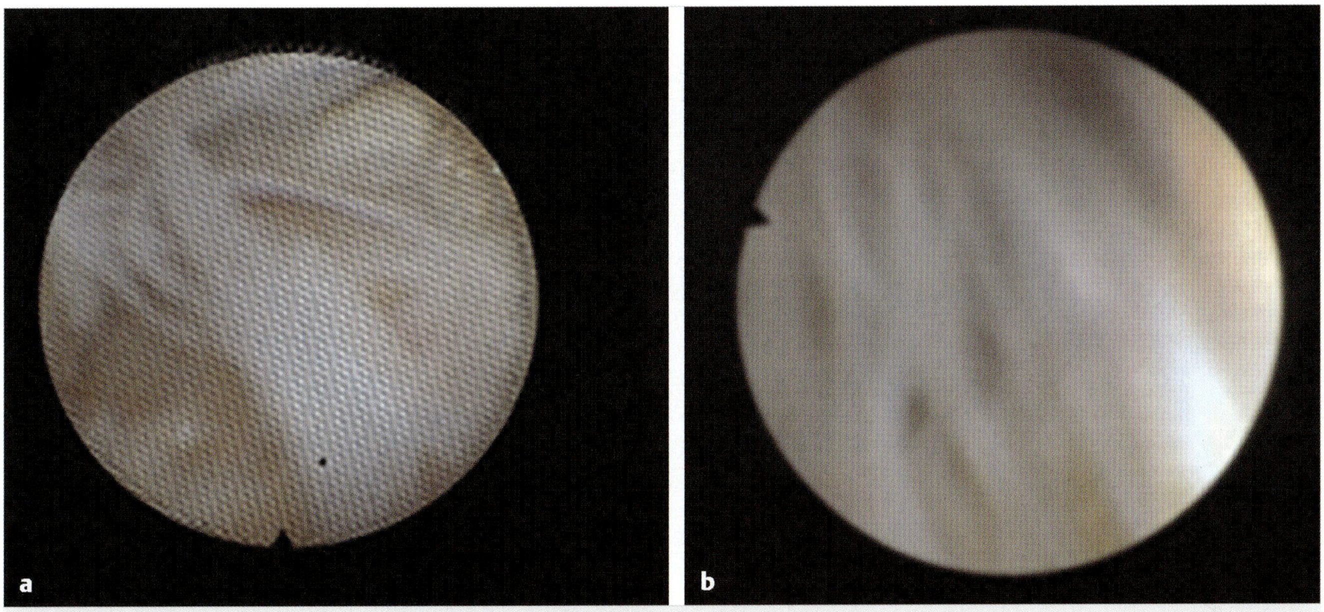

Fig. 12.10 (a,b) Strand-shaped collagen fibers without recognizable vascularization organize into an epidural fibrosis.

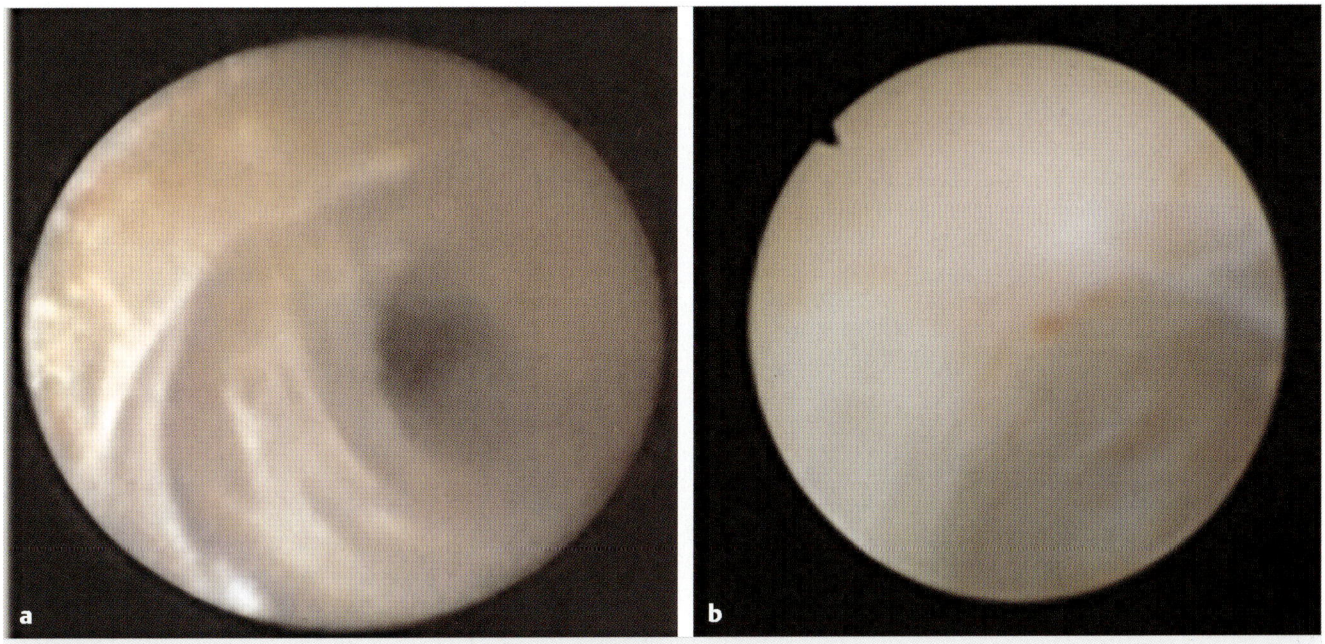

Fig. 12.11 (a,b) Marked space-occupying and stenosing epidural scars.

Osteophytes (bony outgrowths) and protrusions or sequestrations of the nucleus pulposus due to intervertebral disk prolapse and herniation, respectively, can stretch the posterior longitudinal ligament. This compression, in connection with inflammatory changes and changes in the permeability or in the transmural pressure of the endoneural capillaries, can result in edema.

Finally, further pathologic processes lead to disturbance of the epidural blood flow. According to Brown's investigations, the venous blood flow can already be impaired by 5 to 10 mm Hg in the affected segment and the arterial supply reduced by 20 to 30%.[9] Through the disturbance in perfusion due to an applied pressure of 10 mm Hg, nutrient transport to the nerve root can be reduced about 20 to 30%. The inflammatory reaction favors the emergence of an epidural fibrosis.[11,25]

A significant increase of collagen fibers is observed with chronic inflammation, long-standing edema, and the filling of tissue wounds. Pathologic tissue alterations are recognized not only by the growth of collagen fibers but by the abnormal structure of various collagen fiber types. The structural proteins fibrillin or elastin can likewise be pathologically changed.[11]

It should be part of everyday clinical life to apply exclusively preservative-free solutions epidurally or spinally, because preservatives also can induce epidural fibroses and even neurotoxic reactions in the vicinity of the spinal cord.

Epidural scars

A pathologic increase in connective tissue in the epidural space is designated as an epidural fibrosis or epidural scar. (The terms "epidural sclerosis", "callus", and "induration" are also used as synonyms for epidural fibrosis.) Epidural scars play an extraordinarily important role in pain medicine.

The formation of an epidural scar proceeds through the following stages.

1. The first stage, of healing by first intention, is the normally running healing process ending with formation of small amount of granulation tissue.
2. The second stage, of healing by second intention, from the end of the first stage runs until massive granulation tissue formation, by which strong scarring occurs.
3. Scar formation concludes with the stage of wound healing, or repair.

Epidural wound healing is indicated by a strong loss of locally specific cells and can be compensated for by a corresponding scar formation, leading to functional impairment of the tissue.[11]

Wound healing, in which connective tissue encloses nervous structures followed by subsequent scar formation, leads to clinically significant nerve root compression sign and localized back pain as well as diminution of conduction. Krämer correctly compares the connective tissue strands on the spinal dura mater and nerve roots to bell ropes that are actuated by every unthinking movement.[15]

Epidural wound healing is not always a harmless reaction to a reparative inflammatory process; rather, it is often harmful resulting in a specific chronic neuraxial pain syndrome.

Endoscopic Identification

In the endoscopic image, epidural proliferations of connective tissue appear as fibroses, depending on the degree of collagen markedness, in a clear white or gray-white color, and they present themselves as avascular, as a rule. Epidural fibroses range in appearance from strandlike, to space-occupying and in consistency from flexible to stone-hard. Fibroses that are strandlike can be confused with an epidural peripheral nerve. Epidural vessels or nervous structures can also be enclosed in fibroses.

In the pathologic process of epidural space-occupying fibroses, parts of the spinal dura mater and posterior longitudinal ligament or ligamentum flavum are often firmly included. Fibroses in the form of clamplike connective tissue strands can also be seen to pull around the nervous structures and nerve roots and put pressure on them mechanically. The connective tissue strands can also hinder the slippage of the nerve root in the intervertebral foramen.

Epidural fibroses can also be identified as broad flat malformations between the affected nerve root and the osseous environment in the lateral recess.

Along with the epidural fibroses, local chronic inflammatory processes (epiduritis, radiculitis, and arachnoiditis) or a perineural fibrosis, as in FBSS, can also be diagnosed endoscopically in the epidural cavity.

Topographic Aspects

Sacral and lumbar regions

As with adhesions, fibroses are found in the sacral and lumbar regions in most endoscopic investigations. On the basis of a large number of endoscopic investigations of our patients with FBSS, postoperatively arising fibroses, or scars, of varying extent are very frequently encountered. Lumbosacral fibroses often form the basis for spinal stenosis (▶ Fig. 12.12).

To perform successful spinal endoscopies, obstructing fibroses in the sacral or lumbar regions of the epidural space must occasionally be removed by means of microsurgical instruments or the use of a laser.

Thoracic and cervical regions

Fibroses in the thoracic and cervical regions are very frequently clinically relevant to pain. In contrast to the situation in the cervical and lumbar regions, only 0.75% of all intervertebral disk prolapses are diagnosed in the thoracic region, due to lesser biomechanical loading.[8,26]

Our investigations show that fibroses are endoscopically observed in target regions of pathologic processes (regions operated on, with implants, where polymethylmethacrylate [PMMA] bone cement was used, etc.), especially in the ventral cervical and thoracic pain regions (▶ Fig. 12.13).

Marked local to completely stenosing fibroses (in which parts of the longitudinal posterior ligament and/or the spinal dura mater are often firmly included) occur, according to our experience, in the ventral thoracic segments T4–T10(▶ Fig. 12.14). Fibroses occur mainly in the ventral cervical epidural C5–C6, C6–C7 segments.

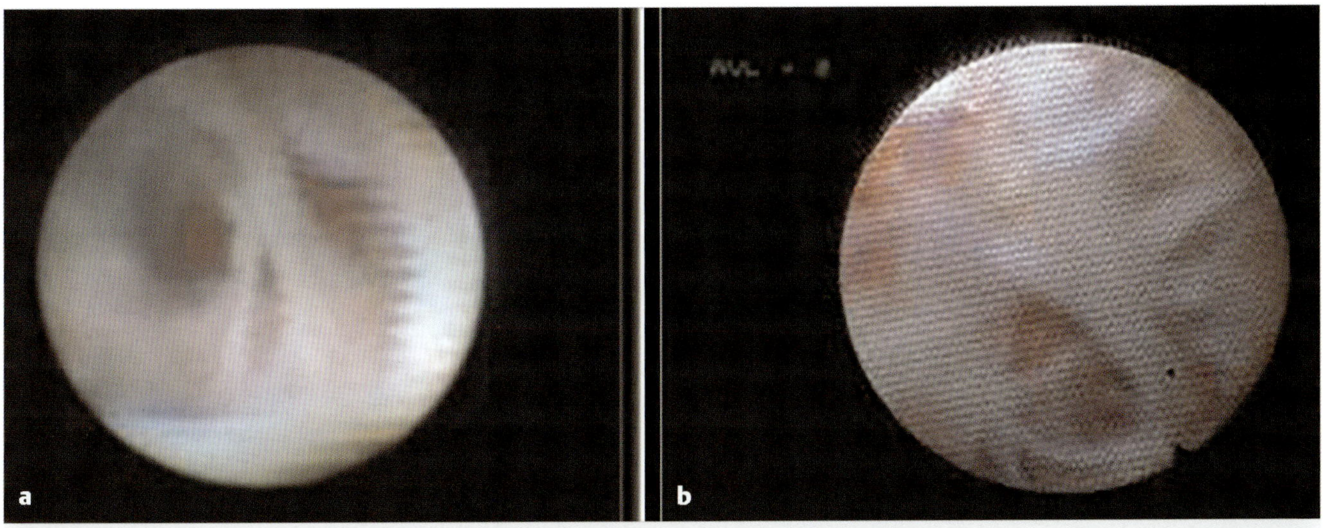

Fig. 12.12 (a,b) Epidural fibrosis in lumbosacral region being investigated.

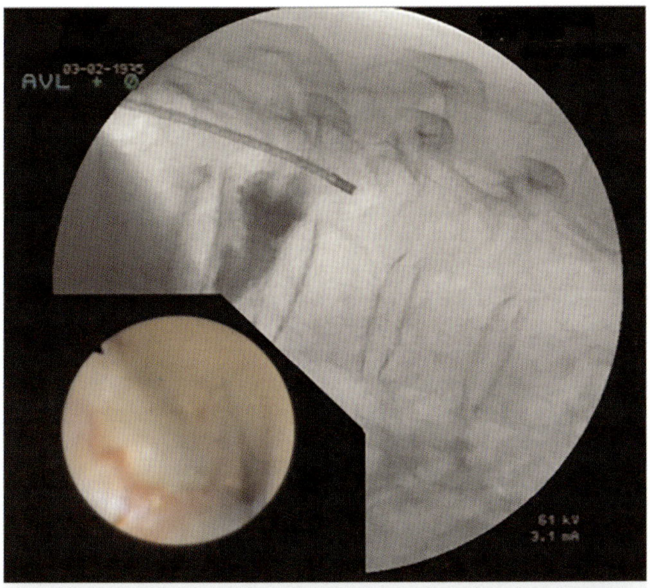

Fig. 12.13 Accidental distribution of polymethylmethacrylate (PMMA) for a vertebroplasty into the epidural space, epiduroscope in the dorsal epidural space presentation of fibrosis between the ligamentum flavum and the spinal dura mater (picture-in-picture technique).

A pain provocation test can be done on endoscopically identified epidural fibroses to determine their relevance to the pain phenomena of a patient.

Fibroses in the Region of Implanted Spinal Cord Stimulation Electrodes and Catheters for Neuromodulation

Spinal cord stimulation electrodes

After long periods of time, epidural fibroses develop in the region of implanted spinal cord stimulation (SCS) electrodes for neuromodulation (▶ Fig. 12.15).

Over the course of time through a strengthened epidural scar formation or marked fibrous enclosure of the tip of an SCS electrode, the electrode can become ineffective. Cicuendez et al reported on spinal cord compression due to epidural scar tissue formation around an SCS electrode.[27,28] In these patients, SCS neuromodulation was ineffective and paraparesis was encountered a year after SCS electrode implantation. After operative decompression, histologic investigation of the epidural tissue around the SCS electrode confirmed an epidural fibrosis from which microbiological investigation excluded a chronic infection (▶ Fig. 12.16).

Epiduroscopy opens the possibility of detecting increased production of fibrous connective tissue or granuloma formation in the region of an SCS electrode tip and intervening microsurgically as necessary in order to restore the effectiveness of neuromodulation after long use, above all without operative electrode replacement.

Case presentation

In two patients with long-standing implanted SCS electrodes (Medtronic, Inc., Minneapolis, MN, USA; St. Jude Medical Inc., St. Paul, MN, USA) in which electrostimulation was no longer effective, an endoscopic investigation with subsequent microsurgical adhesiolysis was performed because of the suspicion of epidural fibrosis of the SCS electrode tip (▶ Fig. 12.17).

Using the FLEX-X[29] flexible epiduroscope (Karl Storz, Tuttlingen, Germany) introduced percutaneously via the sacral access to the dorsal catheter location at levels T9–T10, a white, avascular fibrosis strand embedded in the surrounding fatty tissue was observed. The structure of the endoscopically identified strand corresponded to one of the nerves that passes through the cavity. Via fluoroscopic C-arm imaging in the lateral beam direction, it was confirmed that the SCS electrode tip was contained in the epidural fibrosis.

With the help of special microsurgical instruments advanced through the working channel of the epiduroscope, the SCS electrode tip was laboriously freed from the epidural fibrosis,

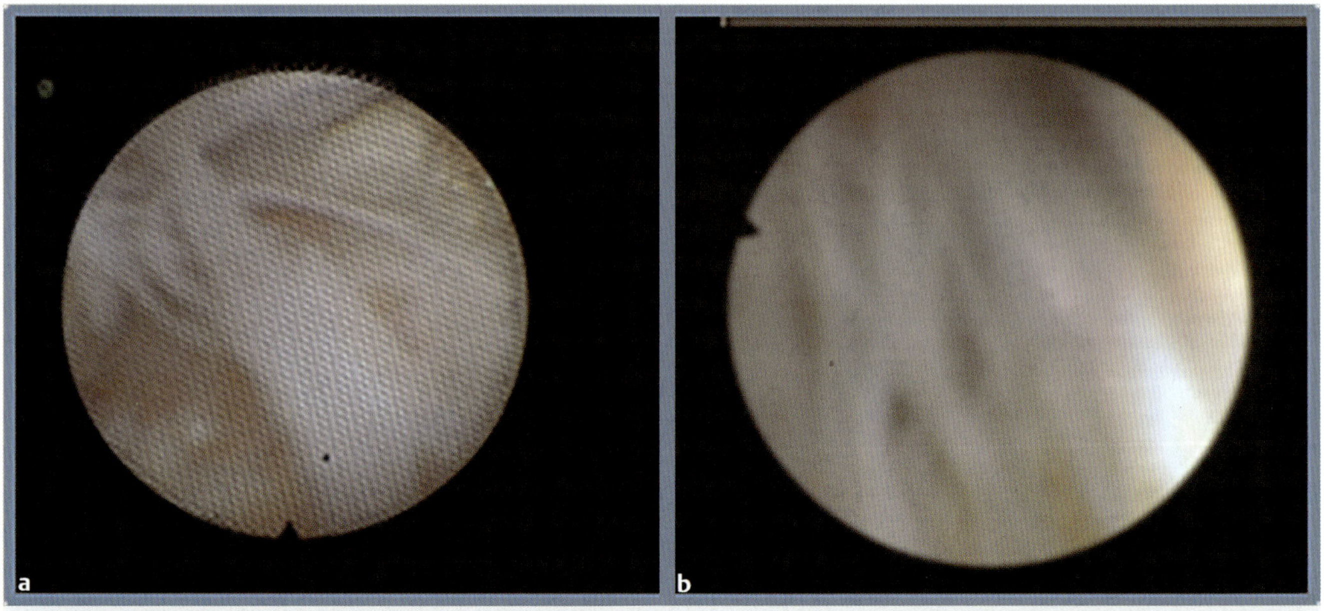

Fig. 12.14 (a,b) Epidural ventral strand-shaped, avascular, connective tissue fibroses in segment L1–L2.

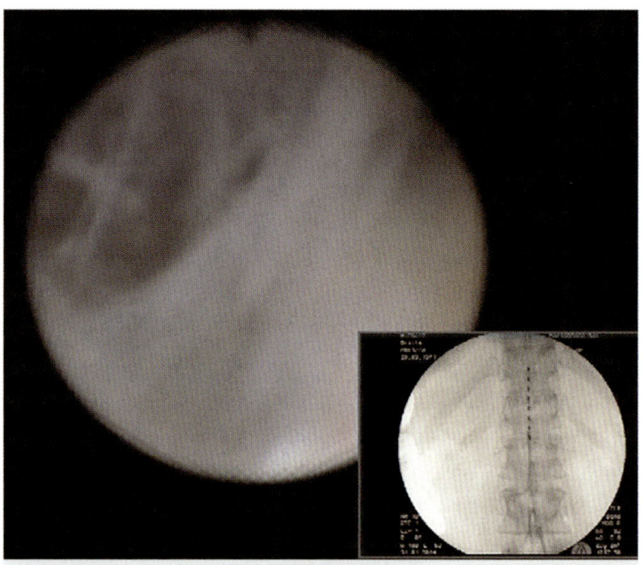

Fig. 12.15 Thoracic epidural fibrosis after spinal cord stimulation electrode implantation (figure-in-figure technique)

restoring the epidural electrostimulation and the accompanying pain reduction.

Catheter

Along with normal catheter positioning, sling and loop formations of the catheter, and folding back of the catheter tip through fibrosis or adhesions could also be observed under endoscopic view in our patients (▶ Fig. 12.18).

It is practically impossible under endoscopic view for a roll-up or a knot or sling formation to occur in placement of a stylet-guided catheter through the working channel of the endoscope. With catheters not placed under endoscopic view, injections in the region of a nerve root can come out in a vein or in a ligament in our experience (▶ Fig. 12.19).

After a long epidural residence period for an implanted catheter, fibroses form along its course. Aldrete[19] found clinical signs of epidural fibrosis between 21 days and 320 days after an epidural catheter was placed.

From our experience, after a residence period of 1 month after placement of a polyurethane catheter, because of marked adhesions and the beginning local fibrosis the catheter was endoscopically recognizable only with effort.

During an epiduroscopic control investigation, we found signs of a nonbacterial inflammatory reaction already after a catheter implantation period of 5 to 7 days (▶ Fig. 12.20).

We confirmed that even epidural catheters placed under endoscopic view or epidurographic control with their tips in the pain-relevant region (a locale with highly developed fibrotic material), after epidural application of local anesthesia, attained a negative analgesia result.

Endoscopic identification and handling for catheter placement or removal can be made easier by strong catheter coloration and corresponding marking aids.

Catheter shear-off

Depending on the catheter implantation technique, the possibility of shearing the catheter off exists, for example, in installation, withdrawal, or removal of the catheter (e.g., through the Tuohy needle). In fact, there are no hints about serious neurologic sequelae of a sheared-off catheter as a foreign body, but some authors advocate surgical removal of it.

To reach the target region for catheter extraction, fibrotic changes around long-implanted catheters in the epidural cavity must be loosened microsurgically. However, the discovery of the catheter after a long residence period in the epidural space can be laborious. On the basis of fibrinous jacketing of the catheter, it is often difficult to distinguish it endoscopically

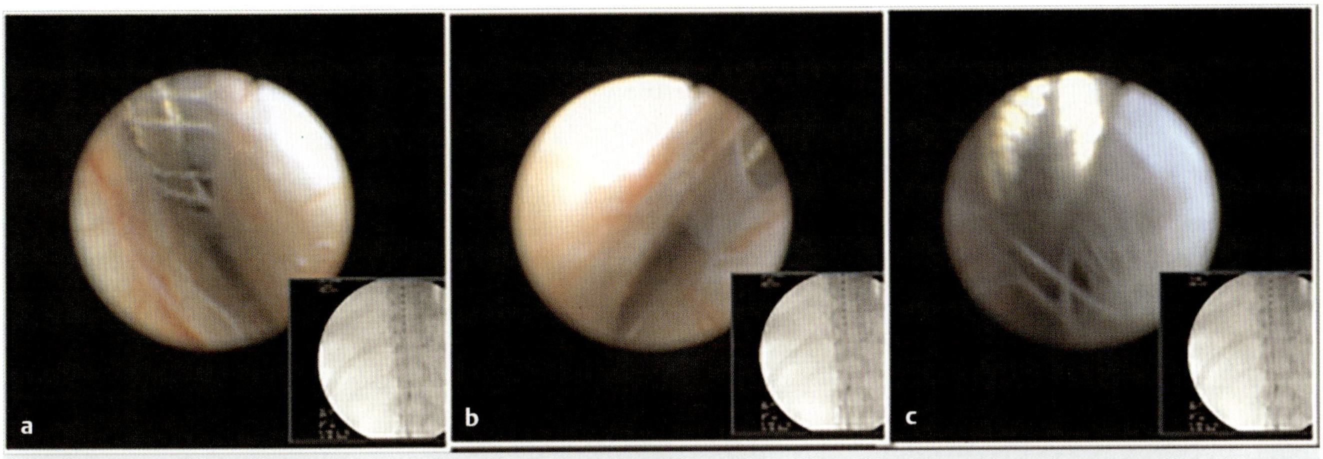

Fig. 12.16 (a–c) Epidural adhesions and fibroses are seen forming around an implanted spinal cord stimulation electrode (picture-in-picture technique).

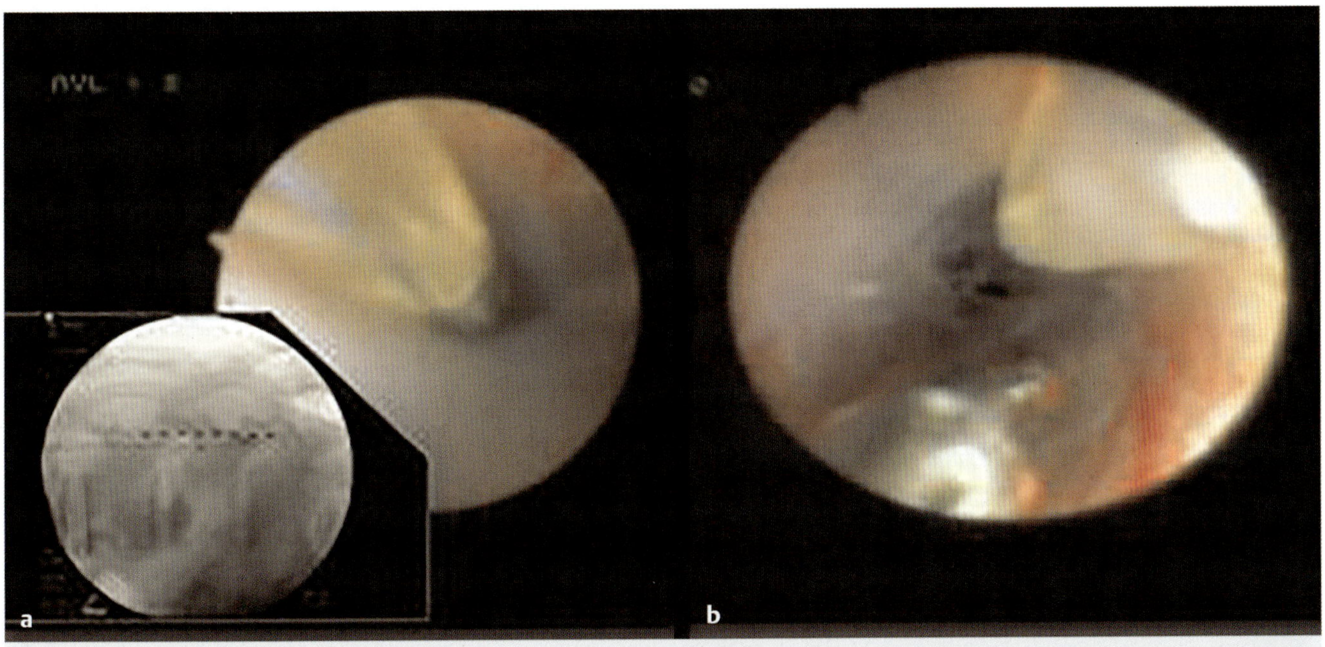

Fig. 12.17 (**a**) Implanted spinal cord stimulation electrode in dorsal thoracic epidural space. (**b**) Grasping forceps plus spinal cord stimulation electrode in the epidural space for mechanical adhesiolysis.

from the remaining epidural fibroses and nerves. We found it necessary to break up the catheter jacketing microsurgically by means of miniature grasping forceps in order to explant the sheared-off catheter part from the epidural cavity or intrathecal space (▸ Fig. 12.21). Because of the catheter material (polyurethane), it is advisable to forgo the use of a laser-supported adhesiolysis.

12.3.3 Inflammatory Processes

Inflammation is a regular consequence of the effect of inflammation-provoking stimuli. Inflammatory processes in the neuraxial region can be the consequence of an infection caused by pathogens, an aseptic inflammation, or iatrogenic reaction after operative interventions on the spinal column.

Acute Inflammatory Reactions

In most forms of inflammation, the escape of fluid and leukocytes from capillary vessels is typical. Depending on the composition of the fluid (exudate or transudate), inflammation is distinguished as serous, fibrinous, purulent, hemorrhagic, or lymphocytic. Purely serous inflammations are mostly nonbacterial.

Necrotizing inflammations are processes provoked by exotoxin-forming microorganisms, which lead to tissue destruction.

Acute bacterial inflammations in the neuraxial region play a special role in medicine. On the basis of the urgency of diagnosis and therapy, patients with bacterial infections may need the services, of course, of emergency and acute medicine staff.

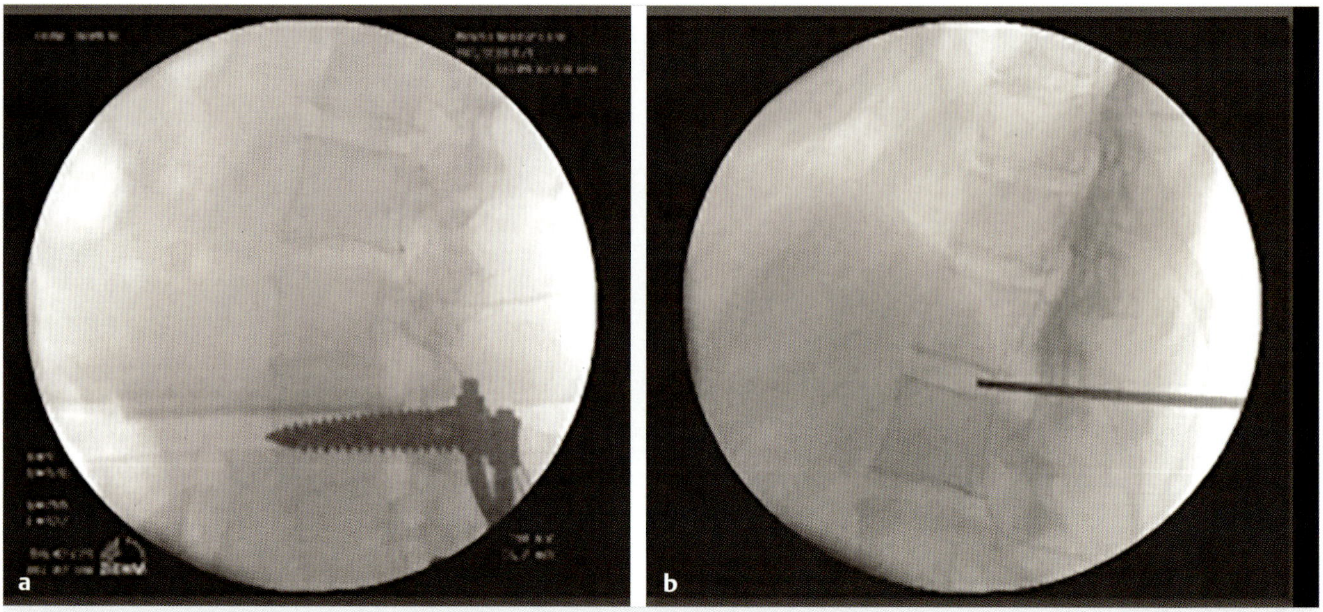

Fig. 12.18 (**a**) Folding back of a conventionally placed spinal catheter, lumbar to caudal. (**b**) Exact thoracic site of the catheter after correction.

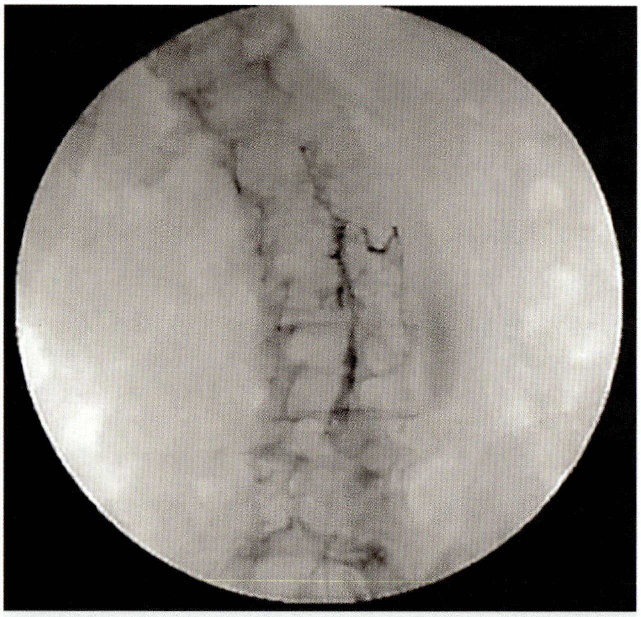

Fig. 12.19 Anteroposterior (AP) epidurogram with a radiopaque catheter following injection of contrast into the lower vena cava.

Chronic Inflammatory Processes

Chronic neuraxial inflammatory processes are frequently associated with extensive destruction of the original tissue structures and finally with wound healing. Macrophages, lymphocytes, plasma cells, and eosinophilic granulocytes play an important role in chronic inflammation. The cytokines secreted by these cells lead to marked and long-lasting changes of the affected tissue.[11,15,30,31]

Epiduritis

Epiduritis is a partial or space-occupying inflammatory process of structures in the epidural space. In an endoscopic image epiduritis presents with the cardinal signs of redness, swelling, and edema and test positive on a pain provocation test.

Choi et al reported on 24 patients who underwent epiduroscopy because of chronic back pains. In all of these patients, endoscopopy revealed inflammation characteristic of epiduritis.[5]

In most cases of patients with chronic inflammatory neuraxial diseases, such as spondylitis, diskitis, spondylodiskitis, radiculitis, or epiduritis, the ventral sections of the spinal canal are involved (▶ Fig. 12.22).

In our more than 2,000 epiduroscopic investigations, we comparatively diagnosed some 42% of patients with chronic inflammatory processes in the sense of a locally limited chronic inflammatory process, with the signs of radiculitis and/or epiduritis (▶ Fig. 12.23).

Laboratory chemical signs of an inflammation, such as an elevated red blood cell sedimentation rate, an elevated level of C-reactive protein or an elevated leukocyte count, cannot be regularly detected in patients in whom a chronic inflammatory process (arachnoiditis, epiduritis, or radiculitis) has been diagnosed.

Endoscopic identification

Chronic inflammatory processes in the epidural space, such as epiduritis, radiculitis, or arachnoiditis, present in endoscopic images as significantly hyperemic, edematous bloated tissue structures.

The often regionally limited epiduritis is diagnosed as a local inflammaory reaction with the cardinal signs of redness, swelling, and edema. The affected epidural structures are strongly hyperemic (reddish to dark red) and edematously bloated. Epiduritis is delimited very significantly from the other healthy epidural tissue by color.

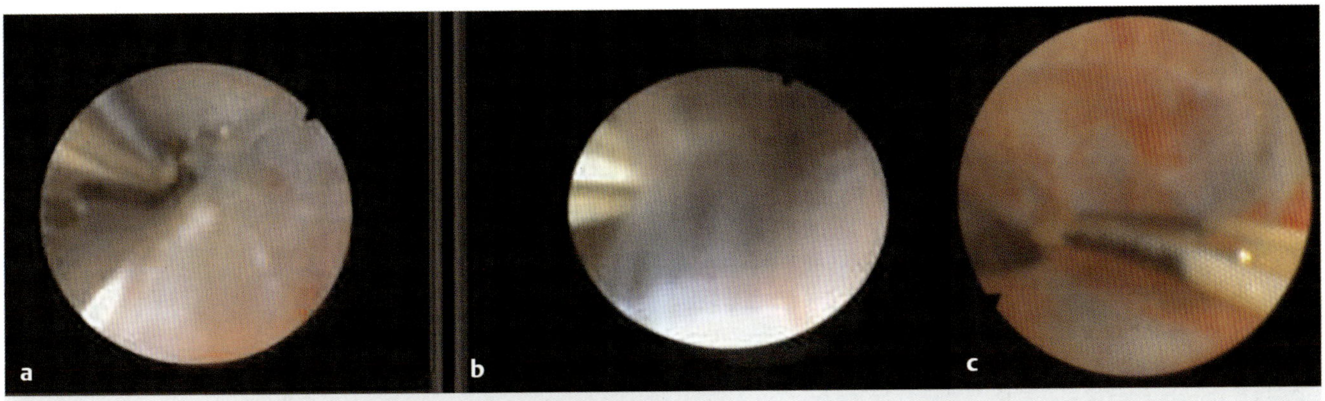

Fig. 12.20 (a) Endoscopically placed epidural catheter. (b) Already after 6 days (control epiduroscopy), the catheter is surrounded by soft epidural adhesions. (c) Marked catheter tip visible in the epidural space.

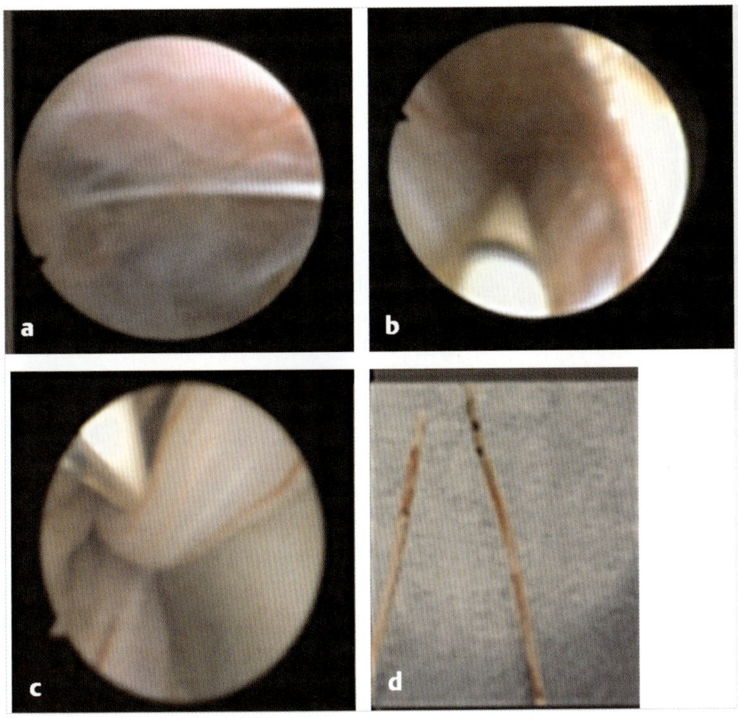

Fig. 12.21 (a–d) Epiduroscopic view of a sheared-off catheter with a partial intrathecal location.

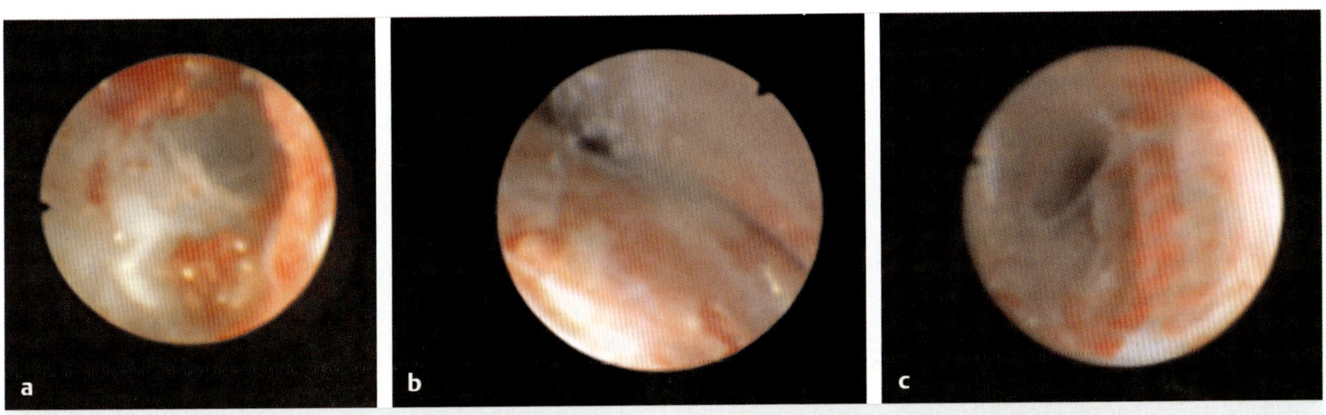

Fig. 12.22 Epiduritis. (a-c) Endoscopic view of a chronic inflammatory process. Along with spread from a point, wide areas of inflammation are also observed. In endoscopic images, inflammation is seen as swelling and redness.

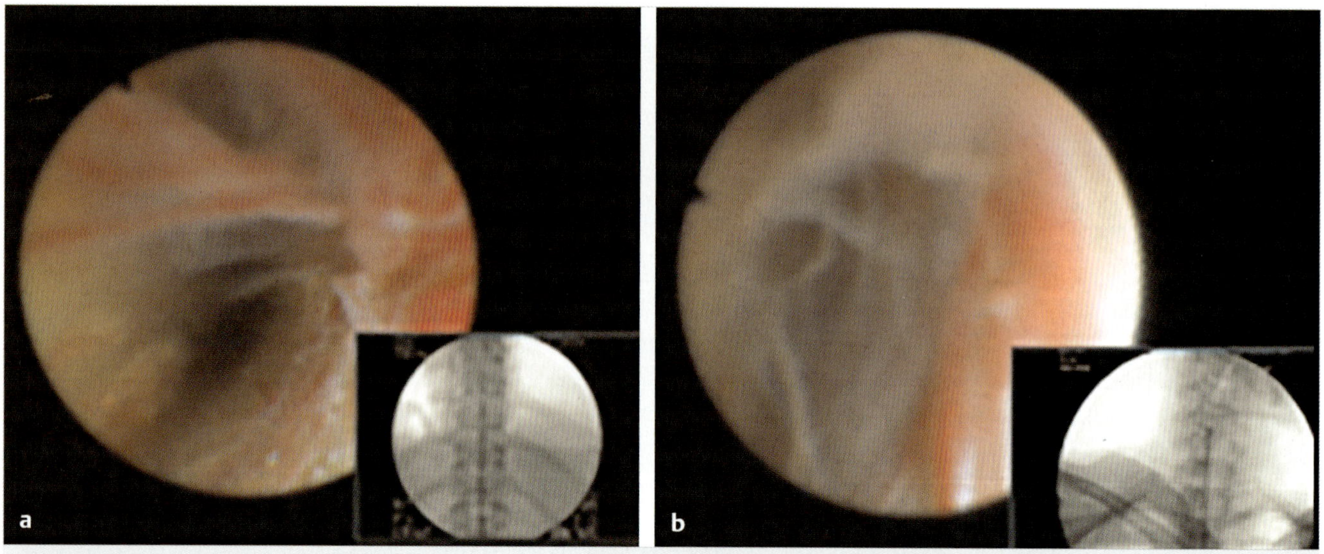

Fig. 12.23 Chronic inflammatory process in the epidural space. (a) Thoracic epiduritis. (b) Cervical epiduritis.

Along with epiduritis that spreads from a point, wide areas of inflammation have also been observed.

Radiculitis

Mechanical compression and stretching, inflammation, and fibroses can irritate the meningeal or mesenchymal tissue of the nerve root, which also can provoke corresponding severe pain symptoms.[32] Endoscopic access to the inflamed nerve root is especially demanding to obtain.[33]

Special pathogenesis

Along with the pathomorphological model of a mechanical spinal root irritation, a secondary inflammatory cause of spinal root irritation is gaining ever more attention.

Prolapsed intervertebral disk tissue as a foreign body in the epidural space can trigger a local inflammatory reaction.[15] The degenerated disk can produce chemical inflammation mediators such as phospholipase A2, interleukin 6, prostaglandin E2, and synovial cytokines and transmit them into the region of the nerve root, thereby touching off and maintaining radiculitis.[23,31,34]

In the environment of the dislocated intervertebral disk tissue, TNF-α and interleukin 1 are released as inflammation mediators. An inflammatory reaction in the environment of the involved nerve root is the result of the mediator release.

These inflammatory reactions are easily seen in epiduroscopic images. Both the granulomatous and the proliferative form of inflammation can be noted. The granulomatous form is distinguished by the development of capillary-rich reparative tissue and the proliferative form through the sprouting of fibroblasts.

For practical pain diagnostics, the experiments of Kizelshteyn et al are significant.[35] In patients in whom a Fogarty catheter had been introduced in the spinal canal, it could be demonstrated that an inflation of the balloon against the healthy dorsal nerve roots caused no pain but rather hypesthesia. If the balloon was inflated in the region of an inflamed nerve, however, the subject felt radicular back pain. This experiment was repeated in different ways, with the same results that a neural inflammation can hold as impetus for pain, hyperalgesia, and allodynia.

Topographic aspects

Endoscopic inspection in the region of the nerve root, which is important for pain medicine, is often difficult.

An endoscopic assessment of vascular supply, pulsation, and contact sensitivity, as well as pathologic alterations such as adhesions, fibroses, edema, and inflammatory changes is possible to date in the vicinity of the ascending and descending branch of the posterior or anterior artery of the vertebral canal, but only within certain limits. The short preganglionic root, the most vulnerable part of the nerve, is poorly visualized endoscopically (▶ Fig. 12.24), and inflammatory processes around this portion of the nerve are also thought to be causes of pain (▶ Fig. 12.25).

Endoscopic identification

In endoscopic images chronically inflamed nerve roots are altered in color and edematous. After long nerve root compression, the nerve root can also present endoscopically as atrophically thin. Occasionally portions of the nerve root become visible at the endoscopic operative site in a reddish or livid bluish color. The pathologically altered nerve roots can probably also appear endoscopically as avascular because do not pulsate, as a rule, due to the surrounding epidural adhesions.

As opposed to contact stimulation (due to endoscope tip, catheter, or microsurgical instruments), inflammatorily altered nervous structures react significantly more sensitively to the pain provocation test than do those in a normal healthy state.

Arachnoiditis

Arachnoiditis is a challenge to pain medicine, diagnostically and therapeutically. Frequently the term arachnoiditis, or arachnopathy, is used for spinal adhesions and also synonymously for epidural malformations. But the term is actually reserved for

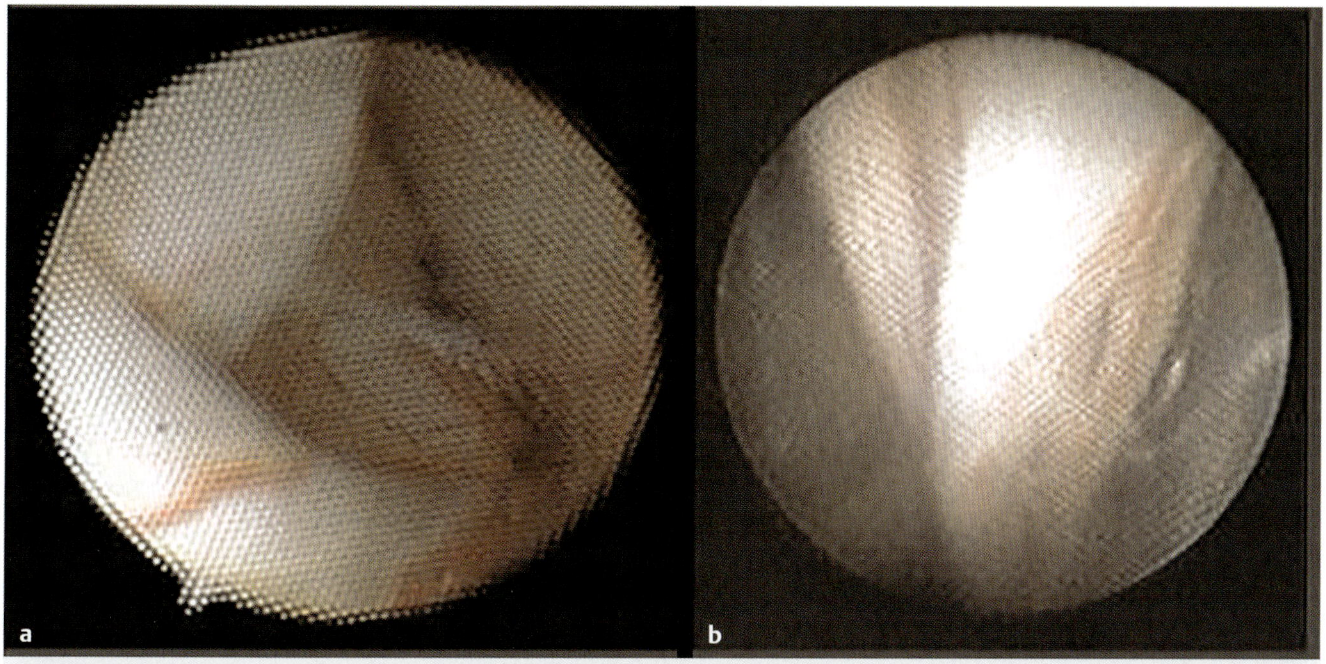

Fig. 12.24 (a) Nerve root with (b) presentation of the vulnerable preganglionic root section.

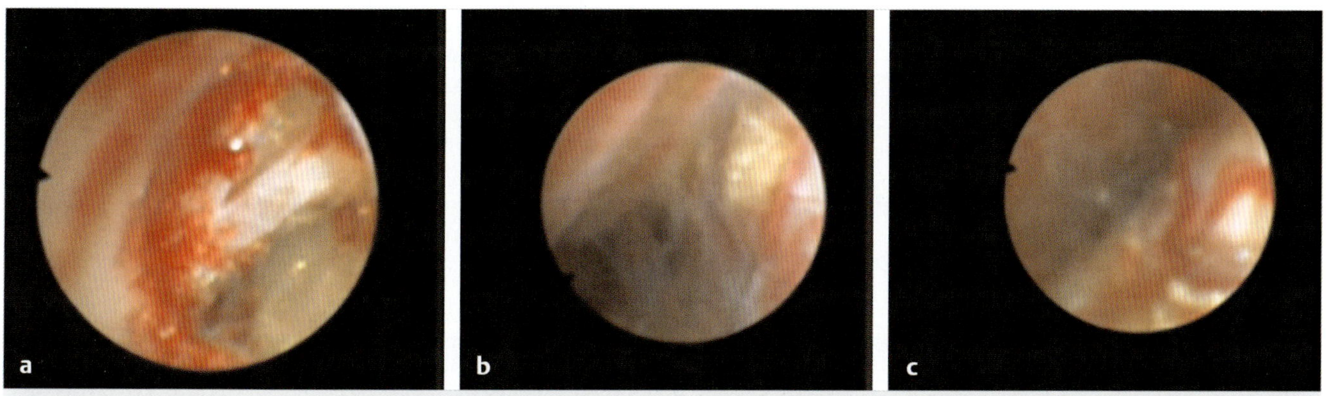

Fig. 12.25 (a-c) Chronic inflammatory processes in the region of the nerve root (radiculitis).

processes that act subdurally on the cauda equina fibers and nerve root sheaths.

The diagnosis of arachnoiditis can be difficult. Computed tomography (CT) or magnetic resonance imaging (MRI) and an electromyelogram (EMG) are helpful for making the diagnosis. A CT with myelogram can eventually identify thickened and hardened nerve roots (and soft tissue, by comparison), which can constrict the spinal canal.

Bacterial and viral spinal infections are possible causes of arachnoiditis (in 7% of cases). Repeated subarachnoidal injections of cytostatic and antimetabolite drugs are also causes for arachnoiditis.[17]

Adhesive lumbar arachnoiditis is a disease with a devastating course.[36]

Special pathogenesis

The inflammatory process of the arachnoid can be triggered mechanically (localized) or chemically (diffuse) and lead to the formation of neuraxial scar tissue with corresponding clinical symptoms. According to Day, arachnoiditis is a complex neuropathic pain event with a multilayered etiology.[37]

In 5 out of 10 cases, an adhesion of the nerve root to the dura mater causes arachnoiditis. Clinically, marked neuropathic pain and neurologic events are the most impressive signs of this disease. Arachnoiditis is often associated with epidural fibroses, but not the other way around.[17]

In the early phase of arachnoiditis (arachnoid adhesions), visible parts of the nerve root present themselves as strongly edematous. Signs of inflammation—hyperemia and swelling of the cauda fibers and nerve roots (radiculitis)—occur. At a later stage, arachnoiditis is characterized by a proliferation of fibroblasts with collagen deposits in the tissue. In the proliferative stage (adhesive arachnoiditis) that follows, the involved nerve roots present as clumped and asymmetric. The swelling of the tissues recedes. The nerve roots are stuck to each other and to the pia mater, and the thick collagen structure has developed into a scar.[17,20,38] According to Warnke, the CSF flow from the distal dural sac is obstructed cranially, toward the brain; thereby the intrathecal pressure rises.[36] Subsequently the scar tissue calcifies in some cases (▶ Fig. 12.26).[17]Adhesive arachnoiditis is the most severe form of arachnoiditis; it is recognized by severely impairing, therapy-resistant back and limb pain and a series of other neurologic problems.

Topographic aspects

Arachnoiditis occurs in connection with spinal column operations in 60% of cases, after neuraxial anesthesia (spinal anesthesia, epidural anesthesia) in 22% of cases, after lumbar puncture in 7% of cases, and after myelography in 3% of cases, as well as following pain treatments and secondary infections.[39]

According to the literature, 87% of arachnoiditis is in the lumbar region, 23% is in the thoracic region, and 34% is in cervical region of the spinal column (▶ Fig. 12.27).[17,40,41,42]

After localization, spinal adhesive arachnoiditis can be found to be distributed at the level of one vertebral body (localized), two or more vertebral bodies(segmental), or two or more contiguous vertebral bodies (continuous). If more than one spinal region (e.g., both the lumbar and thoracic regions) are affected by this chronic inflammatory process, it is called a diffuse distribution of arachnoiditis.[17]

Endoscopic identification

Epiduroscopically in the first phase of development of arachnoiditis, as in perineural nerve sheath fibrosis, the spinal dura mater appears thickened and the surrounding tissue shows increased blood flow.

In the second phase of development of arachnoiditis, there are epiduroscopic indications of scar tissue. Nervous structures are unambiguously grown together with the dura mater.

In the third phase of development, epiduroscopically arachnoiditis demonstrates complete encapsulation of the nerve roots. The compression causes visible atrophic structural changes in the nervous tissue.

Adhesive arachnoiditis, as the most severe form of this disease, can lead to obliteration of the tissue, which is also confirmed by an observed loss of dura mater pulsation as a specific diagnostic sign. In this phase, the spinal dura mater in the affected region can no longer to be distinguished endoscopically from other epidural scar tissue.

In the final phase of adhesive arachnoiditis, the neuraxial inflammatorily altered structures develop into granulation tissue that fills up the spinal canal. This tissue pushes through the very narrow arachnoidal space and can even obliterate it. A constriction-like encasement of the atrophically and ischemically altered nerve roots also occurs.

It is important to note that the performance of an epiduroscopy in segments of marked adhesive arachnoiditis can be technically extraordinarily demanding, and if not done significant limits must be set on the epiduroscopy.

Granulomas

Inflammations in which focal or tuberous collections of inflammatory cells arise are designated as granulomatous inflammations or granulomas. Immediately on the dura mater one observes, according to Berthold,[43] cell- and vessel-poor, collagen-rich skinlike overlays that pass into loose, vessel- and fluid-rich granulation tissue and finally, on the periphery, into a pronounced reticular meshwork and so approach the permanent tissue typical of the epidural space. Granulations in the epidural space were also described in the literature as a cause of paraplegic syndromes.

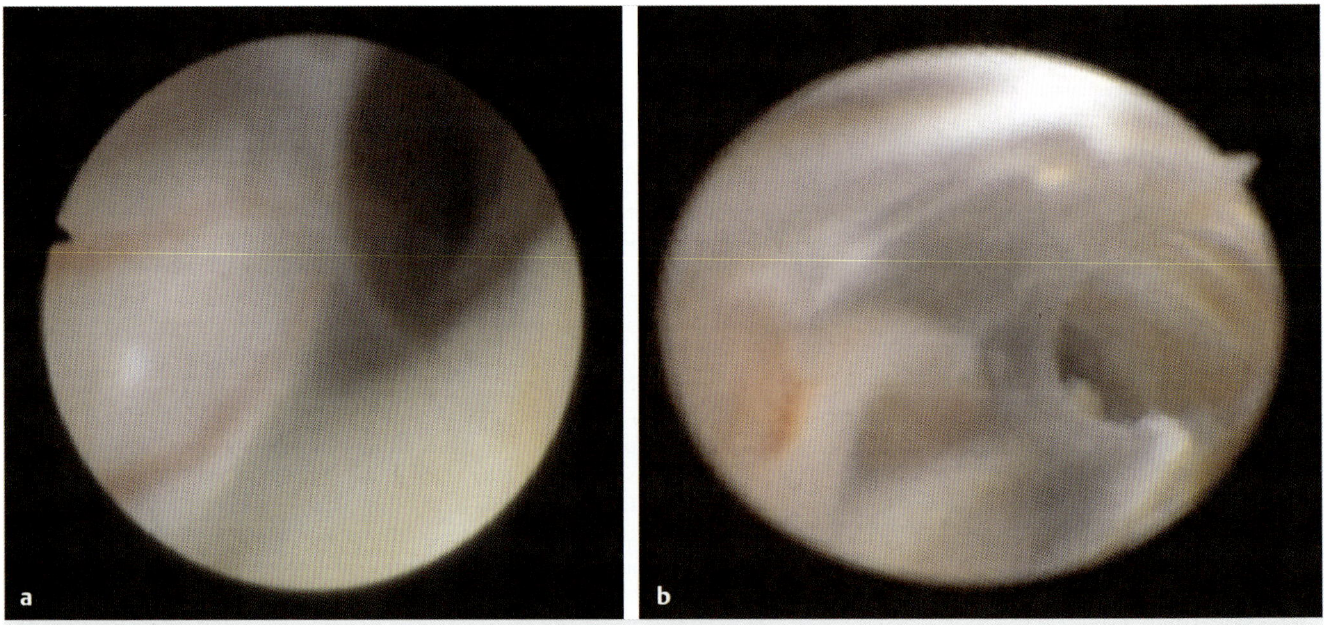

Fig. 12.26 Epiduroscopic view of the spinal arachnoid. **(a)** View into the subarachnoidal space (spinal cord → spinal arachnoid → spinal dura mater). **(b)** View of the filum of the spinal dura mater through the spinal arachnoid (subarachnoidal space).

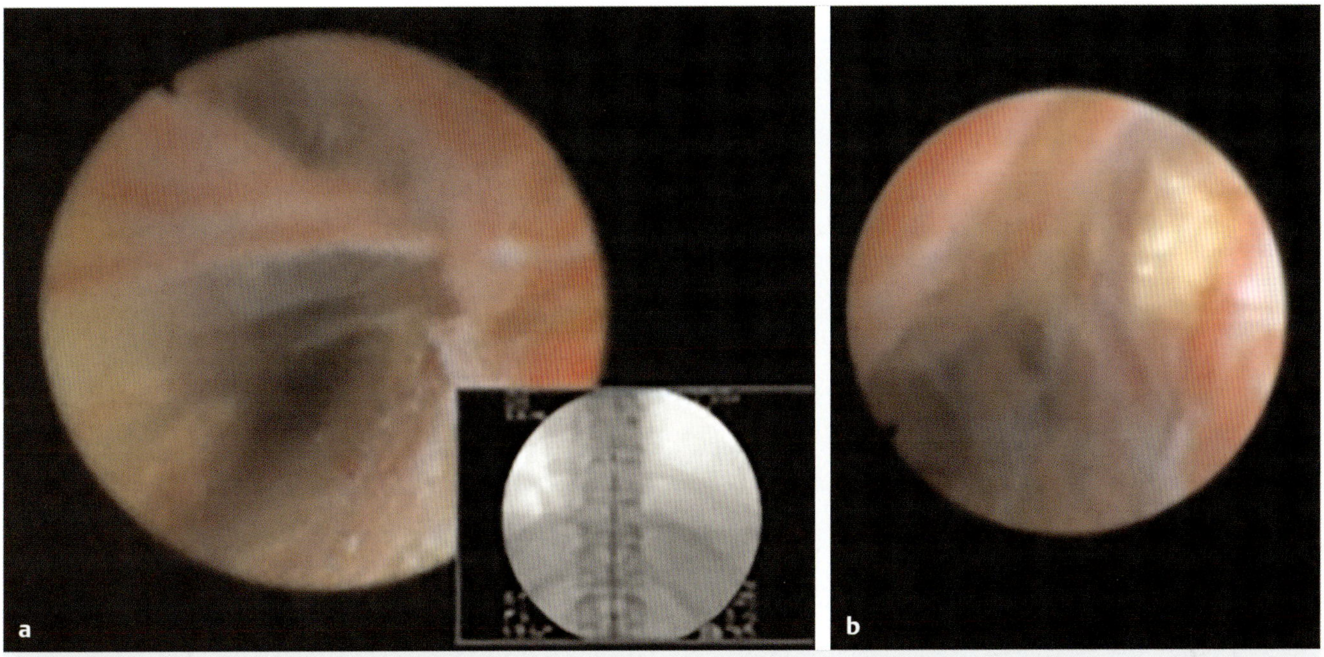

Fig. 12.27 (a,b) Ventral thoracic chronic inflammatory processes (arachnoiditis).

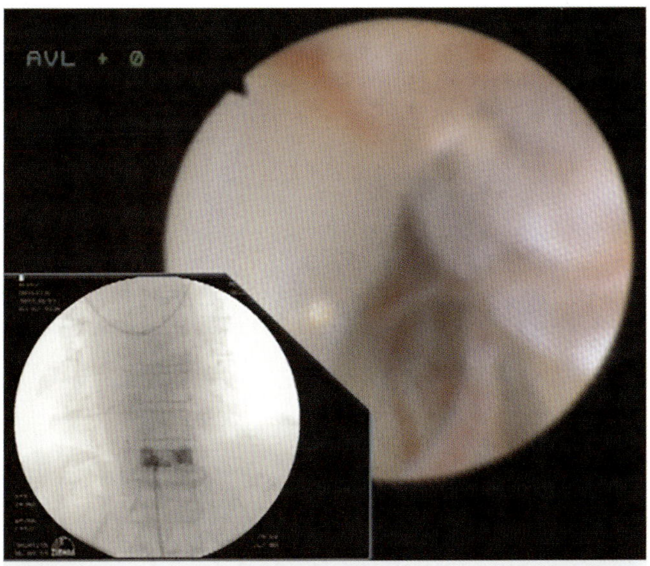

Fig. 12.28 Thoracic epidural fibrosis with connective tissue inclusion of blood vessels after a vertebroplasty with polymethylmetharcrylate (PMMA).

Foreign body granuloma

Foreign body reactions or foreign body granulomas, endoscopically visible wound-healing disturbances, are overshoot formations of granulation tissue. A foreign body granuloma is a typical representative of a histiocytic granuloma in that it's formation is triggered in tissue by reactions to substances native to the body or foreign to the body. Fragments native to the body or foreign to the body that migrate into the epidural cavity can initiate granulomatous reactions. In pain medicine, foreign body reactions (e.g., after electrode and catheter implantation) are a common complication of long-term therapy.

For example, the release of cement (PMMA) into the epidural cavity is deemed a complication of vertebroplasty. Endoscopically, even fibrosis of the PMMA in the epidural space has been observed in some cases (▶ Fig. 12.28).

According to Deer et al, granuloma formations that stay in contact with the dura mater present one of the most difficult side effects of intrathecal therapy (▶ Fig. 12.29).[21] Intraspinal catheter-associated granuloma formation represents a rarity in our patient population.

12.4 Pathologic Structural Changes of Spinal Tissues

12.4.1 Circulatory and Vascular System Diseases

Regional disturbances of the neuraxial circulatory system that can be equated with local stenosis or occlusion of vascular processes or hemorrhages are of special significance.

Vascular diseases, such as those of the arteries, the veins, and the lymphatic vessels, should to be delimited from diseases of the circulation.

Regional Neuraxial Circulatory Disturbances

At the center of the regional disturbances of neuraxial circulation stand the absolute and relative ischemias as well as arterially and venously caused blood circulation disturbances. Arterial thrombi or arterial thromboemboli are the most frequent causes of acute arterial vessel blockages.[29,44] Fatefully, the meeting of a vessel-wall weakness with an acute load can occur

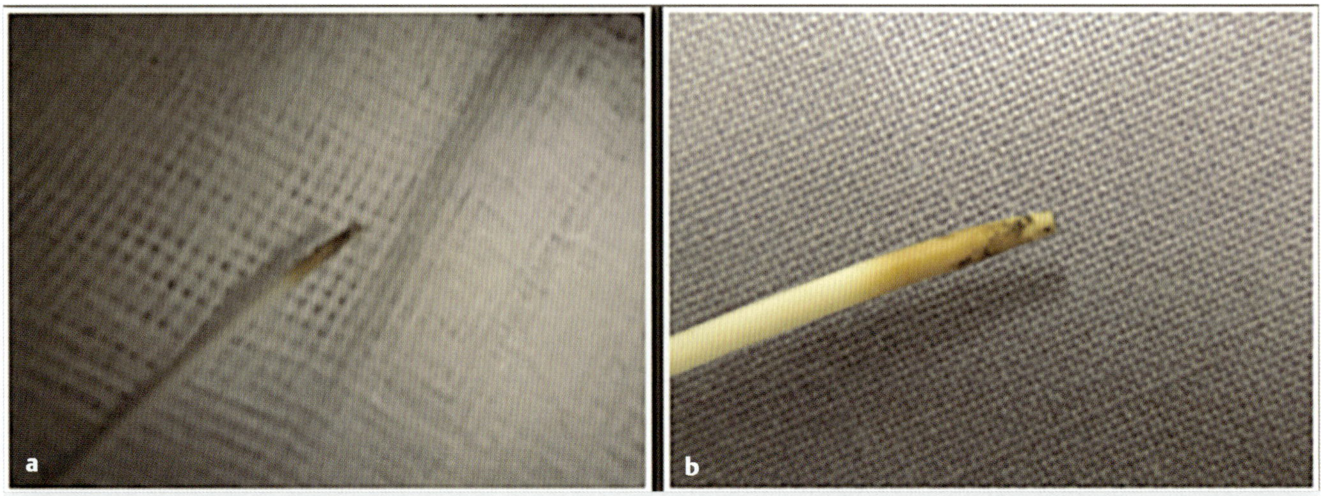

Fig. 12.29 (**a**) Closure of the tip of a spinal catheter through intrathecal granuloma formation at T10 after continuous morphine application for 11 years. (**b**) Intrathecal granuloma after 12-year intrathecal residence of a spinal catheter at T9.

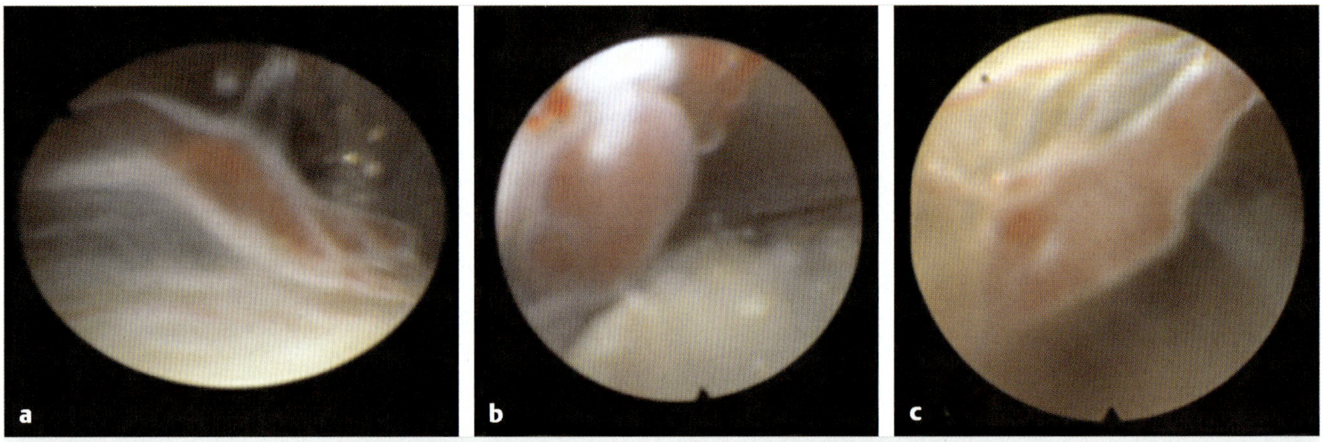

Fig. 12.30 (**a–c**) Pathologic structural alterations of epidural blood vessels.

in spinal endoscopy, catheter or electrode placement, epidural infusion, or microsurgical instrument use, and a previously damaged vessel can rupture particularly easily.

Hemorrhage

Clinical indications of a hemorrhage, for example, as a complication of neuraxial regional anesthesia in the spinal canal, requires an immediate neuro(radio)logic diagnosis. Prompt operative hematoma clearance within the first 8 hours after the occurrence of symptoms leads, as a rule, to a restoration of original conditions.

The danger of an intraoperative hemorrhage in the neuraxial region plays an outstanding role in spinal endoscopy.

When the continuity of a vessel is broken, blood with cellular and plasma components escapes into the immediate epidural, subdural, or intrathecal environment.

Along with bleeding through a tear (rhexis bleeding), bleeding through an intact epithelium (diapedesis bleeding) can occur. In individual cases this form of bleeding in dural segments can be visibly diagnosed endoscopically (▶ Fig. 12.30).

Diseases of Blood Vessels

Diseases of the arteries

In the arterial system, vasculitides often manifest as a primary inflammatory vascular disease. The most common arterial vascular disease is arteriosclerosis, which can also be a cause of arterial aneurysms. An epidural mass due to intervertebral disk degeneration, by an obstruction of the arterial blood flow and the venous plexus can also cause tissue anoxia and stimulate fibrous tissue proliferation.[45] As well, a spinal hematoma, with compression of adjacent neuronal structures as a consequence of a neuraxial puncture,can also occur spontaneously after trauma. Also Hashizume et al observed along with ossification in the region of the posterior longitudinal ligament, adventitious fibrous thickenings of veins and hyaline degeneration of artery walls.[46]

Diseases of the veins

Somewhat as in arteriosclerosis, pressure and flow loads on vessel walls are triggers for a fibrotic or sclerotic conversion of the vein wall and development to phlebectasis or varicoses.

In our epiduroscopic investigations with 120 selected patients with FBSS in 2006, ventral scars with significant indications for congested epidural veins were diagnosed in the epidural segments from S1 to C4 in 12% of cases. In the FBSS patients, veins and venous plexuses presented that were strangulated or changed to (pseudo)varicose by fibroses or scar formations (▶ Fig. 12.31).

Endoscopic identification

Arterial as well as valveless venous vessels pervade the epidural space, and are easily distinguishable by their visible pulsations.

Epidural veins can be subjected to compression, as a consequence of which they appear endoscopically congested and dilated. Congested epidural veins are frequently met with endoscopically, especially in space-demanding, mostly fibrotic segments of the spinal canal. The blood supply thereby diminished leads to reduction of nerve volumes and visible developing perineural fibrosis.

Endoscopically visible nodular and saccular outpouchings and tortuosity of diseased veins due to reduced vessel-wall stability are identified as phlebectasia and varicosities, respectively. Different variants of superficial epidural bleeds that are diapedetically caused—petechiae (pinpoint bleeds), purpura (generalized pinpoint bleeds), ecchymoses (small-area bleeds), and sugillations (large-area bleeds)—can occasionally be observed even during spinal endoscopy, especially in the region near the dura mater.

Hematoma

An epidural hematomas is a rare and extremely dangerous pathologic situation. Hematomas similar to hemorrhages develop as a consequence of damage due to neuraxial punctures or spontaneously after traumas. This space-demanding hemorrhage can lead to a direct myelocompression and reduction of the blood flow to the spinal cord and, further in its development, to irreversible damage at the spinal cord levels.

To avoid clinically relevant and serious neuraxial pathologic anatomical damage during spinal endoscopy, smaller venous and arterial bleeds (e.g., from the taking of tissue samples, removal of pain-relevant fibrotic material, or implantation of catheters) should be treated very promptly by a laser coagulation.

Diseases of Lymphatic Vessels

There are diverse theories about the anatomy and organization of the lymphatic vessels and the course of spinal drainage of CSF. Like blood vessels, the neuraxial lymphatic vessels can suffer disease. Clinically relevant is lymphangiectasia, an unphysiologic expansion of the lymphatic vessels. So far, diseases of the neuraxial lymphatic vessels have played rather a subordinate role in spinal endoscopy and/or epiduroscopy.

12.4.2 Spinal Dura Mater

The spinal dura mater plays an absolutely prominent role in spinal endoscopy. The spinal dura mater serves as endoscopic guardrail and important topographic point of orientation (▶ Fig. 12.32).

For topographically optimized classification and endoscopic assessment of pathologic findings, support from integrated endoscopic-sonographic (ENDO-SONO) systems can be particularly valuable. The supplementary epidural ultrasound investigations reveal, among other things, the pathologic wall strengthening of the spinal dura mater. They also provide further revelations concerning the distribution of the CSF and the topography of epidural pathologic structures in the spinal dura mater near the endoscopic investigation site. Via this integrated system, the possibility exists to generate further morphological information or to do a virtual histologic evaluation.[6,47]

In epidural segments with space-occupying, or stenosing, fibroses, regions of the spinal dura mater are often integrated into the fibrosis. Lesions of the spinal dura mater can result from operative interventions on the spinal column and vary greatly in dependence on the operative technique.[48] Persistent

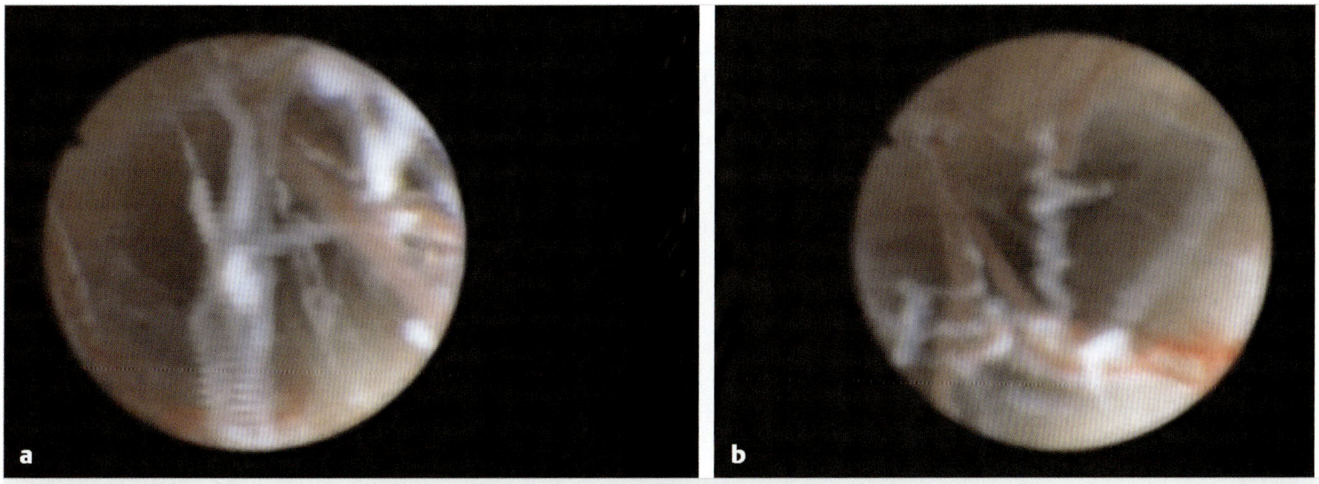

Fig. 12.31 (a,b) Beginning inclusions of epidural adhesions in vascular structures in the epidural space.

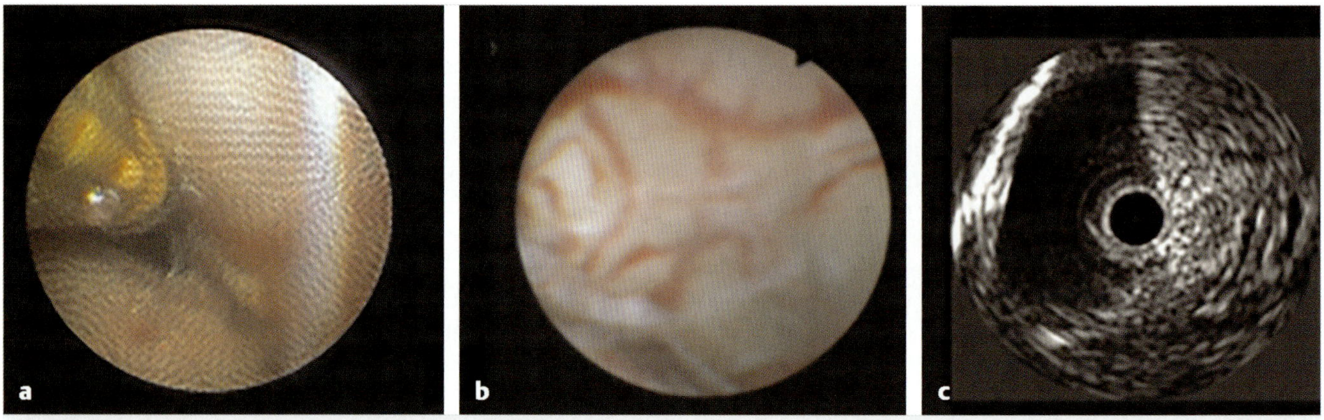

Fig. 12.32 (a) Epiduroscopic placement of an ultrasound catheter for navigation in the region of the spinal dura mater. (b) Endoscopic view of the spinal dura mater. (c) Sonographic view of the epidural cavity and epidural space showing a very strongly marked spinal dura mater and fatty tissue as well as adhesive regions.

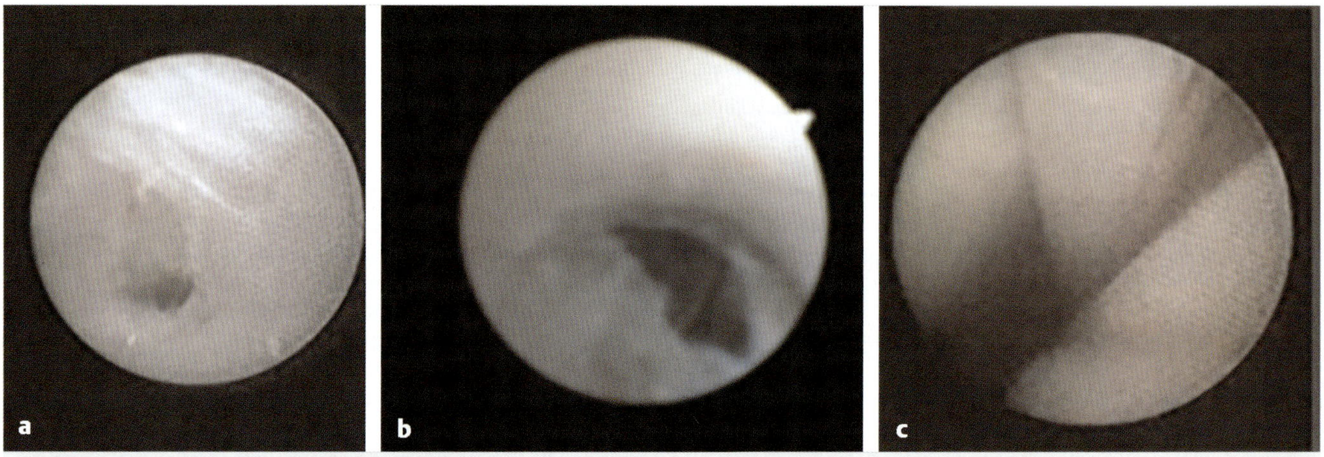

Fig. 12.33 Accidental perforations of the spinal dura mater. (a,b) After an epidural single-shot injection. (c) View of the intrathecal filum terminale through the perforation site.

leakages of CSF can lead to meningeal pseudocysts, to nerve root herniations and the resulting neurologic damage, and to meningitis and arachnoiditis.[49]

Perforation of the Spinal Dura Mater

After regional anesthesia procedures near the spinal cord, injuries to the spinal dura mater due to a puncture needle, a catheter, an endoscope, microsurgical instruments, or laser use can occur, with subsequent consequences such as inflammation, wound healing, or CSF fistulae.

Also with epidural implantation of SCS electrodes for neurostimulation, there is a risk of a dura mater perforation. Accidental dura mater injury is a relatively frequent complication of the placement of an SCS electrode.[28,50]

Endoscopically a dura mater perforation appears as a relatively large tissue defect, depending on the cause (▶ Fig. 12.33).

Care of a spinal dura mater lesion depends on the magnitude of the injury. No studies on the effectiveness of the different treatment methods have been done.[39]

Astonishingly, we have not observed headaches with accidental epiduroscopic dura perforations or thecaloscopies despite relatively larger dura perforations of > 2.5 mm, because they, compared to those perforations due to spinal anesthesia, are also met with the finest needles (29 gauge). On the basis of the fluid in the intrathecal space, the actual endoscopic image is first significantly improved. According to our own very good experiences, with a dura perforation we recommend you apply an epidural blood patch under endoscopic view to seal up the dural defect. The endoscopically applied blood patch is a very reliable treatment (▶ Fig. 12.34).

Under spinal endoscopy, a spontaneous adaptation or bonding of the dural tissue transection has also been observed occasionally with smaller dural injuries.

12.4.3 Arachnoid

Three layers of protective connective tissue surround the spinal cord: the dura mater, the arachnoid, and the pia mater. Between the dura mater and the pia mater lies the intermediate cell layer that consist of arachnoid cells (▶ Fig. 12.26).

Arachnoiditis is a chronic inflammation of the innermost leaf of the spinal dura mater; it is a pathologic reaction of the spinal

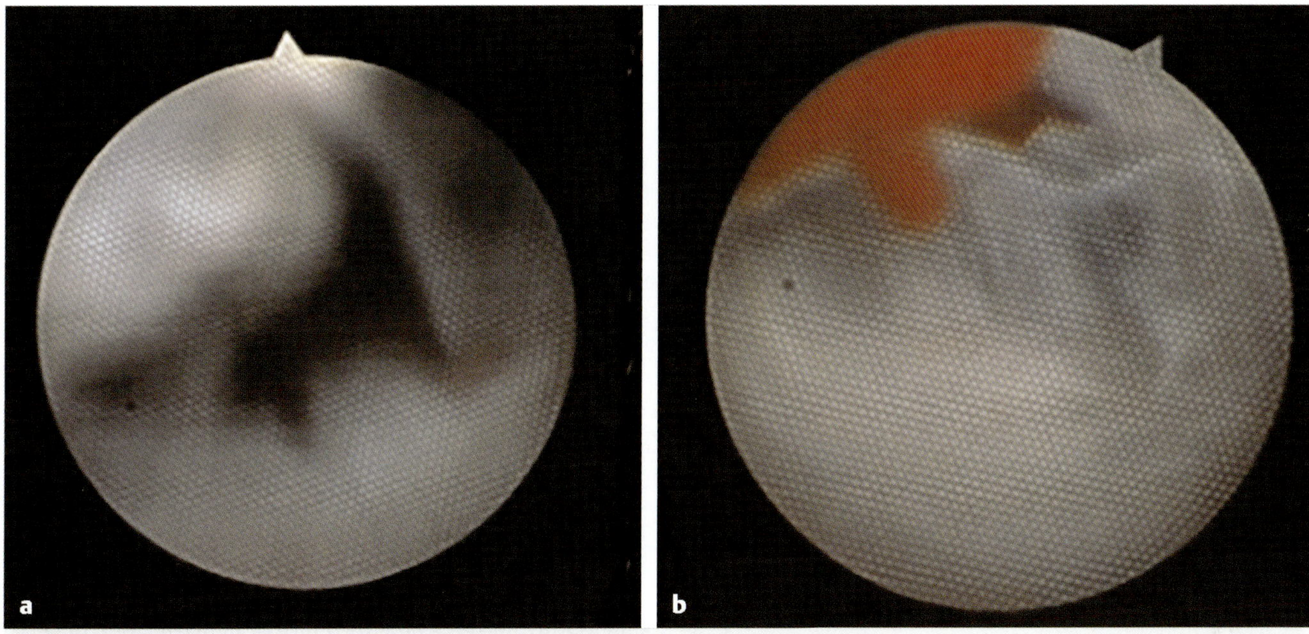

Fig. 12.34 (a) Perforation of the spinal dura mater. (b) Endoscopically applied blood patch.

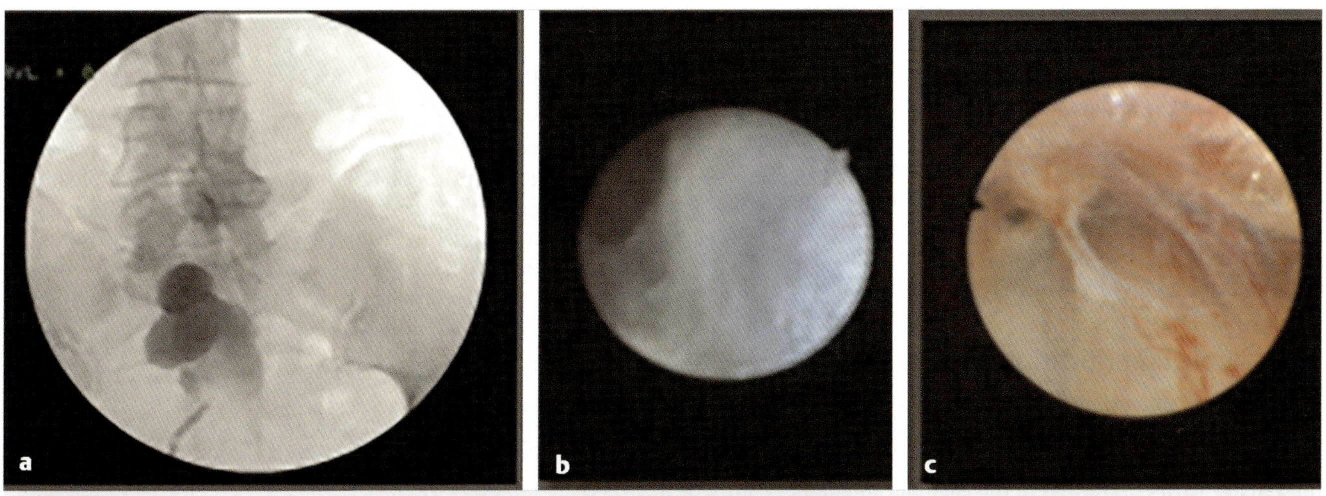

Fig. 12.35 (a) Epiduroscopic image of a conspicuously widened epidural space suspicious for a large cystic change (4 × 2 cm arachnoid cyst) at levels L5–S3. (b) Endoscopic investigation findings show nearly transparent dura-like cyst wall. (c) Above the epidural arachnoid cyst, there are also space-occupying epidural adhesions with significant color change and with indications of a chronic inflammatory process (arachnoiditis) in the region of the L5 cyst.

dura mater against tissue damage with corresponding clinical symptoms (Chapter 2.3.1).

In this context it is significant that, for example, a pull on the arachnoid (intermediate layer) by an epiduroscopic procedure or intervention always also means a pull on the vessels of the spinal cord with all the attendant consequences.

Arachnoid Cysts

Frequently arachnoid cysts are a chance finding in CT or MRI investigations. The authors Weber and Knopf diagnosed at least one arachnoid cyst in 1.7% of 2,536 healthy male candidates for the German air force.[51] Compression of adjacent structures by these cystic intraspinal processes can lead to local pain as well as radicularly radiating pain (▶ Fig. 12.35).

To treat pain due to an arachnoid cyst, it is important to close off the point of communication of the arachnoid cyst with the subarachnoid space, but it is not absolutely necessary to remove the arachnoid cyst totally.

The authors Kyeong-Sik Ryu et al reported on the accidental mechanical formation of a intradural cyst from the advance of the endoscope into the epidural space during epiduroscopy.[52]

Cerebrospinal Fistulas

With the implantation of spinal catheter systems for pharmacologic neuromodulation, CSF fistulas, among other things, can

significantly in the epidural space and does not correlate with the general fat distribution in the rest of the body. Also how it presents endoscopically does not correspond with the common anatomical illustrations.

Epidural fatty tissue forms an ideal pressure pad for the neuraxial structures and an excellent depot for medications that are applied in the epidural space.

In the epidural implantation of electrodes and catheters, the (often scant, however) epidural fat protects against perforations of the spinal dura mater.

Topographic Aspects

The epidural fatty tissue is distributed in a highly variable manner in the epidural cavity per our endoscopic investigations. The principal share of the epidural fatty tissue is located in the dorsal part of the epidural space preferentially in the dorsal middle thoracic and deeper lumbar segments of the spinal column. The ventral epidural fatty tissue distribution is rather local and very sparse based on our observations.

Epidural Lipomatosis

Epidural lipomatosis is recognized by an elevation of the epidural fat content. Epidural lipomatosis can occur idiopathically, but also in patients undergoing long-term steroid therapy or through endogenous steroid hypersecretion.[66]

Kawai reported that after steroid therapy a pathologic increase of the extradural fatty tissue was triggered.[67] In one patient, an idiopathic epidural lipomatosis led to a nerve compression with radicular pain symptoms. Roy-Camille et al reported in 1991 on cases of a symptomatic epidural lipomatosis as a consequence of long-term therapy with epidural steroid injections.[68] The excessive fat deposits around the dural sac led to nerve root compression with the corresponding neurologic symptoms (▶ Fig. 12.41).

Fat Deficit

Igarashi et al showed that the epidural fatty tissue is important for the nerves as a movable layer and is depleted after chronic inflammatory processes.[69] When depleted, epidural nervous structures are drawn into developing scar tissue.

Especially in the ventral section of the epidural space as well as in postoperative situations, only sparse amounts of epidural fatty tissue are found, according to our endoscopic investigations. Epidural fatty tissue was nearly depleted in regions of spinal masses. In patients with an FBSS, in our epiduroscopic investigations, fatty tissue was occasionally significantly reduced or absent in otherwise normal sections of the epidural space.

Endoscopic Identification

The normal lipophilic epidural fatty tissue presents as a fresh, glossy white to yellow–gold in the endoscopic image. The epidural fatty tissue is often threaded with small blood vessels.

In rare lipomatosis, a significant thickening of the dorsal epidural fat layer of thoracic or lower lumbar vertebral column segments is to be observed.

In isolated chronic epidural inflammatory processes and FBSS, epidural fatty tissue frequently appears depleted. Depleted epidural fatty tissue has lost its glossy freshness and appears often rather cloudy and very pallid yellow–white.

12.4.7 Deposits of Surplus Substances

Local Anesthetic Deposits

From the literature, epidural deposits of local anesthetic (e.g., bupivacaine) as well as deposits of X-ray contrast medium (e.g., iodophenylundecylic acid) are known after accidental epidural leakage of contrast medium for myelography.[70]

Korsten et al still found deposits of 10% *n*-butyl-*p*-aminobenzoate (BAB) in epidural segments 36 days after epidural application.[71]

According to Balga et al, white bupivacaine crystals deposited in the thoracic epidural space after long-term epidural bupivacaine infusion were found at patient autopsy.[72] By infrared spectrometry, the epidural deposits also confirmed to consist of bupivacaine. The histologic investigation showed nonreactive fatty tissue necrosis within the bupivacaine crystals. The authors assumed from this that the bupivacaine crystals arose through precipitation.[72]

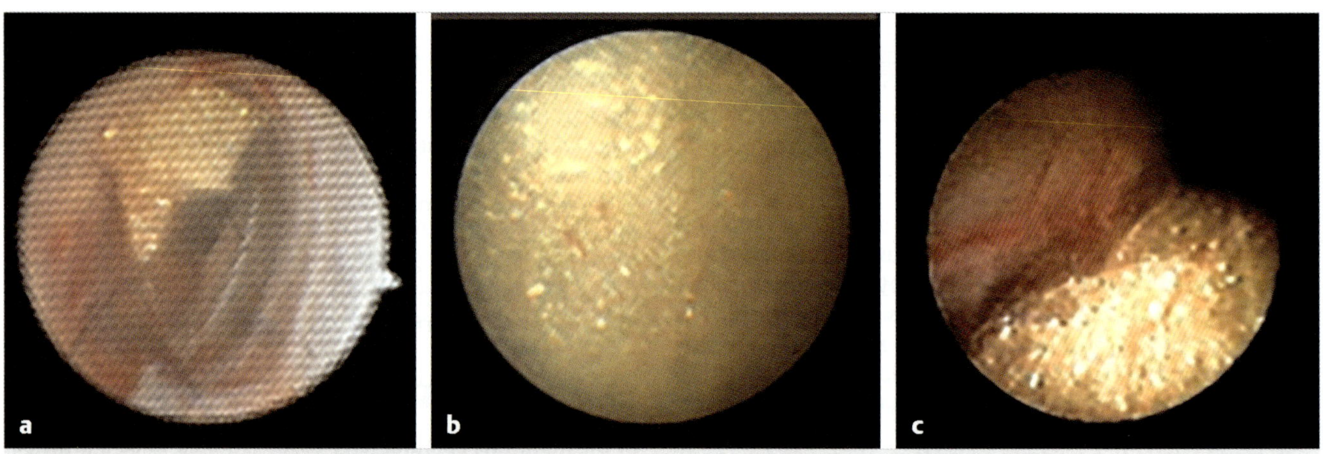

Fig. 12.41 Fatty tissue in the epidural cavity. **(a)** Scattered epidural fatty tissue. **(b)** Profuse epidural fatty tissue. **(c)** Spinal dura mater with normally structured fatty tissue.

Pigment Deposits

Heavner et al reported the presence of yellow pigments (xanthosis) in the epidural space of a patient with a chronic painful radiculopathy.[73] The epiduroscopic findings described by the group of authors in the region of the L5 and S1 nerve roots, associated with the radiculopathy, included diffuse perivascular yellow pigments, an elevated vascularity, and an increase of fibrous tissue.

Pigments in granular or diffuse distribution occur in numerous organs. It should be taken into account that pigments are not necessarily to be classified as pathologic. Pigments are derived frequently from proteins or various decay products of hemoglobin.

We have likewise found similar endoscopic pigment findings in our own patient population as accidental findings, without pain relevance.

12.4.8 Complex Alterations of Epidural Structures

Of outstanding clinical significance in daily pain therapy practice are the effects of complex structural alterations in the region of the spinal dura mater, ligamentum flavum, posterior longitudinal ligament, blood and lymph vessels, nerves, and fatty tissue.

Adhesions, fibroses, and chronic inflammatory processes are the dominant complex pathologic alterations in the epidural space in patients with neuraxial pain syndromes.

Failed Back Surgery Syndrome

The complex pathologic epidural structure alterations are especially to be found endoscopically in so-called failed back surgery syndrome (FBSS).

According to Krämer, all persistent strong discomfort after an intervertebral disk operation that are provoked by segmental instability and growths in the vertebral canal are designated as FBSS, or postdiscotomy syndrome (PDS).[15] Postoperative bleeds, lesions of the spinal dura mater, soft tissue trauma, arachnoiditis, or defects in the fibrinolytic system are discussed as causes. An incidence of 20% is specified for FBSS.

Postoperative scar formation along with segmental instability are the most important factors that are responsible for the development of clinical FBSS symptoms (▶ Fig. 12.42).[7,15,26,74]

In accordance with the few indications from the literature, in our endoscopic investigations of patients with FBSS (C3–S1), along with pronounced pain-relevant epidural scar formation, significantly reduced or absent epidural fatty tissue, unequivocal reduction of epidural vascularization, and significant signs of chronic inflammatory processes are found in the affected epidural segment (C3–S1).

The following endoscopic characteristics are to be ranked as complex epidural pathologic structure alterations found in FBSS:

- Marked stenosis due to fibroses.
- Strongly reduced proportions of epidural fatty tissue.
- Diminution of vascularization; vessel congestion on occasion.
- Chronic inflammatory processes.

Tethered Cord Syndrome

Tethered cord syndrome is a clinically complex disease. Tethered cord is an intraspinal pathology, a taut, thickened and even shortened filum terminale or an intraspinal lipoma that fixes the conus of the spinal cord.[42,75]

The common mechanism by which these intraspinal lesions lead to the clinical symptoms are constriction of the longitudinal freedom of motion of the spinal cord, particularly the conus medullaris. On the basis of the biophysical properties of the spinal cord, a local spinal cord ischemia is the consequence.

As a result of postoperative scar formation, primarily in the region of the filum terminale, a fibrous strand formation is found that has grown together with the spinal dura mater. Because of this pathologic connective tissue process, the position of the conus medullaris shifts to the so-called low conus.[40,75]

A clinical consequence is often a complex of clinical symptoms that comprises severely painful lumbago; neurologic disturbances and failures such as distally accentuated pareses, trophic and sensitivity disturbances of the legs and buttocks; disturbances of bladder function; and orthopedic alterations such as vertebral column malpositionings and foot deformities.[41]

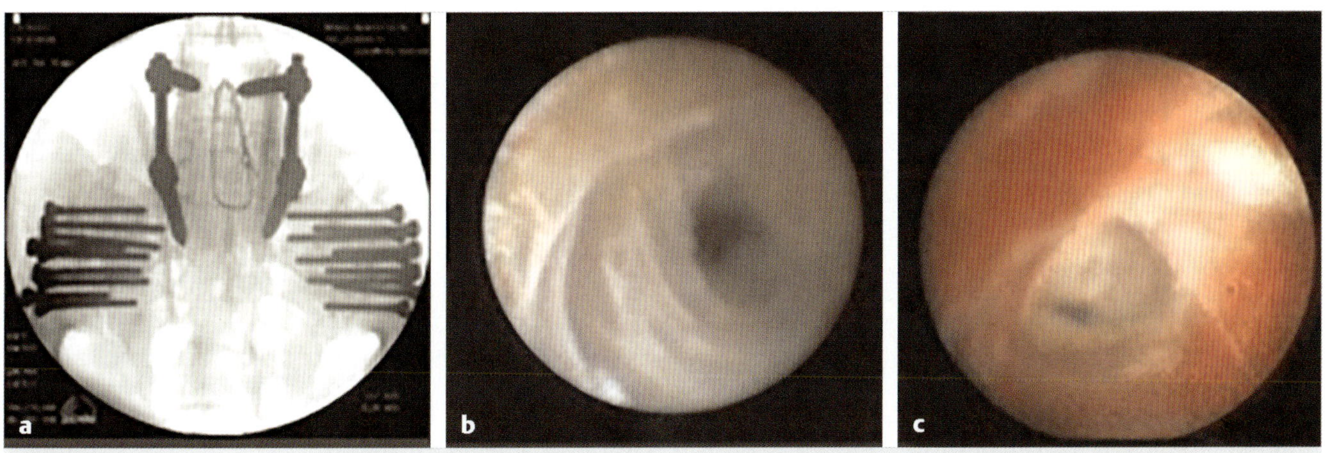

Fig. 12.42 Complex pathologic epidural structural alterations in failed back surgery syndrome. (a) After operative treatment. (b) Fibrosis and stenosis. (c) Chronic inflammatory process.

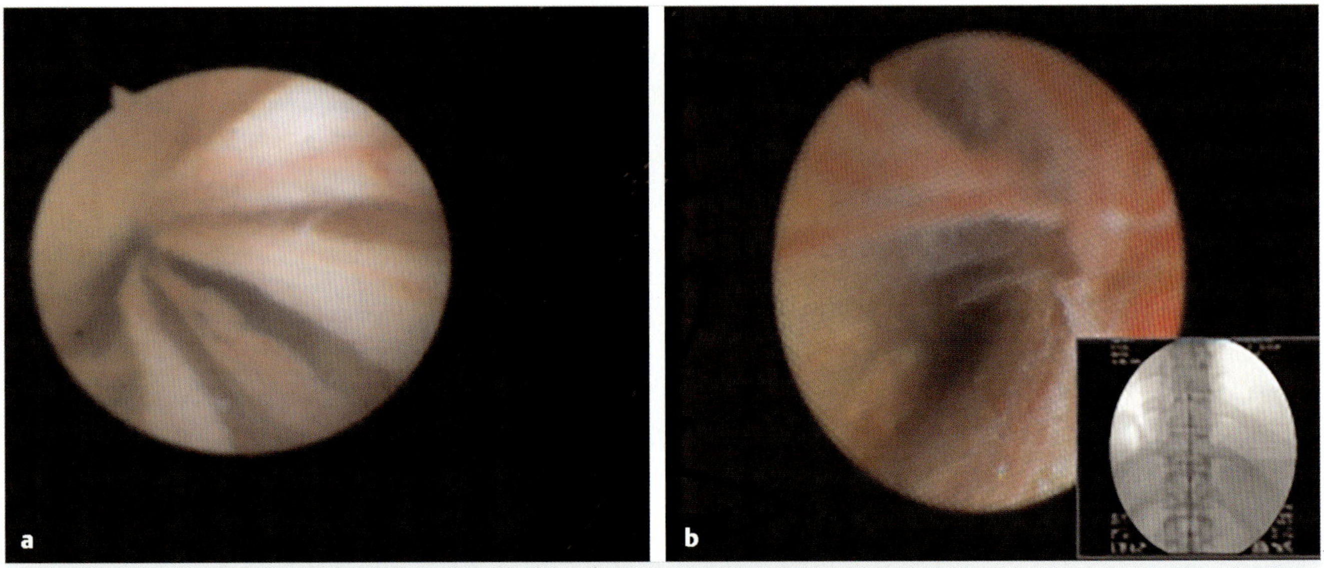

Fig. 12.43 Tethered cord syndrome. (a) Region of the cauda equina with isolated pathologically altered nervous structures after operative corrective measures in the conus medullaris. (b) Thoracic adhesive arachnoiditis with tethered cord in the L5 segment.

The method of demonstration of tethered cord syndrome is by MRI, but also myelogram-CT for special questions.[40]

With spinal endoscopy–thecaloscopy the pathologic anatomical facts of the tethered cord syndrome can be further tested and presented endoscopically. A limitation of the longitudinal range of motion of the spinal cord, particularly the conus medullaris, is recognizable, and can have as a consequence a local ischemia in the spinal cord as mentioned earlier (▶ Fig. 12.43).[40,75] Of course, endoscopic diagnosis of the filum terminale with its pathologic dural alterations is connected with an elevated risk of infection and dura leakage. As a rule, the spinal endoscopy is unproblematic.

Treatment for tethered cord syndrome consists, as a rule, of an operative intervention with performance of a myelolysis and patient-specific pain therapy.

Endoscopic identification

The intraspinal pathology in tethered cord syndrome is identified by an endoscopically visible tautened, thickened, and shortened filum terminale or an intraspinal lipoma that fixes the conus of the spinal cord.

References

[1] Schütze G, Kurtze H. Percutaneous investigation of the epidural space using a flexible endoscope: A contribution to epiduroscopy. Reg Anesth Pain Med. 1993; 18:24

[2] Schütze G. Interventionelles Schmerzmanagement—Bildgestützte Verfahren zur Diagnostik und Therapie rückenmarksnaher Schmerzsyndrome-Spinale Endoskopie, pharmakologische und elektrische Neuromodulation. Bremen, Germany: UNI-MED Verlag AG; 2011

[3] Schütze G. Epiduroscopy [multilingual DVD-ROM]. Tuttlingen, Germany: Karl Storz; 2006

[4] Beltrutti D, Groen GJ, Saberski L, Sandner-Kiesling A, Schütze G, Weber G. Epiduroscopy: consensus decision March 2006. Pain Clin. 2007; 19(2):47–50

[5] Choi YK, Tan MH, Barbella JD, Grubb WR. Epiduroscopic analysis of persistent back and leg pain and the efficacy of epiduroscopy in the treatment of failed back pain syndrome. Reg Anesth Pain Med [Suppl]. 2001; 26(2):90

[6] Schütze G. Epiduroscopy—Spinal Endoscopy. New York, NY: Springer-Verlag; 2008

[7] Faustmann PM. Neuroanatomic basis for discogenic pain. [in German]. Z Orthop Ihre Grenzgeb. 2004; 142(6):706–708

[8] Greenberg M. Handbook of Neurosurgery, 6th ed. New York, NY: Thieme; 2006

[9] Jayson MIV, Keegan A, Million R, Tomlinson I. A fibrinolytic defect in chronic back pain syndromes. Lancet. 1984; 2(8413):1186–1187

[10] Olmarker K, Rydevik B, Holm S. Edema formation in spinal nerve roots induced by experimental, graded compression. An experimental study on the pig cauda equina with special reference to differences in effects between rapid and slow onset of compression. Spine. 1989; 14(6):569–573

[11] Krams M, Frahm S, Kellner U, Mawrin C. Kurzlehrbuch Pathologie. Stuttgart: Thieme; 2013

[12] Kuslich SD, Ulstrom CL, Michael CJ. The tissue origin of low back pain and sciatica: a report of pain response to tissue stimulation during operations on the lumbar spine using local anesthesia. Orthop Clin North Am. 1991; 22 (2):181–187

[13] Omarker K, Myers RR. Pathogenesis of sciatic pain: role of herniated nucleus pulposus and deformation of spinal nerve root and dorsal root ganglion. Pain. 1998; 78(2):99–105

[14] Freund M, Hutzelmann A, Steffens JC, Buhl R, Heller M. MR myelography in spinal canal stenosis. [in German]. Rofo. 1997; 167(5):474–478

[15] Krämer J. Bandscheibenbedingte Erkrankungen. Stuttgart: Thieme; 1994

[16] Roshan M. Topographisch anatomische Grundlagen zur epiduralen perineuralen Injektion an der Lendenwirbelsäule [inaugural dissertation]. Bochum, Germany: Ruhr-Universität Bochum; 2007

[17] Smith S. The Adhesive Arachnoiditis Syndrome. Circle of Friends with Arachnoiditis (COFWA); 2003 http://www.cofwa.org/AASYNDROME-10-03.pdf

[18] Richardson J, McGurgan P, Cheema S, Prasad R, Gupta S. Spinal endoscopy in chronic low back pain with radiculopathy. A prospective case series. Anaesthesia. 2001; 56(5):454–460

[19] Aldrete A. Epidural Fibrosis after Permanent Catheter Insertion and Infusion. Chipley, FL: Pain and Spine Institute; 1995

[20] Burton CV. Lumbosacral arachnoiditis. Spine. 1978; 3(1):24–30

[21] Deer T, Krames ES, Hassenbusch SJ, et al. Polyanalgesic consensus conference 2007: recommendations for the management of pain by intrathecal (intraspinal) drug delivery: report of an interdisciplinary expert panel. Neuromodulation. 2007; 10(4):300–328

[22] Hardwig S. Postoperative Adhäsions-prophylaxe mittels Perfluorcarbon [dissertation]. Berlin, Germany: Medizinischen Fakultät der Charité Universitätsmedizin Berlin; 2011

[23] Laus M, Bertoni F, Bacchini P, Alfonso C, Giunti A. Recurrent lumbar disc herniation: what recurs? (A morphological study of recurrent disc herniation). Chir Organi Mov. 1993; 78(3):147–154
[24] Koch T, Hübler M. Thorakale epidurale Anästhesie und Analgesie. Munich, Germany: Arcis Verlag GmbH; 2003
[25] Brown JH. Pressure caudal anesthesia and back manipulation. Conservative method for treatment of sciatica. Northwest Med. 1960; 59:905–909
[26] Greenberg MS. Handbook of Neurosurgery, 5th ed. New York, NY: Thieme; 2001:314
[27] Cicuendez M, Munarriz PM, Castaño-Leon AM, Paredes I. Dorsal myelopathy secondary to epidural fibrous scar tissue around a spinal cord stimulation electrode. J Neurosurg Spine. 2012; 17(6):598–601
[28] Krames E, Peckham HP, Rezai AR. Neuromodulation. New York: Elsevier; 2009
[29] Alexander GE, Provost TT, Stevens MB, Alexander EL. Sjögren syndrome: central nervous system manifestations. Neurology. 1981; 31(11):1391–1396
[30] Jerosch J, Steinleitner W. Minimal invasive Wirbelsäulen-Intervention. Cologne, Germany: Deutscher Ärzte-Verlag Köln; 2005
[31] Marshall LL, Trethewie ER, Curtain CC. Chemical radiculitis. A clinical, physiological and immunological study. Clin Orthop Relat Res. 1977(129):61–67
[32] Boudelal R. Topographie der lumbalen Nervenwurzeln in Relation zu den Bandscheiben [inaugural dissertation]. Bochum, Germany: Ruhr-Universität Bochum; 2009
[33] Reina MA, De Andrés JA, Hernández JM, Prats-Galino A, Machés F, Peláez J. Transforaminal or translaminar approach for dorsal root ganglion and dorsal nerve root. Anatomical reason for technique decision. Eur J Pain Suppl. 2010; 4 S4:287–297
[34] Obata K, Tsujino H, Yamanaka H, et al. Expression of neurotrophic factors in the dorsal root ganglion in a rat model of lumbar disc herniation. Pain. 2002; 99(1–2):121–132
[35] Kizelshteyn G, Heavner JE, Levine S. Epidural balloon catheter system lysing epidural adhesions. Reg Anesth. 1991; 15 1S:87
[36] Warnke JP, Köppert H, Bensch-Schreiter B, Dzelzitis J, Tschabitscher M. Thecaloscopy part III: first clinical application. Minim Invasive Neurosurg. 2003; 46 (2):94–99
[37] Day PL. Arachnoiditis: A brief Summary of the Literature. Christchurch: New Zealand Health Technology Assessment Clearing House; 2001
[38] Rice I, Wee MY, Thomson K. Obstetric epidurals and chronic adhesive arachnoiditis. Br J Anaesth. 2004; 92(1):109–120
[39] Saxler G, Krämer J, Barden B, Kurt A, Pförtner J, Bernsmann K. The long-term clinical sequelae of incidental durotomy in lumbar disc surgery. Spine. 2005; 30(20):2298–2302
[40] Kothbauer K, Seiler RW. Tethered spinal cord syndrome in adults. [in German]. Nervenarzt. 1997; 68(4):285–291
[41] Maliszewski M, Ładziński P, Majchrzak H. Tethered cord syndrome in adults. [in Polish]. Neurol Neurochir Pol. 2000; 34(6):1269–1279
[42] Yamada S, Lonser RR. Adult tethered cord syndrome. J Spinal Disord. 2000; 13(4):319–323
[43] Berthold H. Epidural granulations as a cause of paraplegic syndromes. [in German]. Dtsch Z Nervenheilkd. 1958; 177(3):209–221
[44] Mok CC, Lau CS, Chan EY, Wong RW. Acute transverse myelopathy in systemic lupus erythematosus: clinical presentation, treatment, and outcome. J Rheumatol. 1998; 25(3):467–473
[45] Kurunlahti M, Kerttula L, Jauhiainen J, Karppinen J, Tervonen O. Correlation of diffusion in lumbar intervertebral disks with occlusion of lumbar arteries: a study in adult volunteers. Radiology. 2001; 221(3):779–786
[46] Hashizume Y, Iijima S, Kishimoto H, Yanagi T. Pathology of spinal cord lesions caused by ossification of the posterior longitudinal ligament. Acta Neuropathol. 1984; 63(2):123–130
[47] Schütze G. Epiduroskopie. Ein praxisorientierter Leitfaden zur epiduroskopischen Diagnostik und Therapie rückenmarksnaher Schmerzsyndrome. Lengerich, Germany: Pabst Science Publishers; 2006
[48] Kurt A. Die klinischen Langzeitfolgen der intraoperativen Duraverletzung in der Wirbelsäulenchirurgie [inaugural dissertation]. Bochum, Germany: Ruhr-Universität Bochum
[49] Black P. Cerebrospinal fluid leaks following spinal surgery: use of fat grafts for prevention and repair. Technical note. J Neurosurg. 2002; 96(2) Suppl:250–252
[50] Raj PP, Lou L, Erdine S, et al. Interventional Pain Management: Image-Guided Procedures. 2d ed. New York, NY: Elsevier; 2009
[51] Weber F, Knopf H. Incidental findings in magnetic resonance imaging of the brains of healthy young men. J Neurol Sci. 2006; 240(1–2):81–84
[52] Ryu KS, Rathi NK, Kim G, Park CK. Iatrogenic intradural lumbosacral cyst following epiduroscopy. J Korean Neurosurg Soc. 2012; 52(5):491–494
[53] Putz R. The detailed functional anatomy of the ligaments of the vertebral column. Ann Anat. 1992; 174:40–47
[54] Kalb S, Martirosyan NL, Perez-Orribo L, Kalani MY, Theodore N. Analysis of demographics, risk factors, clinical presentation, and surgical treatment modalities for the ossified posterior longitudinal ligament. Neurosurg Focus. 2011; 30(3):E11
[55] Smith ZA, Buchanan CC, Raphael D, Khoo LT. Ossification of the posterior longitudinal ligament: pathogenesis, management, and current surgical approaches. A review. Neurosurg Focus. 2011; 30(3):E10
[56] Choi BW, Song KJ, Chang H. Ossification of the posterior longitudinal ligament: a review of literature. Asian Spine J. 2011; 5(4):267–276
[57] Inamasu J, Guiot BH, Sachs DC. Ossification of the posterior longitudinal ligament: an update on its biology, epidemiology, and natural history. Neurosurgery. 2006; 58(6):1027–1039, discussion 1027–1039
[58] Kawauchi Y, Yone K, Sakou T. Myeloscopic observation of adhesive arachnoiditis in patients with lumbar spinal canal stenosis. Spinal Cord. 1996; 34 (7):403–410
[59] Witteler A. Vergleich von kernspin-tomographischen Befunden beim lumbalen Bandscheibenvorfall mit dem histopathologischen Korrelat nach Resektion [inaugural dissertation]. Bochum, Germany: Ruhr-Universität Bochum; 2009
[60] Sairyo K, Biyani A, Goel VK, et al. Lumbar ligamentum flavum hypertrophy is due to accumulation of inflammation-related scar tissue. Spine. 2007; 32 (11):E340–E347
[61] Ohba T, Ebata S, Ando T, Ichikawa J, Clinton D, Haro H. Lumbar ligamentum flavum hematoma treated with endoscopy. Orthopedics. 2011; 34(7):e324–e327
[62] Saetia K, Cho D, Lee S, Kim DH, Kim SD. Ossification of the posterior longitudinal ligament: a review. Neurosurg Focus. 2011; 30(3):E1
[63] Hirai T, Korogi Y, Takahashi M, Shimomura O. Ossification of the posterior longitudinal ligament and ligamentum flavum: imaging features. Semin Musculoskelet Radiol. 2001; 5(2):83–88
[64] Yabuki S, Kikuchi S. Endoscopic surgery for cervical myelopathy due to calcification of the ligamentum flavum. J Spinal Disord Tech. 2008; 21(7):518–523
[65] Bloch J, Hawelski S, Benini A. Cyst of the ligamentum flavum of the lumbar spine: description of 6 cases. [in German]. Schweiz Med Wochenschr. 1997; 127(17):728–732
[66] Payera M, Bach-Kliegel B. Spinal epidural lipomatosis: known or misunderstood? [in German]. Schweiz Med Forum. 2007; 7:921–923
[67] Kawai M, Udaka F, Nishioka K, Houshimaru M, Koyama T, Kameyama M. A case of idiopathic spinal epidural lipomatosis presented with radicular pain caused by compression with enlarged veins surrounding nerve roots. Acta Neurol Scand. 2002; 105(4):322–325
[68] Roy-Camille R, Mazel C, Husson JL, Saillant G. Symptomatic spinal epidural lipomatosis induced by a long-term steroid treatment. Review of the literature and report of two additional cases. Spine. 1991; 16(12):1365–1371
[69] Igarashi T, Hirabayashi Y, Seo N, Saitoh K, Fukuda H, Suzuki H. Lysis of adhesions and epidural injection of steroid/local anaesthetic during epiduroscopy potentially alleviate low back and leg pain in elderly patients with lumbar spinal stenosis. Br J Anaesth. 2004; 93(2):181–187
[70] Jensen JT. Epidural placement of Pantopaque after myelography. Neuroradiology. 1973; 5(4):197–201
[71] Korsten HH, Ackerman EW, Grouls RJ, et al. Long-lasting epidural sensory blockade by n-butyl-p-aminobenzoate in the terminally ill intractable cancer pain patient. Anesthesiology. 1991; 75(6):950–960
[72] Balga I, Gerber H, Schorno XH, Aebersold Keller F, Oehen HP. Bupivacaine crystal deposits after long-term epidural infusion. Anaesthesist. 2013; 62 (7):543–548
[73] Heavner JE, Bosscher H, Dunn D, Lehman T. Xanthosis in the spinal epidural space—an epiduroscopy finding. Pain Pract. 2004; 4(1):39–41
[74] Cosnard G, Cordoliani YS, Sarrazin JL, Soulie D. Imaging studies for failed back surgery syndrome. [in French]. Sem Hop Paris. 1995; 71(25–26):782–792
[75] Khoshhal KI, Murshid WR, Elgamal EA, Salih MAM. Tethered cord syndrome: A study of 35 patients. J Taibah Univ Med Sci. 2012; 7(1):23–28

13 Diagnostic and Functional Epiduroscopic Procedures

Dae Hyun Jo and Jee Youn Moon

13.1 Introduction

Determining the cause of back pain is complicated as it is often multifactorial and also because anatomic abnormalities, which may not be relevant with clinical symptoms, are common in the spine. Epiduroscopy allows spatial and color pictures of epidural anatomic structures, including spinal dura mater, ligamentum flavum, posterior longitudinal ligament, blood vessels, connective tissue, nerve, and fatty tissue. Using epiduroscopy, it is possible to determine a source of pain by identifying the pathology of the epidural space or by making diagnostic blocks onto the target structure.[1,2] Fluoroscopy allows for exact location of the tip of the flexible scope with respect to the bony spinal canal, while direct visualization gives the relative orientation with respect to the surrounding anatomic structures in the epidural space of the lumbosacral spine.

In addition to identifying the lumbar epidural pathology, epiduroscopy can also be used functionally to investigate the outcome of treatment. In a study by Bosscher and Heavner,[3] epiduroscopic treatment predicted its therapeutic outcome based on direct visual information (hyperemia, vascularity, and fibrosis) and mechanical information (pain to touch, contrast spread, and patency). In Bosscher and Heavner's study, a prediction of outcome using diagnostic epiduroscopy was correct in 89 of 114 patients (accuracy of 78%) with sensitivity and specificity of 75 and 82%, respectively. These results suggest that information obtained through epiduroscopy may carry significant diagnostic and prognostic value.

This chapter reviews the diagnostic and functional uses of epiduroscopy.

13.2 Diagnosis of Epidural Adhesions

In the epiduroscopic procedure, adhesion is observed as tissue organized in strings and sheets of white fibers or as a dense white tissue, which appears generally avascular, regardless of their severity (▶ Fig. 13.1). In some patients, adhesions are so solid that the fibrous scar tissue makes mechanical adhesiolysis using epidural impossible (▶ Fig. 13.2). In such cases, it is possible to perform epiduroscopic adhesiolysis with holmium: yttrium-aluminum-garnet laser ablation (▶ Fig. 13.3). If inflammation is visible, then flushing the epidural space with saline and medication can play an important role (▶ Fig. 13.4).

Bosscher and Heavner rated fibrous adhesion identified during the epiduroscopic procedure using the grading system from grade 1 (loose strings and sheets of fibrosis) to grade 4 (dense continuous fibrous material, the epiduroscope cannot be advanced) in accordance with its severity (▶ Fig. 13.5).[4]

13.3 Diagnosis of Inflammation in Epidural Space

In the epiduroscopic procedure, the normal epidural cavity is usually seen with fat tissue (clusters of white or pale-yellow globular tissue with shiny appearance), dura (grey-white, with blood vessels), veins, arteries and fibrous strings or sheets, and ligaments. In acute and chronic inflammation, hyperemia can be observed as a discrete area of the dura root sleeve, peridural membrane, or other epidural structure, with or without adhesions, as compared to normal appearance of the epidural space (▶ Fig. 13.6).

13.4 Diagnosis of Herniated Disks

When the epiduroscope is administered into the anterior epidural space, the disk can be seen. If the disk bulging is large and without annular tear, the disk can push the dura upward (▶ Fig. 13.7). The dura is observed as a white-yellowish band with traverse blood vessels; sometimes, it may look gray-white. If there is an annular tear, a whitish nucleus pulposus can be

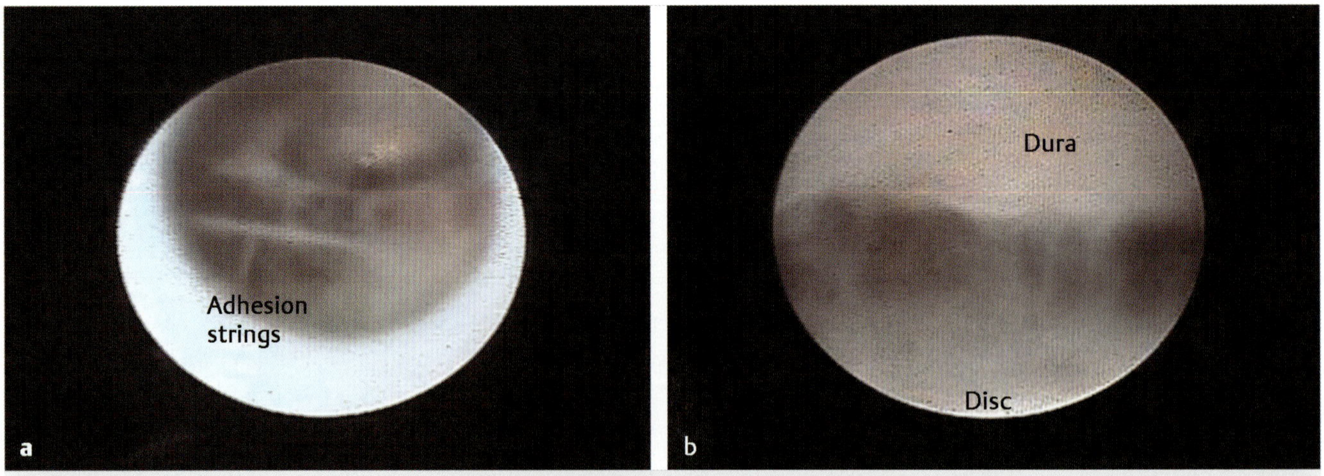

Fig. 13.1 Epidural adhesions. **(a)** The distinguishable epidural structures with fibrous bands. Strings of dense white fibers with no vascular structure are seen. **(b)** Fibrous tissues between the dura (top) and the disk (bottom) are observed.

observed in a red or yellowish inflammatory soup (▶ Fig. 13.8). If disk is migrated downward, the migrated disk is shown as a mass-like lesion (▶ Fig. 13.9).

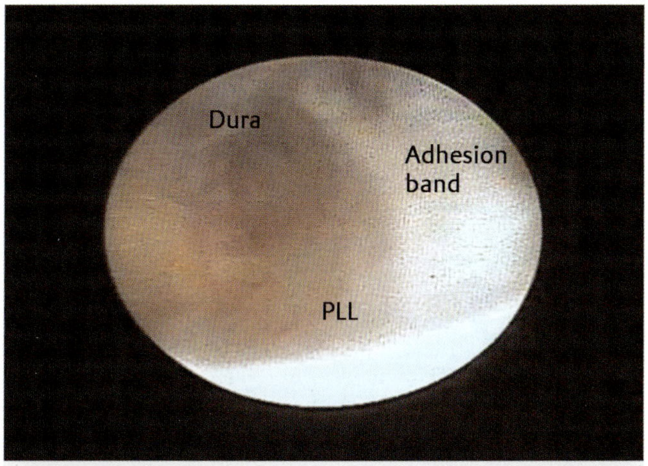

Fig. 13.2 Epidural adhesions. Strings and sheets of fibrous material fill part of the epidural space between the posterior longitudinal ligament and the dura.

13.4.1 Diagnosis of Other Epidural Pathology

Cysts of the ligamentum flavum or the facet joint capsule can be infrequently observed (▶ Fig. 13.10 and ▶ Fig. 13.11). These cysts can lead to nerve root compression. After confirming the cyst with the epiduroscope, targeted laser ablation or mechanical popping with the tip of the epiduroscope can be attempted (▶ Fig. 13.12).

13.4.2 Epidural Pain Provocation Test

In addition to the advantage of visually identifying structures, epiduroscopy can be used to test whether diagnostically significant pain is provoked in the area being examined. Patients can report if the pain elicited by probing epidural structures is concordant with their daily sensation. This functional nature of the test to diagnose painful epidural structure is logical in order to identify the reason for pain. It is also possible to stimulate the nerve roots in the lesion via the epiduroscope to test if the stimulation induces concordant sensations or pain reproduction and to ensure a higher level of accuracy for our diagnosis.

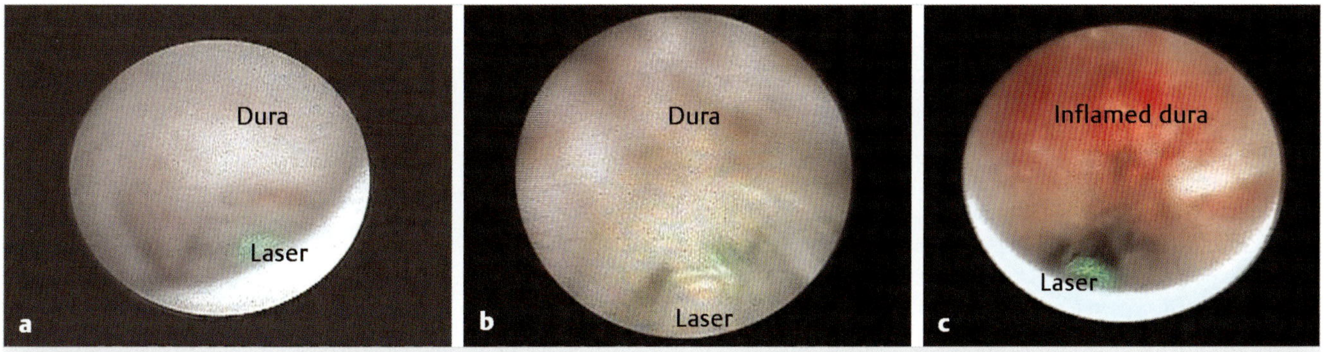

Fig. 13.3 **(a–c)** Epiduroscopic adhesiolysis between the dura and the posterior longitudinal ligament is attempted with holmium:yttrium-aluminum-garnet laser (green light).

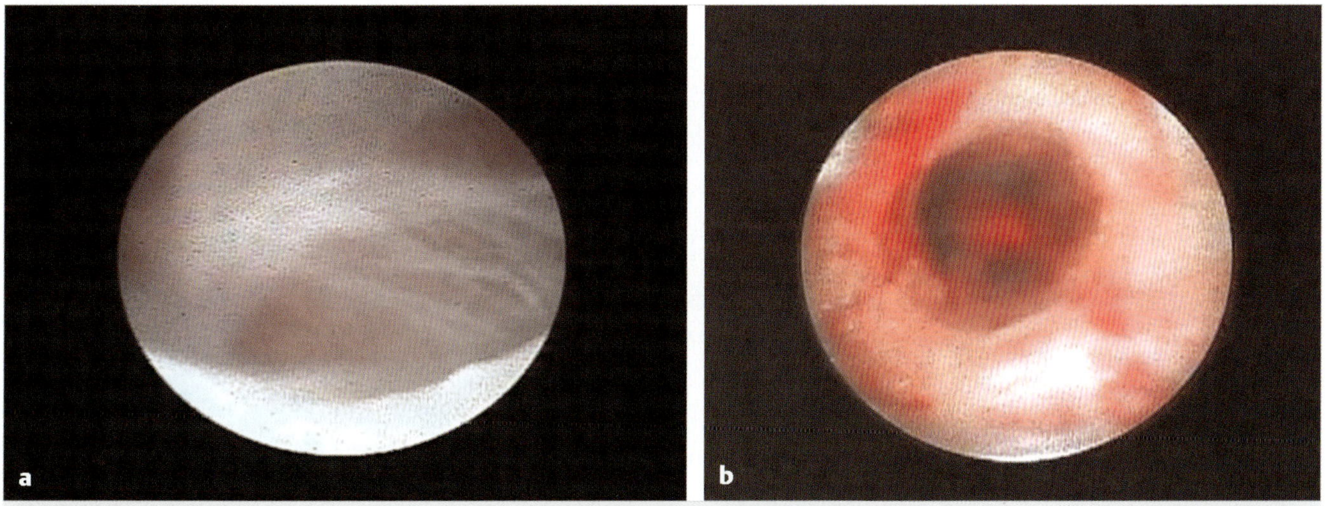

Fig. 13.4 Epidural adhesions with inflammation. **(a)** Strings and sheets of fibrous material form adhesions on the red-colored inflamed dura are observed. **(b)** Sacral epidural space is narrowed due to scar tissue and adhesion after spine surgery.

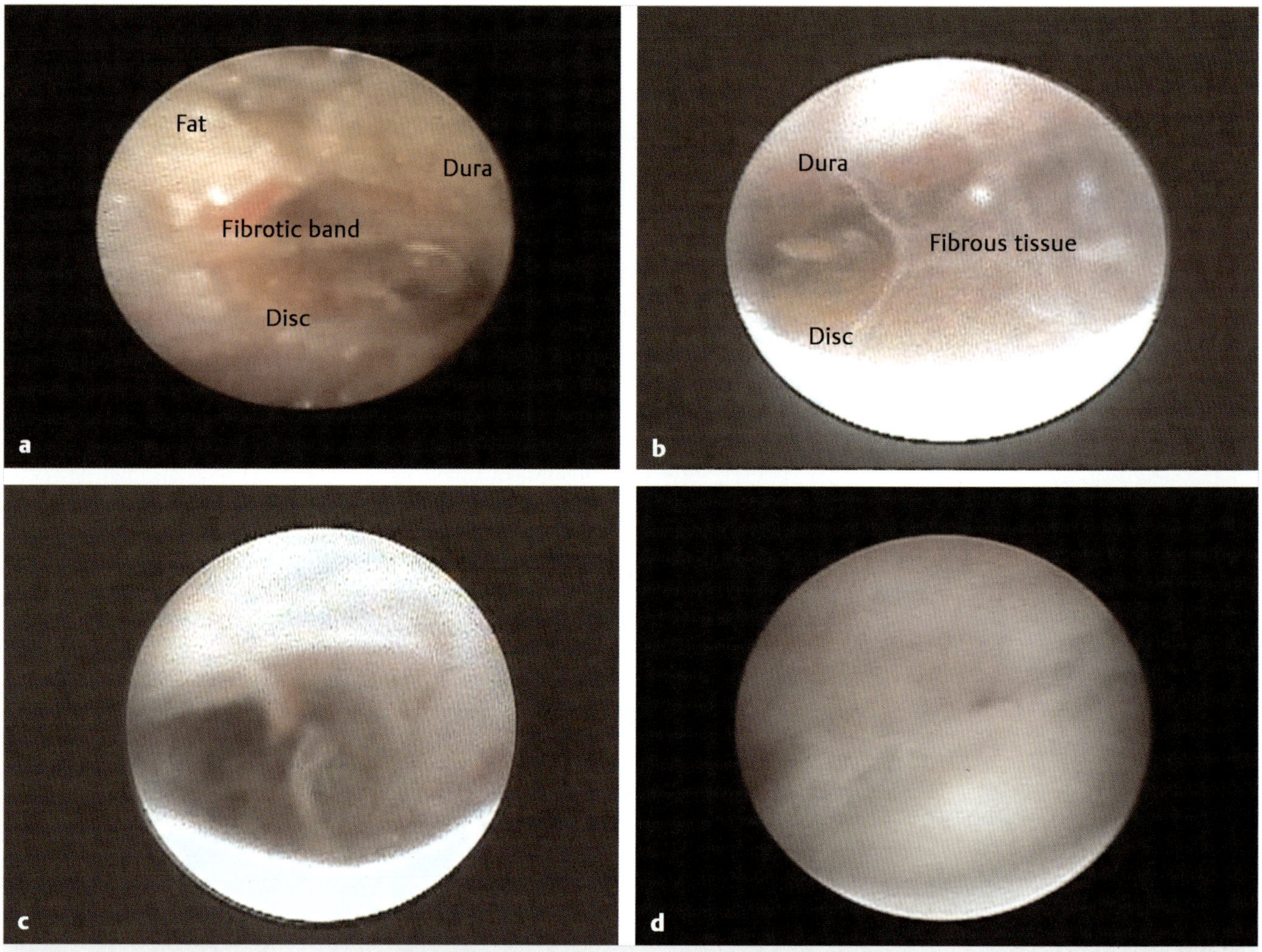

Fig. 13.5 Grades of epidural adhesion. **(a)** Grade 1: early stage of adhesion. Loose strings and sheets of fibrosis are observed. **(b)** Grade 2: more organized, continuous strings and sheets of fibrous material not giving resistance to the scope connect the dura (top) and the disk (bottom). **(c)** Grade 3: dense continuous fibrous material; the scope can only be advanced with difficulty. **(d)** Grade 4: the epidural space is completely obliterated by fibrous tissue that does not allow a scope to penetrate.

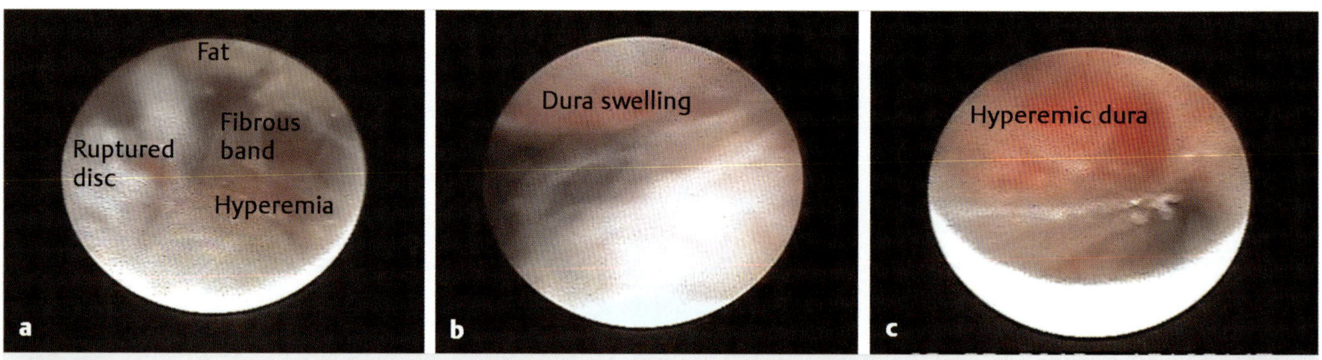

Fig. 13.6 **(a–c)** Inflammation in the anterior epidural space. The ruptured disk is surrounded with hyperemic soft tissues and fibrous band, suggesting the presence of significant inflammation.

It is generally thought that epidural scars themselves are not very painful. Epidural fibrotic tissue can induce mechanical problems and lead to scarred connections with neighboring nerve roots. Changes in length due to movement over the scarred tissue bridges cause pulling and mechanical irritation of the nerve roots that ultimately determine the symptoms of radiculopathy.[1,2] In case of inflamed nerve roots, pain can be provoked (▸ Fig. 13.13). Under epiduroscopy, it is tested whether diagnostically significant pain can be brought about in the area being examined or not.

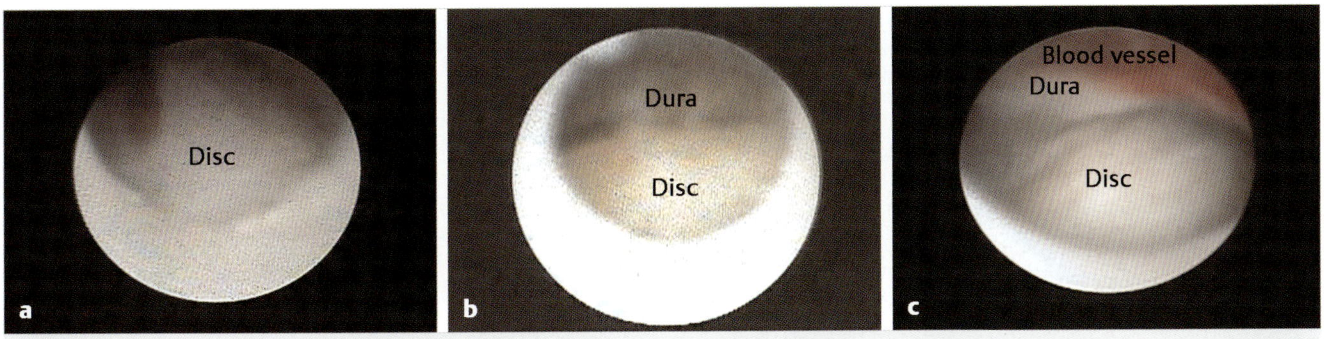

Fig. 13.7 (a–c) Protruded intervertebral disk. A large disk burling pushes the dura upward.

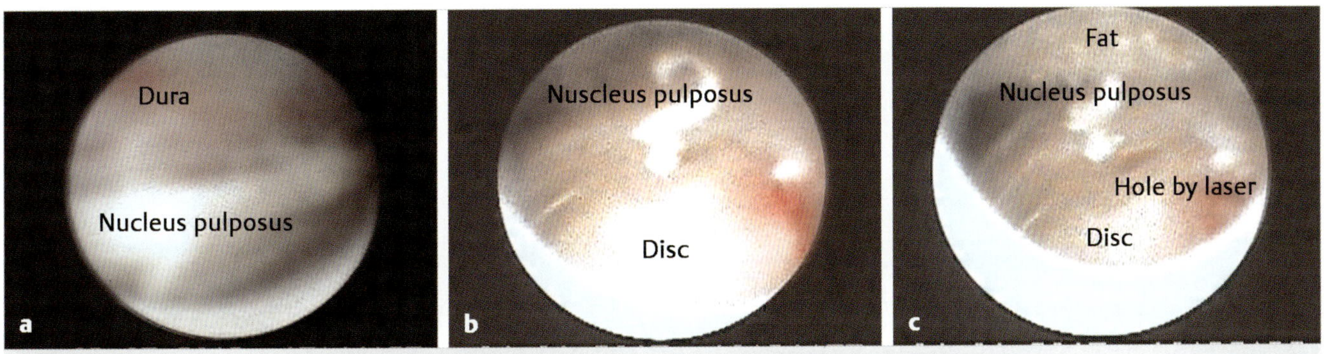

Fig. 13.8 (a–c) Herniated disk. A whitish nucleus pulposus with an annular tear is observed in chronic inflammatory reaction.

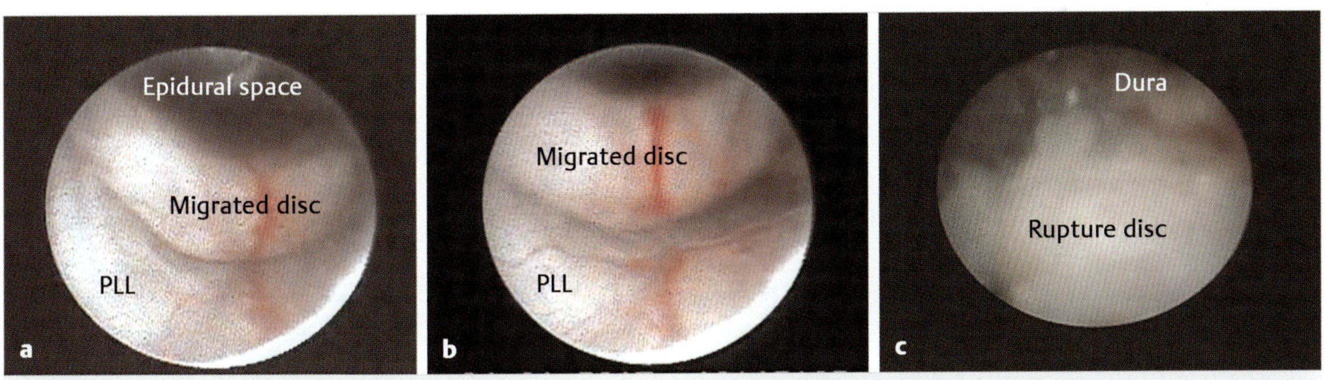

Fig. 13.9 Herniated disk. **(a,b)** Unruptured disk downward in the anterior epidural space. **(c)** The ruptured disk migrates downward, forming an irregular and white masslike lesion.

Alternatively, aided by an epidural laser pain provocation test, epiduroscopy can be used in patients with spinal pain syndrome to examine the extent to which radiologically—or epiduroscopically—identified epidural pathologic and anatomic structures are relevant for the patients' pain. With a defined laser beam, reproducible pain can be stimulated in the targeted epidural structures. The laser emits very short and steep heat impulses that are applied without touching directly, where they activate targeted afferences of the pain system.

In accordance with the study by Bosscher and Heavner, which compared clinical evaluation, magnetic resonance imaging (MRI), and epiduroscopy for determining the vertebral level that was relevant to low back or leg pain, 87% of patients ($n = 125$) reported reproducible pain that was concordant with their practical pain during epiduroscopy.[5] In 28% of patients ($n = 40$), the level determined by clinical evaluation correlated with the level at which pain could be reproduced during epiduroscopy. MRI indicated a specific vertebral level that corresponded to the level at which pain could be reproduced during epiduroscopy in 20% of patients ($n = 28$). Results of Bosscher and Heavner's study indicate that epiduroscopy is more reliable than either clinical evaluation or MRI for determining the vertebral level where clinically significant pathology occurs in patients with low back pain and/or radiculopathy.

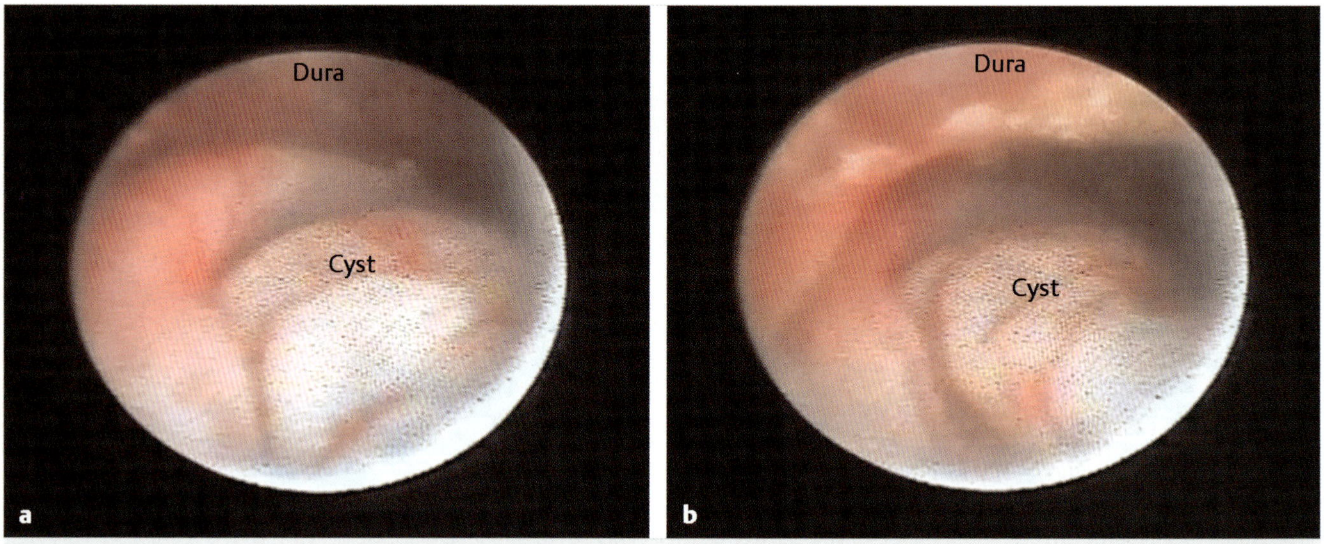

Fig. 13.10 (a,b) Cysts in the ligamentum flavum.

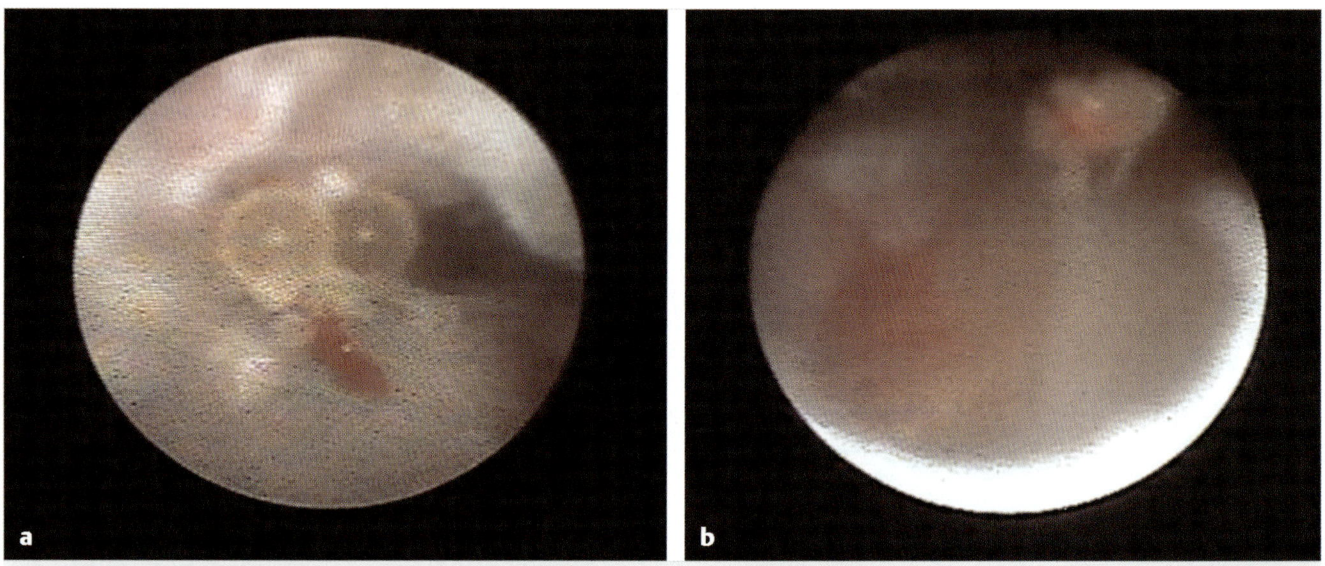

Fig. 13.11 (a,b) Various types of cysts in the epidural space.

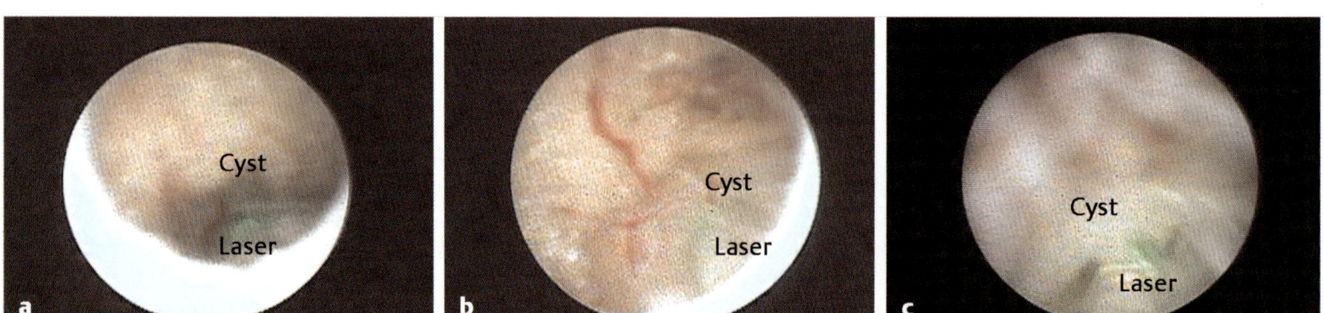

Fig. 13.12 (a–c) Cysts in the facet joint. Targeted laser ablation with holmium:yttrium-aluminum-garnet laser (green light) is attempted.

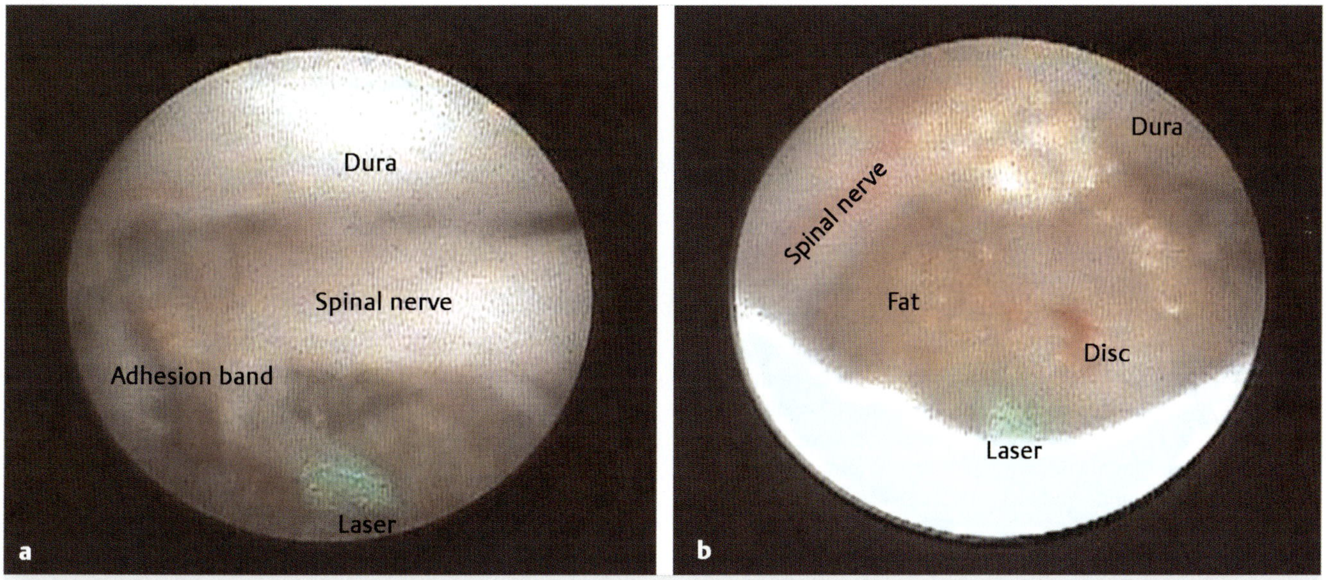

Fig. 13.13 (a,b) In the case of inflamed nerve roots due to herniated disk, pain can be provoked during epiduroscopy. Laser is targeting the adhesion band and the herniated disk.

References

[1] Schutze G. Epiduroscopy – Spinal Endoscopy. Heidelberg: Springer; 2008

[2] Kallewaard JW, Vanelderen P, Richardson J, Van Zundert J, Heavner J, Groen GJ. Epiduroscopy for patients with lumbosacral radicular pain. Pain Pract. 2014; 14(4):365–377

[3] Bosscher HA, Heavner JE. Lumbosacral epiduroscopy findings predict treatment outcome. Pain Pract. 2014; 14(6):506–514

[4] Bosscher HA, Heavner JE. Incidence and severity of epidural fibrosis after back surgery: an endoscopic study. Pain Pract. 2010; 10(1):18–24

[5] Bosscher HA, Heavner JE. Diagnosis of the vertebral level from which low back or leg pain originates. A comparison of clinical evaluation, MRI and epiduroscopy. Pain Pract. 2012; 12(7):506–512

14 Epiduroscopic, Histopathologic, and Microbiological Findings

Günter Schütze

14.1 Introduction

In pain medicine literature, the histologic (fine tissue) analysis of neuraxial tissue samples, taken from the vicinity of the spinal cord with the help of spinal endoscopy, are not at the heart of pain diagnostics currently, but that may change. Different methods are available to investigate and evaluate tissue from the epidural space histologically. Cytologic and histologic light-microscopic laboratory investigations of tissue samples are the basis of routine diagnostics. The analysis of removed cerebrospinal fluid (CSF) and epidural irrigation fluid is also part of cytologic diagnostics.

Infectious inflammatory diseases are a cause, but not the most important, of neuraxial pain in patients. Depending on the age of the patient, the course of the disease, the route of the inflammation and the affected region, conclusions can often be drawn on the causative agent of the infection. But in many cases of infection (including our own), a clear differential diagnosis of the clinical investigation findings only becomes possible through an endoscopically supported biopsy with subsequent histologic investigation. Biopsies also serve for control of the course of therapy and evaluation of the results of therapy. The strategy for the patient's pain therapy can also be optimized utilizing the endoscopic survey findings and the histologic investigation results.

Also, in invasive–interventional pain medicine, every operatively removed piece of tissue must undergo histologic analysis.

A new endoscopic technique, endomicroscopy, is ever more prominently taking center stage in general endoscopic and histologic diagnostics. During endoscopy, endomicroscopy generates histologic-type images of the endoscopically visualized tissue available in real time, making a histologic evaluation possible.[2,3] This cellular-level in vivo diagnostics, or optical biopsy, may play a role in the future as a supplement or alternative to established epidural biopsy.

On the Internet, countless depictions of histologic preparations of the highest quality are available, retrievable at all times, digitized, and also editable with Photoshop (Adobe Systems, San Jose, California), CorelDRAW (Corel Corp., Ottawa, Ontario, Canada), and PowerPoint (Microsoft Corp., Redmond, Washington). These depictions are available for use in comparing with research findings or in creating patient reports and showing comparative images.

14.2 Preanalytics for Histologic and Cytologic Investigation

In diagnostic pain management, the method of obtaining and preparing tissue samples for histologic and cytologic evaluation is significant. These preanalytics include the extraction, workup, storage, and transport of the patient's specimen material. The conditions to which the specimen material is subjected until it enters the laboratory also have a decisive influence on the reliability of histologic and microbiological laboratory findings.

14.2.1 Extraction, Workup, Storage, and Transport of Specimen Material

During spinal endoscopy, neuraxial tissue specimens for biopsy can be extracted through the working channel of the endoscope with the help of a grasping forceps. The tissue specimens secured from the epidural space are first fixed in the operating room (OR) in a 4% formalin solution and sent to the laboratory in special packaging. In the laboratory, the workup of the tissue specimen for microscopic investigation takes place. For most histologic investigations, the light microscope (with a resolution of 200 nm) is used.[4,5]

For histologic investigation, the small tissue specimens from the neuraxial space are embedded in paraffin from which 1.0- to 3.0-µm translucent slices will be prepared. To do this, water is removed from the tissue specimens in the tissue processor. Then the tissue specimens are saturated with paraffin and poured into paraffin blocks. After cooling of the paraffin, the tissue specimens can be found on the upper surface of a paraffin block on the back surface of an embedding cassette. On a microtome cutter, up to 3-µm thin slice preparations can now be made. In a water bath, the tissue specimens are afterward raised on a glass slide. The remaining thin, now translucent tissue slice on the slide can now be automatically or manually stained for visualization of cell nuclei, cell boundaries, and certain tissue structures.[4]

Histopathologic Indications: Stains

The standard staining of histologic preparations is carried out with hematoxylin/eosin (HE).

The stained structures such as the cell nucleus, bacteria, or calcium carbonate are stained blue. Cytoplasm and connective tissue are stained red.

Further staining methods include, for example, Congo red, Ziehl-Neelsen, Prussian blue [Berlin blue], Giemsa, Elastica van Gieson, and silver staining.[4]

With special questions, additional diagnostic methods such as enzymo- and immunohistology as well as molecular biological investigation methods (polymerase chain reaction [PCR] and microarray investigations) come into use.[5]

14.2.2 Removal, Storage, and Transport for Bacteriologic Investigations

For the isolation, differentiation, and cultivation of aerobic and anaerobic bacteria, the necessary test glassware with culture media and incubation systems are available from the industry.[5]

For removal, transport, and conservation of the clinical specimens for cultivation for investigation, sterile, ready-for-use single-use systems are used. The manufacturer's instructions for the use of the single-use systems are helpful.

Spinal Epidural Catheters

Removal, storage, and transport

Some 2 to 3 cm of the catheter tip is cut off with sterile scissors and placed in a sterile test tube. Liquid transport media or blood culture bottles are to be avoided, because a semiquantitative determination is otherwise impossible. The catheter sample should immediately be sent to the laboratory. Storage should be avoided as much as possible. If necessary, the catheter tip can be stored up to 24 hours at 4°C.

Cerebrospinal Fluid and Epidural Irrigation Fluid

Removal, storage, and transport

Before a sample of cerebrospinal fluid (CSF) or epidural irrigation fluid is sent, the laboratory should be contacted to record its upcoming arrival. A strictly aseptic removal of CSF or fluid should be observed. Inoculate a sterile test tube with at least 2 mL of CSF or fluid, depending on the scope of the investigation.

Native cerebrospinal fluid for serology

For CSF antigen detection, blood cultures should be obtained simultaneously witth the CSF sample. CSF diagnostics as the occasion arises are needed with cytology in suspected cases of polyradiculitis or neuroborreliosis. The CSF investigation specimen should be processed in the laboratory within 2 hours of collection without cooling.

Venous Blood

Removal, storage, and transport

Blood samples are needed for basic laboratory evaluations: erythrocyte sedimentation rate (ESR), C-reactive protein (CRP) value, blood count, platelet count, liver and kidney values, and creatine kinase value. An extended blood serology is helpful when there is reasonable suspicion of borreliosis (Lyme disease) or herpes zoster. The blood specimens for a blood culture should be obtained as much as possible in the early phase of the temperature rise, before the beginning of antibiotic therapy or, with current antibiotic therapy, before the next antibiotic dose. The blood draw is to be carried out after skin disinfection with alcohol and drying, using single-use gloves and a blood culture special set (8–10 mL per bottle). The prepared blood cultures are not to be cooled but incubated at 35 to 37°C if an immediate transport to the laboratory is not possible.

Neuraxial Tissue Material

Removal, storage, and transport

The endoscopically, microsurgically obtained neuraxial tissue specimen is to be put into a sterile test tube, stored at room temperature, and promptly transported to the laboratory. When necessary and for storage under 2 hours, the tissue specimen is to be placed in a transport medium such as a 4% formalin solution at room temperature, to be stored and transported.

14.3 Virtual Histology

14.3.1 Optical Biopsy via Endomicroscopy

The integration of a miniaturized confocal microscope in the distal end of a conventional endoscope makes possible a new endoscopic imaging method, endomicroscopy. Endomicroscopy provides a microscopic view of the mucous membrane and mucosal surface during endoscopy in higher resolution by means of laser-supported, confocal fluorescence technology. Cellular, vascular, and connective tissue structures can be differentiated and immediately evaluated during endoscopy via endomicroscopy.

By shifting the color spectrum (visualization tool), even the finest tissue structures can be identified endoscopically. Deepred parts of the visible spectrum are filtered out and the remaining portions of the color spectrum are expanded. Through this technology, such as the Storz Professional Image Enhancement System (SPIES; Karl Storz, Tuttlingen, Germany), the ability to differentiate between tissue types can be facilitated directly on the monitor (▶ Fig. 14.1).

A histologic evaluation of the anatomical structures in real time during endoscopy becomes possible through the enlargement of the endoscopically visualized tissue. Endomicroscopy therefore makes possible cellular level in vivo diagnostics (optical biopsy).[2,6]

On technological grounds, endomicroscopy is not yet ready for clinical use in spinal endoscopy for evaluating cell and tissue.

14.3.2 Sonographically Supported Virtual Histology

Intravascular ultrasound (IVUS) technology offers detailed sonographic information on the composition of anatomical structures of the patient. An ultrasound catheter, placed through the working channel of the epiduroscope, produces sonographic images. The images of virtual histology are processed with the help of the Volcano imaging console and the Eagle Eye Gold IVUS catheter (Volcano Corp., Rancho Cordova, CA, USA) (▶ Fig. 14.2) and are then displayed in real time, so that the pain therapist carrying out the intervention can evaluate them while the patient is still lying on the operating table.[7] This technology offers an automatic measurement tool to simplify the interpretation of the images and uses a predefined color key to show the composition of a specified neuraxial region, such as the spinal dura mater or within the epidural cavity.

Along with the endoscopic imaging, the color presentation of the virtual histology simultaneously delineates four types of tissue components: fibrous, fibrous/fatty, dense calcium, and those with necrotic nuclei.

Immune deficiency syndromes

Acquired immune deficiencies are essentially more frequent than congenital immune deficiencies. The most important acquired immune deficiency is human immunodeficiency virus (HIV) infection (leading to acquired immunodeficiency syndrome, AIDS) and the changes to the immune system due to malignant diseases.

Histopathologic Indications: Immune Deficiency Syndrome

Congenital and acquired immune deficiencies as a rule accompany histologic changes in the lymphatic organs.

Autoimmune diseases

Predispositions and heredity play an important role in the genesis of autoimmune diseases, along with environmental influences, medication factors, and virus infections. Autoimmune diseases are initiated by autoreactive T cells or autoreactive immunoglobulins as humoral factors.

Histopathologic Indications: Autoimmune Disease

In the tissue affected by an autoimmune disease, signs of a chronic lymphocytic inflammation are found. The tissue architecture is destroyed to various degrees.

14.4.3 Inflammation

In the initiation and maintenance of neuraxial pain syndromes, nonbacterial inflammatory reactions plays a crucial role.

Inflammation represents a reaction to tissue damage by chemicals, physical effects, traumatic events, foreign bodies, or microorganisms. During inflammation, a regular sequence of effects due inflammatory stimuli on a cell or tissue occur. A complex reaction of the affected tissue is provoked by various inflammation mediators in which capillary vessels and leukocytes are decisively involved.

The goal of all inflammatory processes is the healing of the tissue damage as well as the elimination of the microorganisms, harmful agents, or foreign bodies, and if no regeneration of the tissue (*restitutio ad integrum*) is possible, then repair of the tissue (*reparation*) results.

Control of Neuraxial Inflammation Processes

Cellular control elements of inflammation

Leukocytes and neutrophil granulocytes are first line cellular control elements of neuraxial inflammation. Macrophages are also activated by serosanguinous and cellular inflammation mediators (arachidonic acid derivatives, histamine, serotonin, interferon, and platelet-activating factor). They permeate from the neuraxial blood vessels into the affected tissue, where they take on a garbage-collection function in the form of phagocytosis and secretion of proteolytic enzymes. From the bloodstream, monocytes also reach the tissue. The monocytes convert in the tissue to macrophages or special tissue macrophages such as histiocytes, Kupffer cells, osteoclasts, and microglia. Epithelioid histiocytes are specially activated forms of macrophages. Macrophages and epithelioid histiocytes can fuse into multinucleated giant cells. Giant cells appear particularly in granulomatous inflammation processes.

The B and T lymphocytes are responsible for the specific immune response. B and T lymphocytes also provide long-term containment of the inflammatory reaction.[5,11] In the case of rare infectiously caused spinal inflammation reactions, the B and T lymphocytes are also decisive for protection against a reinfection.

Humoral control elements of inflammation

Cellular inflammation mediators are produced and secreted by the cells involved in the inflammatory reaction. Humoral inflammation mediators also exist in the blood serum as inactive precursors that are activated by a corresponding stimulus. A release of the cellular inflammation mediators results, for example, actively after an antigen contacts secreting cell (e.g., mast cells) or passively after the destruction of stationary cells. Inflammation mediators are central to the inflammatory process, governing vasodilation, increase of capillary permeability, leukocyte migration, phagocytosis, and the development of general clinical symptoms in the patient.[4,12]

Acute Inflammation

Based on the time course, most inflammations are classified as acute inflammations, that is, having a short time course. Frequent triggers of acute inflammation are exposure to chemicals, physical effects, traumatic events, viruses, or allergens. A characteristic feature is the escape of fluid and leukocytes from the capillary vessels (exudative inflammation).

Depending on the albumin content of the inflammatory fluid, it is classified as an exudate or a transudate, and if as an exudate, by its composition, as a serous, fibrinous, purulent, hemorrhagic, or lymphocytic. It should be noted that serous inflammations are mostly nonbacterial.

The acute inflammation process leads to elimination of the triggering agent. Under certain circumstances, this elimination does not succeed; the inflammation-triggering tissue factors persist. The result is a chronic inflammatory reaction.[4,13,14]

Chronic Inflammation

In chronic inflammation, macrophages, lymphocytes, plasma cells, and eosinophilic granulocytes play a decisive role. The cytokines secreted by these cells lead to marked and long-lasting changes in the affected tissue. Chronic inflammation frequently results in extensive destruction of original organ tissue structure with scarring.

Chronic inflammation processes are classified as chronic lymphocytic, chronic granulating (chronic purulent), or chronic granulomatous inflammations.[13,14] Primary chronic inflammations are distinguished from secondary chronic inflammations by the mechanism of origin of the inflammation.

The form a chronic inflammation takes is an expression of a complex reaction of the vascular and connective tissue apparatus to tissue damage.

In the tissue preparations of patients with neuraxial chronic inflammatory processes (arachnoiditis, epiduritis, radiculitis), the epidural connective tissue structures with their proliferations of fibroblasts become visible as fibrous forms or scar structures in the form of histiocytes and capillary buds. As a consequence of the chronic inflammation, epidural scars develop as the expression of wound healing.

Histopathologic Indications: Chronic Inflammatory Processes

Predominant distinctive findings in histopathologic preparations from tissues with chronic inflammation are:

- Granulation tissue due to cellular proliferation with abundant lymphocytes, histiocytes, and plasma cells or mast cells or polymorphonuclear leukocytes.
- Granulation tissue in the narrower sense, that is, capillary buds and fibroblasts.
- Infiltrating granulation tissue.

14.5 Histologic and Microbiological Findings Indicative of Pathology in Neuraxial Tissue

Relevant histologic findings indicative of neuraxial tissue pathology are:

- Increased collagen formation (adhesions, fibroses).
- Inflammatory processes due to sequestered material.
- Histiocytic granulomas.
- Chronic inflammatory processes (arachnoiditis, epiduritis, radiculitis).
- Bacterial infections and abcesses.

We report here only on histopathologic findings from our own patient population.

14.5.1 Increased Collagen Formation (Epidural Adhesions and Fibroses)

The most frequent histologic and endoscopically observed pathologic anatomical findings in our pain patients are epidural adhesions and fibroses. The terms employed in the literature and in clinical usage for epidural adhesions and fibroses are, among others, "epidural sclerosis," "fibrosis," "callus," "scar," and "induration." They are identical in meaning.[4,15,16]

An injury in the neuraxial region leads first of all to desquamation of the damaged cells as well as to an inflammatory reaction, which is identified by the presence of cellular infiltration, the formation of serosanguineous exudates, and the proliferation of the surrounding cells.[16,17]

Pathogenesis

After operative or interventional procedures near the spinal column and in inflammations in the neuraxial space, tumor necrosis factor alpha (TNFα) and interleukins 1 and 6 (IL1 and IL 6), in particular, are released. These mediators cause a rise in plasminogen-activator inhibitors. These enzymes inhibit the tissue plasminogen activator and the urokinase-type plasminogen activator, which convert inactive plasminogen to plasmin. Plasmin acts to cleave the fibrin gel matrix into individual fibrin monomers.[13,18,19]

As a result, the fibrin compounds persist in connective tissue strands and become converted to epidural adhesions or fibroses. In this phase of development to adhesions or fibroses, the adhesive strands consist for the most part of macrophages, fibroblasts, phagocytes, and mast cells. The number of fibroblasts diminishes over the course of time, and the adhesions differentiate from collagen fibers into vascular strands.[9,20]

Cells of the immune defense that have migrated in as well as the injured cells release cellular mediators (interleukin 8 [IL8], TNFα, IL6, tumor growth factor beta [TGFβ], platelet-derived growth factor [PDGF]). They lead to a migration of further immune defense cells such as monocytes, leukocytes, and granulocytes with segmented nuclei.[11] The monocytes differentiate, as cellular components of the nonspecific immune system, into macrophages and attach themselves onto the epidural wound surface. Through the release of cellular mediators and the expression of thromboplastin (factor III) by the macrophages, the extrinsic part of the coagulation cascade by which a fibrin film can organize itself is started,. The fibrin film also includes leukocytes, erythrocytes, endothelial cells, epithelial cells, mast cells, and cellular debris.[4,12]

If damaged epidural surfaces that are covered with this fibrin film now contact each other, an adhesion forms between the two surfaces. This process can begin immediately postoperatively or up to 5 days postoperatively.[11,18,21]

Normally, the processes of fibrinogenesis and fibrinolysis occur in an exactly balanced physiologic equilibrium. Shifting of this dynamic equilibrium in favor of fibrin formation plays a decisive role in the pathophysiology of epidural adhesions and fibroses. Through the construction of a fibrin network and connective tissue organization, epidural adhesions of the most varied morphology arise.

Fibrinolytic activity is decisively dependent on the activity of immunocompetent cells and inflammation-reducing stimuli, which are, however, diminished in an inflamed region of the epidural space or after interventional manipulation.

An increase in collagen fibers or structural changes in different types of collagen indicates pathologic tissue changes. Structural proteins such as fibrillin or elastin can be pathologically altered.

An increase of collagen fibers per tissue unit is observed in chronic inflammations, in longstanding edemas, and for filling-up tissue defects. In epidural fibrosis, the increase in collagen fibers leads to an extraordinary pathologic hardening (induration) of the affected tissue.

Histopathologic Indications: Adhesions and Fibroses

Microscopically, epidural scars, growths, fibroses, and adhesions consist of collagen fibers with fibroblasts and budding capillaries, colonized by lymphocytes, macrophages, mast cells, and some eosinophilic granulocytes. Sometimes we also found foreign body granulomas from with surrounding giant cells and lymphocytes.

Wound Healing

Wound healing proceeds in distinct phases. The acute fibrinous inflammation reaction corresponds essentially to the exudative phase. In the resorptive phase, the defective tissue is cleared away by macrophages and granulocytes that have migrated to the area. Granulation tissue forms in the reparative phase. In this phase angiogenesis begins, and then stem cells, which differentiate into fibroblasts, migrate to the wound from the bloodstream. In the course of some weeks, the fibroblasts form increasingly stable collagen fibers as scar tissue (wound healing).

Histopathologic Indications: Scar Tissue

The increase of collagen fibers (scar tissue) can be easily identified and, as occasion arises, even quantitively evaluated, particularly by special stains such as Elastica van Gieson (elastic fibers = black; connective tissue, hyaline = red; musculature, fibrin = yellow).

Note: A clinical diagnosis of epidural scarring or scars means connective tissue wound healing.

14.5.2 Intervertebral Disk Prolapse and Sequestration

Fragments of material from advanced degeneration of an intervertebral disk, called a sequestration, can enter the epidural space after rupture of the anulus fibrosus and the posterior longitudinal ligament.

The intact intervertebral disk is composed of collagen fibers and intercellular substance. Its nucleus pulposus includes in its gelatinous substance embedded glycoproteins and high-molecular-weight polysaccharides. The anulus fibrosus consists chiefly of helical collagen fibers intertwined with each other.[12]

As a result of compression or extension, a traumatic rupture of an intervertebral disk can occur.[11,22] The mechanical pressure created by the intervertebral disk fragments (sequestration), when distributed into the epidural space and the resulting biochemical influences on the epidural neural structures, induce an inflammation.

The swelling of the prolapsed intervertebral disk can persist for several weeks on the basis of temporary fluid absorption. After the hydration phase, the dehydration phase of the sequestration sets in. As the volume of fluid decreases, the pressure on the nerve root is reduced.

Small sequestrations are directly resorbed enzymatically through phagocytosis. The breakdown of larger sequestrations is organized via vascularization and connective tissue organization from the surrounding fatty tissue. Nucleus pulposus tissue is broken down by macrophages and T cells.

Histopathologic Indications: Sequestration

Collagen fiber bundles (fibrinoid necrosis), hyaline chondrocytes, multinucleated giant cells, lymphocytes, and neutrophilic granulocytes can be demonstrated histologically in sequestrations.

14.5.3 Histiocytic Granulomas (Epidural Foreign Body)

Granuloma formation serves primarily as a defense against facultative or obligate intracellular bacteria (epithelioid cell granulomas) or in the resorption of foreign material (histiocytic granulomas). Endoscopically visible so-called foreign body granulomas are released in response to the body's own substances or foreign substances in the tissue. A neuraxially placed foreign body in the epidural cavity, such as a catheter, suture material, an electrode, or a broken piece of a needle, is a characteristic cause for formation of this type of granuloma.

In case the foreign body cannot be broken down, the foreign body giant cells persist for years. These cells are characterized histopathologically by an alternate configuration of the nuclei and act as a stimulus for chronic inflammation.

Metabolic diseases and exposure to cell- and tissue-damaging agents can lead to an increased deposition of waste substances in the epidural space as well as to direct cell damage. After an epidural application of 10% *n*-butyl-*p*-amino-benzoate (BAB), deposits were still found in the epidural space after 36 days.[23]

Histopathologic Indications: Histiocytic Granuloma

Histiocytic granulomas are characterized by arrangements of multinucleated giant cells, macrophages, lymphocytes, neutrophilic granulocytes, and proliferating capillaries around the central deposit. The giant cells have often already absorbed the foreign body material, such as catheters, electrodes, or suture material, into the cytoplasm.

Case Presentation

After a percutaneous epidural multiple puncture with a Tuohy needle for pain therapy during epiduroscopy in this female patient, birefringent particles with characteristic multinucleated giant cells of the foreign body type could be demonstrated histologically in the dorsal lumbar epidural tissue specimens (► Fig. 14.3).

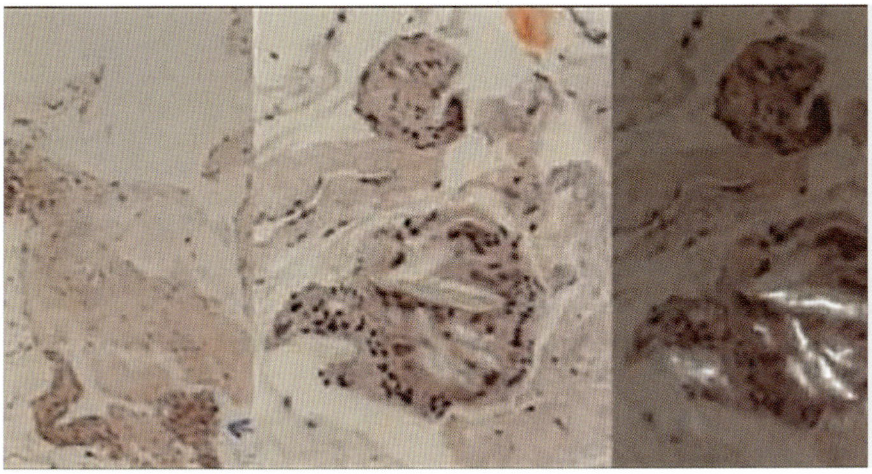

Fig. 14.3 Birefringent particles with characteristic multinucleated giant cells of the foreign-body type occur in this histologic preparation from a patient after multiple epidural punctures. (Courtesy of Prof. Dr. Böhm, Pathologisches Institut, Klinikum Lüdenscheid.)

14.5.4 Chronic Inflammation Processes

Epiduritis

For an endoscopic diagnosis of epiduritis, one would observe a visible inflammation process in the epidural space, which is a reaction of the connective tissue and vascular apparatus to inflammatory stimuli.

Histopathologic Indications: Epiduritis

Strong hyperemia in connection with dry lympholeukocytic infiltrates is characteristic of a microscopic image of epiduritis. The chronic lymphocytic inflammation reaction in epiduritis corresponds to the histopathologic finding of a chronic inflammation.

Corresponding to the course of inflammation, the process can devolve into a proliferative process, that is, fibroblasts can abundantly generate connective tissue fibers that lie on one another densely, hardening the tissue. The cells and even the blood vessels can be pressed together by the fibers, and so the tissue appears pale. The fibers shrink, consequently scars arise that can often be identified epiduroscopically along with the focus of inflammation in the epidural cavity.

Purulent forms of inflammation are caused principally by bacteria. Not infrequently, the appearance of the purulent inflammation is modified in characteristic ways depending on the pathogen type .

Radiculitis

Frequently in patients with a radicular neuropathic pain event, the question of an inflammatory component of the pain presents itself.

Radiculopathy, due to inflammatorily changed nerve roots, can be differentiated into inflammatory neuropathy or metabolic and nutritive/toxic neuropathy. Corresponding to the components of the nerves, the forms of the nerve damage can also be distinguished as neuronal neuropathies, axonal neuropathies, demyelinating neuropathies, and interstitial neuropathies.[24]

Pressure rises due to an epidurally herniated intervertebral disk provoke, for example, a bacterial inflammatory swelling of the root (radiculitis). Also, epidural air pockets, not noticed interoperatively, can lead to spinal cord or nerve root compression and to chronic inflammatory neuropathic damage.[15,25,26] Edema and ischemia (radiculitis) are recognizable histopathologically.[9,26]

Clinical Note

In the absence of a clinical sign of nerve strain, a pain occurring especially at night, resistant to therapy, and that cannot be influenced by lumbar spinal column movement could be a bacterial radiculitis triggered by Borrelia. For a differential diagnosis, a metabolic radiculopathy is also possible, especially in the context of diabetes mellitus.

Histopathologic Indication: Radiculitis

Exudates and proliferation of the epineurium, perineurium, and endoneurium of the affected part of the nerve dominate as histopathologic indications of radiculitis. Edematous changes of the connective tissue fibers, drying out of the myelin sheaths, and immigration of inflammatory cells also make an impression.

In the preparations, inflammatory infiltrates and necroses as well as hemorrhages can be visible.

Histopathologic Indication: Metabolic/Nutritive/Toxic Radiculitis

The loss of myelinated nerve fibers in some segments and occasionally axonal disintegration are important histopathologic indications of toxic nerve damage.

Arachnoiditis

The arachnoid is a trabecular structure of collagen fibers clad in squamous epithelium and forms a lamellar cell layer on the dura mater.[27] Arachnoiditis is a chronic inflammatory process of the spinal dura mater and in particular the arachnoid, diagnosed endoscopically. As a special pathologic reaction to neuraxial tissue damage, arachnoiditis progresses with impressive clinical symptoms and an unfavorable prognosis.

Macroscopically, arachnoiditis means a chronic inflammation of the external leaf of the arachnoid with epidural (peridural, extradural) fibroses, that is, fibroses located on the outside of the spinal dura mater. Cases of arachnoiditis are often associated with epidural fibrosis, but not the other way around.[13]

In the literature the terms "epidural fibrosis," "epidural adhesion," and "spinal scar" are frequently employed for the clinical diagnosis of arachnoiditis.

Histopathologic Indication: Arachnoiditis

Chronic granulating or granulomatous inflammation with a network of capillaries from macrophages, lymphocytes, plasma cells, and fibroblasts are the distinctive histopathologic marks of arachnoiditis.

Microscopic investigations show characteristic findings of arteritis in the blood vessels of the chronically inflamed arachnoid.

14.5.5 Bacterial Infections

In all procedures near the spinal canal, such as endoscopy, punctures and injections, spinal cord stimulation (SCS), neuraxial interventions, and operative interventions on the vertebral column, strict asepsis must be observed.

Extraspinal subcutaneous, paravertebral intraarticular, or intraspinal infections or abscesses can occur in spite of carefully performed injections. In epidural interventions, intraspinal infections (e.g., abscesses, meningitis) can arise primarily or be a rare consequence of a secondary spread of infection into the spinal canal.

Staphylococcus aureus has been shown to be the most common causative agent.[25,28] The coagulase-positive, gram-positive staphylococci can colonize the patient's skin and mucous membranes temporarily or for a long time; they are also important causes of nosocomial infections (about 12%), particularly postoperative wound infections.[11,29,30]

Catheter and Electrode Infections

Between 1 and 60 days can elapse from the conventional placing of a catheter to the occurrence of the first neurologic symptoms of infection.[15,31] Neurologic complications after epidural catheter infections are rare events; they can range from irritation of individual nerve roots or cauda equina syndrome to complete paralytic syndrome. The incidence of life-threatening complications in perioperative therapy with epidural catheters ranges, according to the literature, from 1 in 900 to 1 in 3,000.[31] In long-term catheter therapies it is to be assumed that the incidence of complications increases. Along with catheter-caused infections, infection secondary to catheter malpositioning and therapy mistakes are the basis for the increase in incidence.

With the use of implanted percutaneous leads for SCS, infections occur in 3 to 6% of the cases.[32]

When there is a suspicion of an epidural catheter or electrode infection, the catheter or electrode should be removed immediately and a further diagnostic workup, such as an MRI with contrast medium, should be performed. Bacteriologic investigation of the catheter tip and blood draws for blood cultures are useful for diagnosis or in the case of suspicion of a catheter or electrode infection. Each infection must be surgically treated as the occasion arises according to the clinical findings.

Extraordinary Findings

A 45-year-old female patient was admitted to our clinic for performance of a spinal endoscopy because of failed back surgery syndrome (FBSS). The patient reported that a multifunction catheter had been placed in her via the sacral hiatus days before for pain therapy. In the initial preoperative clinical examination a profound epithelial wound, with exuded fibrinogen and visible fibrin plaques, was found in the sacral access region of the sacral hiatus. Elevated formation of granulation tissue with epidermatization proceeding from the wound edge was also documented (▶ Fig. 14.4).

Epidural Abscesses

Bacterial inflammations and epidural abscesses are extremely rare but serious complications of neuraxial procedures.

An abscess can be associated with a regional anesthetic procedure, but can also occur independently of it. More frequently than after neuraxial blockades or interventions, however, epidural abscesses occur spontaneously.[30] A hematogenous spread of germs from the neuraxial puncture site, a migration of germs, for example, along the needle or catheter, or a contamination of medications can be the cause of an abcess.[33]

Staphylococcus aureus and *S. epidermis* are the principal infectious agents. Reihsaus et al analyzed 915 patients from the literature with a spinal epidural abscess and summarized that the first symptoms, which were most highly suspicious for the

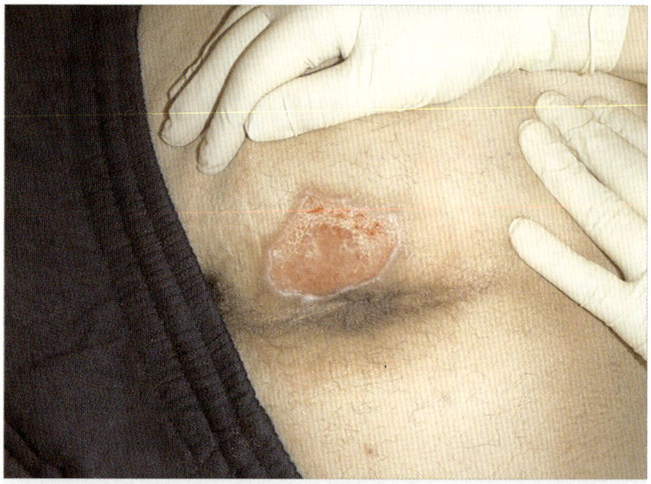

Fig. 14.4 Deep and widespread epithelial wounds after placement of a multifunction catheter through the sacral access to the epidural space.

existence of a spinal epidural abscess, were spinal back pain, local sensitivity, signs of meningitis, and signs of acute inflammation.[30]

Immediate or emergency diagnostic measures (MRI) and therapeutic measures (laminectomy and antibiotics) are necessary to avoid neurologic deficits or long-term damage.

Meningoencephalitis

Bacterial meningitis in connection with a neuraxial invasive procedure that is not its cause is very seldom observed.[30]

Bacterial meningitis occurs more frequently after intrathecal analgesia procedures, but is also can occur after epidural anesthesia. When bacterial meningoencephalitis occurs after neuraxial analgesia, as a rule, an injury to the spinal dura mater is to be assumed[22,28] and an infection due to a spread of hematogenous germs into the spinal canal.[20,34]

In postinterventional meningoencephalitis, α-hemolytic streptococci are frequently demonstrated. The occurrence of meningitis as a consequence of a surgical catheter implantation for spinal administration of medications is found on average in 3% of cases.[5]

14.6 Antibiotic Therapy for Neuraxial Pain Syndromes

14.6.1 Antibiotic Prophylaxis

A perioperative application of antibiotic in the form of a single-dose short intravenous infusion represents useful antibiotic prophylaxis for spinal endoscopy. It is performed 1/2 to 1 hour before the interventional procedure of the spinal endoscopy for anesthetic introduction for example. The single dose is sufficient for routine epiduroscopic neuraxial interventions according to our experience.

Patients with high risk (patients with artificial heart valves, bioprostheses, or previous bacterial endocarditis) especially and those with moderate risk (congenital heart defects, rheumatic or other acquired valve defects, hypertrophic cardiomyopathy, mitral valve prolapse with mitral insufficiency) of infection should recieve antibiotic prophylaxis.[5]

14.6.2 Antibiotic Therapy

Selection of an antibiotic is guided by the expected spectrum of pathogenic agents and individual patient factors, such as risk factors, previous hospitalization and antibiotic therapy, type and duration of operative intervention, and so on. The most frequent infectious agents, *S. aureus* and *S. epidermis*, should be covered by any antibiotic treatment. In clinical practice, the use of a cephalosporin such as cefuroxime has proved its worth for EDS antibiotic prophylaxis.

Antibiotic therapy alone can be contemplated for a bacterial infection in the absence of neurologic deficits in the affected patient. In all other cases a surgical decompression with focal rehabilitation must take place. A change in antibiotic therapy should only be done with unequivocal proof of the infectious agent from clinically relevant material and utilizing a current antibiogram.

For special indications and dosages of various antibiotics or combinations of anitbiotics, the most current literature and the manufacturer's prescribing information should be considered.

14.7 Conclusion

For the histologic diagnosis of neuraxial pain syndromes, a goal-oriented biopsy under endoscopic view is highly recommended. Experience is required to carefully interpret the corresponding pathologic endoscopy findings and histopathologic investigation results and to determine the best individualized treatment plan.

In 77.8% of our patients undergoing epiduroscopies up to the year 2015 (more than 2,000 epiduroscopies), the histopathologic findings match the clinical endoscopic findings Since the clinical introduction of spinal endoscopy in our pain clinic in 1992, including epiduroscopic catheter management and endoscopically supported interventional procedures, no neuraxial infections have been reported! Of < 1,500 smears from patients obtained endoscopically from the epiduroscopically investigated neuraxial region (sacral–cervical, dorsal–ventral) of our patients with pain syndrome and microbiologically cultured (Hans Günther Wahl, Mikrobiologisches Institut, Klinikum Lüdenscheid, Germany), not a single pathogen has grown! These results can be ascribed to the everyday implementation of very strictly controlled hygiene and quality management in the, and the clinic. In our pain clinic as in others, histopathology has a special place in the diagnosis of neuraxial pain syndromes; it serves not only epiduroscopic quality control but also the diagnostic safety of patients.

Looking forward to the future of histopathologic diagnosis in the pain clinic, we mention a promising new tool, endomicroscopy. In addition to the analysis of the surface structure of tissue, endomicroscopy, an endoscopic method of molecular imaging currently under development, makes possible microscopic analysis of cell structures of mucous membranes in vivo. Admittedly, it cannot yet be used for diagnosis in the pain clinical, but clinical trials to date on humans and animal have already demonstrated the possibilities of endomicroscopy.[2,3,6]

References

[1] Schütze G. Interventionelles Schmerzmanagement–Bildgestützte Verfahren zur Diagnostik und Therapie rückenmarksnaher Schmerzsyndrome-Spinale Endoskopie, pharmakologische und elektrische Neuromodulation. Bremen, Germany: UNI-MED Verlag AG; 2011

[2] Kiesslich R, Burg J, Vieth M, et al. Confocal laser endoscopy for diagnosing intraepithelial neoplasias and colorectal cancer in vivo. Gastroenterology. 2004; 127(3):706–713

[3] Rey JW, Kiesslich R, Hoffman A. New aspects of modern endoscopy. World J Gastrointest Endosc. 2014; 6(8):334–344

[4] Krams M, Frahm SO, Kellner U, Mawrin C. Pathologie. Stuttgart, Germany: Georg Thieme Verlag; 2013

[5] Theuretzbacher U, Seewald M. Mikrobiologie im klinischen Alltag. Stuttgart, Germany: Verlag W. Kohlhammer; 1999

[6] Goetz M, Watson A, Kiesslich R. Confocal laser endomicroscopy in gastrointestinal diseases. J Biophotonics. 2011; 4(7–8):498–508

[7] Schütze G. Epiduroscopy–Spinal Endoscopy. New York, NY: Springer-Verlag; 2008

[8] Curran R, Crocker J. Atlas der Histopathologie. 2000. Berlin, Germany: Springer-Verlag; 2000

[9] Leonhardt H. Histologie und Zytologie des Menschen. Stuttgart, Germany: Georg Thieme Verlag; 1971

[10] Rice I, Wee MYK, Thomson K. Obstetric epidurals and chronic adhesive arachnoiditis. Br J Anaesth. 2004; 92(1):109–120
[11] Rosenow D, Tronnier V, Göbel H. Neurogener Schmerz. Heidelberg, Germany: Springer Medizin Verlag; 2005
[12] Krämer J. Bandscheibenbedingte Erkrankungen. Stuttgart, Germany: Georg Thieme Verlag; 1994
[13] Smith S. The Adhesive Arachnoiditis Syndrome. Circle of Friends with Arachnoiditis (COFWA); 2003 http://www.cofwa.org/AASYNDROME-10-03.pdf [Accessed in October 2003.]
[14] Thompson J. Pathogenesis and prevention of adhesion formation. Dig Surg. 1998; 15(2):153–157
[15] Jerosch J, Steinleitner W. Minimal invasive Wirbelsäulen-Intervention. Cologne, Germany: Deutscher Ärzte-Verlag; 2009
[16] Zühlke HV, Lorenz EM, Straub EM, Savvas V. Pathophysiology and classification of adhesions. [in German] Langenbecks Arch Chir Suppl II Verh Dtsch Ges Chir 1990;1990:1009–1016
[17] Kamel RM. Prevention of postoperative peritoneal adhesions. Eur J Obstet Gynecol Reprod Biol. 2010; 150(2):111–118
[18] diZerega GS. Biochemical events in peritoneal tissue repair. Eur J Surg Suppl. 1997(577):10–16
[19] Thompson JN, Whawell SA. Pathogenesis and prevention of adhesion formation. Br J Surg. 1995; 82(1):3–5
[20] Liakakos T, Thomakos N, Fine PM, Dervenis C, Young RL. Peritoneal adhesions: etiology, pathophysiology, and clinical significance. Recent advances in prevention and management. Dig Surg. 2001; 18(4):260–273
[21] Koch T, Hübler M. Thorakale epidurale Anästhesie und Analgesie. Munich, Germany: Arcis Verlag GmbH; 2003
[22] Eckard A. Praxis LWS-Erkrankungen. Berlin, Germany: Springer Verlag; 2011
[23] Korsten HH, Ackerman EW, Grouls RJ, et al. Long-lasting epidural sensory blockade by n-butyl-p-aminobenzoate in the terminally ill intractable cancer pain patient. Anesthesiology. 1991; 75(6):950–960
[24] Unterharnscheidt F. Pathologie des Nervensystems VI.B. Heidelberg, Germany: Springer- Verlag; 1993
[25] Botwin KP, Brown LA, Sakalkale D, Savarese R. Side effects and complications of injection procedures: anticipation and management. In: Slipman CW, Derby R, Simeone FA, Mayer TG, eds. Interventional Spine—An Algorithmic Approach. Philadelphia, PA: Saunders-Elsevier; 2008:213–227
[26] Kennedy TM, Ullman DA, Harte FA, Saberski LR, Greenhouse BB. Lumbar root compression secondary to epidural air. Anesth Analg. 1988; 67(12):1184–1186
[27] Kurt A. Die klinischen Langzeitfolgen der intraoperativen Duraverletzung in der Wirbelsäulenchirurgie [inaugural dissertation]. Bochum, Germany: Ruhr-Universität Bochum; 2006
[28] Coscia MF, Trammell TR. Pyogenic lumbar facet joint arthritis with intradural extension: a case report. J Spinal Disord Tech. 2002; 15(6):526–528
[29] Michel H, Steffen P, Weichel T, Seeling W. Epiduritis after long-term pain therapy with an epidural catheter—review of the literature with a current case report. [in German]. Anaesthesiol Reanim. 1997; 22(3):69–79
[30] Reihsaus E, Waldbaur H, Seeling W. Spinal epidural abscess: a meta-analysis of 915 patients. Neurosurg Rev. 2000; 23(4):175–204, discussion 205
[31] Gerdesmeyer L, Wagner K, Birkenmaier C. Epineurolyse mittels Katheter nach Racz. In: Jerosch J, Steinleitner W, eds. Minimal invasive Wirbelsäulen-Intervention. Cologne, Germany: Deutscher Ärzte-Verlag; 2009:67–78
[32] Tronnier V. Neuromodulation bei chronischen Schmerzzuständen. Bremen, Germany: UNI-MED Verlag AG; 2005
[33] Wang LP, Hauerberg J, Schmidt JF. Incidence of spinal epidural abscess after epidural analgesia: a national 1-year survey. Anesthesiology. 1999; 91 (6):1928–1936
[34] Weingarten TN, Hooten WM, Huntoon MA. Septic facet joint arthritis after a corticosteroid facet injection. Pain Med. 2006; 7(1):52–56

15 Anesthetic Management for Epiduroscopy

Shiraz Yazdani and Salahadin Abdi

15.1 Introduction

Epiduroscopy is a procedure that requires careful anesthetic management. Although the instrumentation involved in this procedure is a small fiberoptic endoscope, the size of the endoscope is much larger than instruments usually introduced into the epidural space. This can lead to significant complications if care is not taken when directing the instrument. A properly sedated patient is the key to avoiding complications. Too deep of an anesthetic can lead to the patient being unable to verbalize when the scope may be causing damage to critical structures. Too light of an anesthetic can lead to patient movement, making it difficult and potentially more dangerous to maneuver the scope in the epidural space.

15.2 Preoperative Evaluation

Patients selected for epiduroscopy have had pain for longer than 3 months and have failed a variety of traditional treatments, including interventional procedures such as epidural steroid injections. Conscious sedation is often provided for interventional procedures. Many of these patients have also failed medication management options, potentially including moderate- to high-dose opioid therapy. This leads to challenging intraoperative management of these patients. It is critical to perform a thorough preoperative evaluation of patients scheduled for epiduroscopy. History taking should include the following:

- Review of previous anesthetic records:
  - Previous sedative medications used and patient response.
  - Hemodynamic response to various medication classes.
  - Oxygen supplementation required during previous cases.
  - Postanesthesia care unit status including postanesthetic recovery score and any unusual occurrences.
  - Any particular pressure points that required extra padding in the prone position.
  - Type of head pillow required for patient comfort during the procedure.
- Airway evaluation:
  - Mallampati score.
  - Respiratory status in supine, seated, and prone positions.
  - Ease of previous airway device placement including endotracheal tube or supraglottic device such as a laryngeal mask airway.
  - Removable appliances such as dentures or bridges.
- Medication use:
  - Opioid use and dosing, including last dose.
  - Benzodiazepine use, which is common due to the high prevalence of anxiety among patients with chronic pain.
  - Psychoactive medication use such as selective serotonin reuptake inhibitors and tricyclic antidepressants.
  - Allergies to medications and responses.
- Medical comorbidities:
  - Properly controlled hypertension is critical to minimize the potential risk of bleeding in the epidural space.
  - Diabetic control may limit the use of corticosteroids by the interventional pain physician during the procedure.
  - Diabetic autonomic neuropathy can lead to abnormal cardiovascular responses to stimulation and anesthetic medications.
  - Coronary artery or peripheral vascular disease may preclude the use of certain sedative medications, which have a more pronounced effect on hemodynamics.
  - Chronic liver disease may affect biotransformation of anesthetic medications, leading to increased likelihood of accumulation.
  - Chronic kidney disease leads to decreased clearance; dose adjustments must be made and some medications should be avoided entirely.
- Surgical changes:
  - Previous cervical, thoracic, or lumbar surgery may lead to special considerations when patient positioning is determined.
  - Previous limb surgery or lymph node dissection may limit the available sites for intravenous access and sphygmomanometry.

Physical examination should focus on the following:

- Neurologic examination including sensation, strength, and reflexes to be tested postoperatively in order to test for any compromise.
- Respiratory status including oxygen supplementation, continuous positive airway pressure device use, predisposition for obstructive sleep apnea, reactive airway disease, and chronic obstructive pulmonary disease.[1,2,3]
- Cardiac evaluation for any clicks, murmurs, or gallops indicative of new or worsening cardiac disease.
- Musculoskeletal and integumentary examination to evaluate for any skin changes or ulcers that may be present, especially at the operative site, of which the pain physician must be made aware.

15.3 Patient Positioning

Epiduroscopy requires prolonged prone positioning (▶ Fig. 15.1) and, as such, all appropriate precautions should be taken:

- Preoperative and intraoperative evaluation should be made of all pressure points.[4,5,6]
- Soft padding must be applied to all of these pressure points.
- The patient should be asked about comfort at these points prior to any significant sedative or opioid administration.
- Appropriately sized prone head pillows should be selected based on patient head circumference and review of previous records.

Patients should be secured to the bed to minimize the risk of significant patient movement:

- Arms are typically flexed 90 degrees at the shoulder and elbow should be on arm boards.[7,8,9]
- Care must be taken to avoid nerve compression at the elbow with foam pads.

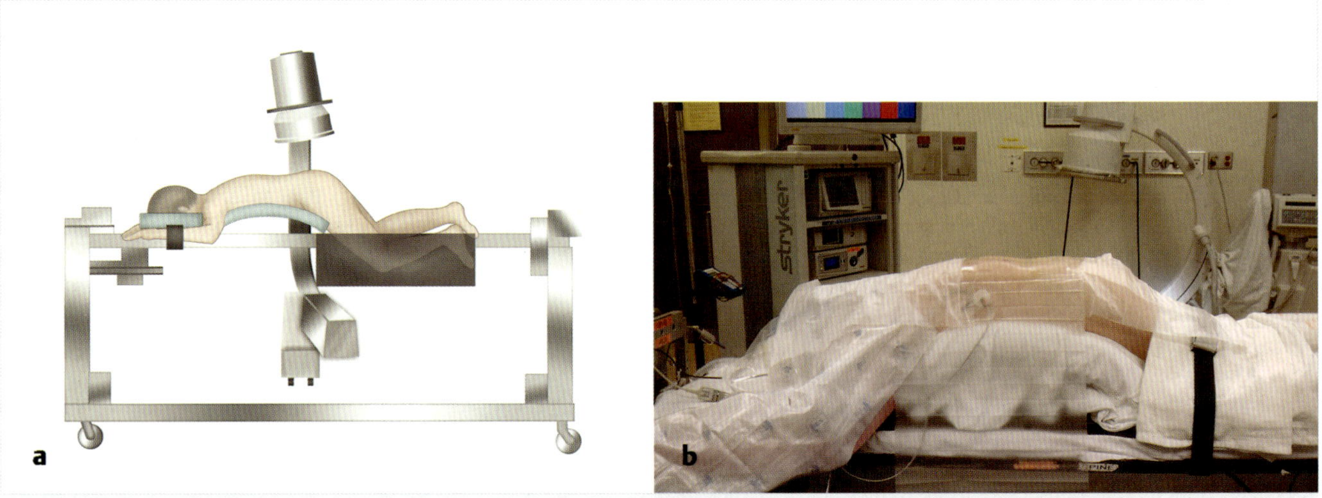

Fig. 15.1 (a,b) Patient positioning. The patient is placed in a prone position, often with a Wilson frame. C-arm fluoroscopy may be used for localization.

- Intravenous lines must be checked for patency, especially if placed in the antecubital fossa, once the patient is properly positioned.

15.4 Patient Monitoring

- As with all monitored anesthesia care/conscious sedation cases, standard monitors should be applied as recommended by the American Society of Anesthesiologists.[1,2,3]
- Monitors include continuous pulse oximetry, three- or five-lead electrocardiogram, intermittent noninvasive sphygmomanometry, and capnography.
- Oxygen supplementation may be required and is usually delivered via a nasal cannula, although a face mask may be required for some patients in order to deliver a higher concentration of inspired oxygen.
- The most important monitor is an appropriately sedated patient who is able to communicate with both the anesthesia care team as well as the physician performing the procedure, which is critical for evaluation and diagnosis of pain generators as well as for monitoring any neurologic changes intraoperatively.

15.5 Medications

15.5.1 Benzodiazepines

- Benzodiazepines exert their action by enhancing the effect of γ-aminobutyric acid (GABA) at the $GABA_A$ receptor.
- They cause sedative, anxiolytic, antiepileptic, anterograde amnestic, and muscle relaxant effects.
- Most commonly, they are used to treat preoperative procedure anxiety.
- Their effects are more pronounced in elderly patients, and dose adjustments are critical in this patient population.[7,8,9]
- Altered liver and/or renal function necessitates dose reduction strategies to avoid accumulation.
- In a small percentage of patients, they can cause disinhibition, leading to an uncooperative patient during the procedure.
- Short-acting variants such as midazolam are most commonly used.

15.5.2 Opioids

- Opioids exert their actions at various endogenous opioid receptors.
- They cause analgesia, sedation, and respiratory depression.
- Their effects are potentiated by the concurrent use of benzodiazepine.
- As with benzodiazepines, short-acting variants are most commonly used.
- Caution must be taken as potent analgesia can prevent the spinal endoscopist from properly identifying pain generators.

15.5.3 Propofol

- Propofol is a sedative–hypnotic medication.
- It exerts its action through potentiation of GABA-A receptors, although sodium channel activity has also been studied.[10,11]
- It is used during conscious sedation cases and commonly used to induce general anesthesia.
- It does not have analgesic properties.
- It has significant respiratory depressant properties.
- As it commonly leads to alterations in consciousness, even at subinduction doses, its use is generally avoid during spinal endoscopy due to the need for a responsive patient.
- Small amounts may be used during insertion of the dilators and endoscope into the sacral hiatus as this is usually the most uncomfortable part of the procedure and the short duration of action of propofol is ideal for such short times of stimulation.

15.5.4 Ketamine

- Ketamine is classified as a dissociative anesthetic.
- Ketamine exerts its action by antagonizing the *N*-methyl-D-aspartate (NMDA) receptor, although other sites of action are also proposed including opioid receptors and monoamine transport sites.

Table 15.1 Spinal/epidural analgesic and pharmacologic therapy

Medication	Class	Mechanism of action	Indication	Contraindication	Notes
Lidocaine	Local anesthetics	Sodium channel blockade	Nerve block, analgesia	Allergy to components	High concentration, risk of transient neurologic symptoms
Bupivacaine	Local anesthetics	Sodium channel blockade	Nerve block, analgesia	Allergy to components	R-enantiomer with higher risk of cardiotoxicity
Ropivacaine	Local anesthetics	Sodium channel blockade	Nerve block, analgesia	Allergy to components	Lower risk of cardiotoxicity than bupivacaine
Clonidine	Sympatholytic	Alpha-2 receptor agonist	Analgesia	Bradycardia, hypotension	Augments effects of local anesthetics
Ketamine	Anesthetic	NMDA antagonist	Analgesia	Tachycardia, hypertension	Potentially neurotoxic
Fentanyl	Opioid	Mu-receptor agonist	Analgesia	Sedation, respiratory depression	Short duration of action, lipophilic
Morphine	Opioid	Mu-receptor agonist	Analgesia	Sedation, respiratory depression	Long duration of action, hydrophilic
Hydromorphone	Opioid	Mu-receptor agonist	Analgesia	Sedation, respiratory depression	Lacks second peak effect present with morphine
Dexamethasone	Glucocorticoid	Transactivation via glucocorticoid receptor	Anti-inflammatory, antinociception	Uncontrolled diabetes or hypertension, pregnancy, severe osteoporosis	Nonparticulate
Methylprednisolone	Glucocorticoid	Transactivation via glucocorticoid receptor	Anti-inflammatory, antinociception	Uncontrolled diabetes or hypertension, pregnancy, severe osteoporosis	Little particulate aggregation
Triamcinolone	Glucocorticoid	Transactivation via glucocorticoid receptor	Anti-inflammatory, antinociception	Uncontrolled diabetes or hypertension, pregnancy, severe osteoporosis	Particulate size similar to Methylprednisolone
Hyaluronidase	Enzyme	Degrade hyaluronan	Breakdown of scar tissue	Allergy to components	Often used for postsurgical scar tissue formation
Ozone	Gas mixture	Reduces free radicals and proinflammatory mediators	Anti-inflammatory, antinociception	Diskitis, infection	Use in epidural space is experimental
Hypertonic saline	Solution	Adhesiolysis, neurolysis	Anti-inflammatory, antinociception	Allergy to components	Used as a bolus or infusion for lysis of adhesions

- It can be used in the induction of general anesthesia and, at lower doses, for sedation.
- It has analgesic, sedative, and amnestic properties.[10,11]
- Neurologic consequences such as hallucinations and nightmares can occur.
- Respiratory drive is minimally affected, making ketamine an option in patients with preexisting respiratory disease.

15.5.5 Alpha-2 Agonists

- Clonidine and dexmedetomidine fall into the category of alpha-2 agonists used for sedation.
- Dexmedetomidine is several-hundred-fold more selective for the alpha-2 receptor than clonidine, making it a popular medication in recent years for sedation.
- Cardiovascular effects can be pronounced, including bradycardia and hypotension.[10,11]
- Respiratory drive is maintained, making these medications ideal for use in sedation cases.
- They do possess mild analgesic properties, but usually not enough to cause interference in the diagnostic portion of the spinal endoscopy procedure.
- Bolus dosing should be avoided due to the aforementioned cardiovascular effects, and titration by intravenous infusion is recommended.

▶ Table 15.1 summarizes the medications.

References

[1] Eilers H. Intravenous anesthetics. In: Miller RD, Pardo MC, eds. Basics of Anesthesia. Philadelphia, PA: Elsevier Saunders; 2007:99-113.

[2] Sweitzer BJ. Preoperative evaluation and medication. In: Miller RD, Pardo MC, eds. Basics of Anesthesia. Philadelphia, PA: Elsevier Saunders; 2007:165-186.

[3] Lee JW, Cassorla L. Patient positioning and associated risks. In: Miller RD, Pardo MC, eds. Basics of Anesthesia. Philadelphia, PA: Elsevier Saunders; 2007:300-316.

[4] Reeves JG, Glass P, Lubarksy DA, et al. Intravenous anesthetics. In: Miller RD, Eriksson LI, Fleisher LA, et al., eds. Miller's Anesthesia. Philadelphia, PA: Churchill Livingstone Elsevier; 2005:720-758.

[5] Fischer SP, Bader AM, Sweitzer BJ. Preoperative evaluation. In: Miller RD, Eriksson LI, Fleisher LA, et al., eds. Miller's Anesthesia. Philadelphia, PA: Churchill Livingstone Elsevier, 2005:1001-1063.

[6] Cassorla L, Lee JW. Patient positioning and anesthesia. In: Miller RD, Eriksson LI, Fleisher LA, et al., eds. Miller's Anesthesia. Philadelphia, PA: Churchill Livingstone Elsevier; 2005:1151-1169.

[7] Barash PG, Cullen BF, Stoelting RK, et al., eds. Intravenous anesthetics. In: Handbook of Clinical Anesthesia. Philadelphia, PA: Wolters Kluwer Health/Lippincott Williams & Wilkins; 2013:248-268.
[8] Barash PG, Cullen BF, Stoelting RK, et al, eds. Preoperative patient assessment and management. In: Clinical Anesthesia. Philadelphia, PA: Wolters Kluwer Health/Lippincott Williams & Wilkins; 2013:569-595.
[9] Barash PG, Cullen BF, Stoelting RK, et al, eds. Patient positioning and potential injuries. In: Clinical Anesthesia. Philadelphia, PA: Wolters Kluwer Health/Lippincott Williams & Wilkins; 2013:793-813.
[10] Butterworth JF, Mackey DC, Wasnick JD, eds. Intravenous anesthetics. In: Morgan & Mikhail's Clinical Anesthesiology. New York, NY: McGraw-Hill; 2013:175-188.
[11] Butterworth JF, Mackey DC, Wasnick JD, eds. Preoperative assessment, premedication, & perioperative documentation. In: Morgan & Mikhail's Clinical Anesthesiology. New York, NY: McGraw-Hill; 2013:295-308.

16 Step-by-Step Epiduroscopic Techniques

Shiraz Yazdani and Salahadin Abdi

16.1 Introduction

Chronic, radicular lower back pain remains one of the most challenging symptoms to properly diagnose and treat. Although imaging modalities such as computed tomography (CT) and magnetic resonance imaging (MRI) have made significant advancements in the past several decades, they cannot offer direct visualization of epidural scarring and fibrotic tissue.

Although myeloscopy has been described since the 1930s,[1] only since the 1990s has flexible fiberoptic epiduroscopy been regularly described in the literature. It is a technique in which a small-diameter flexible endoscope is introduced into the epidural space. This allows direct visualization of the epidural space along with structures such as the dura, nerve roots, and neuroforamina. It also allows the practitioner to diagnose pathology such as epidural scarring or fibrosis via direct visualization of such structures.

The injection port of the epiduroscope can be, and oftentimes is, used in a therapeutic manner to inject local anesthetics, corticosteroids, hyaluronidase, etc. As this is a relatively new technique in the field of percutaneous interventional epidural procedures, it is important to review the anatomy and the step-by-step techniques involved in epiduroscopy.

16.2 Patient Selection

- Epiduroscopy is most commonly performed for patients with postlaminectomy syndrome.
- Patients should have chronic pain of more than 3 months' duration.
- The primary pain complaint in patients should be radicular pain.
- As epiduroscopy is used primarily for diagnosis and treatment of epidural scarring and fibrosis, it follows that patients who have had at least one surgical spine procedure are candidates for this procedure.[2]
- Patients with largely nonradiating back pain should be approached with caution as determination of pain generators in these patients using a technique such as epiduroscopy has not been fully established.[3]
- Conservative and traditional interventional treatments should first be attempted as per the standard of care in the treating location including, but not limited to, physiotherapy, pharmacotherapy, epidural corticosteroid injections, and epidural lysis of adhesions.
- Patients should be able to tolerate the prone position with minimal sedation so as not to confound the diagnostic value of the procedure.
- As with all interventional procedures, epiduroscopy is contraindicated in patients with surgical site/systemic infections or bleeding diathesis.

16.3 Preoperative Imaging

- Patients should have imaging with CT or MRI to rule out any other treatable pathology prior to epiduroscopy.
- Radicular symptoms may be discordant to findings on imaging, owing to the poor visualization of epidural scarring and fibrosis on these imaging modalities.
- Severe central canal stenosis or cauda equine crowding may be a relative contraindication as introducing an endoscope into an epidural space that is close to being compromised may cause neural compression owing to the epiduroscopic technique as well as the size, albeit relatively small, of the epiduroscope.

16.4 Patient Preparation

- As with all procedures, all risks of the procedure should be discussed with the patient and informed consent obtained.
- Although an interlaminar technique for epiduroscopy has been described,[4] most approaches in the literature support access to the epidural space via the sacral hiatus. The transforaminal approach has also been increasingly used and its steps have been covered in this book.
- With either approach, the patient is placed in the prone position with appropriate padding of bony prominences. Tolerance of this position with minimal sedation is required.
- Preoperative antibiotics within 30 minutes of skin violation have been described due to the invasiveness of the procedure and can be accomplished with 1,000 to 2,000 mg of intravenous cefazolin or 600 to 900 mg of intravenous clindamycin for those who are penicillin-allergic (current institutional standards should be utilized).
- Sterile preparation of the operative site should be achieved with either chlorhexidine or povidone-iodine as per current institutional standards.
- Full surgical gown and gloves after thorough, standard hand scrub should be donned by those involved in the procedure.
- Sterile draping of the lumbosacral spine as well as the operative fluoroscope should be performed using strict aseptic technique.

16.5 Epiduroscope Insertion

- Small, flexible, fiberoptic endoscopes vary in size, and those used for epiduroscopy are typically 2.5 to 3 mm in diameter.
- The tip of the endoscope is typically bidirectional and rotates in either direction up to 170 degrees.
- The working port of the endoscope is smaller, approximately 1 mm in diameter, and allows the introduction of injectate as well as tools such as forceps and laser fiber. ▸ Fig. 16.1 demonstrates the equipment required to perform epiduroscopy.
- The skin entry site and trajectory to the sacral hiatus are identified with anteroposterior (AP) and lateral fluoroscopy. Local anesthetic is injected into the overlying skin.
- The caudal space is accessed using a Tuohy needle (▸ Fig. 16.2, ▸ Fig. 16.3, and ▸ Fig. 16.4).
- Entry into the space is confirmed on AP and lateral fluoroscopy with injection of radiopaque contrast dye.

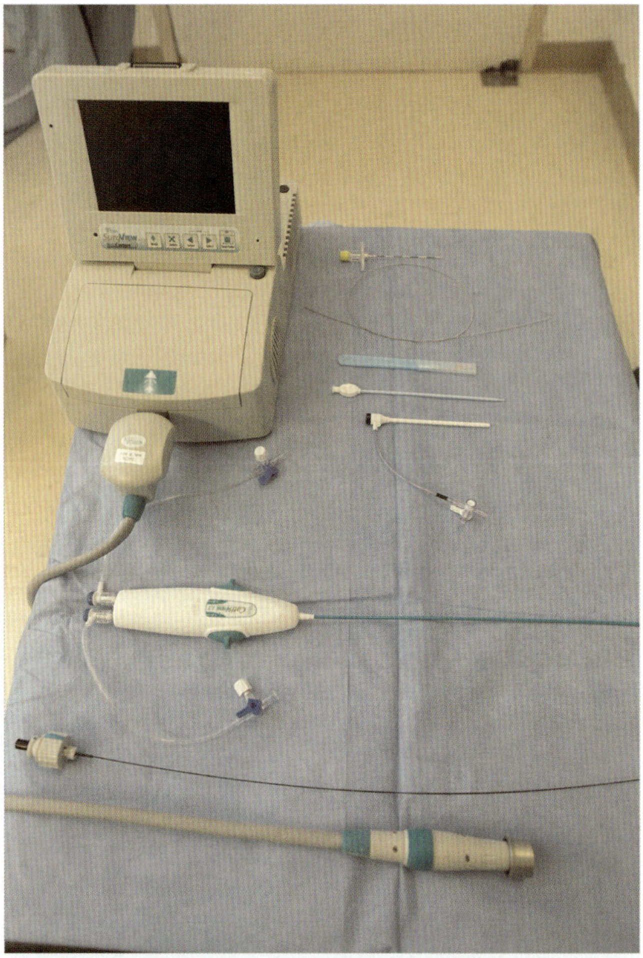

Fig. 16.1 Equipment used for epiduroscope introduction. Top to bottom: Tuohy needle, 11 blade, dilator, sheath, dual-channel endoscope introducer, microendoscope, and connector to the imaging system and light source. Top left of the Mayo stand shows the imaging system and light source.

- The Tuohy needle stylet is then removed and the guidewire is passed through the needle into the epidural space (▶ Fig. 16.5) and is confirmed with AP and lateral fluoroscopy (▶ Fig. 16.6 and ▶ Fig. 16.7).
- The needle is then removed (▶ Fig. 16.8) and the skin entry site enlarged with a blade (▶ Fig. 16.9).
- The dilator is introduced over the guidewire (▶ Fig. 16.10).
- The dilator and sheath are then introduced over the guidewire into the epidural space (▶ Fig. 16.11).
- The dilator and guidewire are removed, leaving the sheath in place (▶ Fig. 16.12).
- Gravity irrigation is connected to the sheath, and the endoscope introducer is passed into the epidural space (▶ Fig. 16.13).
- Depending on patient anatomy, it may be necessary to use stepwise dilation of the sacral hiatus prior to introduction of the endoscopic sheath.
- The endoscope is then advanced through the sheath into the epidural space (▶ Fig. 16.14 and ▶ Fig. 16.15).

16.6 Epiduroscope Advancement and Visualization of Pathology

- The epiduroscope should advance freely in the epidural space without resistance.
- Intermittent normal saline infusion via the working channel of the epiduroscope should be used to expand the epidural space.
- No data have been established as to the maximum total volume of normal saline that can be injected, but this is typically set at approximately 30 mL and should be titrated based on patient discomfort caused by rapidly or excessively increasing epidural pressure.
- Direct visualization should be maintained at all times, and live fluoroscopy should be utilized whenever advancing or redirecting the scope.

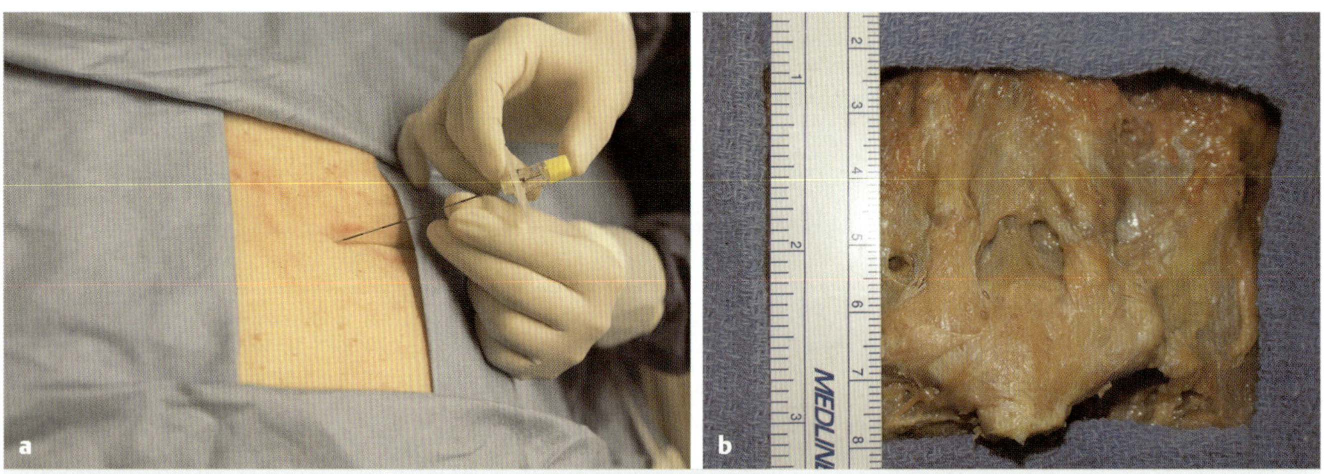

Fig. 16.2 (a) The puncture site and area around the sacral hiatus is first injected with a local anesthetic. The Tuohy needle is inserted until the sacral hiatus is punctured. The initial needle angle is approximately 45 degrees. **(b)** Cadaveric dissection of the sacral hiatus. The range of width of the base is 7 to 27 mm, with a mean of 17 mm. The average anteroposterior diameter of the sacral canal is 6 mm, with a range of 0 to 11 mm. The average distance from the apex of the sacral hiatus to the dura is 47 mm, with a range of 19 to 75 mm. The sacral hiatus is closed by the coccygeal ligament. This ligament must be perforated in order to gain access to the epidural space.

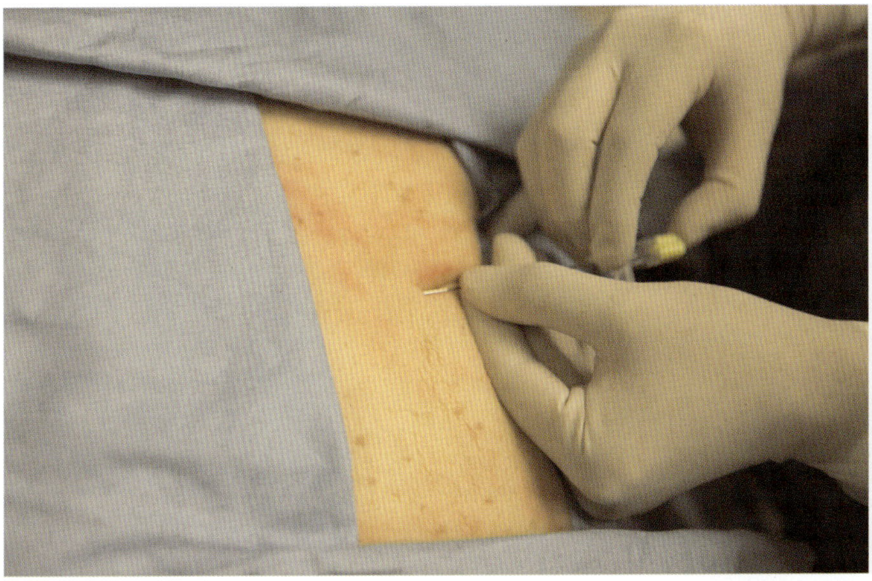

Fig. 16.3 Upon skin entry, the Tuohy needle is advanced towards the sacral hiatus under the fluoroscope guidance.

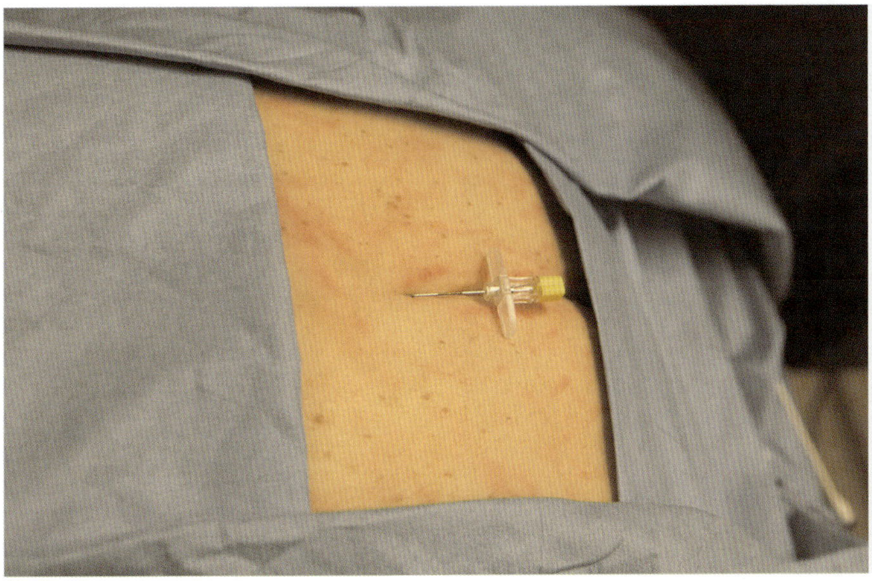

Fig. 16.4 Once the sacrococcygeal ligament has been perforated, the needle is aligned with the vertebral canal. The needle position can be confirmed with fluoroscopy.

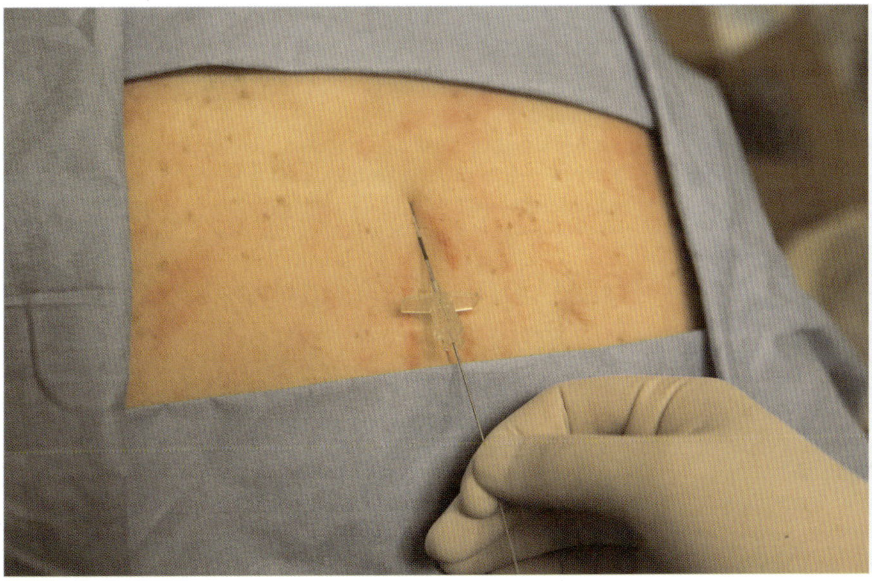

Fig. 16.5 The stylet from the Tuohy needle is removed and the guidewire introduced into the epidural space.

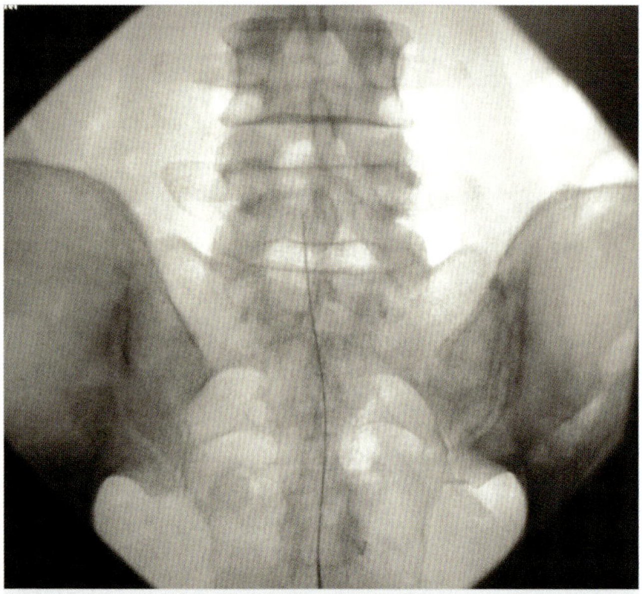

Fig. 16.6 Anteroposterior fluoroscopy demonstrating the guidewire through the sacral hiatus in the epidural space.

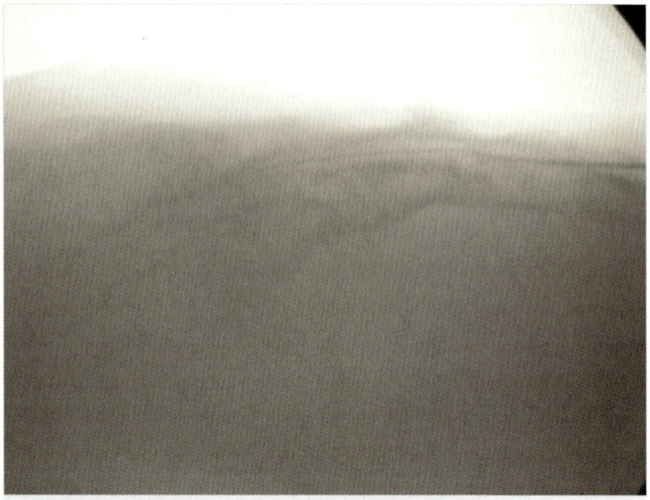

Fig. 16.7 Lateral fluoroscopy demonstrating the spinal needle through the sacral hiatus in the epidural space. The guidewire is passed through the spinal needle in the epidural space.

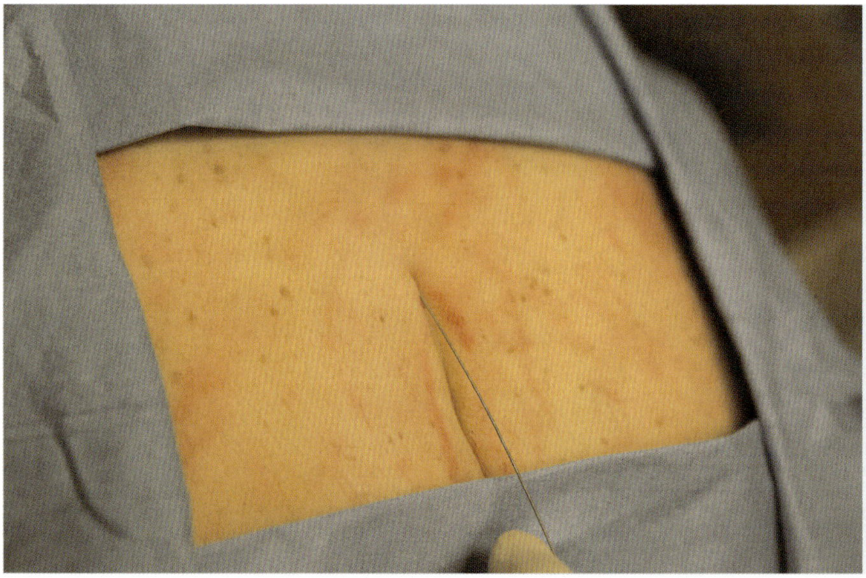

Fig. 16.8 While keeping the guidewire in position, the Tuohy needle is removed.

- The cephalad limit of advancement is typically at L1–L2 as, beyond this level, the epidural space narrows considerably due to the increasing diameter of the conus medullaris and spinal cord.
- Normal and pathologic structures, including the dura, periosteum, epidural fat, blood vessels, nerve roots, and epidural scar tissue or fibrosis, should be analyzed on images obtained through the epiduroscope.
- Fibrous tissue can often be found following spinal surgical intervention and is thought to cause pain from its adhesion to neural structures such as nerve roots (▶ Fig. 16.16).
- Grading of epidural fibrosis has been described and ranges from grade 1 (loose strings and sheets of fibrous tissue) to grade 4 (dense, continuous fibrous material that does not allow passage of the epiduroscope).
- Further diagnosis can be carried out by injecting radiopaque contrast dye in order to evaluate for filling defects at suspected pathologic levels.

16.7 Contact Diagnosis and Therapeutic Injection

- As with most diagnostic procedures, it is prudent to begin the contact diagnostic portion of the procedure on the nonsymptomatic side to act as a control.
- Direct contact of neural structures and/or fibrous tissue with adhesion to these structures should be performed.
- The patient should be able to answer questions clearly and appropriately.

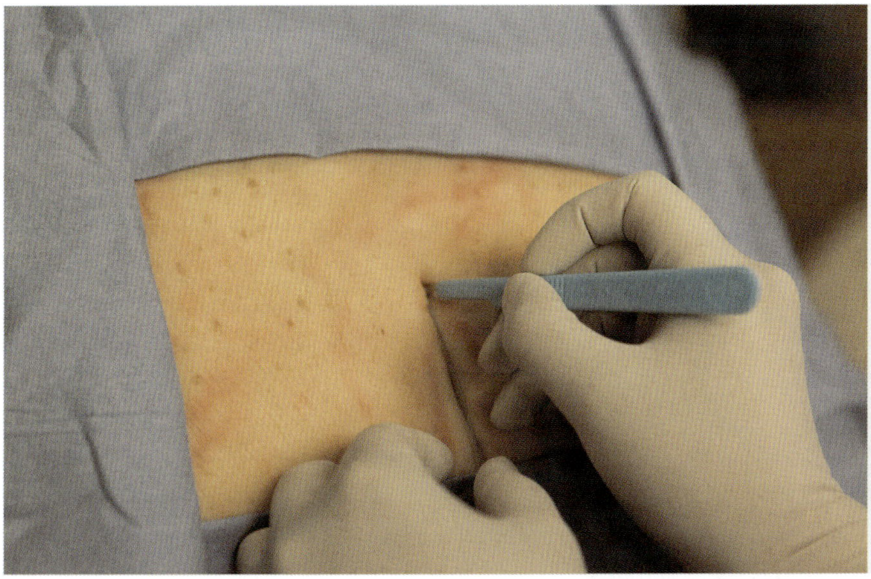

Fig. 16.9 The skin entry site is enlarged to accommodate the dilator.

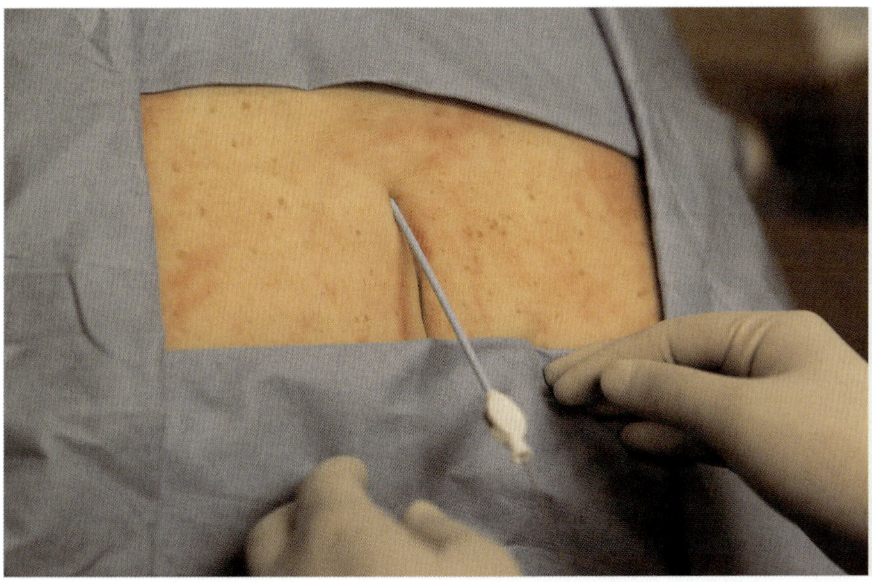

Fig. 16.10 The dilator is passed over the guidewire.

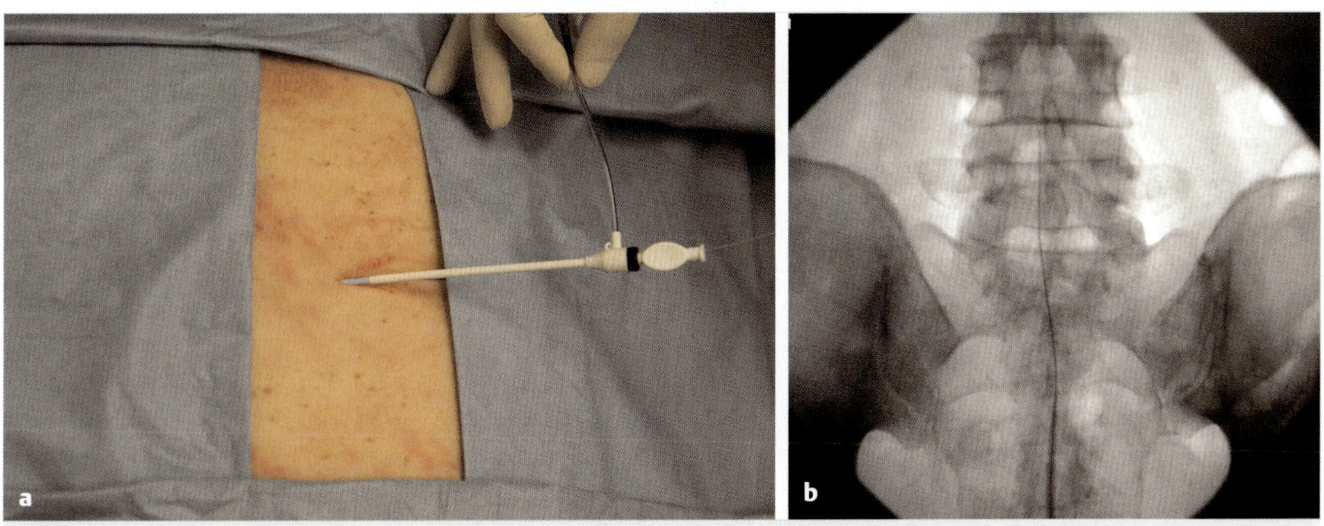

Fig. 16.11 (**a**) The dilator and sheath are then passed over the guidewire into the epidural space. (**b**) Anteroposterior fluoroscopy demonstrating the guidewire through the sacral hiatus in the epidural space. The dilator and sheath are passed over the guidewire into the epidural space.

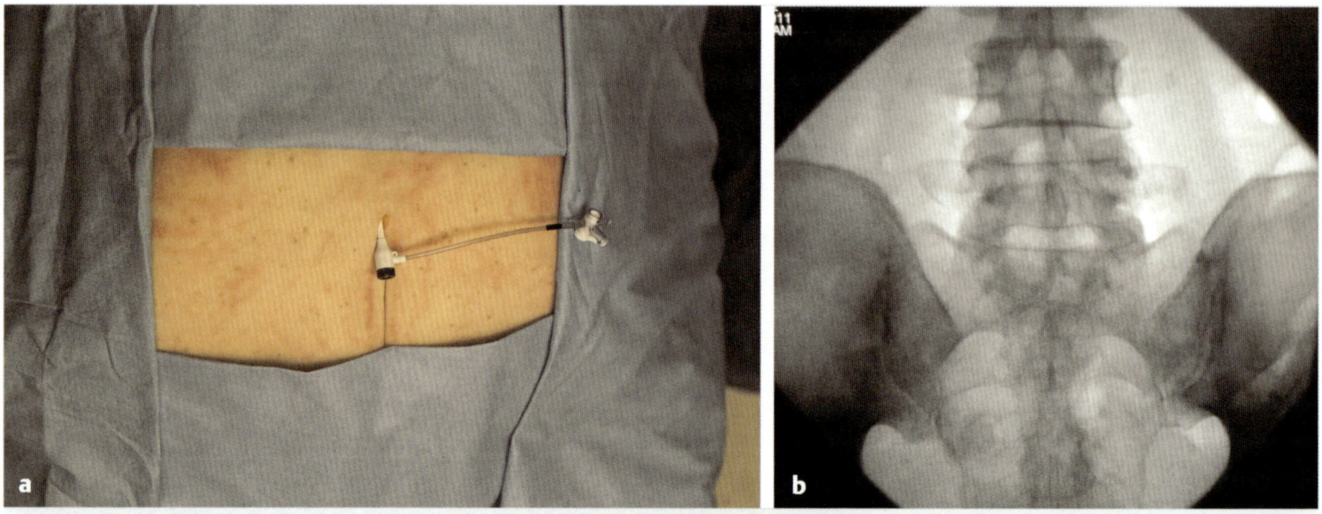

Fig. 16.12 (**a**) The dilator is removed, leaving the sheath in the epidural space. (**b**) Anteroposterior fluoroscopy demonstrating the sheath through the sacral hiatus in the epidural space after the guidewire and dilator have been removed.

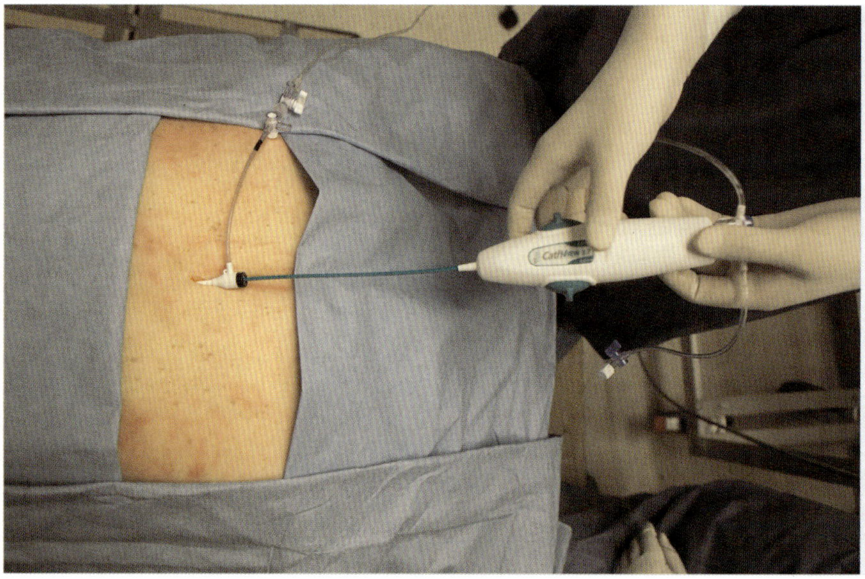

Fig. 16.13 Gravity irrigation is connected to the sheath. The dual-channel microendoscope introducer is passed into the epidural space.

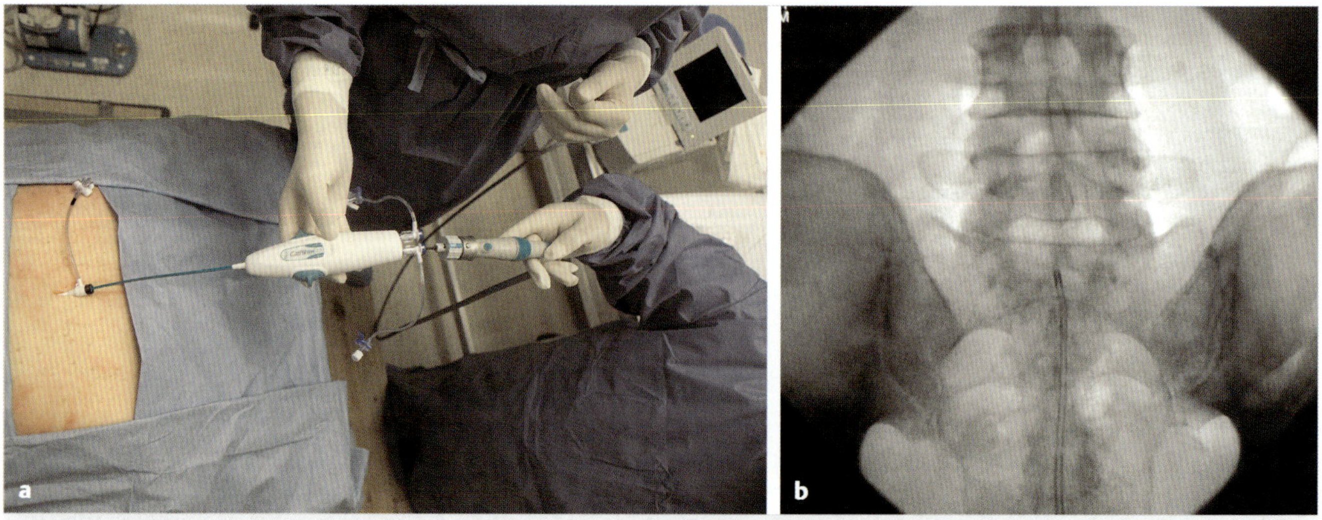

Fig. 16.14 (**a**) The microendoscope is passed through the introducer into the epidural space. (**b**) Anteroposterior fluoroscopy demonstrating the sheath through the sacral hiatus and passage of the epiduroscope into the epidural space.

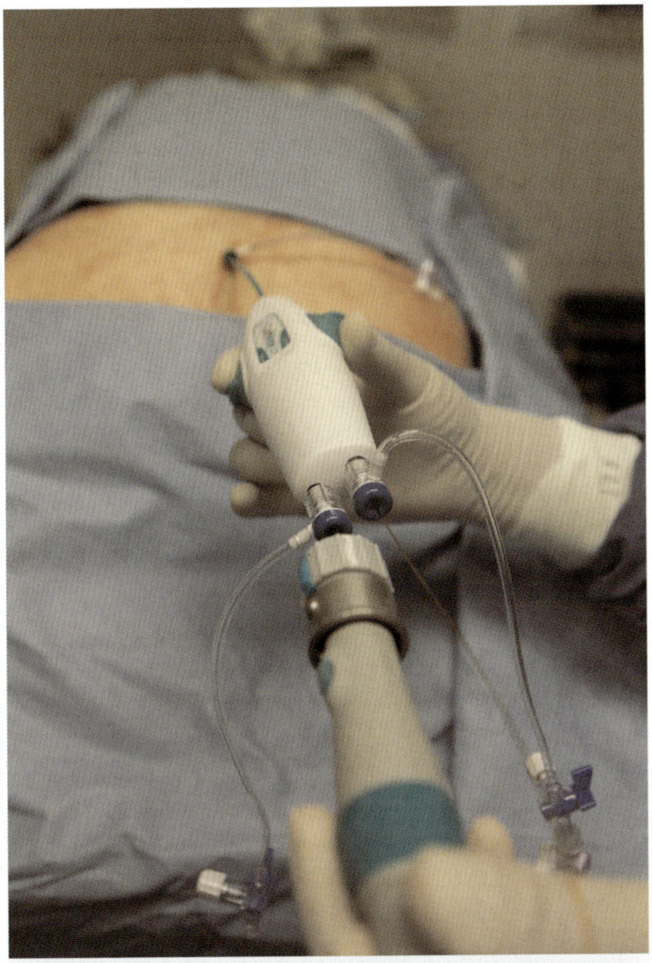

Fig. 16.15 The microendoscope is passed through the introducer into the epidural space. Confirmation of the endoscope in the epidural space is by direct visualization of epidural anatomic structures on the display.

- Direct contact of the epiduroscope with epidural structures may cause pain, and reports of such pain should be compared to the patient's usual pain complaint in order to establish concordancy or discordancy with clinical examination and imaging results.
- Once a concordant painful structure is identified, the practitioner may elect to perform a directed injection at that area.
- Injectates can include, but are not limited to, local anesthetic, corticosteroid, hypertonic saline, and hyaluronidase.
- If lysis of adhesions is being performed and a preinjection contrast study demonstrated a filling defect, contrast injection may be repeated postinjection to evaluate for appropriate adhesiolysis.

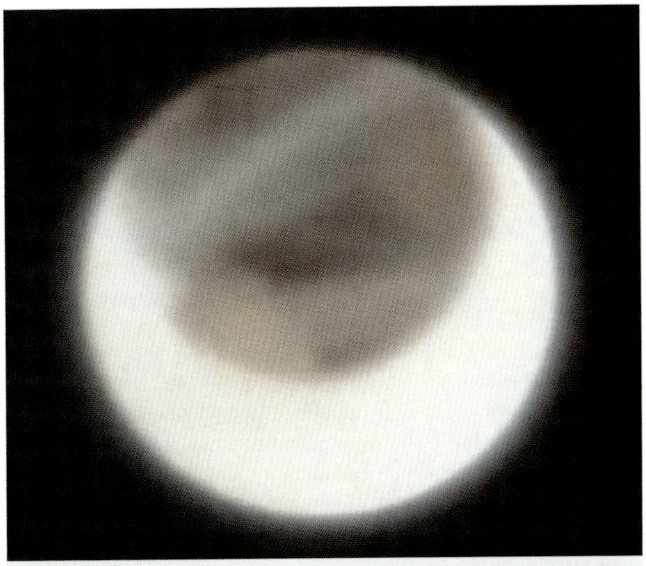

Fig. 16.16 Epiduroscopic view demonstrates significant fibrous bands in the epidural space.

16.8 Epiduroscope Removal and Procedure Completion

- Once the procedure has been completed, the epiduroscope should be returned to the neutral position and slowly withdrawn under fluoroscopic guidance.
- The sheath should then be carefully removed due to its size.
- Superficial bleeding should be treated with pressure occlusion.
- If a significant skin defect is present from insertion of the dilator, it may be approximated using dermal adhesive and/or adhesive skin strips.
- Patients should be monitored postoperatively for any signs of neurologic compromise, as well as surgical site bleeding and dural puncture headache.
- Epiduroscopy is typically an outpatient procedure and patients are discharged the same day.

References

[1] Burmann MS. Myeloscopy or the direct visualization of the spinal cord. J Bone Joint Surg. 1931; 13(4):695-696

[2] Bosscher HA, Heavner JE. Incidence and severity of epidural fibrosis after back surgery: an endoscopic study. Pain Pract. 2010; 10(1):18–24

[3] Vanelderen P, Van Boxem K, Van Zundert J. Epiduroscopy: the missing link connecting diagnosis and treatment? Pain Pract. 2012; 12(7):499–501

[4] Avellanal M, Diaz-Reganon G. Interlaminar approach for epiduroscopy in patients with failed back surgery syndrome. Br J Anaesth. 2008; 101 (2):244–249

17 Spinal Epiduroscopic Mechanical Adhesiolysis

Kent H. Nouri

17.1 Indications

Endoscopic adhesiolysis is built on the principle that by safely utilizing flexible fiberoptic catheters to access the epidural space via the sacral hiatus, a three-dimensional visualization of the space will allow for targeted mechanical rupture of pain-causing adhesions. The procedure allows for specific optical investigation of nerve roots and its pathology, and thus allows for adhesiolysis if pain is attributed to the visualized scarring noted. Adhesiolysis may be achieved by utilizing the fiberoptic catheter to mechanically disrupt connective tissue with injectate. Conditions for which endoscopic adhesiolysis is indicated include failed back surgery syndrome/postlaminectomy pain syndrome, epidural adhesions, and disk herniation. Endoscopic adhesiolysis is thus indicated in patients with an appropriate diagnostic evaluation with chronic lower back pain and radiculopathy who have failed to respond to more conservative treatment modalities.

17.2 Pathophysiology

- Epidural fibrosis is an inflammatory reaction of the arachnoid, a nonvascular elastic tissue that encases the central nervous system.
- Etiologies include hematoma, infection, annular tear, and surgical trauma, which lead to an inflammatory cascade including edema, venous congestions, and fibrosis.[1]
  - A retrospective review of 182 surgical revisions in patients with failed back surgery found that more than 60% were due to epidural fibrosis, which did not respond well to surgery.[2]
- In turn, the fibrosis and scarring cause a mechanical pressure on the nerve roots with a decrease in cerebrospinal fluid–mediated nutrient delivery, causing them to be hyperesthetic and hypersensitive to compression.
  - As approximately 75% of the nutrition to the nerve roots depends upon cerebral spinal fluid flow,[3] in disease states such as fibrosis, this may lead to a critical cause of neural damage.
  - A study testing sensory nerve function before and after epiduroscopic adhesiolysis found a significant improvement in A-beta and A-delta nerve conduction values.[4]
- Nucleus pulposus application via herniation may also lead to rapid thrombus formation within the intraneural capillaries, causing an impairment of blood flow and thus abnormalities in nerve conduction and pain generation.[5]

17.3 Anatomy

- The epidural space lies between the dural sac and the ligamentous boundaries of the vertebral canal (▶ Fig. 17.1).
- This space is filled with areolar connective tissue, which surrounds the dural sac while lining the laminae and pedicles. This epidural membrane lines the vertebral bodies and then passes deep into the posterior longitudinal ligament. The membrane is also drawn laterally to form a sheath around the dural sleeve of the nerve roots.[6]

17.4 Technique

- The most commonly used method of epidurosopic adhesiolysis include the approach via the sacral hiatus; however, the transforaminal approach may be utilized for a more targeted approach on specific nerve roots, which may be inaccessible utilizing a posterior epidural approach.
  - The advantages of the straight entry via the caudal approach include ease of access with passage of the endoscopic device into the epidural space, with less risk of complications including dural puncture and damage to the device itself (▶ Fig. 17.2).
- Prior to undergoing endoscopic adhesiolysis, more conservative treatment modalities including medical management and epidural steroid injections should be exhausted.
  - Appropriate laboratory studies should be performed to rule out bleeding disorders, and anticoagulants should be stopped to avoid bleeding postoperatively.
- Utilizing sterile precautions, the patient is placed in a prone position on the operating room table and the fluoroscope is adjusted over the lumbosacral regions to provide for both anteroposterior and lateral views.
- After infiltration of local anesthetic, an epidural needle is introduced within the epidural space and a lumbar epidurogram is performed using approximately 2 to 5 mL of contrast.

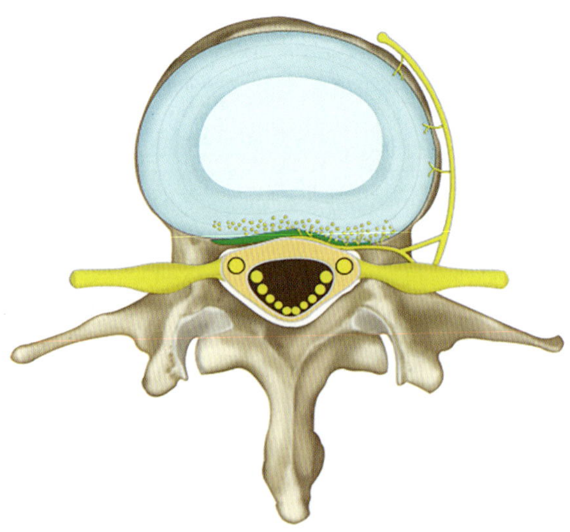

Fig. 17.1 Superior view at the level of the L4–L5 intervertebral disk. Normal relationship between the intervertebral disk and L4 exiting nerve root and L5 traversing nerve root in the epidural space is observed. The epidural space is indicated by the arrow.

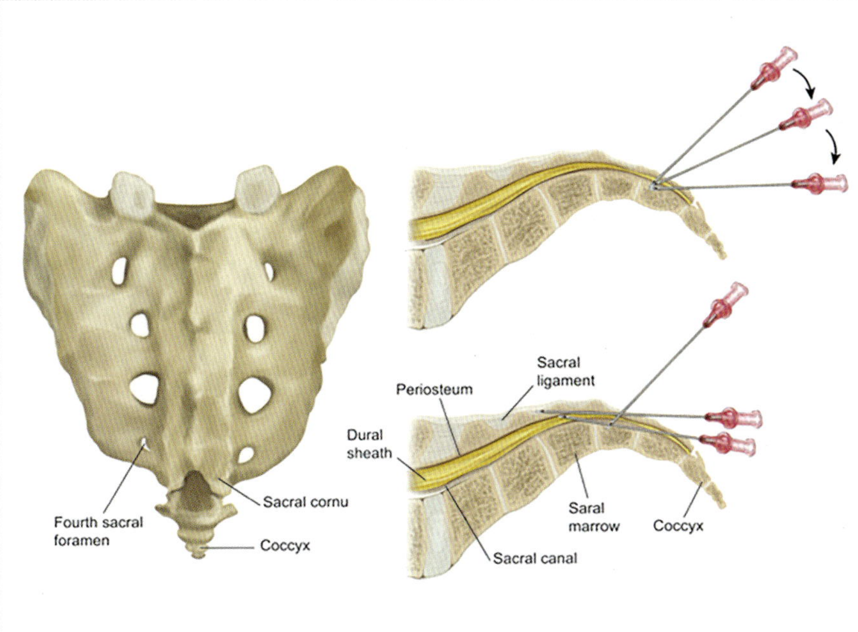

Fig. 17.2 Entry into the sacral canal for epiduroscopy is through the sacral hiatus. Entry of the spinal needle into the sacral canal requires perforation of the sacrococcygeal ligament. The initial entry angle of the needle is approximately 45 degrees. Once the sacrococcygeal ligament is perforated, the needle is aligned to the axis of the vertebral canal. Proper entry is confirmed with fluoroscopy and contrast injection.

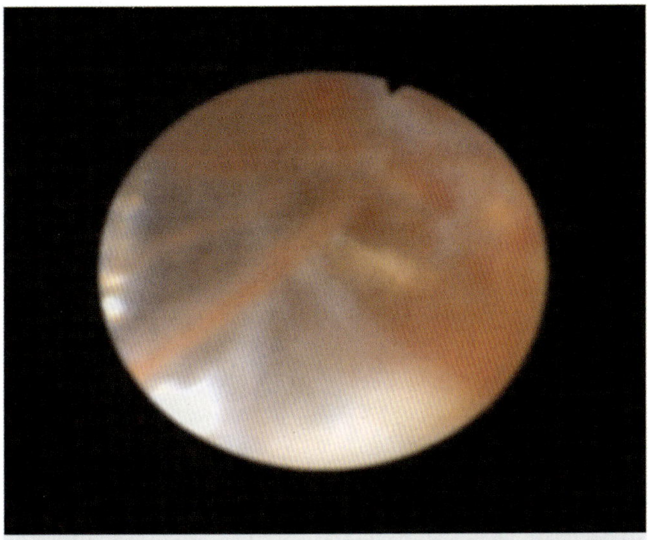

Fig. 17.3 Significant adhesions secondary to arachnoiditis is seen on epiduroscopy.

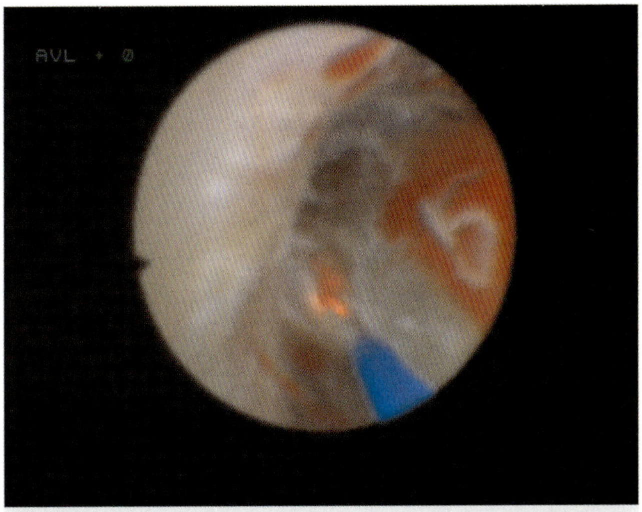

Fig. 17.4 The forward-firing laser is used for adhesiolysis.

 - The epidurogram is performed to examine the contrast flow to find filling defects.
- A guidewire is then inserted through the needle and advanced to the level of suspected pathology.
- At this point, a small incision is made and an introducer with dilator is advanced over the guidewire to the epidural space using the Seldinger technique.
- The dilator and guidewire are then removed from the introducer and replaced with a fiberoptic spinal endoscope.
- Utilizing gentle irrigation with normal saline to distend the epidural space and provide an adequate visual field, the fiberoptic myeloscope is gently advanced and rotated in multiple directions with direct visualization to the affected nerve roots.
 - In cases of arachnoiditis, diffuse inflammation with the presence of fibroinflammatory tissue has been observed (▶ Fig. 17.3).[7]
 - Stenosis may be observed as difficulty in separating dura mater of the different structures with fluid infusion, with separation of the two layers on catheter advancement but immediate resumption of contact on epiduroscope withdrawal.[7]
- Confirmation of the impaired nerve root may be performed by reproduction of the pain by instrumental contact.
- Adhesiolysis is performed by mobilization of the adhesions that are in contact with the affected nerve roots, using the tip of the spinal endoscope, a dissector through the working channel, laser coagulation (▶ Fig. 17.4), or balloon dilation (▶ Fig. 17.5).

- Some practitioners inject a solution containing anesthetic and steroid to the affected areas, utilizing the pocket created by the mechanical and hydrostatic adhesiolysis.
- A confirmatory epidurogram is then performed showing contrast flow through the previously blocked epidural space to verify that adhesiolysis has been accomplished.
- Once the catheter is removed and bandages applied, the patient is monitored for the absence of neurologic and anesthetic complaints and thus may be discharged accordingly.
- Common complications may include pain at the procedure site, steroid side effects, dural puncture with headache and cerebrospinal leak, infection, paresthesia, and transient subarachnoid blockade if anesthetics were utilized in the injectate.
- More severe complications reported in the literature include visual impairment secondary to retinal hemorrhage, due to increased cerebrospinal fluid pressure with bolus fluid injections.[8]

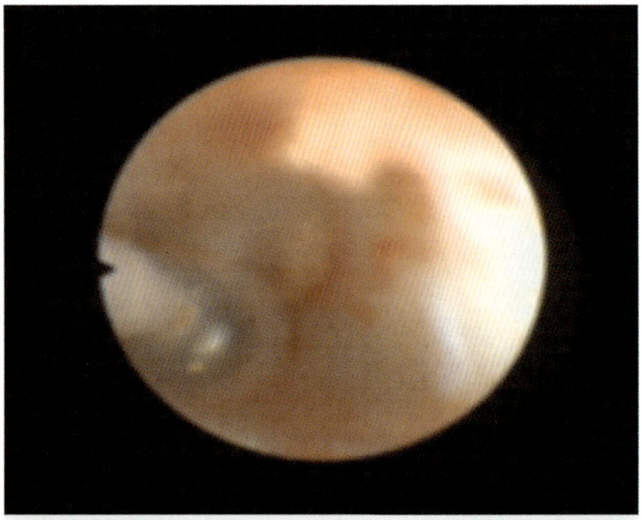

Fig. 17.5 The Fogarty balloon is inflated for adhesiolysis.

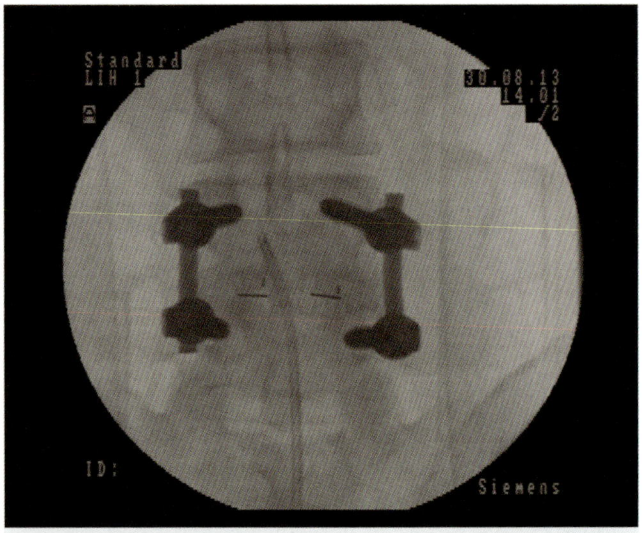

Fig. 17.7 Anteroposterior fluoroscopic view of the lumbar spine of a patient with a previous L4–L5 interbody fusion with posterior instrumentation with failed back surgery syndrome. The epiduroscope can be seen at the level of the left L4 pedicle.

17.5 Example Cases

- A patient with a previous L4–L5 interbody fusion with posterior instrumentation (▶ Fig. 17.6 and ▶ Fig. 17.7). Epiduroscopy demonstrated adhesions in the epidural space (▶ Fig. 17.8). Adhesiolysis is performed under direct vision with a dissector (▶ Fig. 17.9).

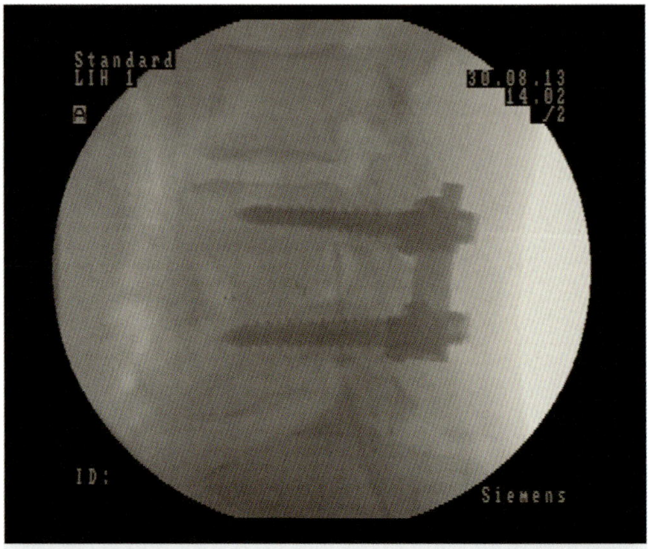

Fig. 17.6 Lateral fluoroscopic view of the lumbar spine of a patient with a previous L4–L5 interbody fusion with posterior instrumentation with failed back surgery syndrome. The epiduroscope can be seen in the ventral aspect of the spinal canal.

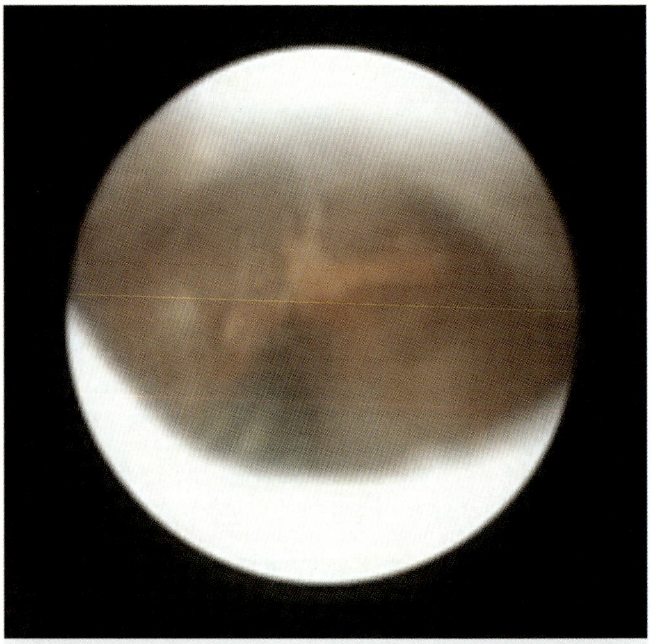

Fig. 17.8 Epiduroscopic view demonstrates significant adhesions in the epidural space.

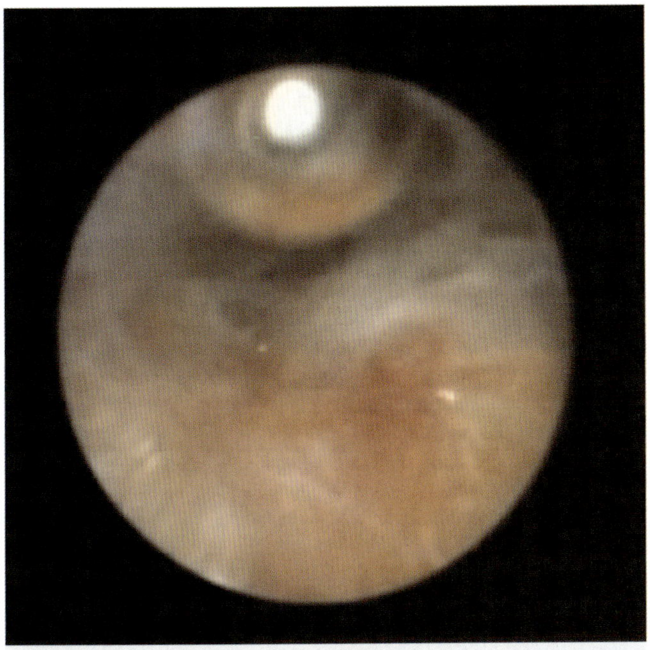

Fig. 17.9 Adhesiolysis in the lumbar epidural space.

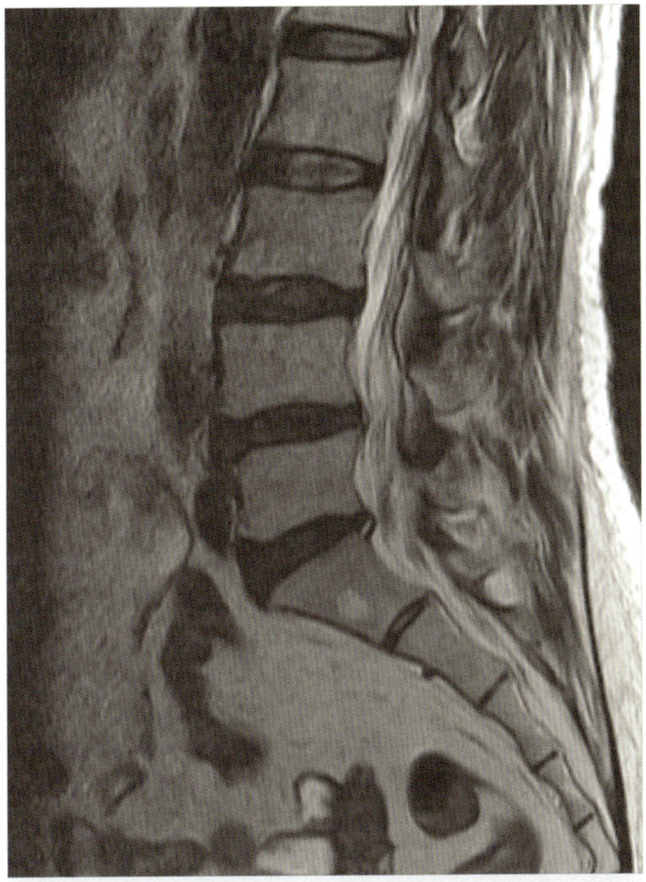

Fig. 17.10 Midsagittal T2 MRI of the lumbar spine of a 59-year-old woman with a 4-month history of left leg pain.

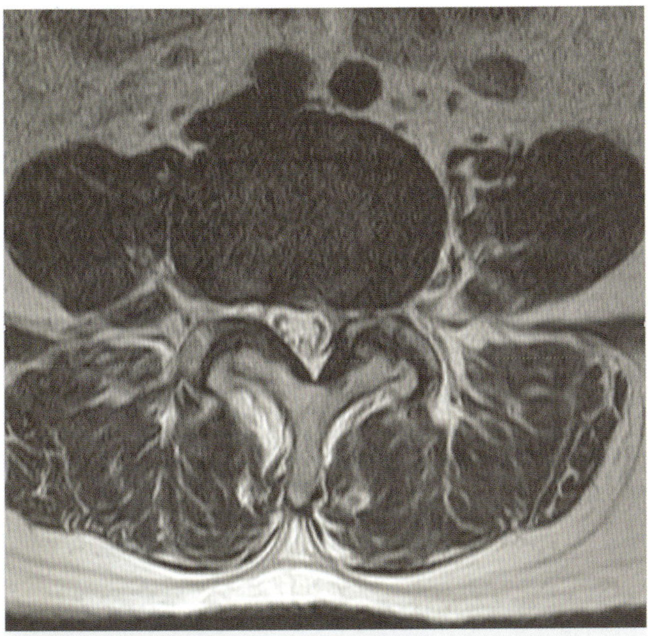

Fig. 17.11 Axial T2 MRI of the lumbar spine at L3–L4 of the same patient in ▶ Fig. 17.10.

- A 59-year-old woman with a 4-month history of pain in her left leg. Midsagittal T2 MRI (▶ Fig. 17.10) and axial T2 MRI of the lumbar spine (▶ Fig. 17.11) demonstrate central and lateral recess stenosis, respectively. Epiduroscopy demonstrated significant adhesions in the epidural space (▶ Fig. 17.12 and ▶ Fig. 17.13). Adhesiolysis was performed with laser coagulation.

Fig. 17.12 Epiduroscopic view demonstrates significant fibrous bands in the epidural space.

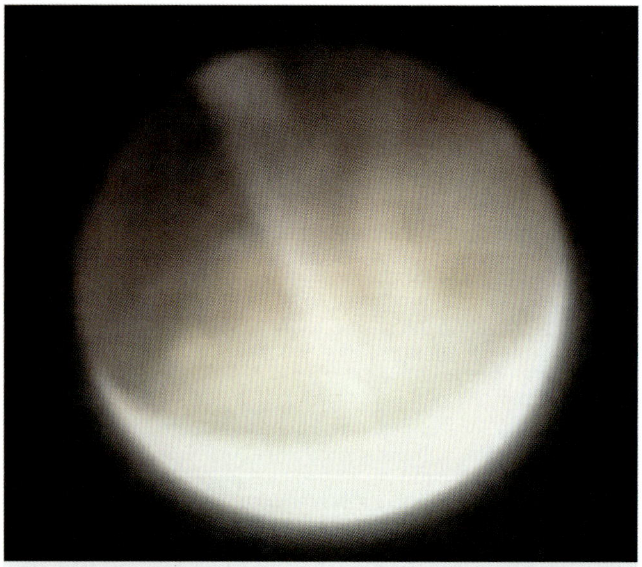

Fig. 17.13 Epiduroscopic view demonstrates significant fibrous bands in the epidural space.

References

[1] Manchikanti L, Saini B, Singh V. Spinal endoscopy and lysis of epidural adhesions in the management of chronic low back pain. Pain Physician. 2001; 4 (3):240–265

[2] Fritsch EW, Heisel J, Rupp S. The failed back surgery syndrome: reasons, intraoperative findings, and long-term results: a report of 182 operative treatments. Spine. 1996; 21(5):626–633

[3] Rydevik B, Brown MD, Lundborg G. Pathoanatomy and pathophysiology of nerve root compression. Spine. 1984; 9(1):7–15

[4] Sakai T, Aoki H, Hojo M, Takada M, Murata H, Sumikawa K. Adhesiolysis and targeted steroid/local anesthetic injection during epiduroscopy alleviates pain and reduces sensory nerve dysfunction in patients with chronic sciatica. J Anesth. 2008; 22(3):242–247

[5] Kayama S, Konno S, Olmaker K, Yabuki S, Kikuchi S. Incision of the annulus fibrosus induces nerve root morphologic, vascular and functional changes. An experimental study. Spine. 1996; 21:2539–2543

[6] Wiltse LL, Fonseca AS, Amster J, Dimartino P, Ravessoud FA. Relationship of the dura, Hofmann's ligaments, Batson's plexus, and a fibrovascular membrane lying on the posterior surface of the vertebral bodies and attaching to the deep layer of the posterior longitudinal ligament. An anatomical, radiologic, and clinical study. Spine. 1993; 18(8):1030–1043

[7] Avellanal M, Diaz-Reganon G. Interlaminar approach for epiduroscopy in patients with failed back surgery syndrome. Br J Anaesth. 2008; 101(2):244–249

[8] Gill JB, Heavner JE. Visual impairment following epidural fluid injections and epiduroscopy: a review. Pain Med. 2005; 6(5):367–374

18 Spinal Epiduroscopic Laser Therapy

Jae Do Kim and Gun Woo Lee

18.1 Introduction

Epiduroscopic procedure for the management of spine-related problems is one of the most commonly performed interventions; it is a minimally invasive approach and has theoretically several advantages, including: (1) adhesiolysis of pathologic lesions such as adhesion and scarred tissue; (2) chemical delusion by saline irrigation; and (3) additional anti-inflammatory effect using steroid. However, only a moderate proportion of patients show improvement in pain and functional level, and these relatively unsatisfactory outcomes suggest that additional methods should be sought to improve its efficacy, for which the additive use of laser can be an ideal option adapting with conventional procedure.

Clinical adaptation of laser has been shown to lead to favorable outcomes in spine territory as laser can reduce pressure by vaporizing a certain volume of disk material, which reduces the pressure between the nucleus pulposus and the peridiskal tissue, causing retraction of the herniation away from the dura and nerve root.[1] Various types of laser for managing spinal disorders have been reported at the clinical or experimental level, including the neodymium:yttrium-aluminum-garnet (Nd:YAG) laser, frequency-doubled Nd:YAG laser, holmium:YAG (Ho:YAG) laser, and semiconductor laser. Of them, the Ho:YAG laser has a wavelength of 2.1 µm and reaches a depth of ≤ 0.5 mm in the tissue.[2,3,4,5,6,7] Tissue permeability is low, tissue vaporization is excellent, and the damage of surrounding tissue is minimal.[4] In the author's thought, the Ho:YAG laser may cause less damage to the surrounding tissues such as dura or nerve root in small enclosed spaces compared with other types of lasers, and thus we utilized it.

18.2 Procedures

18.2.1 Step 1: Position and Anesthesia

Each patient was instructed to lie on a radiologic testing table in the prone position; the skin in the region of the sacral hiatus was swabbed with disinfectant for sterilization (▶ Fig. 18.1 and ▶ Fig. 18.2), and local anesthesia with 1% lidocaine (Xylocaine, AstraZeneca) was administered (▶ Fig. 18.3).

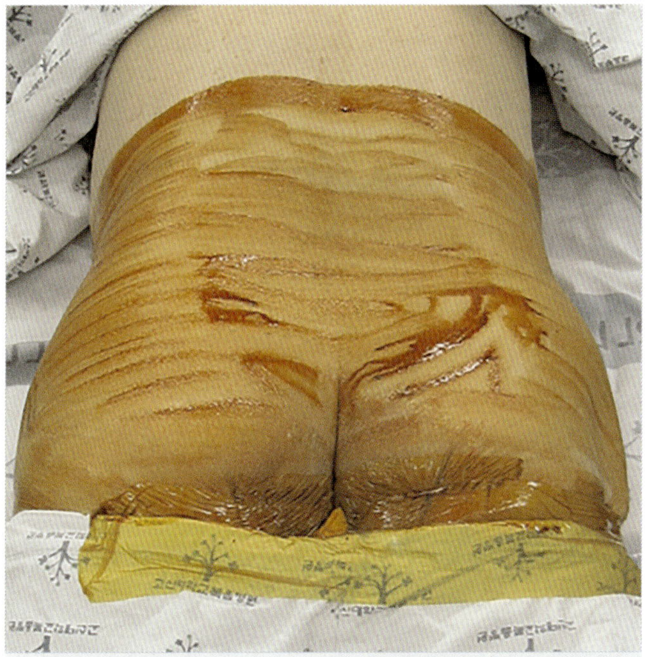

Fig. 18.1 Patient position.

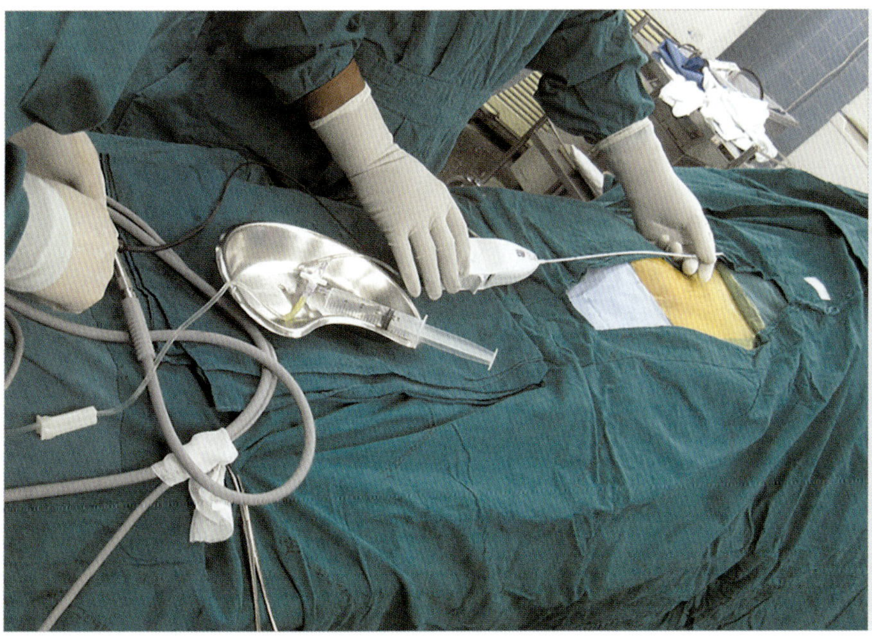

Fig. 18.2 Preparation.

During the procedure, no other sedatives were administered except the intravenous injection of meperidine hydrochloride (Demerol 25 mg, Sanofi-aventis U.S. LLC) to help ease the patients' pain that occurred during procedure and to facilitate verbal communication between the surgical staff and patients throughout the procedure.

18.2.2 Step 2: Making a Portal for Procedure

An 18-gauge Tuohy epidural needle was used for puncturing the sacral hiatus and was advanced into the sacral canal under C-arm guidance (▶ Fig. 18.4, ▶ Fig. 18.5, and ▶ Fig. 18.6). Then, a guidewire was inserted into the opening via the needle that was removed (▶ Fig. 18.7 and ▶ Fig. 18.8). After applying 1-cm longitudinal skin incision at insertion site of guidewire, a dilator was then inserted through the opening to achieve the needed bony space (▶ Fig. 18.9), which was removed once the space was secured. After that, the epiduroscopic procedure should be initiated.

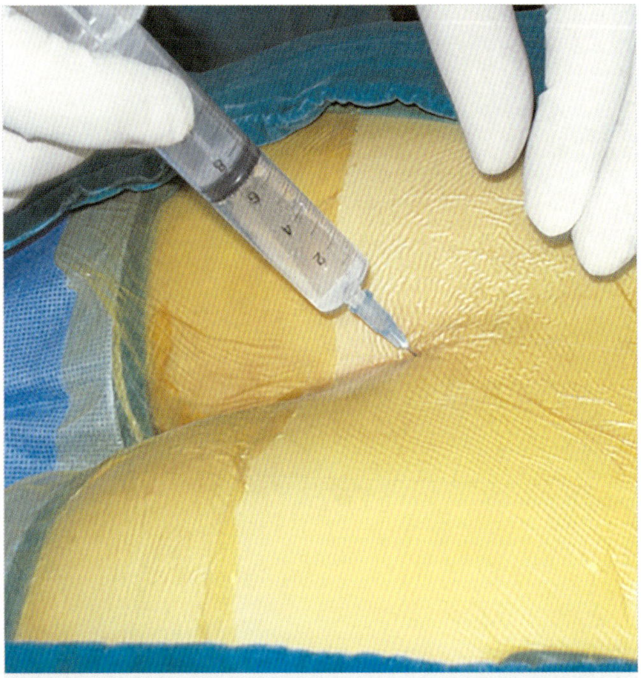

Fig. 18.3 Local anesthesia at incision site around sacral hiatus.

18.2.3 Step 3: Advancing to the Target Lesion

The introducer was inserted, followed by a flexible endoscope (Myelotec, model 75298871) into the epidural space through the introducer (▶ Fig. 18.10). The flexible endoscope was then maneuvered to reach the target lesion at either the ventral or the dorsal side of cauda equine (▶ Fig. 18.11). Epidural saline solution was used to irrigate and clear the area visualized, exert pressure to expand the epidural space, and improve clarity through the endoscopic video screen. This allowed visualization of adhesion status with bands, inflammatory tissues, fibrous connective tissues, and adipose tissue around the dura and nerve root, as well as anterior and posterior stenotic lesions such as thickened ligamentum flavum, hypertrophied facet joint, and protruded intervertebral disk (▶ Fig. 18.12).

In addition to securing epidural visibility endoscopically, a contrast agent (Omnipaque, Nycomed) was injected to perform epidurography to identify the adhesive lesion clearly under fluoroscopic image. We even stimulated the nerve roots in the lesion via the endoscope. This was to test if the stimulation induced concordant sensations or pain reproduction in the patient and to ensure a higher level of accuracy for our diagnosis, if it did. Next, we manipulated the flexible endoscope inside the epidural root lesion and performed adhesiolysis and foraminoplasty (▶ Fig. 18.13).

18.2.4 Step 4: Laser Ablation at Pathologic Lesions

The guidewire for a placement of laser catheter was introduced under endoscopic visualization directly into the

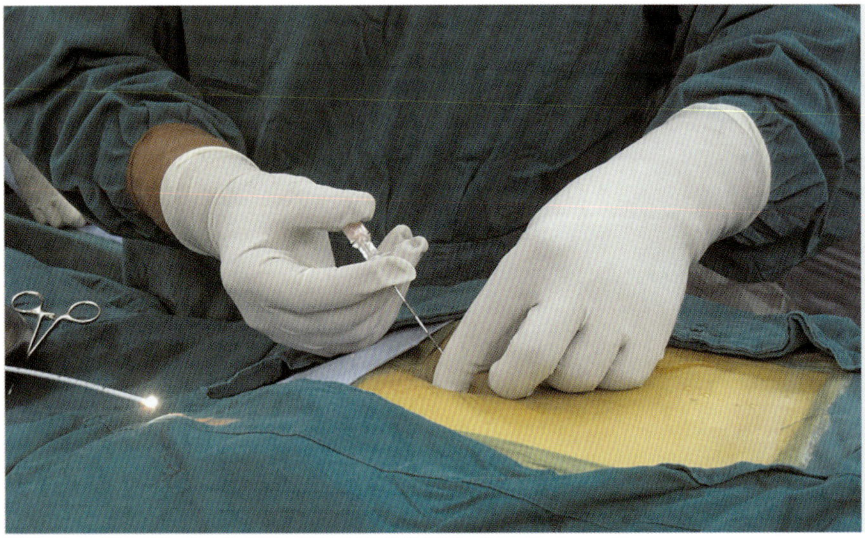

Fig. 18.4 18-gauge Tuohy needle insertion at the site.

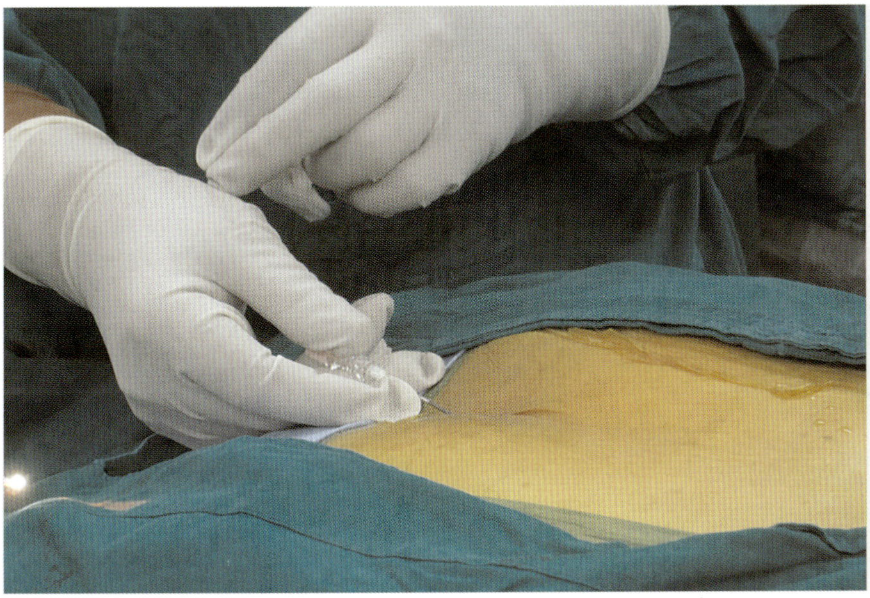

Fig. 18.5 Advancing the needle and tilting it caudally.

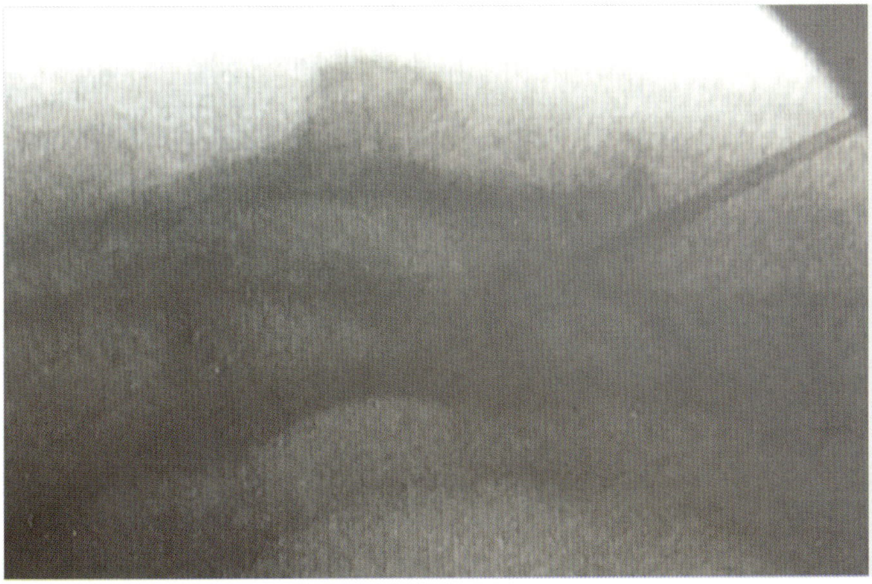

Fig. 18.6 Confirming the needle placement into the sacral hiatus.

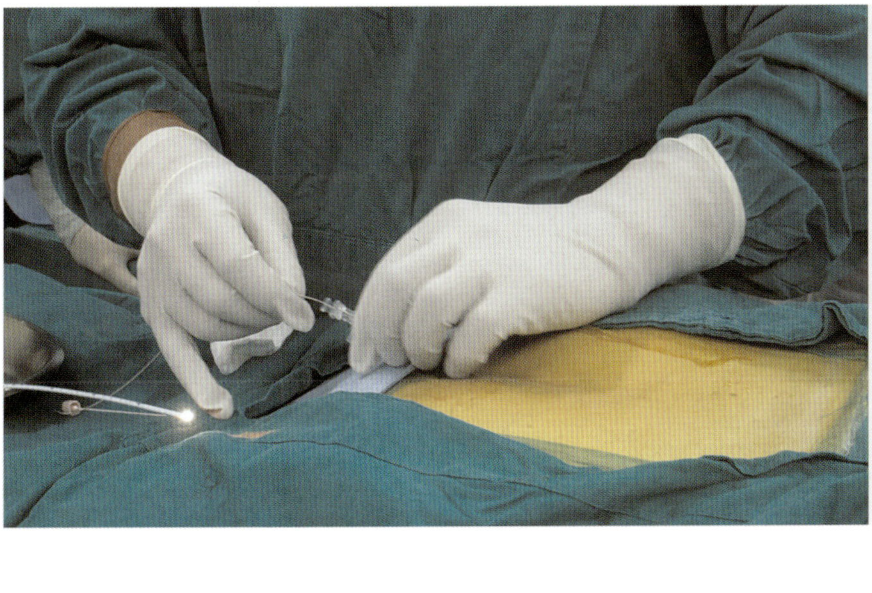

Fig. 18.7 Guidewire insertion for epiduroscope through the 18-gauge Tuohy needle.

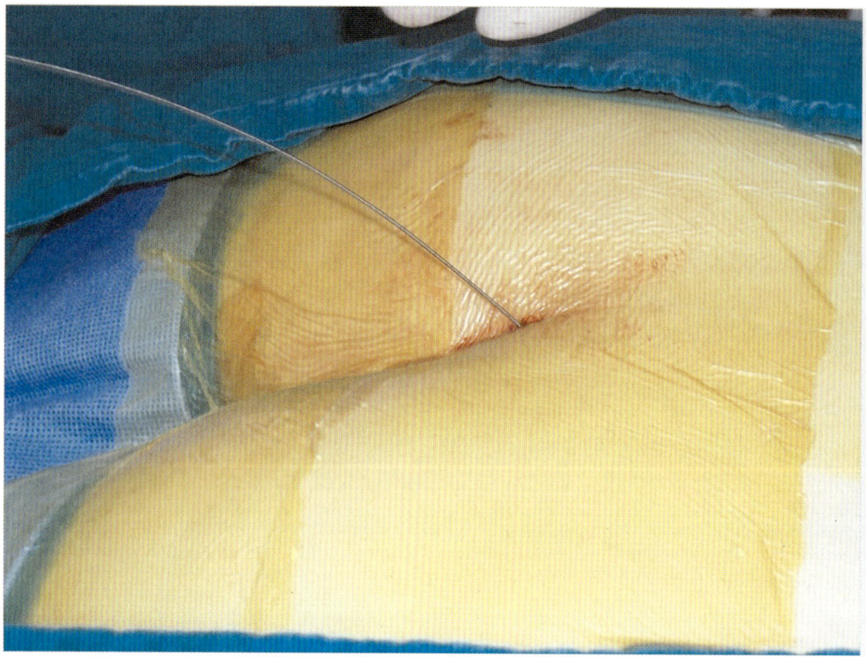

Fig. 18.8 Remained guidewire after removing the needle.

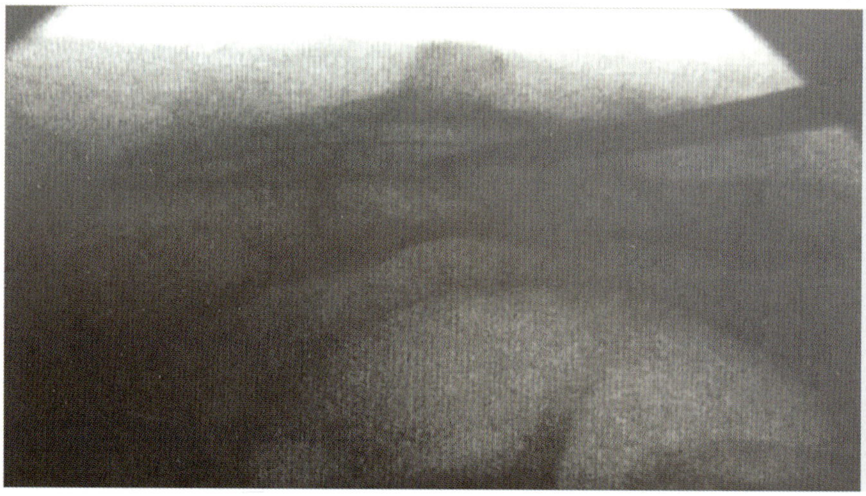

Fig. 18.9 Dilator insertion through the opening to achieve the needed bony space, which was confirmed under fluoroscopic images to locate inside the sacral canal.

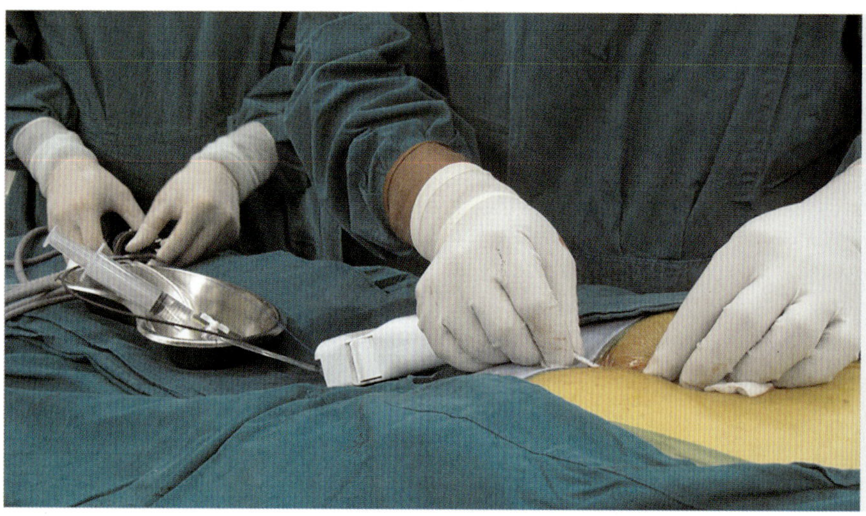

Fig. 18.10 Initiation of epiduroscopic procedure.

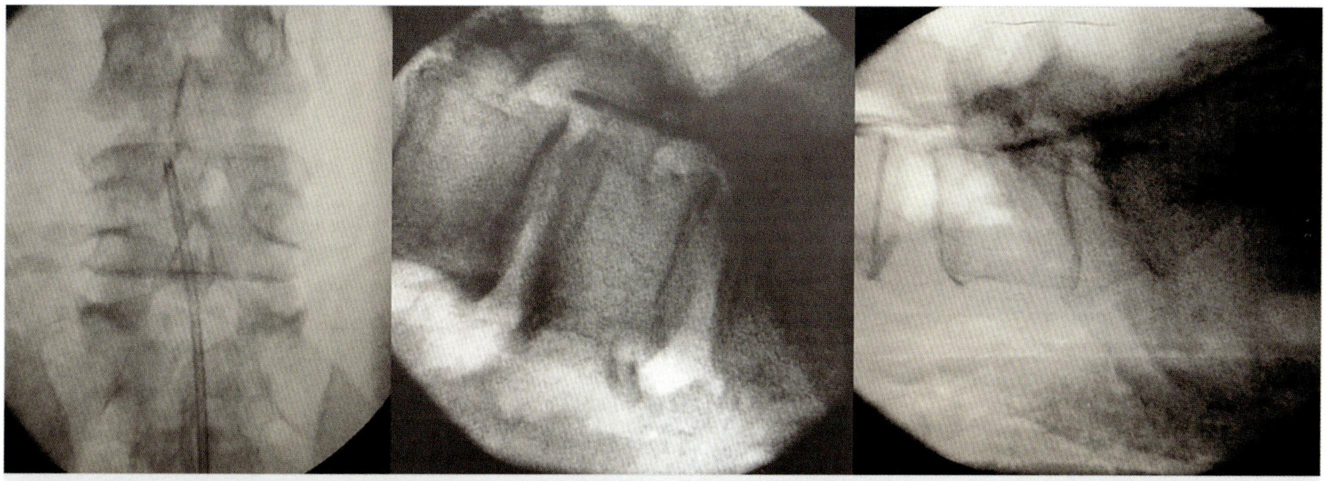

Fig. 18.11 Approaching the target lesion under fluoroscopic guidance.

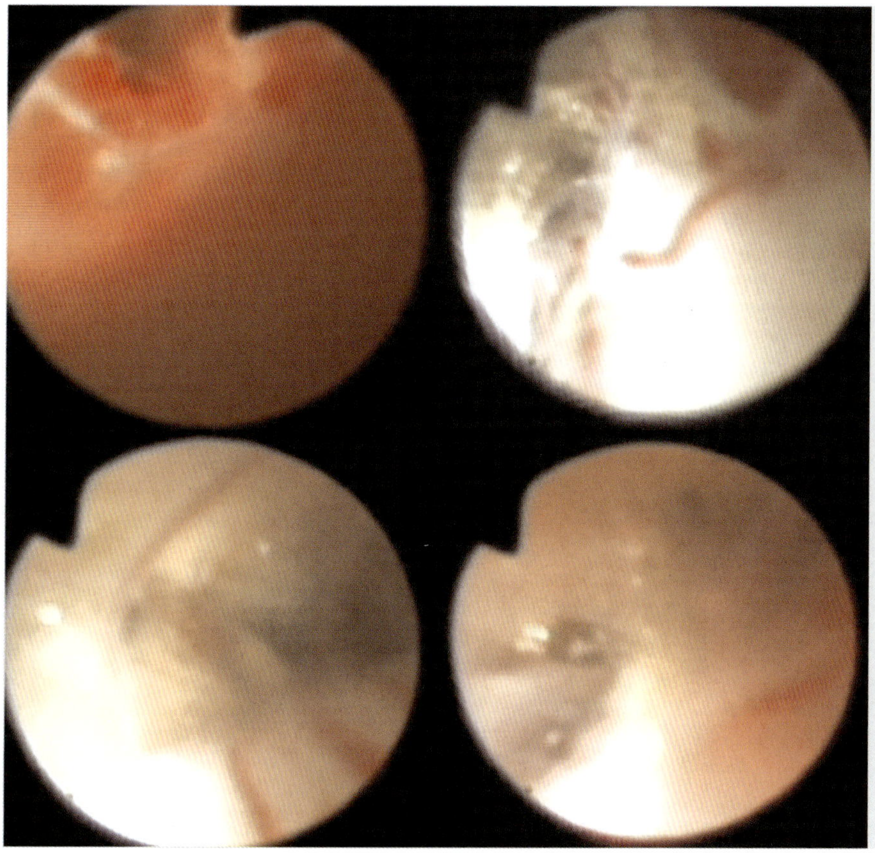

Fig. 18.12 Various pathologic lesions.

causative factors via the working channel. When the contact of guidewire to the pathologic lesion was observed under endoscopic visualization, laser catheter was advanced to the lesion. When the pathologic lesions at the ventral or dorsal side of dura were cauterized using laser ablation via the endoscopic camera, the lesions were shrunk, and dura and nerve roots were decompressed (▶ Fig. 18.14). After the procedure, a triamcinolone 80 mg (Kenalog, Bristol Myers Squibb Co.) was injected directly in the treatment area to maximize the treatment effects.

18.3 Important Notes during Laser Ablation of Epiduroscopic Procedures

Additive laser ablation with epiduroscopic procedure has several advantages, including evaporation of herniated disk materials, destroying inflammatory materials and pathologic lesions, and shrinkage of pathologic soft tissues such as hypertrophied ligamentum flavum, annulus fibrosis, and facet joint capsule, which

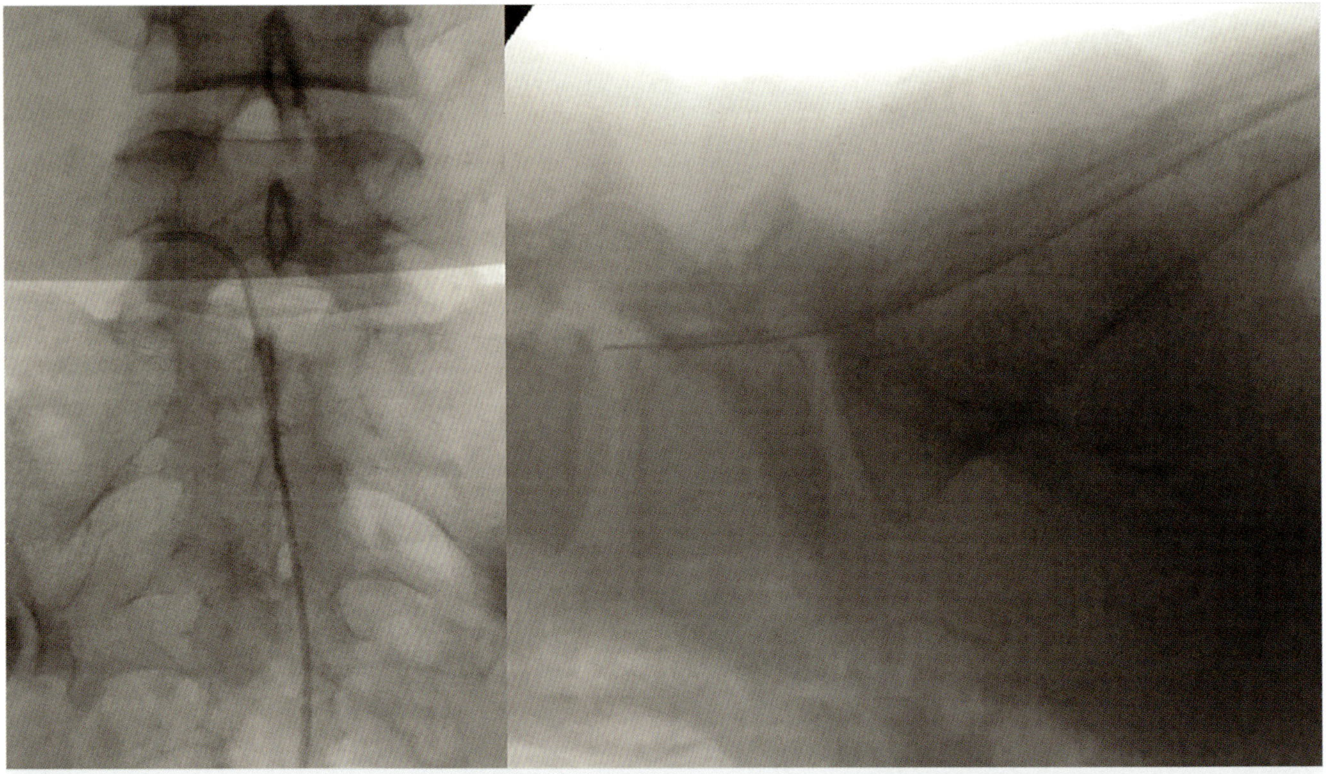

Fig. 18.13 Manipulation of the flexible endoscope inside the epidural root lesion, and performing adhesiolysis and foraminoplasty under fluoroscopic guidance.

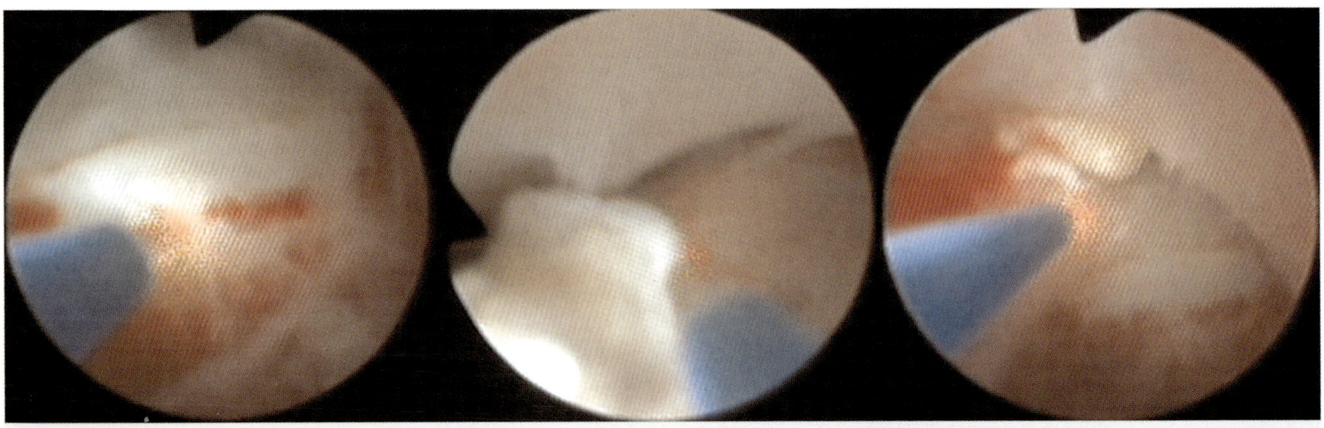

Fig. 18.14 Cauterization of the pathologic lesions using laser ablation.

would be the superiority of this procedure to the conventional epiduroscopic procedure discussed in previous articles.[1,3,7]

However, the laser ablation can also be related to its nature complications, including thermal injury of dura or nerve root and ablation to the normal structures. To avoid those problems, the authors suggest the following:

- Laser ablation should be performed with saline irrigation to reduce thermal injury.
- Prior to laser ablation, optimal location and direction of the catheter tip should be definitely proven by the use of guidewire.
- The surgeon should confirm a location of the catheter tip, which is away from the neural structure.
- The catheter tip is located as close as possible at the pathologic lesion.

18.4 Clinical Outcomes in Epiduroscopic Treatment with or without Laser

18.4.1 Patients

A total of 20 patients in group A (epiduroscopic neural decompression, END) and 78 patients in group B (epiduroscopic neural decompression with laser, ELND) were enrolled in this study and evaluated. Demographic data are not significantly different between the groups in terms of age, gender, height, weight, past illness, and others.[1,3]

18.4.2 Outcome Measures

The primary posttreatment study end point was the pain intensity at the lower back and the radiating pain on the lower extremity. The pain intensity was evaluated with visual analog scale (VAS) scores, for which patients were instructed to make a mark on a horizontally oriented, 10-point VAS sheet, with "no pain" at the far left of the sheet and "greatest pain possible" at the far right of the sheet. The scores were obtained preoperatively, weekly for the first month after surgery, then at 3 and 6 months, and then at 1 year.

Secondary end points included the following: (1) clinical outcomes based on the Roland Morris Disability Questionnaire (RMDQ), (2) surgical outcomes, and (3) procedure-related complications. The surgical outcome was evaluated with surgery-related detailed information, including estimated surgical time, corrected surgical time per level, the amount of blood loss, and the length of hospital stay. In addition, postoperative problems and complications were recorded in detail, including nerve root or cord injury during procedure, postoperative neurologic symptoms or deterioration, and postoperative infection or metal failure.

18.4.3 Results

In the END group, the mean VAS score for low back pain was 8.5 ± 1.3 preoperative, 3.8 ± 0.4 at 1 month after surgery, and 6.7 ± 0.9 at 2 years after surgery ($p=0.12$), exhibiting a "V-shaped" upward tendency, and the mean VAS score for leg pain was 6.7 ± 0.8 preoperative, 3.1 ± 0.7 at 1 month after surgery, and 5.2 ± 0.6 at 2 years after surgery ($p=0.15$). The mean RMDQ score was 12.6 ± 1.2 preoperative, 8.6 ± 0.6 at 1 month after surgery, and 10.4 ± 0.5 at 2 years after surgery ($p=0.09$), also exhibiting V-shaped upward trend in which the score at final follow-up was almost identical to that before the surgical treatment.

In the ELND group, the mean VAS score for low back pain was 8.1 ± 0.7 preoperative, 3.1 ± 0.3 at 1 month after surgery, and 4.4 ± 0.5 at 2 years after surgery ($p=0.01$), demonstrating a relatively consistent improvement during follow-up, and the mean VAS score for leg pain was 6.2 ± 0.9 preoperative, 2.8 ± 0.4 at 1 month after surgery, and 4.7 ± 0.6 at 2 years after surgery ($p=0.07$). The mean RMDQ score was 13.2 ± 1.6 preoperative, 7.2 ± 0.5 at 1 month after surgery, and 8.1 ± 0.3 at 2 years after surgery ($p=0.03$), also demonstrating a relatively consistent improvement during follow-up with regard to functional activity. There was a statistical significant improvement in the VAS and RMDQ score after 6 months after the ELND procedure in comparison to the END procedure.

Posttreatment 1-year outcomes based on the MacNab criteria were positive in 82% of the ELND group, compared to 45% of the END group ($p=0.02$).

We experienced several epiduroscopy-related complications, including transient headache ($n=3$), pain over the site of the scope insertion ($n = 3$), focal infection of the entry site ($n = 2$), and mild meningitis ($n = 1$). However, all symptoms improved after bed rest and conservative treatment with medications. One case of foot-drop-related laser complication occurred, which was recovered within 6 months.

References

[1] Lee GW, Jang SJ, Kim JD. The efficacy of epiduroscopic neural decompression with Ho:YAG laser ablation in lumbar spinal stenosis. Eur J Orthop Surg Traumatol. 2014; 24 Suppl 1:S231–S237

[2] Epstein JM, Adler R. Laser-assisted percutaneous endoscopic neurolysis. Pain Physician. 2000; 3(1):43–45

[3] Kim JD, Jang JH, Jung GH, et al. Epiduroscopic laser disc and neural decompression. J Neurosurg Rev. 2012; 1:14–19

[4] Imhoff AB. The use of lasers in orthopaedic surgery. Oper Tech Orthop. 1995; 5:192–203

[5] Abrisham SMJ, Kermani-Alghoraishi M, Ghahramani R, Jabbari L, Jomeh H, Zare M. Additive effects of low-level laser therapy with exercise on subacromial syndrome: a randomised, double-blind, controlled trial. Clin Rheumatol. 2011; 30(10):1341–1346

[6] Yeldan I, Cetin E, Ozdincler AR. The effectiveness of low-level laser therapy on shoulder function in subacromial impingement syndrome. Disabil Rehabil. 2009; 31(11):935–940

[7] Ruetten S, Meyer O, Godolias G. Application of holmium:YAG laser in epiduroscopy: extended practicabilities in the treatment of chronic back pain syndrome. J Clin Laser Med Surg. 2002; 20(4):203–206

19 Epiduroscopic Procedure Using the 1,414-nm Nd:YAG Laser (Lutronic Laser)

Jun Young Kim, Il Sup Kim, and Jae Taek Hong

19.1 Introduction

Laser spine surgery has been in use for the treatment of spinal diseases since the 1980s. Laser therapy in spinal diseases is mostly carried out using the percutaneous and endoscopic approach. In this method, skin incision, as well as damage to the muscles, blood vessels, and nerves around the spine, is minimized. It is also advantageous as operations are done under local anesthesia and the satisfaction of patients is high owing to the short operation time.

In February 1986, Choy et al[1] implemented percutaneous laser disk decompression (PLDD) on a lumbar herniated disk. Since then, the development of laser therapy for spinal diseases has focused on lumbar disk disease, and PLDD became an active treatment following U.S. Food and Drug Administration (FDA) approval in 1991.[2] Also, its application range has been expanding in recent times to include treatment of lumbar spinal stenosis and endoscopic treatment of ruptured cervical disk.[3,4]

19.2 The Types of Laser and Their Respective Advantages

Lasers have been used in medicine since the early 1960s. There are several types of lasers in use for the lumbar spine, with the most common being the holmium:yttrium-aluminum-garnet (Ho:YAG) and neodymium:YAG (Nd:YAG) lasers.[5] The Ho:YAG laser (2,090 nm), owing to its close approximation to the intense 2-μm absorption band of water, appears to be a viable candidate for clinical trials of laser diskectomy.[6] Also, the Ho:YAG laser is a pulsed laser, in contrast to the continuous-wave near-infrared lasers, and therefore has the advantage of producing minimal amounts of heat in adjacent tissues.

From the standpoint of spinal surgery, Choy et al,[1] in 1987, reported the first results of the Nd:YAG in laser-assisted disk decompression, yielding initial resolution of sciatica in 9 out of 12 patients. It is postulated that due to vaporization of water, internal disk pressure decreases, which, in turn, results in disc decompression as well as decompression of neighboring neural elements. Ongoing investigations of laser systems in the mid-infrared range have led to more refined and appropriate optical wavelengths for use in the medical arena. Specific wavelengths applicable for use in the human intervertebral disk and currently approved by the FDA include potassium titanyl phosphate at 532 nm, Nd:YAG at 1,064 and 1,440 nm, CO_2 at 10,600 nm, and Ho:YAG at 2,100 nm wavelength. With the exception of the CO_2 laser, each of these laser systems can be delivered via flexible fiberoptics, making them suitable for use in the intervertebral disk.

Currently, Ho:YAG is the most commonly used laser for ablation therapy. The Ho:YAG laser has been accepted as the safest laser because of low tissue penetration. The Nd:YAG laser, which was introduced before the Ho:YAG laser, was not used for deep tissue penetration. However, the Nd:YAG laser can also provide shallow tissue penetration by adjusting its wavelength to that of the Ho:YAG laser.[7] The advantages of the Nd:YAG include easier laser control, a smaller manufacturing fee, and equivocal effectiveness.[8,9] Moon et al[10] reported that the 1,414-nm Nd:YAG laser is effective and safe for decompressing herniated intervertebral disks under the guidance of a spinal epiduroscope in vivo porcine models and in human cadaveric models.

19.3 Percutaneous Endoscopic Laser Annuloplasty

Lee and Kang[11] introduced the percutaneous endoscopic laser annuloplasty (PELA) procedure, which cauterizes the granulation tissue related to the damaged annulus fibrosus. The main advantage of the PELA procedure is to preserve a healthy nucleus pulposus by minimizing damage to the center and front of the disk. They reported significant improvement of lumbago and statistically significant improvements in line with Macnab's criteria following the PELA treatment. However, according to Carragee et al,[12] only approximately 30% showed improvement. The frequency of specific complications or infection was not high. It is likely that damage to other parts was minimized by the short treatment time and use of the endoscope.

19.4 Epiduroscopic Laser Neural Decompression

Epiduroscopic laser neural decompression (ELND) is conducted using an extradural endoscope. For most patients, a nerve root block or palliative surgery is conducted. However, if these measures fail and there is no improvement after surgery, ELND is considered. It allows communication between a doctor and a patient under local anesthesia and exfoliation of adhesion in the epidural area. ELND also removes damaged tissue and inflammation and a protruded disk to lessen pressure on the nerves. Additionally, steroids and a topical anesthetic can be injected directly around a lesion. Hence, ELND can be a good therapy method for patients with various lesions. Jo et al[13] implemented ELND using Ho:YAG lasers. In their procedure, a 2,100-nm wavelength was selected for the tissue to be removed. The laser was then inserted at a depth of 0.3 to 0.5 mm and the tissue was cauterized, thus minimizing damage to the surrounding tissue. However, since there are cases where the nerve roots are damaged or diskitis occurs, caution is required for this procedure.

19.5 Transforaminal Epiduroscopic Laser Annuloplasty

Transforaminal epiduroscopic laser annuloplasty (TELA) has been recently introduced and considered as a new method for

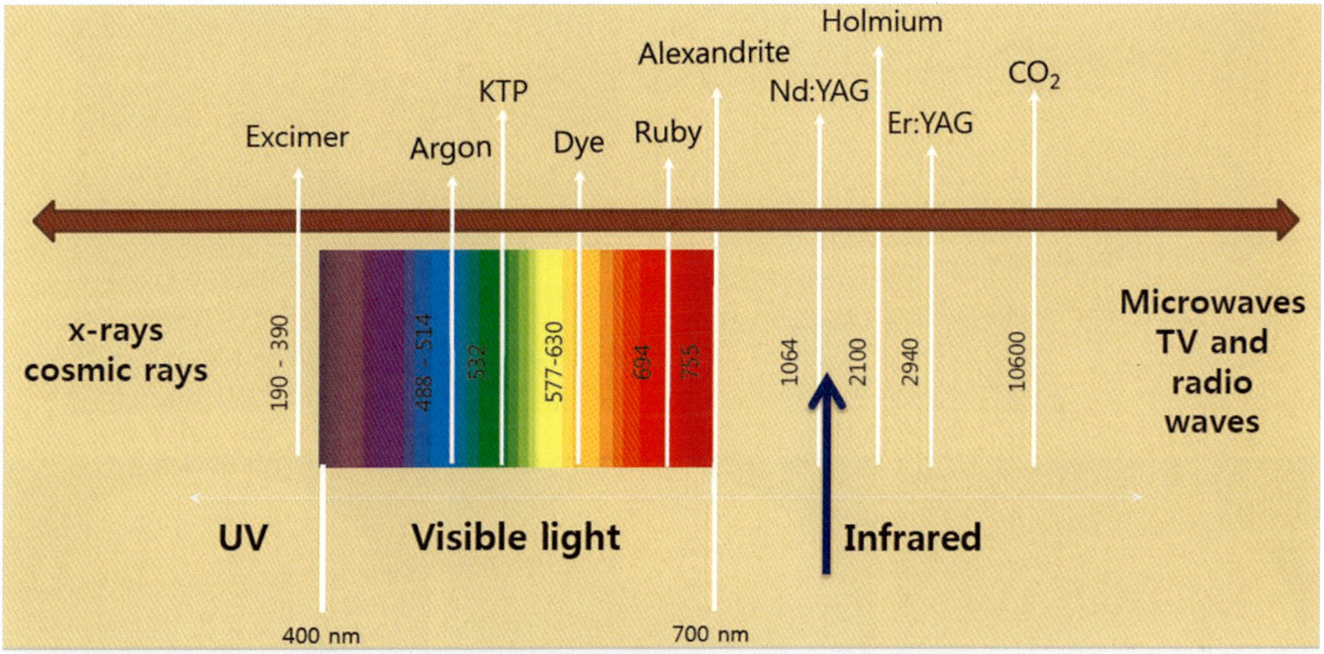

Fig. 19.1 1,414-nm laser (transforaminal epiduroscopic laser annuloplasty; Lutronic).

the treatment of not only herniated lumbar disk, spinal stenosis, and postlumbar surgery syndrome but also chronic refractory low back pain, which is not well responded by other interventional treatments. It is a minimal invasive procedure for spine pain using epiduroscope and laser. For laser irradiation, an Nd: YAG laser (Lutronic) with a wavelength of 1,414 nm operating in the 200 to 600 mJ range was used (▶ Fig. 19.1).

We can observe the epidural space by epiduroscope as well as remove the lesions such as herniated disk, fibrosis, and adhesion by laser. The advantages of TELA compared with back surgery include relatively noninvasiveness, short operating time, reduction of risks related to general anesthesia, possibility of communication with the patient during the procedure (leading to reduced risks of accidental nerve damage), and easier access into the ventral epidural space.

The TELA procedure was done under local anesthesia with the patient in the prone position. The intraoperative fluoroscopy (C-arm) was used to select and approach the exact surgical site. Because of the use of local anesthesia, the surgeon could communicate with the conscious patient and confirm the patient's response. The usual skin entry point was 10 to 12 cm lateral from the midline. The surgeon advanced the needle on a trajectory that approached the facet joint, and the usual axial angle of needle was 20 to 30 degrees from horizon. The 18-gauge spinal needle is advanced to the target annulus lateral to medial pedicular line (foraminal zone). And then, the guidewire is inserted through the spinal needle. Next, a 14-gauge Tuohy needle is inserted through the guide under fluoroscopic guidance until the tip reaches the ventral epidural space (▶ Fig. 19.2).

A skin incision is made with a no. 15 scalpel where the guidewire meets the skin surface. Next, a dilator is placed over the guidewire and gently advanced into the ventral epidural space. After inserting the dilator into the epidural space, epidurography was used to outline neural tissues of the thecal sac and exiting nerve root so that they might be avoided to prevent neural injury (▶ Fig. 19.3). Next, the introducer sheath and dilator set is inserted over the guidewire into ventral epidural space. After that, the NeedleView CH (BioVision) is inserted into epidural space. And, annular probing or diskography is performed. Surgical instruments including BioVision LLC endoscopic graspers and ACCUPLASTI LASER system (1,414-nm Nd:YAG laser) are used to complete the disk repair (▶ Fig. 19.4).

The endoscope used projects fiberoptic images using a high-definition video platform, allowing excellent spatial resolution and real-time imagery. The forward-firing and side-firing laser fiberoptics are then used to perform the thermal treatment of the damaged annulus fibrosus (▶ Fig. 19.5). The laser power level was set at 200 to 300 mJ (10–20 Hz) for denervation and coagulation and at 300 to 600 mJ (10–20 Hz) for vaporization and ablation. Intraoperative images demonstrate the use of the side-firing laser for annuloplasty (▶ Fig. 19.6) and diskectomy (▶ Fig. 19.7).

In some cases when larger herniated fragments were observed during the treatment, the surgeon could use endoscopic forceps in the same working channel as well as the laser (▶ Fig. 19.8). Upon completion of the disk repair, the endoscope and cannula are removed to allow primary closure of the wound with suture and steri-Strips.

The authors studied heat distributions of the cadaveric lumbar intervertebral disks during laser annuloplasty using 1,414-nm Nd:YAG laser. Laser annuloplasty was performed on fresh human cadaveric lumbar spine with 20 intact intervertebral disks in a 37°C circulating water bath using newly developed 1,414-nm Nd:YAG laser (ACCUPLASTI) (▶ Fig. 19.9). Five thermocouples (HYPO-30-1/2-T-G60-SMPW-M, OMEGA) were attached to different locations on the disk, and, at the same time, in each location, the temperature during annuloplasty was measured and analyzed (▶ Fig. 19.10 and ▶ Fig. 19.11).

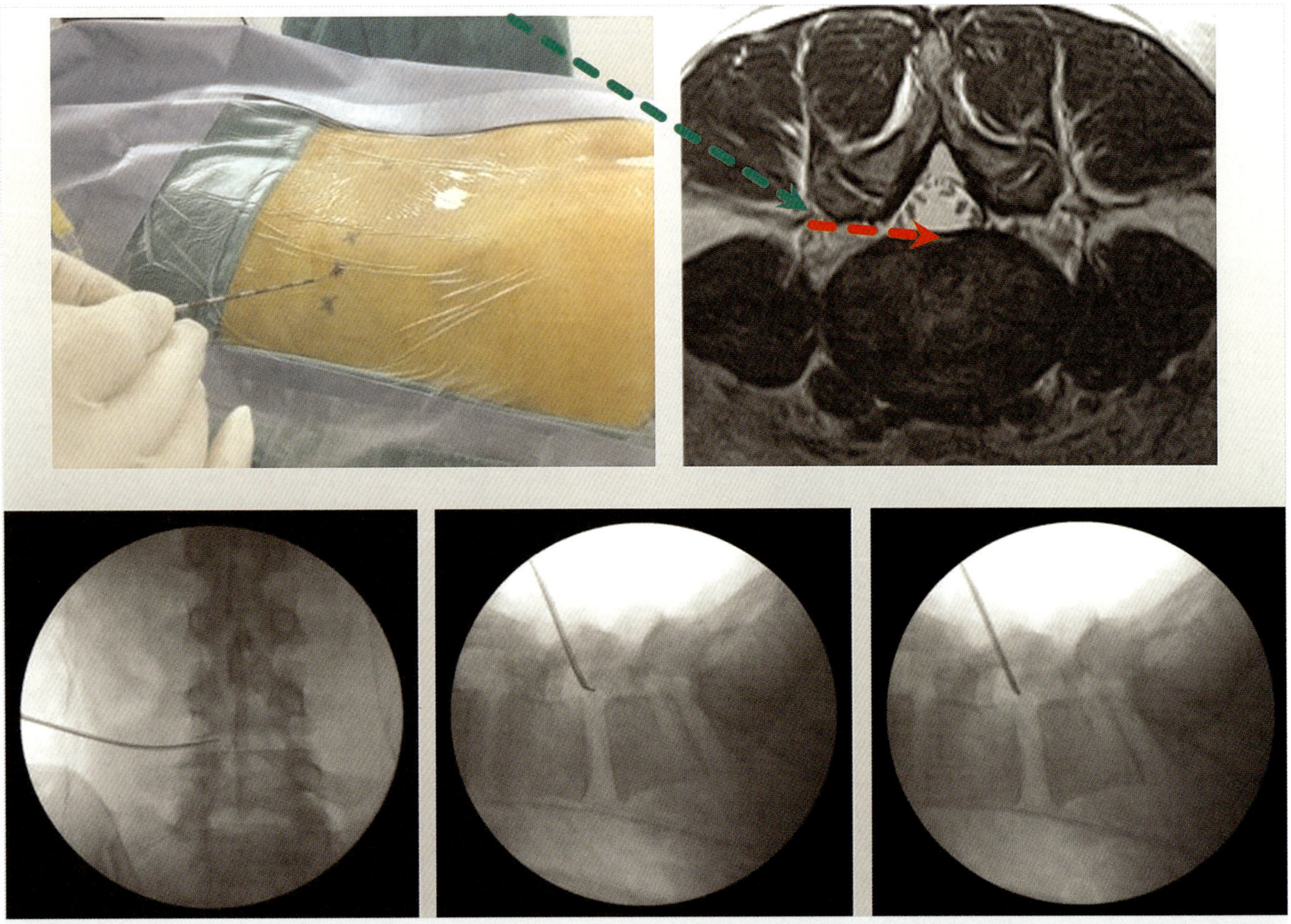

Fig. 19.2 The usual skin entry point was 10 to 12 cm lateral from the midline. The 14-gauge Tuohy needle is inserted through the guide under fluoroscopic guidance until the tip reaches the ventral epidural space.

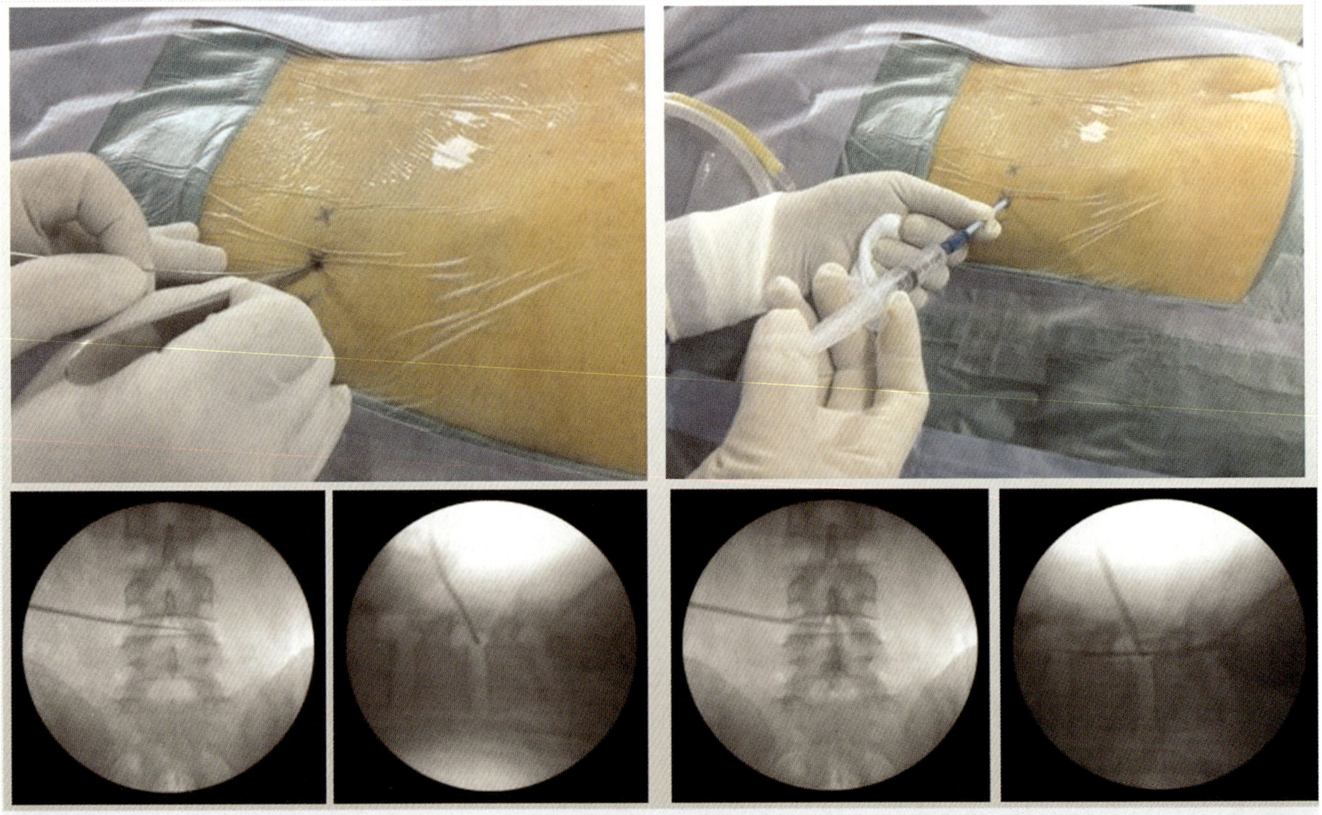

Fig. 19.3 The dilator is placed over the guidewire and gently advanced into ventral epidural space. After inserting the dilator into the epidural space, epidurography was performed.

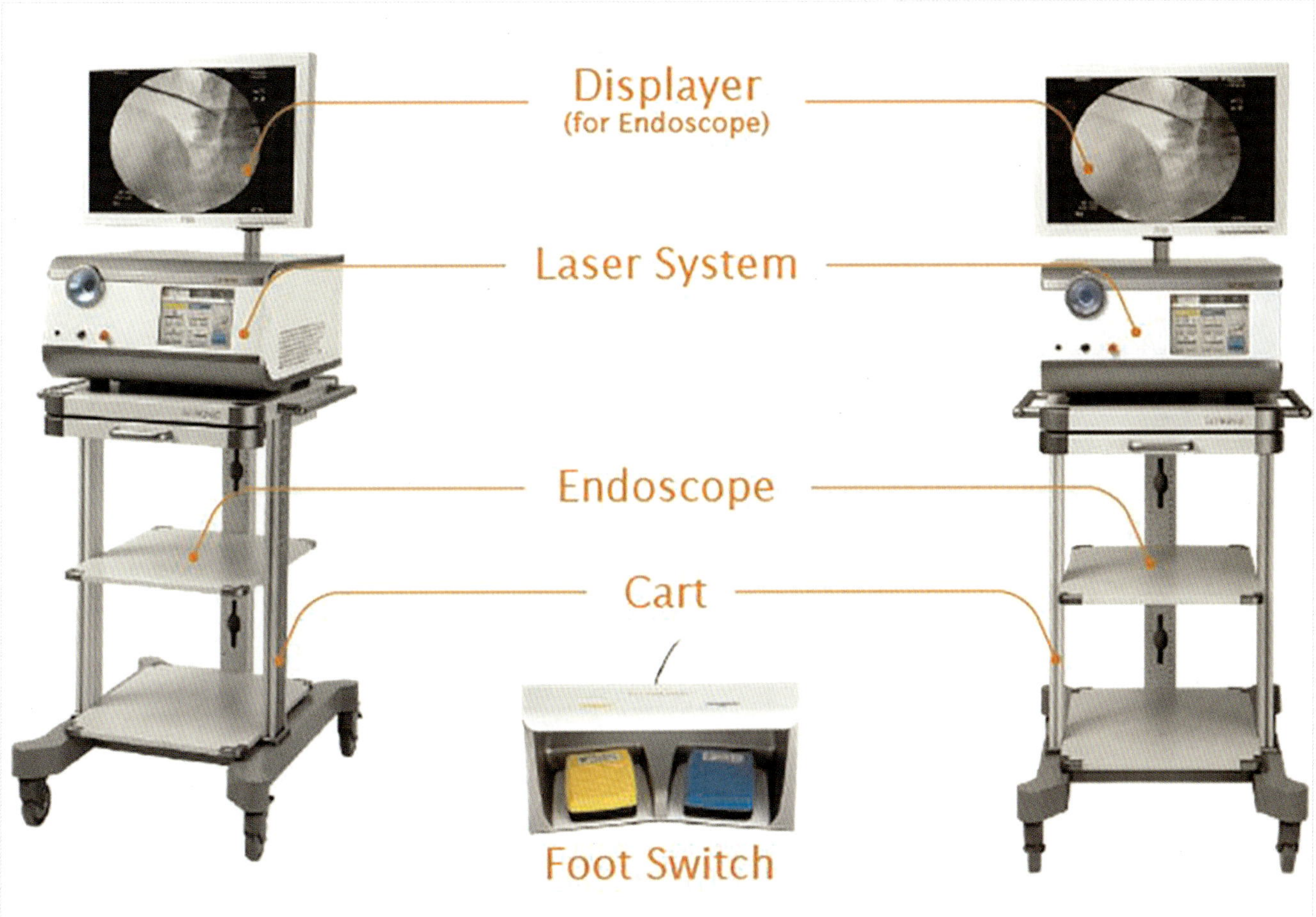

Fig. 19.4 Lutronic ACCUPLASTI LASER system.

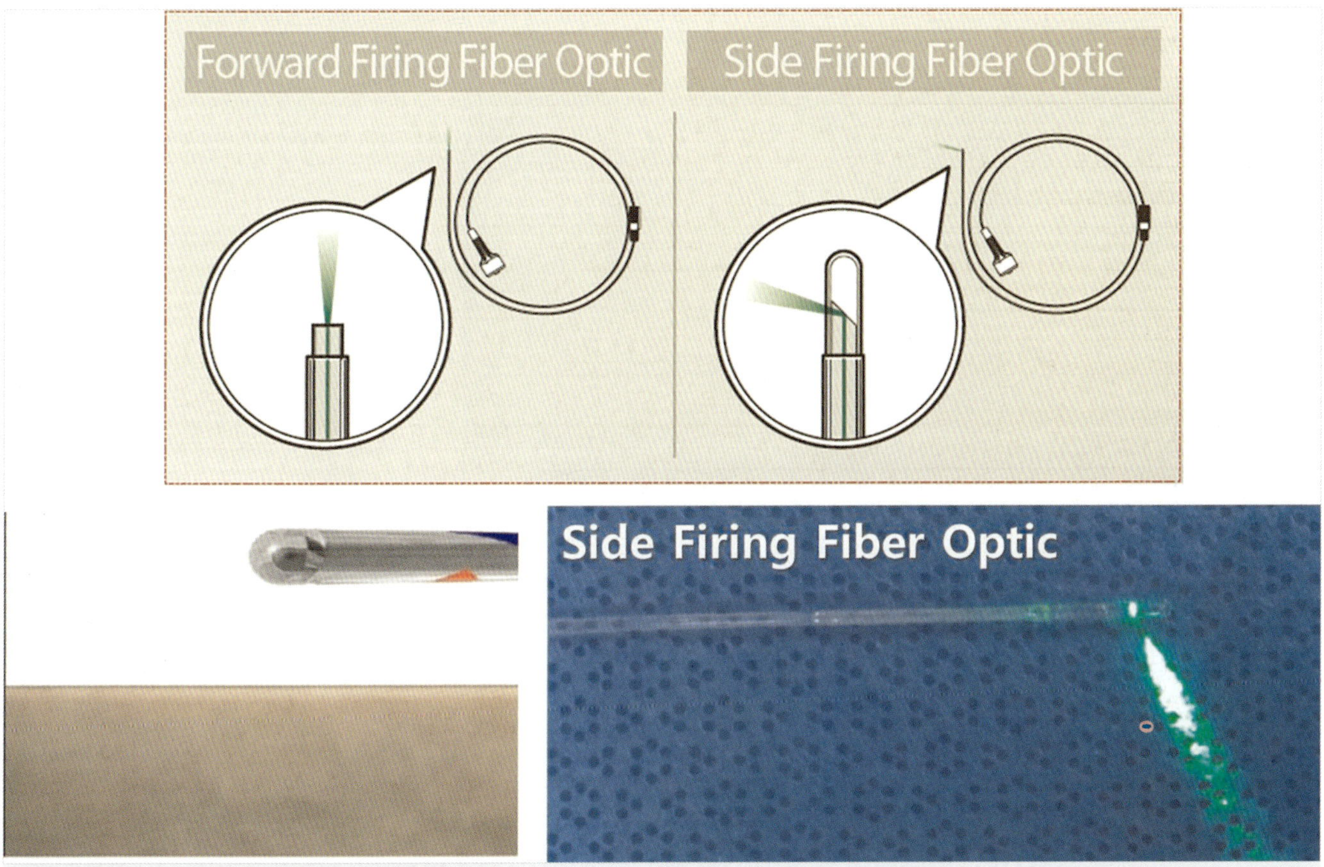

Fig. 19.5 Forward-firing and side-firing fiberoptics.

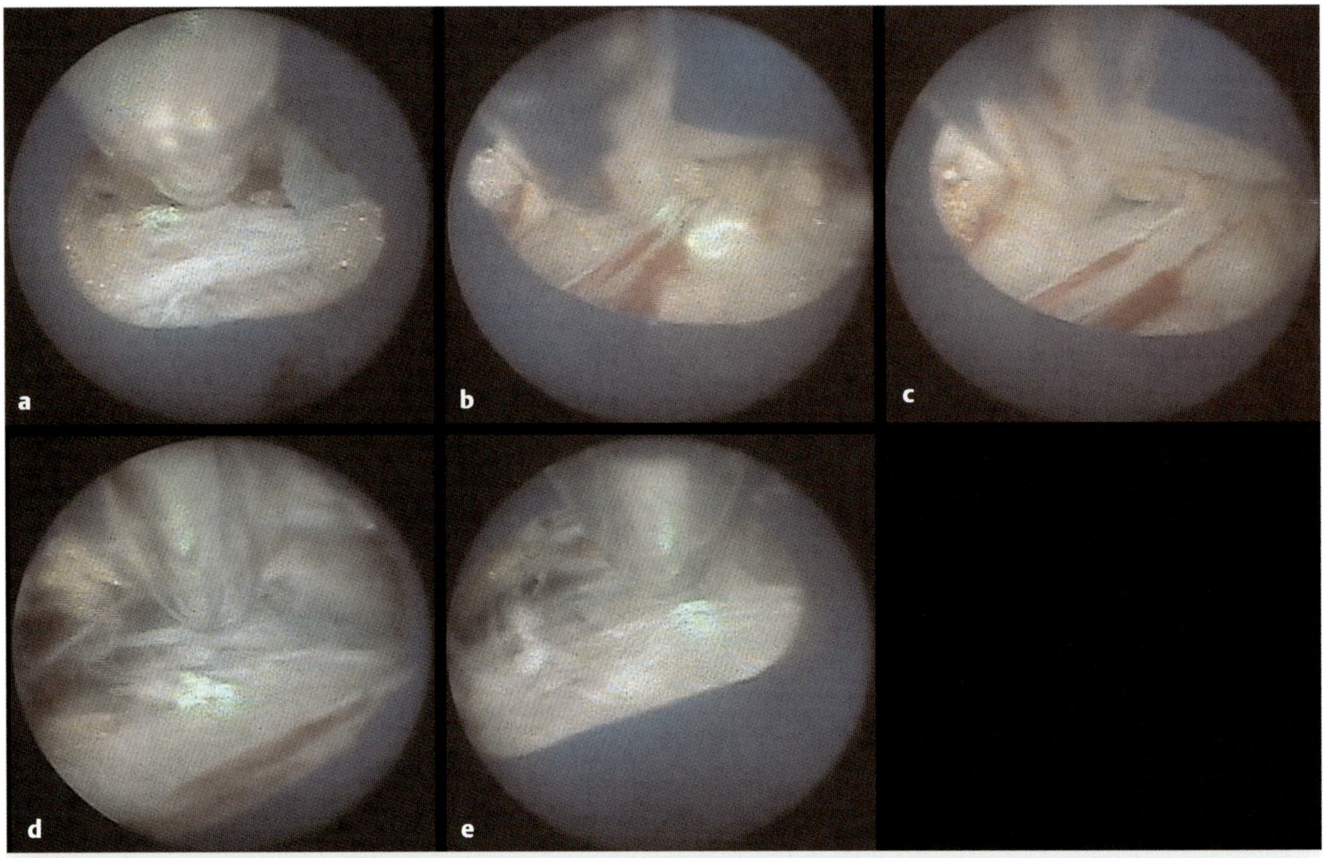

Fig. 19.6 (a) Foraminoplasty can be performed with the side-firing laser. **(b)** Annuloplasty is done with the side-firing laser. Blood vessels are coagulated in the process of laser annuloplasty. **(c)** Annuloplasty is done with the side-firing laser. Blood vessels are coagulated in the process of laser annuloplasty. The blood vessels can be seen to be coagulated after laser application. **(d)** The traversing nerve root can be retracted with the distal end of the laser as the annuloplasty is performed. **(e)** The midline annulus in the ventral epidural space can also be reached with the side-firing laser.

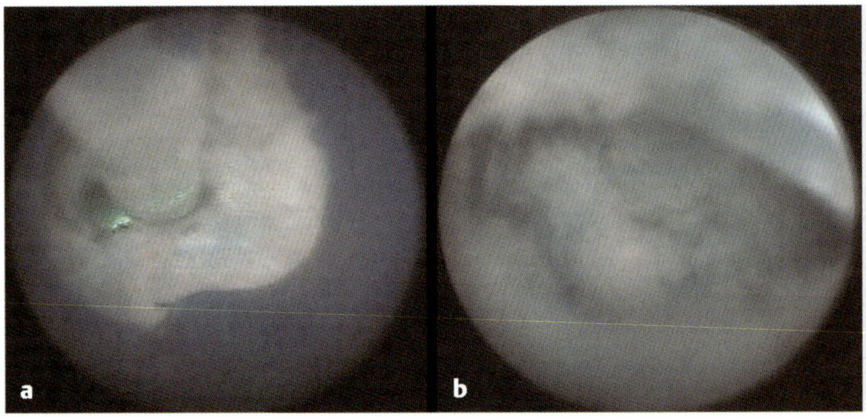

Fig. 19.7 (a) The intradiskal contents are coagulated and shrunken with the side-firing laser. **(b)** View of the intervertebral disk after laser coagulation and diskectomy.

Thermal probe's temperature was higher in locations closer to laser fiber end tip and on lateral locations, rather than farther and in depth locations. The average maximum temperature in locations 3 mm deeper to the laser fiber end tip was 47 ± 4.3°C, measured at 13.6 ± 0.8 seconds, and in 6 mm depth, the temperature was 38.7 ± 0.9°C, measured at 24.8 ± 3 seconds. In 3, 6, and 9 mm lateral to the laser fiber end tip, the temperatures were 69.5 ± 13.3°C, 37.8 ± 0.7°C, and 37.2 ± 0.3°C, respectively, measured at 10.6 ± 0.5, 26.6 ± 3.4, and 38.8 ± 4.4 seconds, respectively. In accordance with the laser fiber end tip and the depth, temperature above 45°C, which triggering nociceptive ablation in 16 levels among 20 levels (80%), was measured in 3 mm depth, in accordance with the laser fiber end tip and laterality, every measurement had above 45°C, and also was measured temperature over 60°C, which can trigger collagen denaturation at 16 levels among 20 levels (80%).

There are many potential benefits of TELA for symptomatic lumbar disk disease. First, this technique allows for easier access into the ventral epidural space and clear visualization by curved scope and is familiar to surgeons as it is similar to PELD. Also,

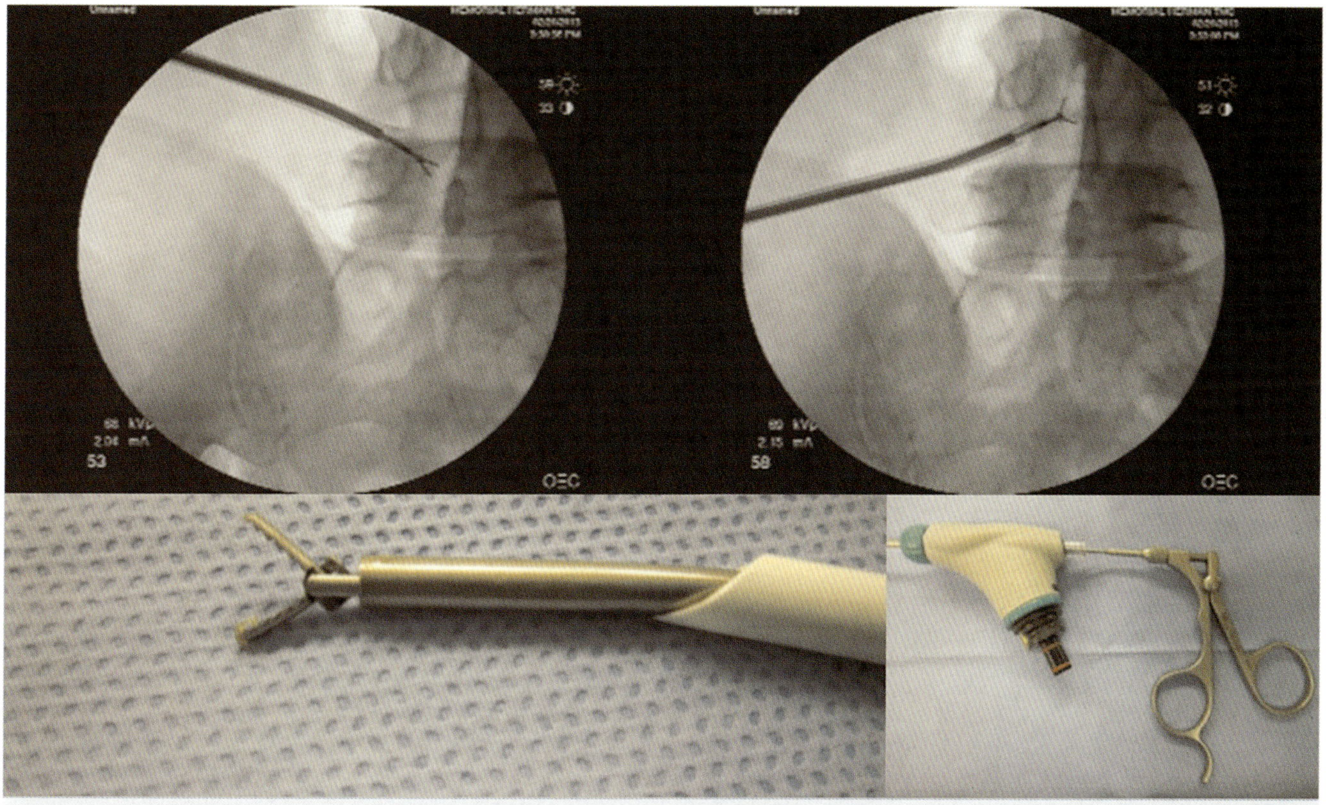

Fig. 19.8 The endoscopic forceps is used for the removal of larger herniated fragments.

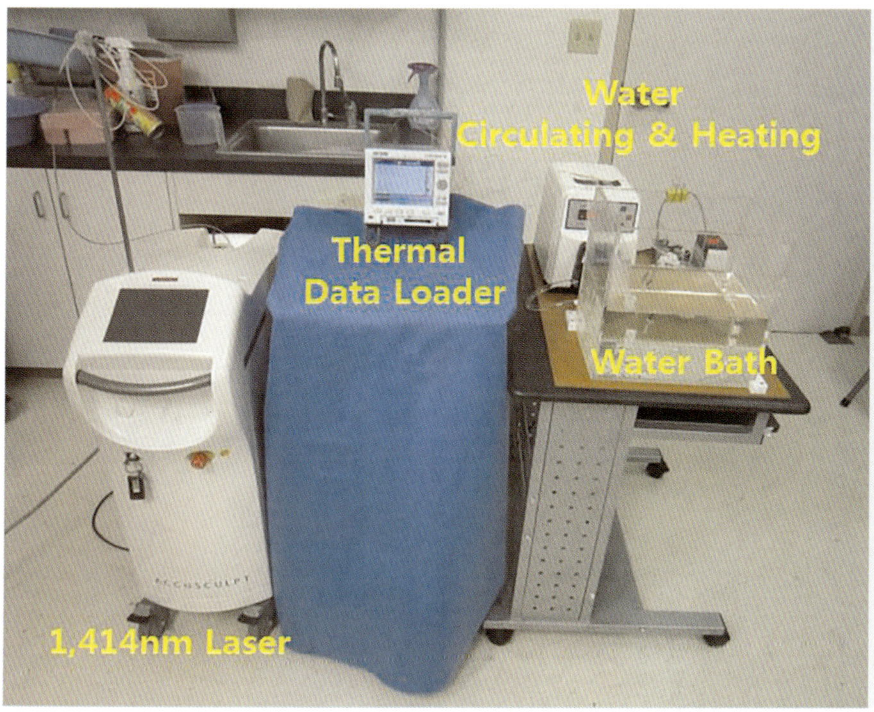

Fig. 19.9 Experimental setup.

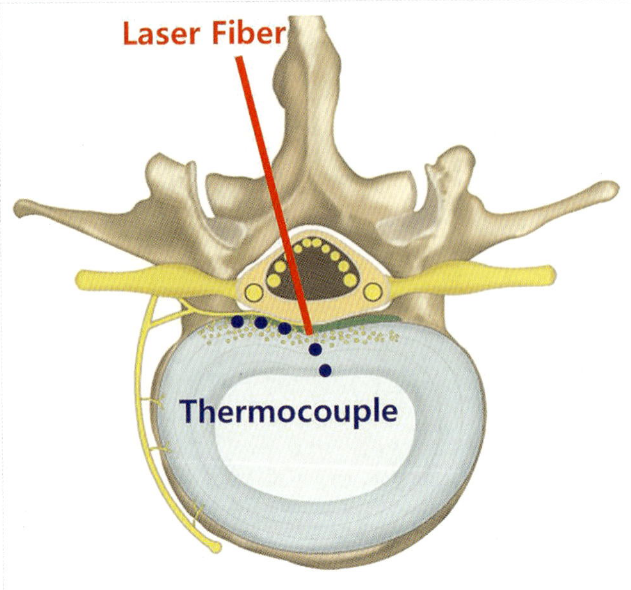

Fig. 19.10 The location of thermocouples.

Fig. 19.11 Measurement of temperatures during laser annuloplasty.

TELA has advantages over palliative operation treatment in terms of less invasive procedures, shorter treatment periods, and faster recovery times. Future studies will assess the clinical effectiveness of this novel procedure to treat lumbar disk disease.

References

[1] Choy DS, Case RB, Fielding W, Hughes J, Liebler W, Ascher P. Percutaneous laser nucleolysis of lumbar disks. N Engl J Med. 1987; 317(12):771–772

[2] Choy DS. The true story of percutaneous laser disc decompression. J Clin Laser Med Surg. 2001; 19(5):231–233

[3] Choy DS, Ngeow J. Percutaneous laser disc decompression in spinal stenosis. J Clin Laser Med Surg. 1998; 16(2):123–125

[4] Deukmedjian AJ, Cianciabella A, Cutright J, Deukmedjian A. Cervical Deuk Laser Disc Repair(®): A novel, full-endoscopic surgical technique for the treatment of symptomatic cervical disc disease. Surg Neurol Int. 2012; 3:142

[5] Raj PP. Intervertebral disc: anatomy-physiology-pathophysiology-treatment. Pain Pract. 2008; 8(1):18–44

[6] Quigley MR, Shih T, Elrifai A, Maroon JC, Lesiecki ML. Percutaneous laser discectomy with the Ho:YAG laser. Lasers Surg Med. 1992; 12(6):621–624

[7] Jayasree RS, Gupta AK, Bodhey NK, Mohanty M. Effect of 980-nm diode laser and 1064-nm Nd:YAG laser on the intervertebral disc—in vitro and in vivo studies. Photomed Laser Surg. 2009; 27(4):547–552

[8] Marguet CG, Sung JC, Springhart WP, et al. In vitro comparison of stone retropulsion and fragmentation of the frequency doubled, double pulse Nd:YAG laser and the holmium:YAG laser. J Urol. 2005; 173(5):1797–1800

[9] Choy DS, Altman P, Trokel SL. Efficiency of disc ablation with lasers of various wavelengths. J Clin Laser Med Surg. 1995; 13(3):153–156

[10] Moon BJ, Lee HY, Kim KN, et al. Experimental evaluation of percutaneous lumbar laser disc decompression using a 1414 nm Nd:YAG laser. Pain Physician. 2015; 18(6):E1091–E1099

[11] Lee SH, Kang HS. Percutaneous endoscopic laser annuloplasty for discogenic low back pain. World Neurosurg. 2010; 73(3):198–206, discussion e33

[12] Carragee EJ, Lincoln T, Parmar VS, Alamin T. A gold standard evaluation of the "discogenic pain" diagnosis as determined by provocative discography. Spine. 2006; 31(18):2115–2123

[13] Jo DH, Yang HJ. The survey of the patient received the epiduroscopic laser neural decompression. Korean J Pain. 2013; 26(1):27–31

20 Spinal Epiduroscopic Radiofrequency Therapy

Paulo Pereira and Pedro Monteiro

20.1 Introduction

Although the mechanisms of recurrent/persistent postoperative radicular pain and its relation with epidural fibrosis remain incompletely understood,[1] a series of patients have been reported to benefit with targeted approaches, namely, transforaminal epidural steroids, epiduroscopy, and adhesiolysis.[2] While more common after spinal surgery, epidural adhesions can also occur in patients not operated, in response to the release of proteoglycans from the nucleus pulposus into the epidural space through tears in the annulus fibrosus, or due to the presence of infection, hematoma, or contrast dye in the epidural compartment.[2]

Several techniques have been used for epiduroscopic adhesiolysis, namely, mechanical (performed with the tip of the endoscope, a Fogarty catheter or similar), electrothermal (using a laser or radiofrequency [RF] energy), and chemical (hyaluronidase).[3] One of the advantages of using RF for epidural adhesiolysis is to vaporize and coagulate the epidural fibrous bands keeping the tissue temperature below 50°C.[4]

20.2 Step 1: Sacral Hiatus Epiduroscopic Approach

- Endoscopic approach to the spinal epidural space through the sacral hiatus (▶ Fig. 20.1) has been described in Chapter 9. Through the spinal needle, the RF probe can be passed into the epidural space and maneuvered to the area of interest (▶ Fig. 20.2).
- Conscious sedation and systemic analgesia are strongly recommended when adhesiolysis is planned during epiduroscopy to render the procedure more comfortable for the patient.
- Epidurography is useful to identify areas not filled with the contrast dye, usually corresponding to extensive epidural fibrosis, blocking the epidural or periradicular space, where the epiduroscope is primarily directed.

20.3 Step 2: Irrigation of the Epidural Space with Saline

- Small boluses of isotonic saline are repeatedly flushed in the epidural space, accomplishing several purposes:
  - Opening the epidural space and cleaning the visual field.
  - Tearing mild fibrosis bands.
  - Controlling minor bleedings.
  - Cooling down the neural structures when performing epidural RF.
  - Washing out inflammatory mediators (such as phospholipase A2, tumor necrosis factor-α, interleukin [IL]-1β, IL-6, or IL-8) from the epidural space.
- Care must be taken concerning the volume of fluid injected in the epidural space and the speed and pressure of injection to avoid increase of the epidural and subarachnoid pressure and the consequences that may arise related thereto. We monitor but do not limit the volume of saline solution used for irrigation. Instead, we prefer not using a valve in the portals of the endoscope and injecting the saline in small boluses, which allows the fluid to exit the spinal canal through the portals and the intervertebral foramina. Moreover, in the rare case the patient reports any relevant symptom, we hold the procedure for a few minutes.[2]

20.4 Step 3: Mechanical Adhesiolysis

- Mechanical adhesiolysis shall precede the RF therapy and might be all that is required in cases with less thick fibrosis.[2,5]
- The tip of a Fogarty catheter may be used to dissect mild adhesions (▶ Fig. 20.3). The balloon is usually inflated and moved back and forth to stretch and tear denser areas of fibrosis and provides more available space in the local epidural area (▶ Fig. 20.4). It can also be useful for mechanical hemostasis.
- A 1-mm flexible endoscopic grasping forceps can be used to remove epidural scar tissue if no blood vessels are identified in the vicinity (▶ Fig. 20.5).

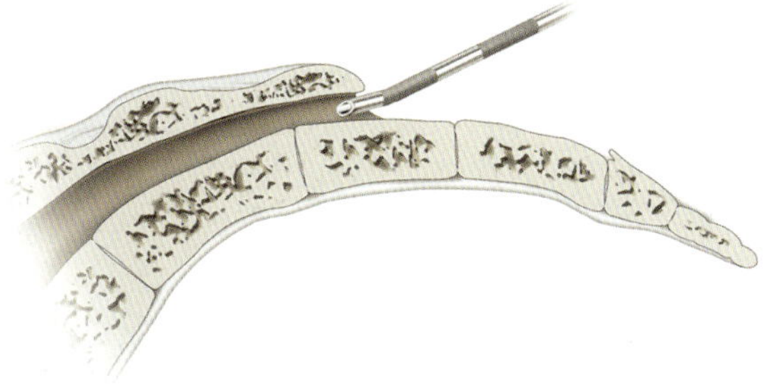

Fig. 20.1 Percutaneous insertion of the spinal needle through the sacral hiatus into the epidural space.

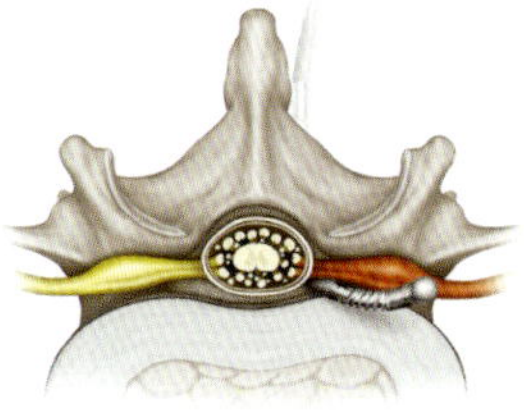

Fig. 20.2 The radiofrequency probe in the epidural space can be maneuvered to treat focused targets.

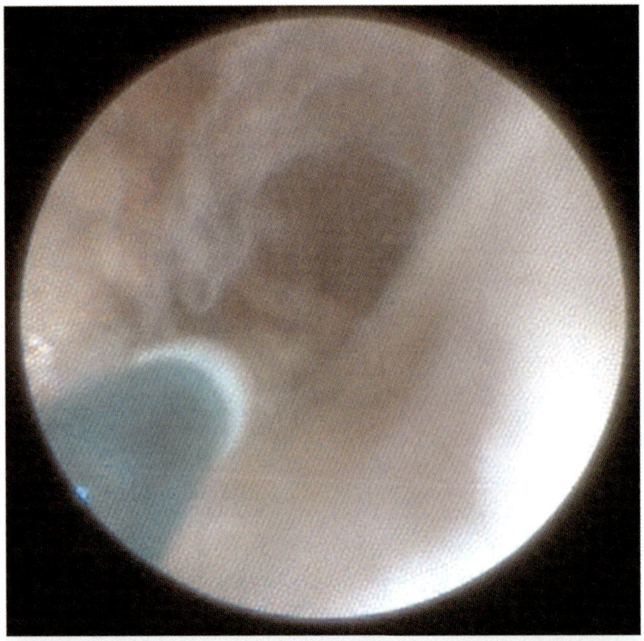

Fig. 20.3 Epidural adhesiolysis with the tip of a Fogarty catheter.

Fig. 20.4 The balloon of the Fogarty catheter is inflated and moved back and forth for adhesiolysis.

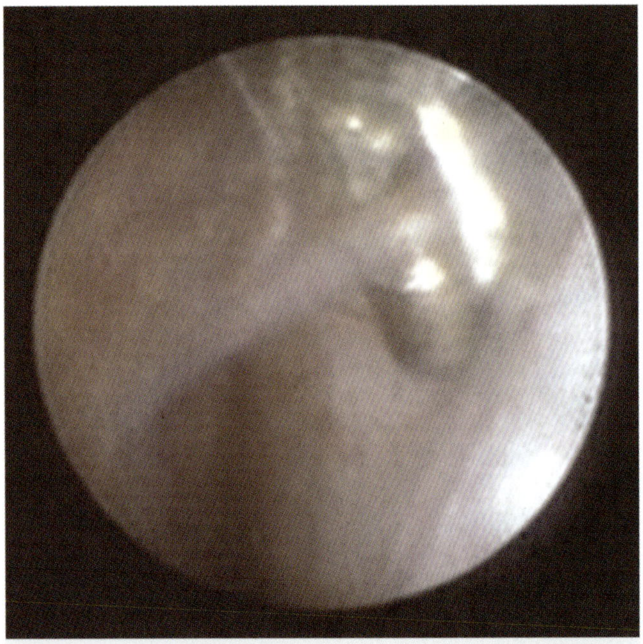

Fig. 20.5 An endoscopic forceps can be used to grasp and tear epidural scar tissue if no blood vessels are present at its surface.

20.5 Step 4: Radiofrequency Adhesiolysis

- The thickest and hardest fibrotic areas and *septa* can be partially destroyed using an RF catheter. We use a disposable monopolar electrode, with a plastic insulation of the shaft and an active ball tip with a diameter of 0.8 mm connected to a magnetic resonance generator, which employs frequencies of 4, 8, 12, and 16 MHz (Resaflex and Resablator, MRT [Molecular Resonance Technology], AMS Group S.R.L) (▶ Fig. 20.6). It allows cauterization and coagulation of epidural scar tissue without increasing the local temperature above 50°C (▶ Fig. 20.7).[4]
- In cases where compact fibrosis is found, preventing the progression of the epiduroscope, an opening is performed with the RF catheter and then the Fogarty catheter is inserted into the opening and the balloon is inflated in order to expand the passage.[2,5]
- After adhesiolysis, epidural hemostasis must be checked. Epidural bleeding from small vessels can be managed with the RF catheter (▶ Fig. 20.8).

20.6 Step 5: Repeated Epidurogram

- When filling defects were present on the initial epidurogram and adhesiolysis techniques were used, a control epidurogram is useful to assess the efficacy of the procedure.
- The syringe containing the saline is replaced by another one filled with a nonionic contrast (e.g., Ultravist 240, Bayer Schering Pharma A.G.) and 5 to 10 mL is injected in the epidural space, caudally to the intervention.
- After the epidurogram, the epidural space is irrigated with saline to scatter the contrast dye.

20.7 Step 6: Pharmacologic Therapy

- After adhesiolysis, a number of therapeutic drugs can be injected in the epidural space, directly around the

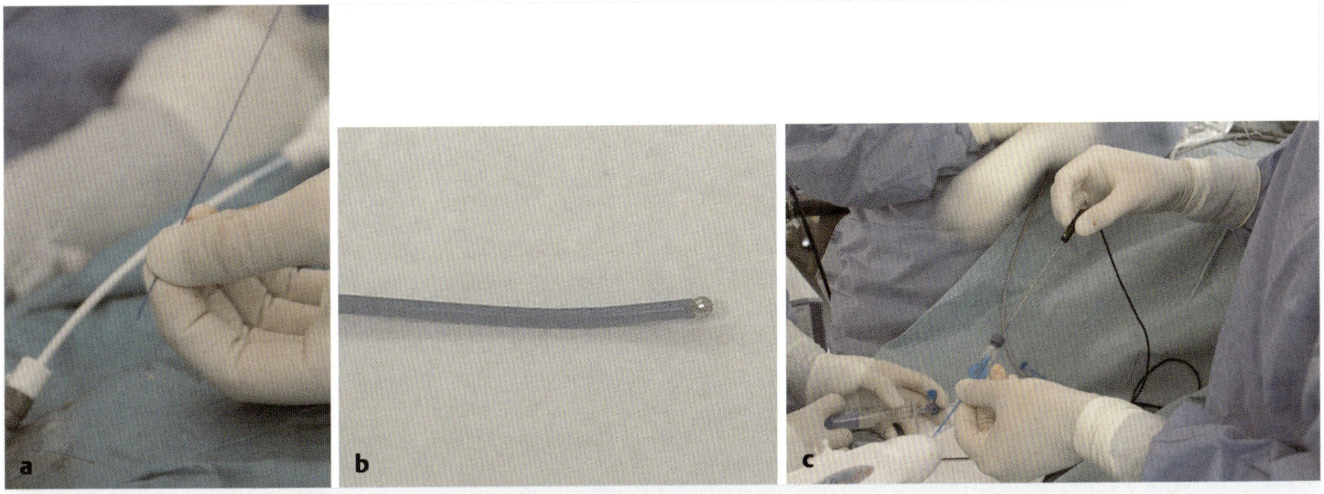

Fig. 20.6 (a–c) An insulated radiofrequency catheter with an active tip is inserted through the portal of the epiduroscope.

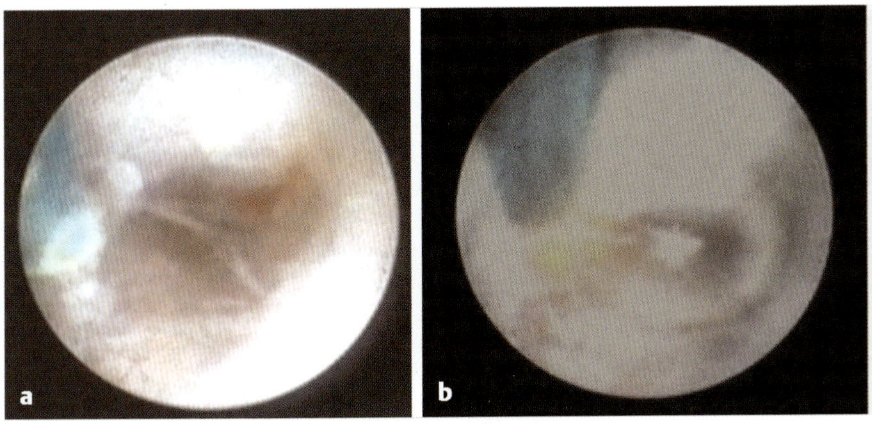

Fig. 20.7 (a,b) Adhesiolysis using the radiofrequency catheter.

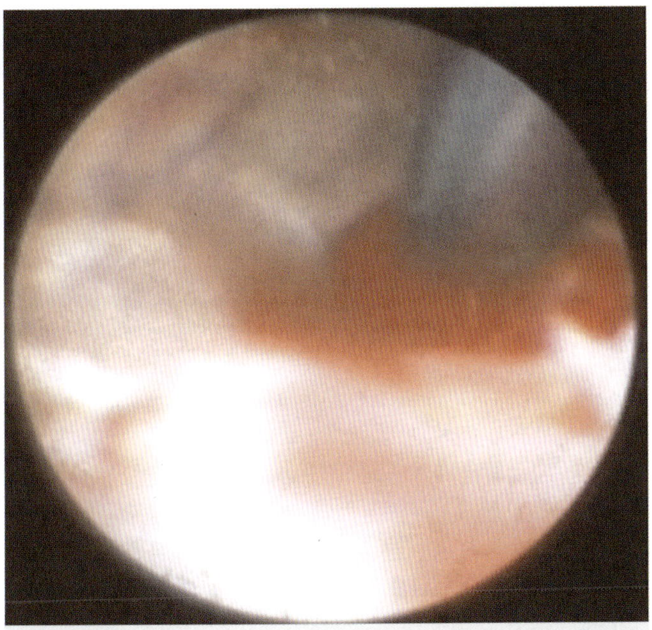

Fig. 20.8 The radiofrequency catheter used for the coagulation of small epidural blood vessels.

intervention and the previously identified painful area, including steroids, local anesthetics, clonidine, and hyaluronidase. We use 5 mL of bupivacaine hydrochloride 0.5% and 12 mg of betamethasone (betamethasone sodium phosphate 6 mg + betamethasone acetate 6 mg/2 mL, Celesdepot, Schering-Plough, Merck & Co.) (▶ Fig. 20.9).

- During the drug injection and the removal of the epiduroscope, care must be taken to avoid the efflux of the therapeutic agents through the portals of the epiduroscope or the sacral hiatus (▶ Fig. 20.10).

20.8 Step 7: Postoperative Care

- After the procedure, nursing care includes monitoring of vital signs, headache, neck and low back pain, neurologic symptoms, and surgical dressing.
- A postoperative analgesic regimen is established.
- Patients are kept in bed for a few hours after the procedure, and then they progressively begin ambulation and daily activities. We discharge the patients the following morning.

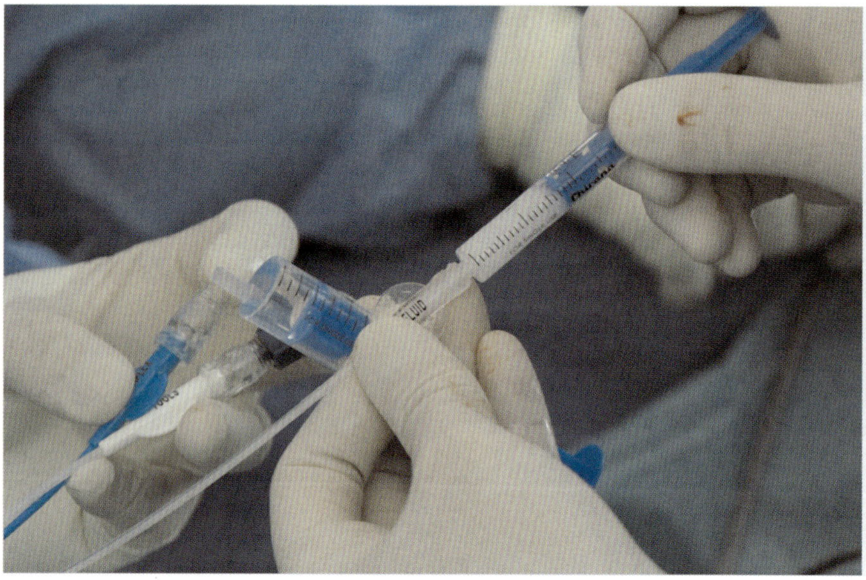

Fig. 20.9 Injection of steroid mixture in the epidural region where the intervention has been performed.

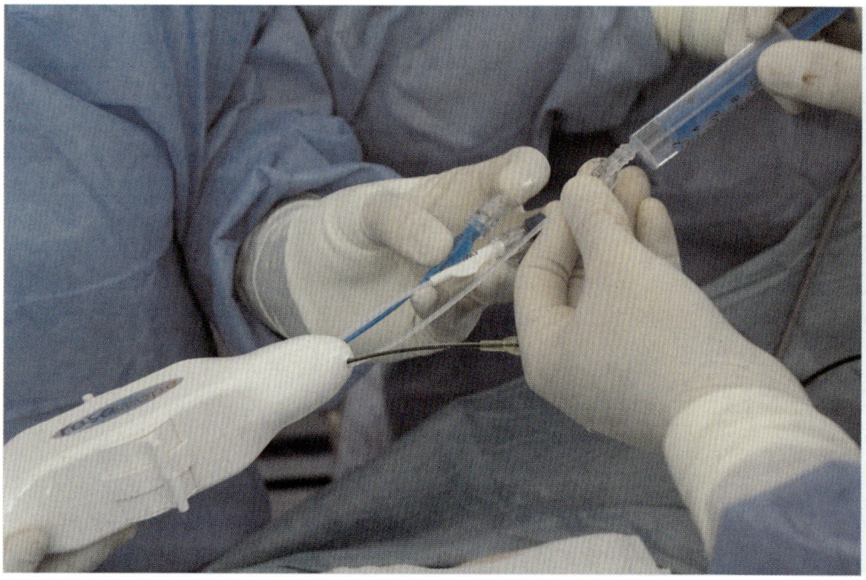

Fig. 20.10 During the injection of the medication in the epidural space, the portals are closed to avoid efflux of the drugs.

References

[1] Pereira P, Avelino A, Monteiro P, Vaz R, Castro-Lopes JM. New insights from immunohistochemistry for the characterization of epidural scar tissue. Pain Physician. 2014; 17(5):465–474

[2] Pereira P, Severo M, Monteiro P, et al. Results of lumbar endoscopic adhesiolysis using a radiofrequency catheter in patients with postoperative fibrosis and persistent or recurrent symptoms after discectomy. Pain Pract. 2016; 16 (1):67–79

[3] Schütze G. Epiduroscopy – Spinal Endoscopy. London: Springer; 2008

[4] Raffaeli W, Righetti D, Andruccioli J, Sarti D. Periduroscopy: general review of clinical features and development of operative models. Acta Neurochir Suppl (Wien). 2011; 108:55–65

[5] Raffaeli W, Righetti D. Surgical radio-frequency epiduroscopy technique (R-ResAblator) and FBSS treatment: preliminary evaluations. Acta Neurochir Suppl (Wien). 2005; 92:121–125

21 Transforaminal Epiduroscopic Laser Annuloplasty for Diskogenic Pain

Victor P. Lo, Jongsun Lee, Ashley E. Brown, Alissa Redko, and Daniel H. Kim

21.1 Introduction

Pain in the lower back can affect up to 85% of the population at some point in life.[1] In most cases, low back pain is self-limiting; however, it can become chronic and disabling in 5% of patients.[2] The precise anatomic cause can often be difficult to identify. It has been estimated that approximately 40% of chronic low back pain originates from the intervertebral disk.[3] Histologic analysis of intervertebral disk revealed significant sensory innervation in the posterolateral aspect of the annulus fibrosis.[4] Direct stimulation of the outer annulus fibrosis in vivo demonstrated concordant pain.[5]

Treatment of chronic diskogenic low back pain has been challenging. Conservative measures often fail to reduce the pain or improve function. Lumbar arthrodesis for diskogenic pain reported only a 46% satisfactory clinical outcome.[6] Successful fusion by elimination of a painful motion segment has not demonstrated a significant improvement in pain and functional status.[7] In addition, the surgery is associated with its complication risks, morbidity, and prolonged recovery. This has led to the development of minimally invasive intradiskal therapeutic approaches to open surgical procedures, including intradiskal electrothermal therapy (IDET), radiofrequency ablation (RFA), cryotherapy, percutaneous endoscopic laser diskectomy (PELD), and percutaneous endoscopic laser annuloplasty (PELA).[8,9] The proposed mechanism of action of intradiskal therapy is a combination of destruction of the annulus nociceptors and shrinkage of the intervertebral disk.

Given that the pain-generating regions of diskogenic pain are located in the posterolateral aspect of the annulus, an extradikcal epidural approach can also be utilized for assessment and treatment. An extradiskal epidural approach has the added benefit of direct visualization of the epidural space and its structures with a flexible endoscope (epiduroscopy). In addition, the epidural structures can be assessed to determine whether it is concordant with the patient's clinical symptoms. Lumbosacral epiduroscopy has been demonstrated to be more accurate in identifying vertebral level pathology than clinical assessment or magnetic resonance imaging (MRI).[10] In addition, epiduroscopic findings were noted to be predictive of treatment outcomes.[11] The epiduroscopic approach has been previously reported to treat lumbar stenosis and back–leg pain syndrome.[12,13,14,15,16,17,18] However, the traditional approach of epiduroscopy with its entry site through the sacral hiatus may be limited by bony stenosis of the hiatus, lumbar stenosis, or epidural scaring from previous surgery.

In combination with epiduroscopy, the use of a laser can enhance the treatment efficacy of diskogenic pain. Systematic review of the literature for use of laser in lumbar disk decompression demonstrated positive results, with 75% of patients treated reporting significant pain relief for 12 or more months.[19] Laser disk decompression has also been demonstrated to be comparable to diskectomy.[20] Various types of lasers have been used in managing spinal disorders.[21,22,23] One type, the neodymium:yttrium-aluminum-garnet (Nd:YAG) laser, has been demonstrated to be effective in managing spine disorders in several clinical studies.[24,25,26,27]

In this chapter, we describe an approach for the management of diskogenic pain utilizing a novel curved spinal endoscope and introduction technique in combination with an Nd:YAG laser. The curved endoscope provides transforaminal access where current rigid spinal endoscopes or anatomic barriers may limit passage (▸ Fig. 21.1). Transforaminal epiduroscopic laser annuloplasty (TELA) provides a minimally invasive and direct approach to the evaluation and treatment of diskogenic pain.

21.2 Indications for Transforaminal Epiduroscopic Laser Annuloplasty

- Internal disk disruption.
- Herniated nucleus pulposus (HNP) with predominately axial back pain.
- Annular tear.
- Adhesions secondary to failed back surgery syndrome.
- Diskal cyst.
- Mild to moderate neuroforaminal stenosis.

21.3 Contraindications for Transforaminal Epiduroscopic Laser Annuloplasty

- Large HNP with radiculopathy.
- Severe neuroforaminal stenosis.
- Spinal instability.
- Modic changes.
- Patients with high iliac crest and L5–S1 level pathology.

21.4 NeedleView CH Endoscope System

- The endoscopic system is a disposable semirigid fiberoptic based microendoscope with a single working channel (NeedleView CH, BioVision Technologies). The endoscope has a 160-mm working length with a 3.4-mm outer diameter. There is a 1.85-mm diameter working channel and a built-in 0.7-mm fiberoptic channel with 17,000-pixel resolution (▸ Fig. 21.2).
- The distal third of the endoscope can be bent to a desired angle to facilitate entry into the ventral epidural space from a transforaminal approach (▸ Fig. 21.3).

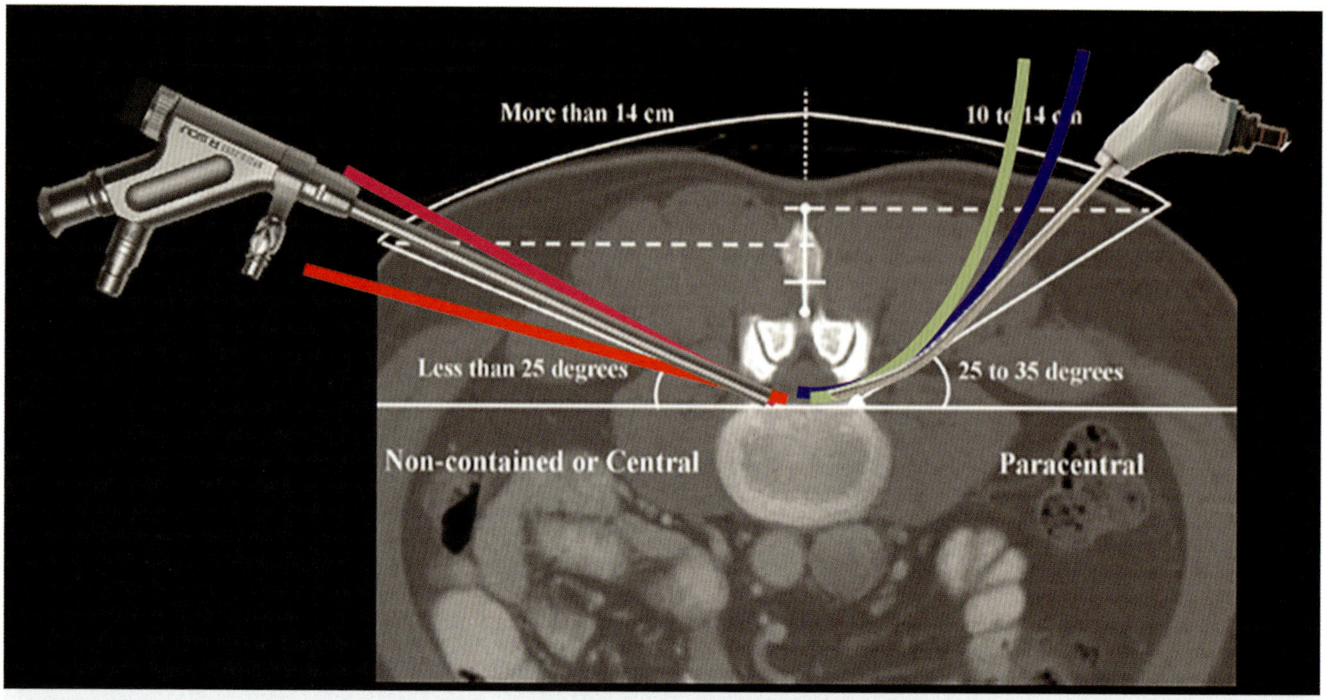

Fig. 21.1 The standard endoscope (left) has a fixed trajectory to the neural foramen from the skin entry site. A curved endoscope (right) has a similar skin entry site and can navigate into the neural foramen with a less steep trajectory, thus bypassing potential anatomic barriers.

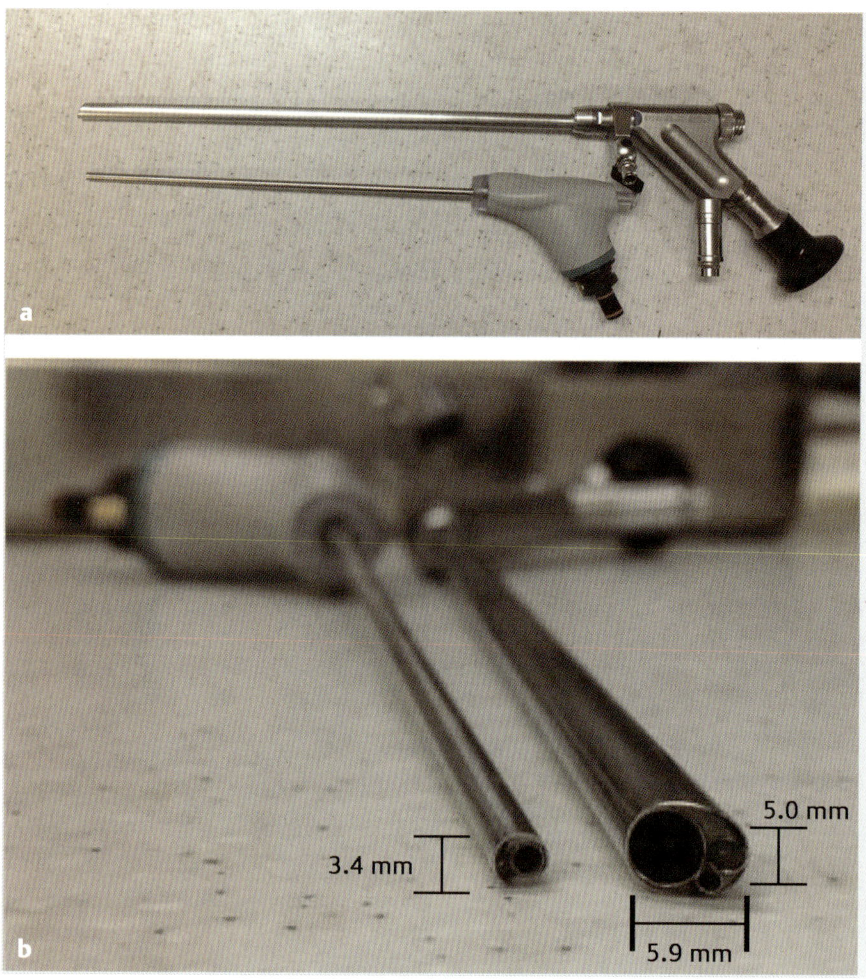

Fig. 21.2 (**a**) NeedleView CH microendoscope compared with a Vertebris endoscope (Richard Wolf). (**b**) NeedleView CH's outer diameter is 3.4 mm compared to Vertebris' outer diameter of 5.9 × 5.0 mm.

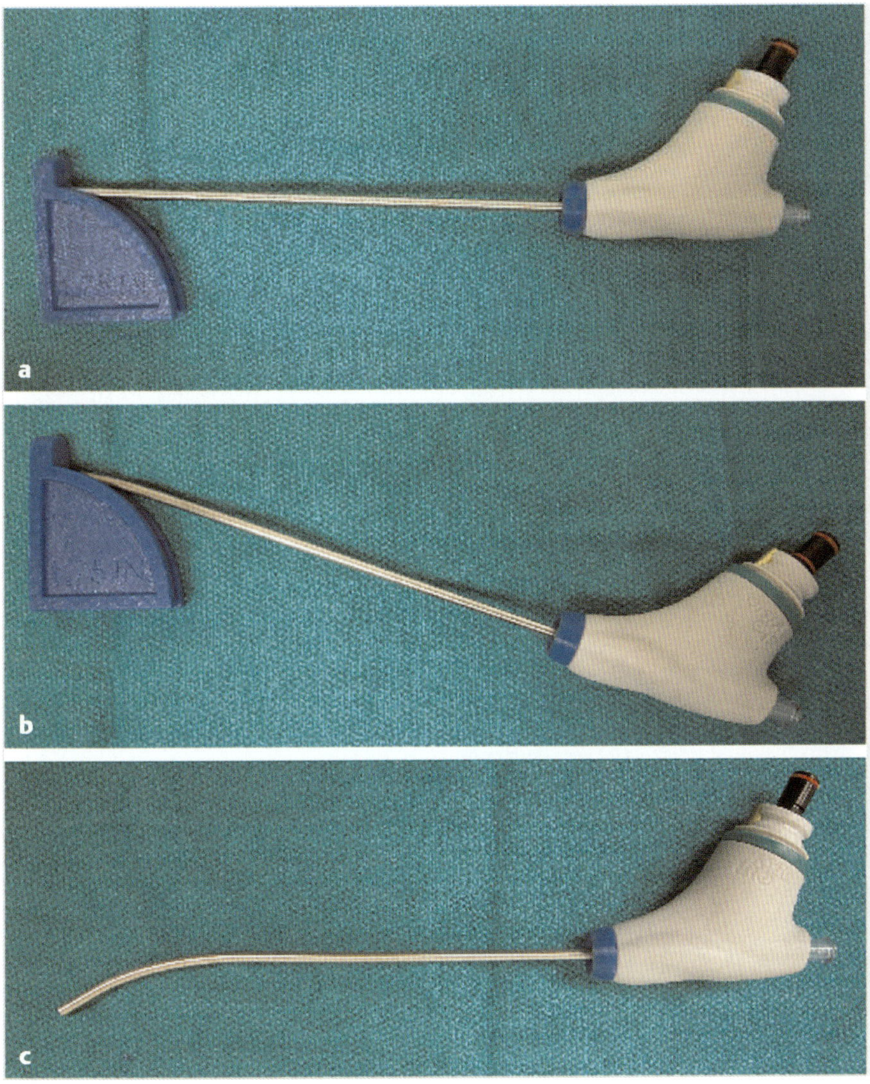

Fig. 21.3 The endoscope is bent to the desired entry angle. **(a)** The endoscope is placed on the molding frame. **(b)** A curve in the distal end is made to the desired angle. **(c)** The final configuration of the bent endoscope.

21.5 NeedleCam HD Visualization System

- The NeedleCam HD Visualization System (BioVision Technologies) incorporates a light-emitting diode light source and high-resolution camera in a single compact unit.
- The light source and video images are transmitted through a single cable. The video output is connected to a high-definition display with 1,920 × 1,080 resolution.

21.6 Equipment

- A pulsed Nd:YAG laser with a wavelength of 1,414 nm is transmitted through 3-m fiber (ACCUPLASTI, Lutronic).
- The laser is delivered through a 550-μm side-firing opening (▶ Fig. 21.4).
- Equipment required for endoscope introduction includes the following (▶ Fig. 21.5):
- 18-gauge spinal needle.
- 21-gauge spinal needle.
- 14-gauge × 127-mm Tuohy needle.
- 18-gauge × 152-mm Tuohy needle.
- 12-Fr cannula and 12-Fr dilator.
- 70-cm guidewire.
- Endoscope bender.
- No. 15 blade.

21.7 Anesthesia

- The procedure is performed under conscious sedation.
- Local anesthetic is administered at the skin entry site and trajectory of the endoscope.

21.8 Positioning

- The patient is placed in prone position on a Wilson frame and radiolucent operating table.
- The spine is flexed intraoperatively with the Wilson frame.
- The C-arm fluoroscopy unit is positioned for anteroposterior (AP) and lateral images.

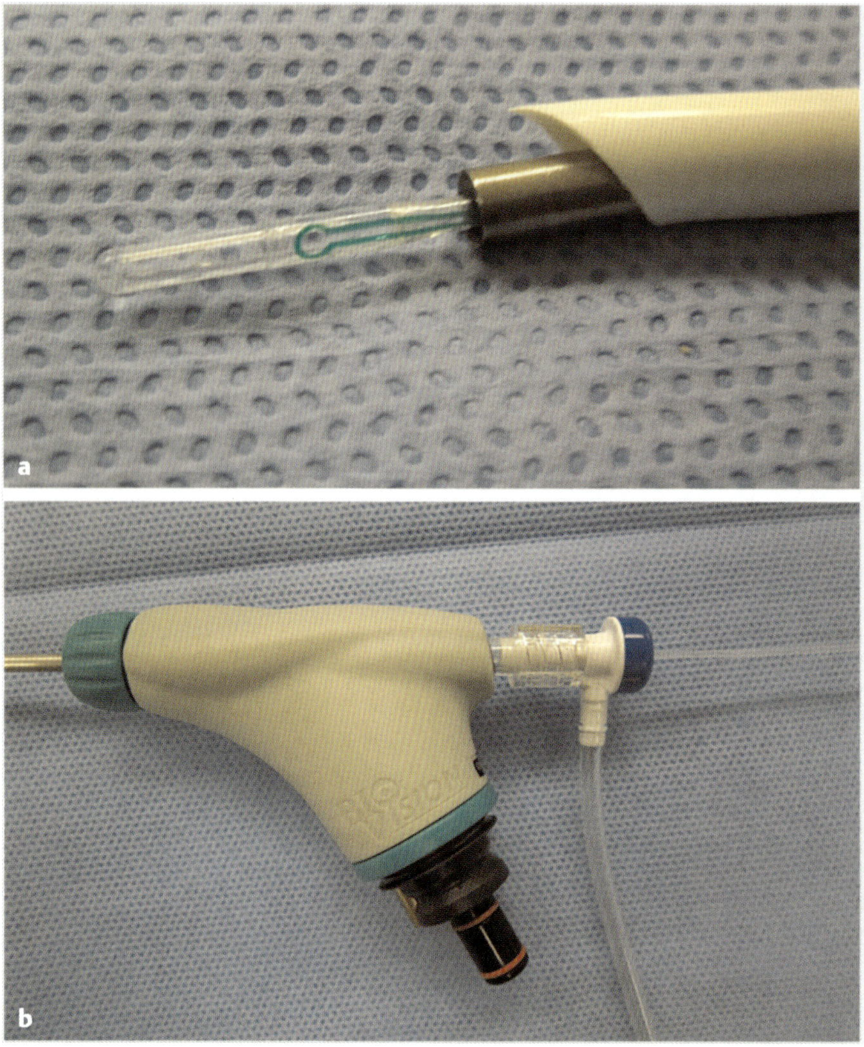

Fig. 21.4 **(a)** Side-firing Nd:YAG laser tip. **(b)** Insertion of laser fiber through the endoscope.

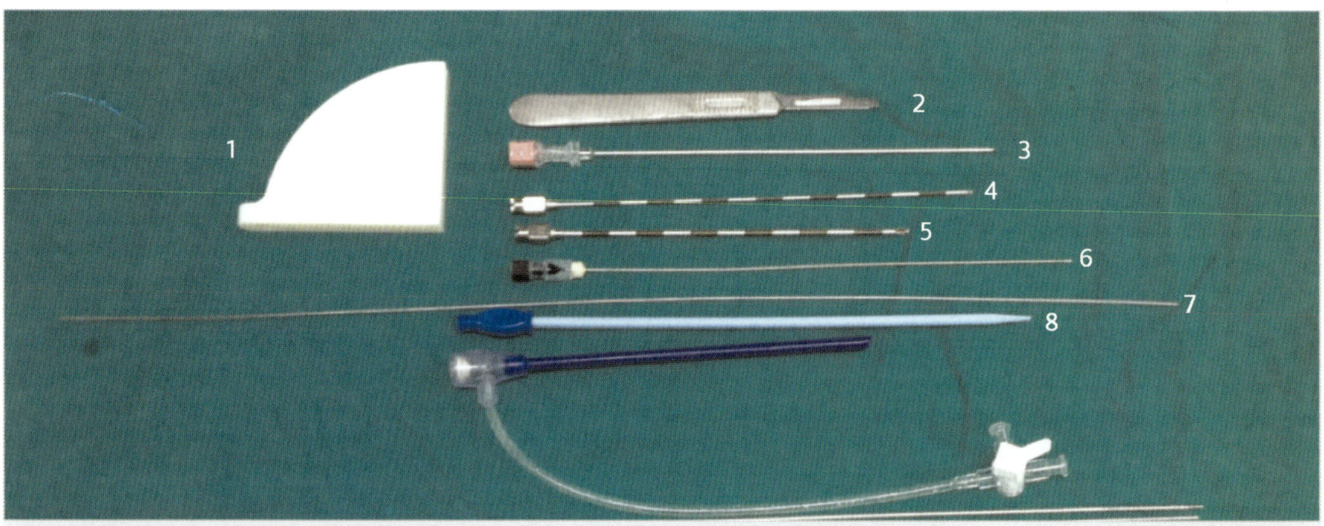

Fig. 21.5 Instruments required for introduction of the curved endoscope: (1) endoscope bender; (2) no. 15 blade for skin incision; (3) 18-gauge spinal needle; (4) 18-gauge Tuohy needle; (5) 14-gauge Tuohy needle; (6) 21-gauge spinal needle; (7) guidewire; and (8) flexible dilator and sheath.

21.9 Technique for Transforaminal Epiduroscopic Laser Annuloplasty

- Axial MRI or computed tomographic images are used to calculate the distance of the skin entry site from midline, with the trajectory to the neural foramen. The typical distance is 9 to 13 cm from midline.
- Inject the skin entry point with local anesthetic.
- The 18-gauge spinal needle is inserted until the epidural space is reached (▶ Fig. 21.6a,b). This is confirmed with AP and lateral fluoroscopy (▶ Fig. 21.6c,d).
- The stylet is then removed and the guidewire is then passed in the epidural space at the lateral recess (subarticular zone). Confirm the epidural location of the guidewire with AP and lateral images.
- The spinal needle is then removed and the 14-gauge Tuohy needle is inserted over the guidewire into the epidural space. The guidewire is then advanced into the ventral epidural space (▶ Fig. 21.7).
- Remove the 14-gauge needle and confirm location of the guidewire on fluoroscopy.
- Slide the 12-Fr dilator and sheath over the guidewire (▶ Fig. 21.8) and confirm ventral location in the epidural space with contrast (▶ Fig. 21.9a,b).
- After the removal of the dilator, the ventral location of the sheath in the epidural space is confirmed with AP and lateral fluoroscopy followed by contrast injection (▶ Fig. 21.9c,d).
- The NeedleView CH microendoscopic camera is bent to the desired angle and inserted into the sheath, into the epidural space (▶ Fig. 21.10), and confirmed with fluoroscopy (▶ Fig. 21.11). Ventral epidural anatomy is identified on endoscopy (▶ Fig. 21.12).
- At this point, the epidural region can be probed (▶ Fig. 21.13a,b) or diskography (▶ Fig. 21.13c,d) can be

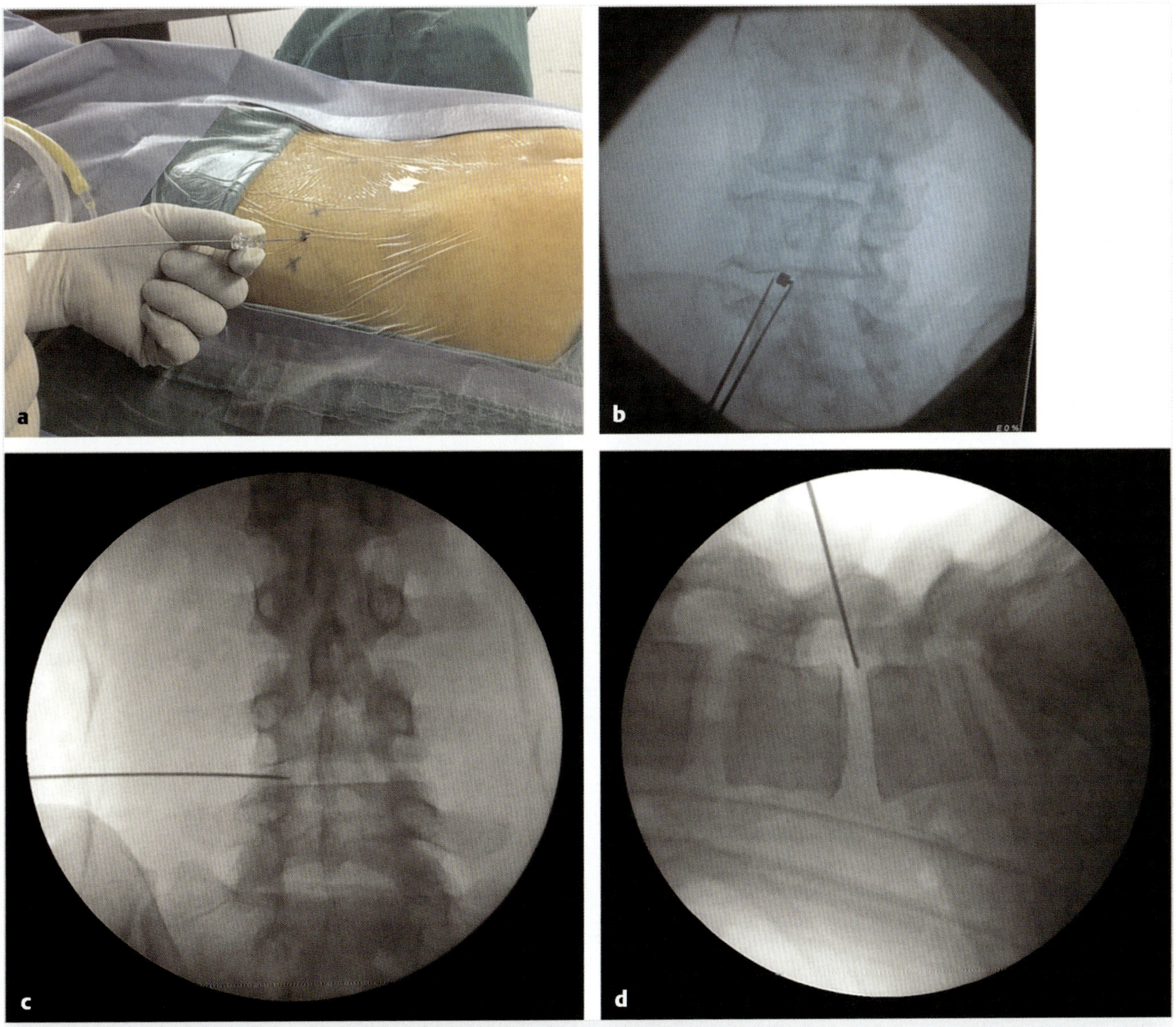

Fig. 21.6 **(a)** Insertion of spinal needle. **(b)** Oblique fluoroscopic projection to confirm trajectory to disk space. **(c)** Anteroposterior fluoroscopic image to confirm needle position at neural foramen. **(d)** Lateral image to confirm needle position at the neural foramen.

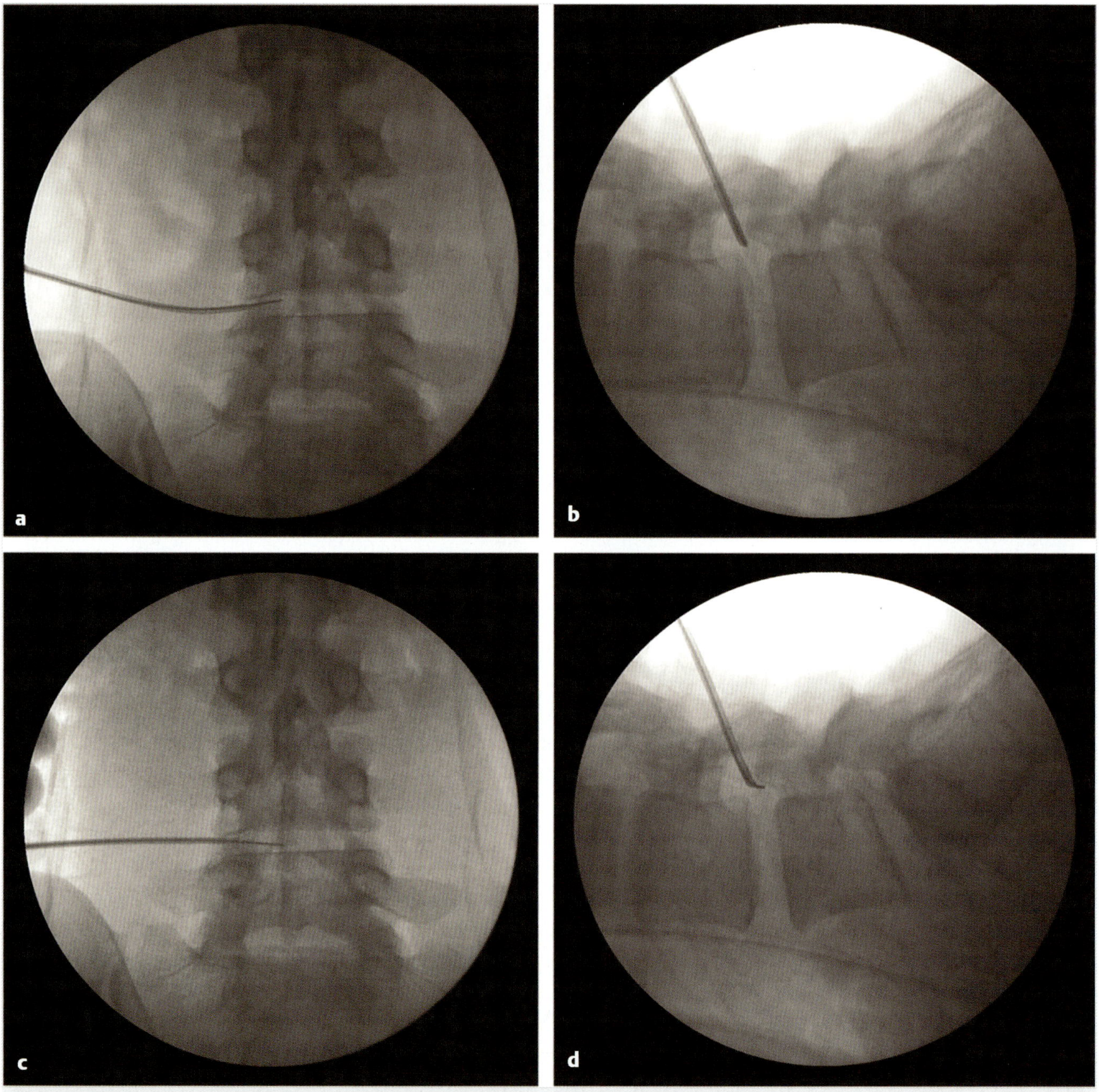

Fig. 21.7 **(a)** Fluoroscopic anteroposterior (AP) view of 14-gauge Tuohy needle with guidewire in lateral recess. **(b)** Fluoroscopic lateral view of 14-gauge Tuohy needle with guidewire through the foramen in lateral recess. **(c)** Fluoroscopic AP view after advancement of the guidewire into the ventral epidural space. **(d)** Fluoroscopic lateral view after advancement of guidewire into the ventral epidural space.

performed. Endoscopic forceps can also be introduced for the removal of free fragments (▶ Fig. 21.14 and ▶ Fig. 21.15).

21.10 Laser Annuloplasty

- The side-firing Nd:YAG laser is then introduced through the working channel into the epidural space.
- The application of the laser, 0.25 W, 150 mJ at 20 Hz, with 0.5- to 1.0-second pulses and interval of 1 to 2 seconds, to a total energy deliver of ~500 J (▶ Fig. 21.16).
- The laser setting can be adjusted either for thermal ablation of the nociceptive nerve ending or for coagulation for annular shrinkage.

21.11 Potential Complications

- Dural or neural injury.
- Annular damage.
- Headaches from elevated intracranial hypertension.

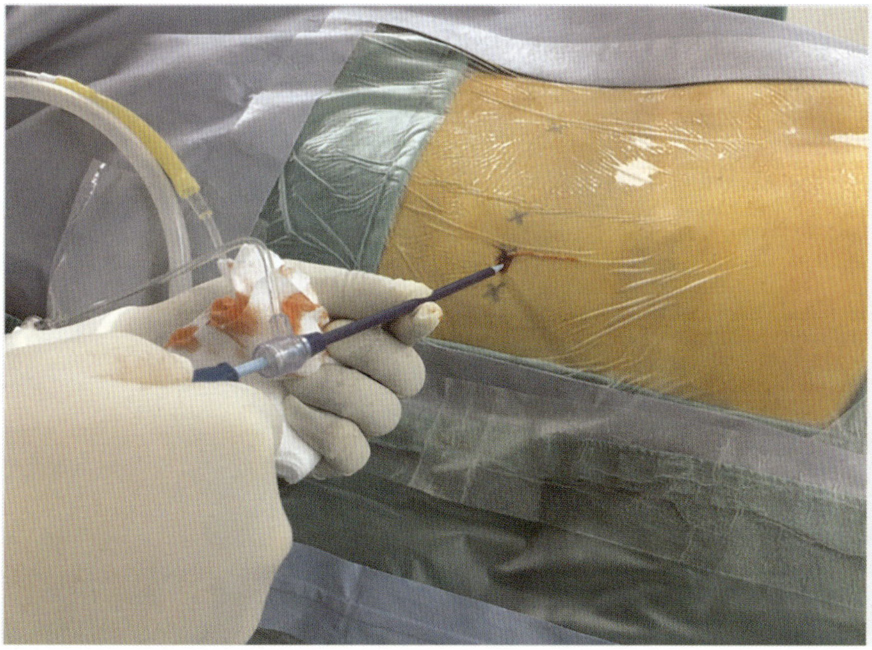

Fig. 21.8 Passage of dilator and sheath over guidewire into the epidural space.

21.12 Conclusion

One of the main mechanisms of diskogenic pain is damage to the posterior annulus.[28] Growth of nociceptive free nerve endings in the annulus fibrosis represents the source of pain-generating signals.[29,30] Initial management options include medications and physical therapy. One would consider a procedure after failure of conservative management. With the lack of demonstrated treatment efficacy of open surgical procedures for diskogenic pain, minimally invasive techniques such as IDET, RFA, PELD, and PELA offer alternative options.[6,8,9]

Minimally invasive treatment options with intradiskal approaches, such as intradiskal electrothermal therapy, RFA, or PELA, require the creation of a new annular opening to allow entry of the treatment device. This can potentially lead to new diskogenic pain symptoms or disk herniation. An extradiskal epidural approach would eliminate the risk for further damaging the intervertebral disk during the treatment process. In addition, with intradiskal procedures, one cannot visualize the entire annular surface within the spinal canal to assess for other sites of pathology. In order to inspect the annular surface within the spinal canal, an endoscope can be used to visualize the epidural space. The standard spine endoscope, such as the one used in PELD, is effective for diskectomy, with less tissue disruption in comparison to open surgery[31]; however, its rigid construction limits visualization of the entire disk surface. In addition, the position of the working channel is critical to the success of the procedure.[32] To see the entire annular portion of the disk, access into the ventral lumbar epidural space is required.

Epiduroscopy is an effective approach to access the epidural space. In addition, epiduroscopy has the benefit of direct visualization of structures within the epidural space and the ability to probe in order to assess whether symptoms can be reproduced. Epiduroscopy has reported to treat lumbar stenosis and back–leg pain syndrome with an approach through the sacral hiatus.[12,13,14,15,16,18] Introduction of the epiduroscope through the sacral hiatus to lumbar disk pathology requires a relatively long distance to navigate the endoscope. In addition, bony stenosis of the sacral hiatus or lumbar epidural space may make passing the endoscope to the target impossible.

We have described a novel approach, TELA, to the ventral epidural space through the lumbar neural foramen and passage of a laser for treatment. This is accomplished by passing a bendable endoscope with an outer diameter of 3.4-mm in combination with a flexible sheath with a total outer diameter of 4 mm. In addition, the presence of a working channel allows the passage of endoforceps or laser. TELA technique passes a 1,414-nm side-firing Nd:YAG laser into the epidural space for thermal ablation of the free nerve endings in the annulus as well as for annular shrinkage for decompression.

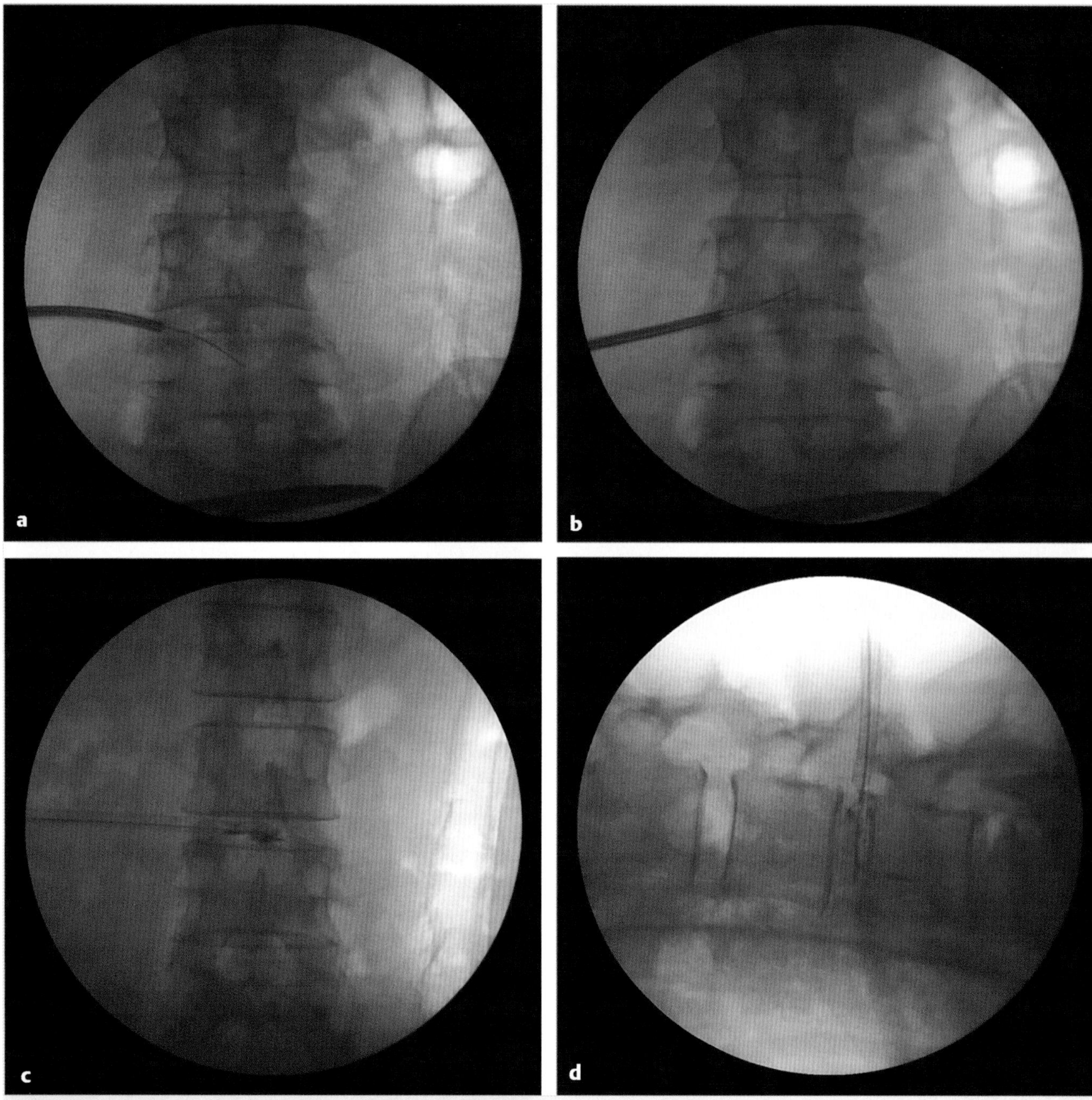

Fig. 21.13 (a) Anteroposterior (AP) fluoroscopy of a probe maneuvered caudally in the ventral epidural space. **(b)** AP fluoroscopy of a probe maneuvered rostrally in the ventral epidural space. **(c)** AP fluoroscopy of a diskogram. **(d)** Lateral fluoroscopy of a diskogram.

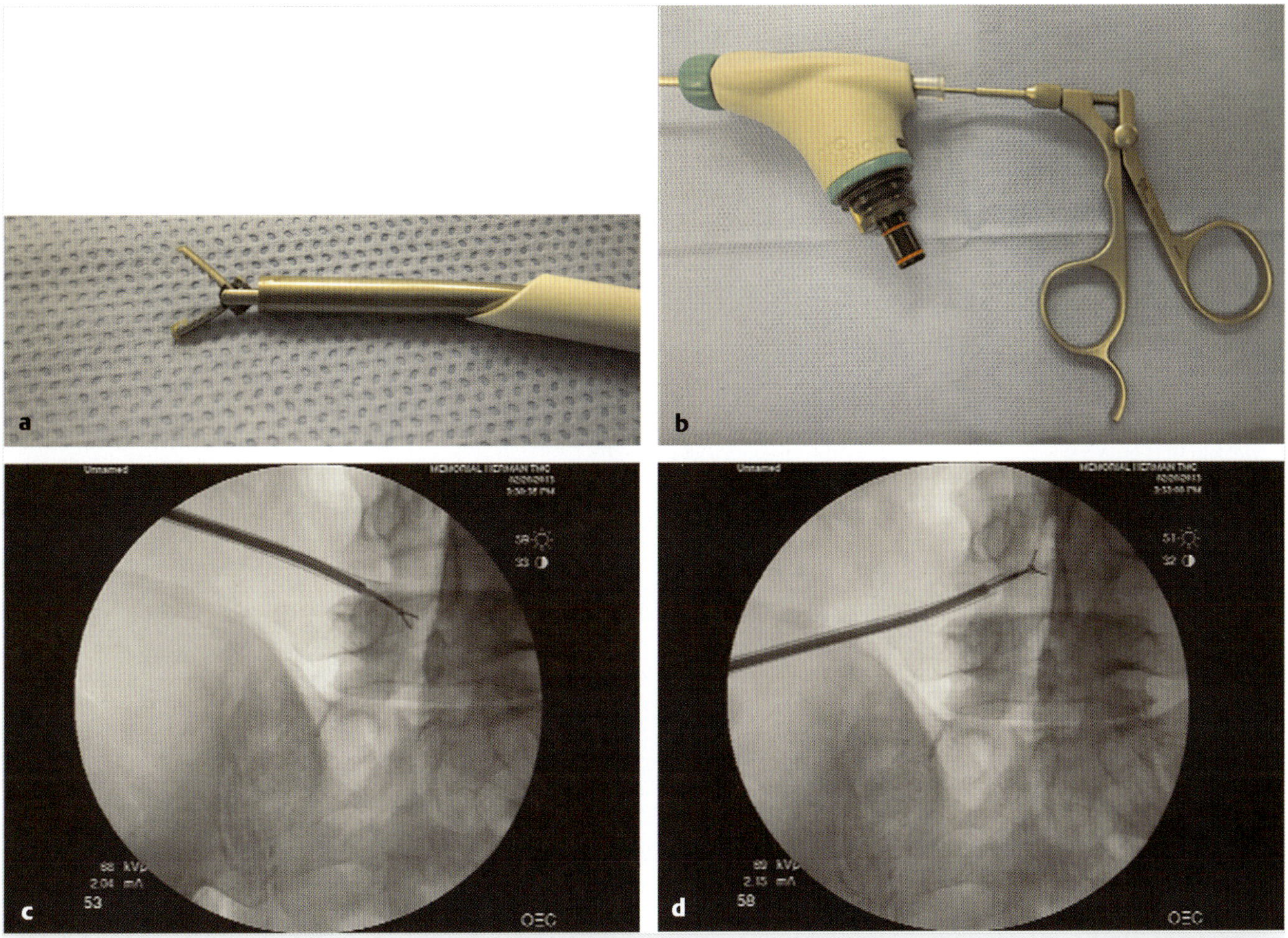

Fig. 21.14 (a) Endoscopic forceps, 1.5 mm. (b) Endoscopic forceps through the working channel. (c) Anteroposterior (AP) fluoroscopy of endoscopic forceps in the ventral epidural space reaching caudally. (d) AP fluoroscopy of endoscopic forceps in the ventral epidural space reaching rostrally.

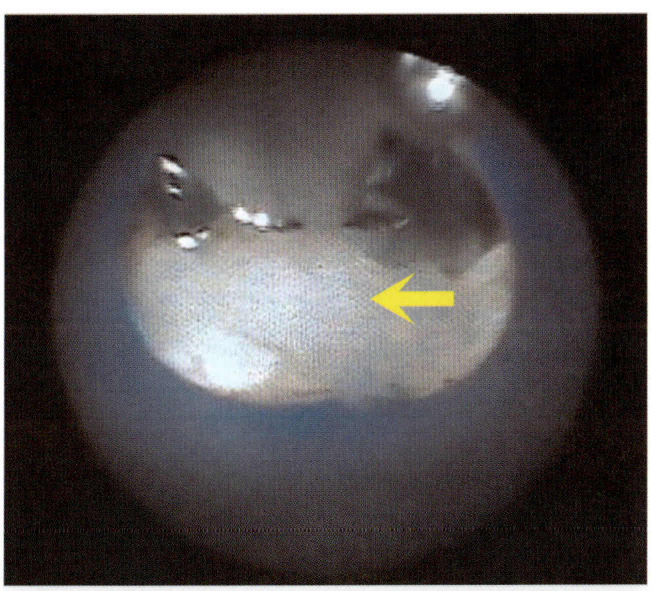

Fig. 21.15 Endoscopic forceps removing intervertebral disk material (*arrow*).

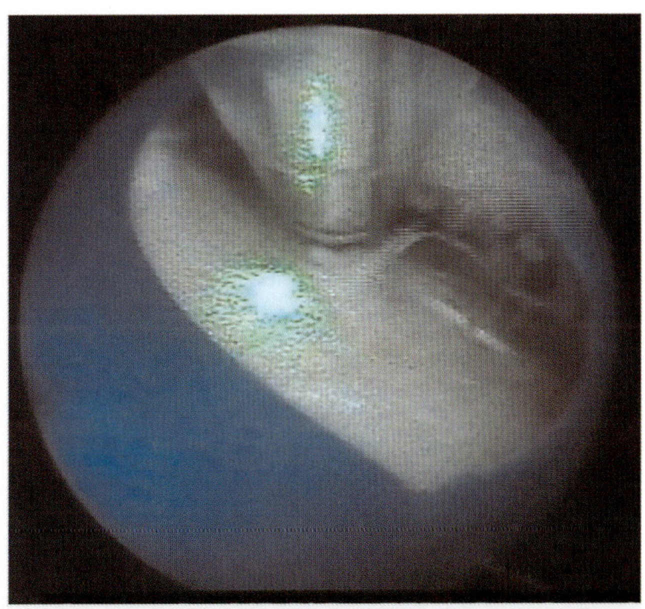

Fig. 21.16 Annuloplasty with side-firing Nd-YAG laser.

References

[1] Andersson GB. Epidemiological features of chronic low-back pain. Lancet. 1999; 354(9178):581–585

[2] Andersson GB, Svensson HO, Odén A. The intensity of work recovery in low back pain. Spine. 1983; 8(8):880–884

[3] Schwarzer AC, Aprill CN, Derby R, Fortin J, Kine G, Bogduk N. The prevalence and clinical features of internal disc disruption in patients with chronic low back pain. Spine. 1995; 20(17):1878–1883

[4] Yoshizawa H, O'Brien JP, Smith WT, Trumper M. The neuropathology of intervertebral discs removed for low-back pain. J Pathol. 1980; 132(2):95–104

[5] Kushlich SD, Ulstrom CL, Michael CJ. The issue origin of low back pain and sciatica: a report of pain response to tissue stimulation during operations on the lumbar spine using local anesthesia. Orthop Clin North Am. 1991; 22 (2):181–187

[6] Wetzel FT, LaRocca SH, Lowery GL, Aprill CN. The treatment of lumbar spinal pain syndromes diagnosed by discography. Lumbar arthrodesis. Spine. 1994; 19(7):792–800

[7] Mirza SK, Deyo RA. Systematic review of randomized trials comparing lumbar fusion surgery to nonoperative care for treatment of chronic back pain. Spine. 2007; 32(7):816–823

[8] Singh K, Ledet E, Carl A. Intradiscal therapy: a review of current treatment modalities. Spine. 2005; 30(17) Suppl:S20–S26

[9] Lee SH, Kang HS. Percutaneous endoscopic laser annuloplasty for discogenic low back pain. World Neurosurg. 2010; 73(3):198–206, discussion e33

[10] Bosscher HA, Heavner JE. Diagnosis of the vertebral level from which low back or leg pain originates. A comparison of clinical evaluation, MRI and epiduroscopy. Pain Pract. 2012; 12(7):506–512

[11] Bosscher HA, Heavner JE. Lumbosacral epiduroscopy findings predict treatment outcome. Pain Pract. 2014; 14(6):506–514

[12] Lee GW, Jang SJ, Kim JD. The efficacy of epiduroscopic neural decompression with Ho:YAG laser ablation in lumbar spinal stenosis. Eur J Orthop Surg Traumatol. 2014; 24 Suppl 1:S231–S237

[13] Ruetten S, Meyer O, Godolias G. Endoscopic surgery of the lumbar epidural space (epiduroscopy): results of therapeutic intervention in 93 patients. Minim Invasive Neurosurg. 2003; 46(1):1–4

[14] Jo DH, Kim ED, Oh HJ. The comparison of the result of epiduroscopic laser neural decompression between FBSS or not. Korean J Pain. 2014; 27(1):63–67

[15] Kallewaard JW, Vanelderen P, Richardson J, Van Zundert J, Heavner J, Groen GJ. Epiduroscopy for patients with lumbosacral radicular pain. Pain Pract. 2014; 14(4):365–377

[16] Igarashi T, Hirabayashi Y, Seo N, Saitoh K, Fukuda H, Suzuki H. Lysis of adhesions and epidural injection of steroid/local anaesthetic during epiduroscopy potentially alleviate low back and leg pain in elderly patients with lumbar spinal stenosis. Br J Anaesth. 2004; 93(2):181–187

[17] Avellanal M, Diaz-Reganon G. Interlaminar approach for epiduroscopy in patients with failed back surgery syndrome. Br J Anaesth. 2008; 101(2):244–249

[18] Sakai T, Aoki H, Hojo M, Takada M, Murata H, Sumikawa K. Adhesiolysis and targeted steroid/local anesthetic injection during epiduroscopy alleviates pain and reduces sensory nerve dysfunction in patients with chronic sciatica. J Anesth. 2008; 22(3):242–247

[19] Manchikanti L, Abdi S, Atluri S, et al. An update of comprehensive evidence-based guidelines for interventional techniques in chronic spinal pain. Part II: guidance and recommendations. Pain Physician. 2013; 16(2) Suppl:S49–S283

[20] Brouwer PA, Brand R, van den Akker-van Marle ME, et al. Percutaneous laser disc decompression versus conventional microdiscectomy in sciatica: a randomized controlled trial. Spine J. 2015; 15(5):857–865

[21] Gottlob C, Kopchok GE, Peng SK, Tabbara M, Cavaye D, White RA. Holmium: YAG laser ablation of human intervertebral disc: preliminary evaluation. Lasers Surg Med. 1992; 12(1):86–91

[22] Quigley MR, Shih T, Elrifai A, Maroon JC, Lesiecki ML. Percutaneous laser discectomy with the Ho:YAG laser. Lasers Surg Med. 1992; 12(6):621–624

[23] Sato M, Ishihara M, Arai T, et al. Use of a new ICG-dye-enhanced diode laser for percutaneous laser disc decompression. Lasers Surg Med. 2001; 29 (3):282–287

[24] Choy DSJ, Case RB, Fielding W, Hughes J, Liebler W, Ascher P. Percutaneous laser nucleolysis of lumbar disks. N Engl J Med. 1987; 317(12):771–772

[25] Choy DSJ, Ascher PW, Ranu HS, et al. Percutaneous laser disc decompression. A new therapeutic modality. Spine. 1992; 17(8):949–956

[26] Gangi A, Dietemann JL, Ide C, Brunner P, Klinkert A, Warter JM. Percutaneous laser disk decompression under CT and fluoroscopic guidance: indications, technique, and clinical experience. Radiographics. 1996; 16(1):89–96

[27] Yonezawa T, Onomura T, Kosaka R, et al. The system and procedures of percutaneous intradiscal laser nucleotomy. Spine. 1990; 15(11):1175–1185

[28] Moneta GB, Videman T, Kaivanto K, et al. Reported pain during lumbar discography as a function of anular ruptures and disc degeneration. A re-analysis of 833 discograms. Spine. 1994; 19(17):1968–1974

[29] Bogduk N, Tynan W, Wilson AS. The nerve supply to the human lumbar intervertebral discs. J Anat. 1981; 132(Pt 1):39–56

[30] Bogduk N, Windsor M, Inglis A. The innervation of the cervical intervertebral discs. Spine. 1988; 13(1):2–8

[31] Pan L, Zhang P, Yin Q. Comparison of tissue damages caused by endoscopic lumbar discectomy and traditional lumbar discectomy: a randomised controlled trial. Int J Surg. 2014; 12(5):534–537

[32] Choi KC, Lee JH, Kim JS, et al. Unsuccessful percutaneous endoscopic lumbar discectomy: a single-center experience of 10,228 cases. Neurosurgery. 2015; 76(4):372–380, discussion 380–381, quiz 381

22 Spinal/Epidural Analgesic and Pharmacologic Therapy

Shiraz Yazdani and Salahadin Abdi

22.1 Introduction

Neuraxial drug administration describes a method of delivering pharmacologic therapies directly to and around the spinal cord and nerve roots. This is commonly used in the field of anesthesiology as the primary or adjuvant anesthetic for patients undergoing surgery. In the field of pain medicine, medications are commonly delivered into the epidural or intrathecal space for the treatment of chronic pain. As many common causes of chronic pain can be traced to spinal nerve roots, targeted delivery of medications to these areas can be a valuable tool in the management of chronic pain.

Using epiduroscopy, these medications can be directed at structures suspected to be pain generators by visualization of pathology, contact diagnosis, or diagnostic/therapeutic injection. The epiduroscope can also be used as an aid in percutaneous catheter placement for continuous/intermittent infusion of medications to pain-generating structures. What follows is the most commonly used local anesthetics, steroids, and other agents.

22.2 Local Anesthetics

See ▶ Table 22.1 for the pharmacologic properties of commonly used local anesthetics.

- Potency related to lipid solubility.
- Duration of action related to protein binding.
- Rapidity of onset related to acid dissociation constant (pKa).

22.2.1 Lidocaine

- Hydrophilic.
- Low protein binding.
- Fast onset of action.
- Medium duration.
- Higher relative risk of transient neurologic symptoms than other local anesthetics.

22.2.2 Bupivacaine

- High protein binding.
- Medium onset of action.
- Long duration of action.
- Racemic mixture:
    - R-enantiomer is more likely to cause central nervous system toxicity and cardiotoxicity.

22.2.3 Ropivacaine

- High protein binding.
- Medium onset of action.
- Long duration of action.
- Less proconvulsant and arrhythmogenic than equivalent doses of bupivacaine.

22.3 Clonidine

- Alpha-2 agonist.
- Effective as an analgesic intrathecally.
- Used as an adjuvant to local anesthetics and opioids in the epidural space:
    - Reduces concentration of local anesthetic required for sensory blockade.
    - Increases degree of motor blockade caused by local anesthetics.
- May cause hypotension, bradycardia, and sedation.

22.4 Ketamine

- *N*-methyl-D-aspartate receptor antagonist.
- Use in neuraxial analgesia remains under investigation.
- Potentially neurotoxic if not used in a preservative-free solution.
- Combination with local anesthetics and opioids in the epidural space may provide improved analgesia.[1]

22.5 Opioids

- Cause analgesia via regional spinal cord receptors, cerebral receptors, and peripheral/central effects due to vascular absorption.
- Lipophilic opioids (e.g., fentanyl) have a rapid onset but short duration:
    - Narrower segmental analgesia.
- Hydrophilic opioids (e.g., morphine) have a slow onset but longer duration:
    - Wider segmental analgesia.

Table 22.1 Pharmacologic properties of commonly used local anesthetics

Local anesthetic	pKa	Lipid solubility (relative)	Potency (relative)	Percentage protein bound	Onset of action	Duration of action
Procaine	8.1	1	1	6	Slow	Short
Lidocaine	7.7	150	2	65	Fast	Medium
Bupivacaine	8.1	1,000	8	95	Medium	Long
Ropivacaine	8.1	400	6–8	94	Medium	Long

Source: Used with permission from Schug SA, Saunders D, Kurowski I, Paech M. Neuraxial drug administration: a review of treatment options for anaesthesia and analgesia. CNS Drugs 2006;20(11):917–933.

Table 22.2 Pharmacologic properties of some commonly used glucocorticoids

Glucocorticoid	Equivalent potency (mg)	Liquid type	Particulate aggregation	Maximum particle size (µm)
Betamethasone	0.75	Suspension	Some	500
Dexamethasone	0.75	Solution	None	0.5
Methylprednisolone	4	Suspension	Few	> 500
Triamcinolone	4	Suspension	Extensive	> 500

Source: Data from MacMahon PJ, Eustace SJ, Kavanagh EC. Injectable corticosteroid and local anesthetic preparations: a review for radiologists. Radiology 2009;252(3):647–661.

22.6 Glucocorticoids

See ▶ Table 22.2 for the pharmacologic properties of commonly used glucocorticoids.

- Mimic mechanism of cortisol.
- Potent anti-inflammatory effects.
- Direct neural antinociceptive effects.
- Must be cautious with adverse effects of high-dose glucocorticoids.
- Solubility plays a role in duration of action and potential for particular aggregation:
  - Dexamethasone is nonparticulate.
  - Particulate steroids may aggregate into clusters larger than red blood cells, leading to vascular occlusion.[2]

22.7 Hyaluronidase

- Enzyme that degrades hyaluronan, an element of the extracellular matrix.
- Can help break down scar tissue that may form and be a cause for pain.
- Scar tissue may develop in the form of epidural adhesions following invasive neurosurgical procedures.
- Epiduroscopy can be used to direct injection of hyaluronidase into the areas of concentrated scar tissue.

22.8 Ozone

- Administered as an oxygen–ozone gas mixture at 1 to 40 µg of ozone per mL of oxygen.
- Increased glycolysis, improved immunomodulation, analgesic, and anti-inflammatory effects have been described.[4]
- May be used intradiskally.
- Use in epidural space still being investigated.
- Complications rare but include retinal hemorrhage, septic diskitis, and epidural abscess.

22.9 Isotonic/Hypertonic Saline

- Isotonic saline is often mixed with other medications to increase delivery area.
- Saline alone may have analgesic effects by "washing out" inflammatory cytokines.
- 10% hypertonic saline can be used for adhesiolysis when scar tissue formation is thought to directly cause pain or prevent the spread of medications to the affected nerve root.[5]
- Can be given as a bolus or infused continuously via a catheter.
- Epiduroscopy can be used to direct hypertonic saline at visualized areas of scar tissue formation.

22.10 Conclusion

Neuraxial administration of medications via direct needle entry, catheter placement, or the injection port of an epiduroscope can be an effective way to treat chronic pain from neuraxial structures in a targeted fashion. Studies are ongoing for many medications, but the commonly used local anesthetics, glucocorticoids, and opioids have a fair amount of data behind their use. Delivering these medications to the pain generators directly potentially minimizes adverse systemic effects and allows their use at the lowest effective dose. In the case of epidural adhesions, directed delivery may be the only method to treat the underlying cause of pain.

References

[1] Subramaniam K, Subramaniam B, Steinbrook RA. Ketamine as adjuvant analgesic to opioids: a quantitative and qualitative systematic review. Anesth Analg. 2004; 99(2):482–495

[2] Schilling LS, Markman JD. Corticosteroids for pain of spinal origin: epidural and intraarticular administration. Rheum Dis Clin North Am. 2016; 42 (1):137–155, ix

[3] Rahimzadeh P, Sharma V, Imani F, et al. Adjuvant hyaluronidase to epidural steroid improves the quality of analgesia in failed back surgery syndrome: a prospective randomized clinical trial. Pain Physician. 2014; 17(1): E75–E82

[4] Gazzeri R, Galarza M, Neroni M, Esposito S, Alfieri A. Fulminating septicemia secondary to oxygen-ozone therapy for lumbar disc herniation: case report. Spine. 2007; 32(3):E121–E123

[5] Manchikanti L, Rivera JJ, Pampati V, et al. One day lumbar epidural adhesiolysis and hypertonic saline neurolysis in treatment of chronic low back pain: a randomized, double-blind trial. Pain Physician. 2004; 7(2):177–186

23 Epiduroscopy-Assisted Spinal Cord Stimulator Electrode Implantation

Billy K. Huh and Carrie E. Johnson

23.1 Introduction

Epiduroscopy can be instrumental for spinal cord stimulator (SCS) electrode implantation in anatomically challenging patients. Epiduroscopy also allows precise visualization of the epidural space (▶ Fig. 23.1) and maneuvering of the electrode tip to its optimal position (▶ Fig. 23.2) when conventional technique of pushing electrode through Tuohy needle is not viable. When electrode advancement is significantly hindered by adhesion (▶ Fig. 23.3) or scar tissues, epiduroscopy can be used to lyse adhesions, remove scar tissue, and facilitate lead advancement.[1,2]

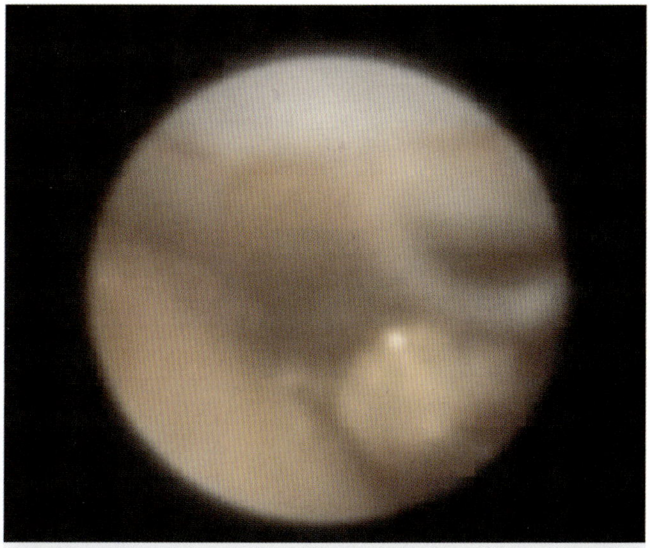

Fig. 23.1 The epiduroscope in the ventral epidural space can clearly identify anatomic structures for lead placement.

Over time, increased scarring or fibrous growths that envelop the tip of an SCS electrode can cause the stimulation to become ineffective over time (▶ Fig. 23.4). Epiduroscopy allows this extraneous fibrous tissue or scarring near the implanted SCS electrode tip to be removed in order to restore the efficacy of neuromodulation after long-term use without having to replace the electrode.

Electrode tips are also vulnerable to migration over time. Epiduroscopy can help with repositioning of displaced electrodes, thereby restoring efficiency and efficacy without replacing the electrode. This reduces the risks associated with removal and reimplantation of the SCS system.

Epiduroscopically assisted electrode implantation lowers risk of injury to the spinal dura mater injury, epidural bleeding, spinal trauma, and intraoperative pain. In addition, unnecessary stimulation tests or displaced electrodes can be avoided or can enable early recognition of potential high-risk situations.[1,2]

23.2 Indications

Difficult SCS electrode implantation can be due to the following:

- Epidural lesion.
- Epidural adhesion.
- Epidural fibrosis.
- Epidural scarring.
- Spinal stenosis
- Postlaminectomy pain or failed back syndrome.

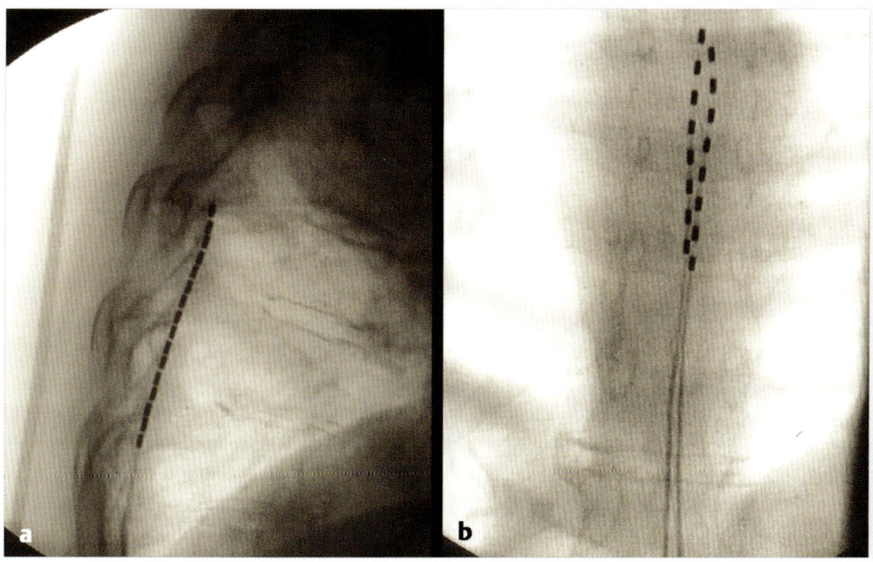

Fig. 23.2 **(a)** Lateral fluoroscopy demonstrating the stimulator lead position in the epidural space. **(b)** Anteroposterior fluoroscopy demonstrating the stimulator lead position in the epidural space near midline.

Fig. 23.3 Epiduroscopy can identify adhesions and fibrous bands in the epidural space that could potentially limit stimulator lead placement. In addition to identification, intervention (e.g., adhesiolysis) can be performed to facilitate lead positioning and placement. (Reproduced with permission from Ruetten S, Komp M, Merk H, Godolias G. Full-endoscopic interlaminar and transforaminal lumbar discectomy versus conventional microsurgical technique: a prospective, randomized, controlled study. Spine 2008;33(9):931–939.)

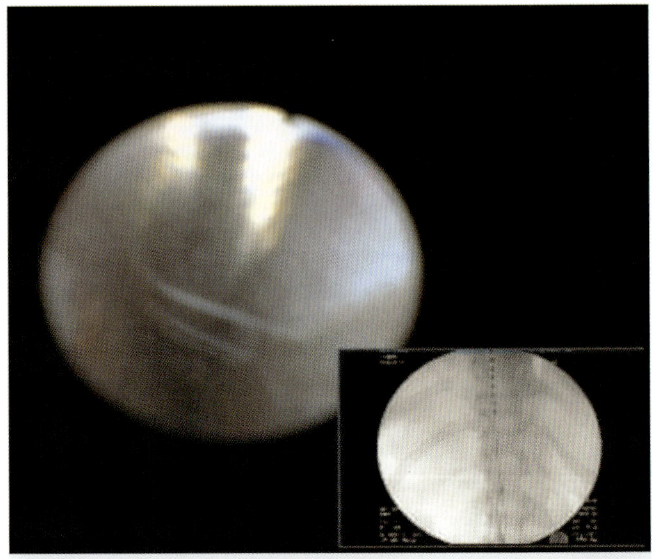

Fig. 23.4 Epiduroscopic view of scar tissue around the distal spinal cord stimulator leads. (Reproduced with permission from Ruetten S, Komp M, Merk H, Godolias G. Full-endoscopic interlaminar and transforaminal lumbar discectomy versus conventional microsurgical technique: a prospective, randomized, controlled study. Spine 2008;33 (9):931–939.)

23.3 Advantages of Epiduroscopy-Assisted SCS Electrode Implantation

- Facilitates electrode placement with adhesiolysis of scar tissues.
- Allows precise positioning of electrode with multidirectional push/pull capability.
- Lowers the risk of injury to the spinal dura mater.
- Reduces the incidence of epidural bleeding, spinal trauma, and intraoperative pain.
- Avoids unnecessary stimulation tests for displaced electrodes.[1,2]
- Enhances chance of successful implant and patient satisfaction.
- Avoids pathologic and anatomic obstacles.
- Eliminates the need to replace the electrodes after migration.
- Reduces X-ray exposure.

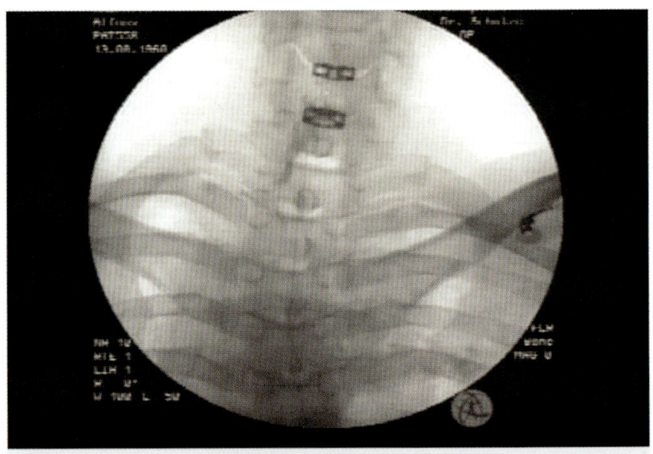

Fig. 23.5 Contrast injection in the cervical epidural space demonstrates limited spread, indicating adhesions and fibrosis. Adhesiolysis can be performed to open up the epidural space. (Reproduced with permission from Ruetten S, Komp M, Merk H, Godolias G. Full-endoscopic interlaminar and transforaminal lumbar discectomy versus conventional microsurgical technique: a prospective, randomized, controlled study. Spine 2008;33(9):931–939.)

23.4 Technique

A 17-gauge Tuohy needle is introduced into epidural space via an interlaminar approach one vertebral level below the tip of electrode using loss of resistance technique. Once the position of needle is confirmed under fluoroscopy, an epidurogram is performed with Omnipaque (GE Health Inc., Princeton, NJ, USA) to identify the degree and severity of epidural adhesions. Then, a guidewire is slowly introduced through the Tuohy into the epidural space to avoid dura puncture. The puncture site is dilated with a dilator, and an introducer sheath is passed over the guidewire. After removing the guidewire, epiduroscope is passed through introducer sheath. The epiduroscopy visualizes epidural space and electrode. The steerable catheter is rotated side by side to perform adhesiolysis if required. The epidural space is irrigated intermittently with normal saline. Note that the quantity of normal saline should not exceed 100 mL. Epidurogram will confirm the extent and success of adhesiolysis (▶ Fig. 23.5). Using fluoroscopic guidance and endoscopic visualization, microinstruments or balloon catheters can be used to remove or lyse adhesions (▶ Fig. 23.6). The epiduroscopy has

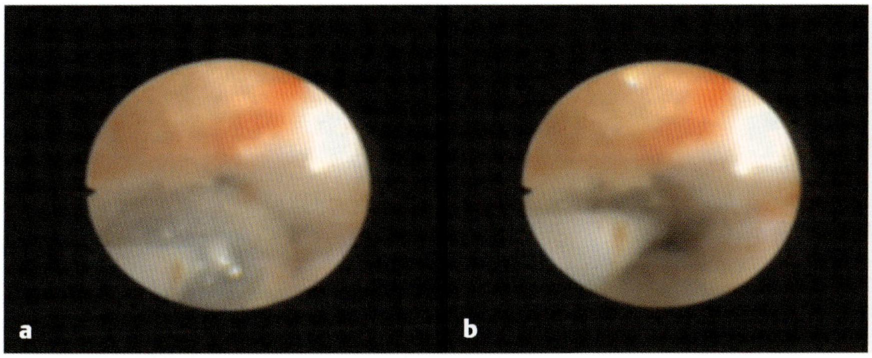

Fig. 23.6 (a) Epiduroscopic view of the Fogarty balloon catheter inflated for adhesiolysis. (b) Increased spinal epidural space is created after deflation of the balloon catheter. This can facilitate placement of stimulator leads.

dual effect: first, direct adhesiolysis by steerable catheter and, second, electrode repositioning by device pincer, especially when the usual push technique fails to advance the electrode tip to the desired location due to adhesion.[3]

23.5 Potential Adverse Effects

- Temporary pain.
- Bruising.
- Soreness.
- Headaches.
- Infection.
- Bleeding.
- Nerve damage.
- Worsening of symptoms.
- Allergic reactions to the injected medications and electrode.

23.6 Contraindications

- Allergy to any of the medications to be injected.
- Blood thinning medications.
- Active infection.

References

[1] Schütze G. Epiduroscopy – Spinal Endoscopy. New York, NY: Springer-Verlag; 2008

[2] Schütze G. Epiduroscopically guided percutaneous implantation of spinal cord stimulation electrodes. Management of pain a world perspective II. In: Raj P, Erdine S, Niv, eds. Management of Pain. Bologna: Monduzzi Editore S.p. A; 1996:301–304

[3] Baheti DK. Interventional pain management. Bombay Hospital Journal. Available online at: http://www.bhj.org.in/journal/2002_4403_jul/review_414.htm; accessed May 5, 2015

24 Epiduroscopic Ozone Therapy

Siddarth Thakur and Salahadin Abdi

24.1 Introduction

- Ozone is an allotropic form of oxygen, administered as an oxygen–ozone gas mixture at concentrations ranging from 1 to 40 µg per milliliter of oxygen and made in an ozone generator.[1,2] Ozone is formed by an atom of oxygen and a molecule of oxygen, as represented in the mechanism: $O + O_2 = O_3$.
- It has been used for the treatment of back pain by minimally invasive techniques over the past two decades, particularly in Europe and South America.[1,3,4]
- Ozone may be utilized with steroids, intradiskal radiofrequency, local anesthetics, collagenases, or antibiotics and administered intradiskally, intramuscularly (into the paraspinal muscles), or into the epidural space.[1,5]
- Proposed mechanisms of action include mechanical and chemical adhesiolysis, reduction of inflammation, and shrinkage of intervertebral disk content.[2,3,6]
- Ozone is a strong oxidizing agent that quickly reacts with collagens and proteoglycans in the nucleus pulposus of intervertebral disks causing resorption of water and reduction of disk volume, which, in turn, results in decreased mechanical irritation of nerves in the disk annulus and relief of nerve compression from herniated disks (▶ Fig. 24.1).[2,3]
- Ozone improves perineural circulation and tissue oxygenation and also serves as an anti-inflammatory agent. Ozonized proteins inhibit proinflammatory signals, decrease free radical production, and block the expression of COX-2 enzymes and products such as prostaglandins.[2,3,5]
- Epidurscopic ozone therapy consists of endoscopic adhesiolysis followed by instillation of ozone–oxygen gas mixture.

24.2 Indications

- Most commonly it used in patients suffering from failed back surgery syndrome (FBSS); the etiology of pain may be inflammatory changes in nerve roots, local ischemia,[3] structural changes in vertebral column, lumbar degenerative disease, epidural fibrosis. and adhesions (▶ Fig. 24.2) which may compress nervous structures.[4]
- Also it is used for refractory low back pain secondary to degenerative disease with neurogenic claudication or radiculopathy, spondylolisthesis, spinal stenosis, or lumbar disk herniation, specifically in patients who fail conservative measures (physiotherapy and analgesics) and routine interventional procedures (epidural steroid and facet injections).[1,3,4,5,6]
- It may serve as an alternative to additional open back surgery.
- Contraindications include allergy to ozone, glucose-6-phosphate dehydrogenase deficiency (favism), and hyperthyroidism, in addition to the contraindications for epiduroscopy (anticoagulant use, active infection, and compromised cardiac or pulmonary function).[4]

24.3 Technique

- Spinal endoscopy is carried out under local anesthesia with light sedation. The patient is awake to avoid unintentional pressure effects in the epidural space.[4]
- The patient is placed prone with pillows underneath hips for patient comfort and in order to straighten the lumbar.[5]
- Prophylactic antibiotics are administered.

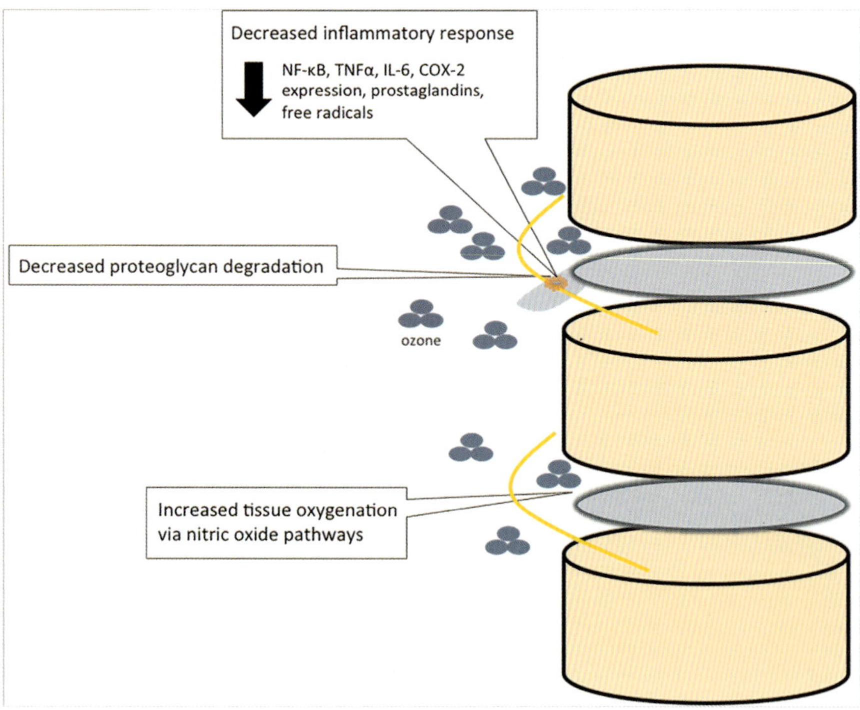

Fig. 24.1 Mechanisms of ozone on inflammation reduction.

- The procedure is performed with aseptic technique and utilizes fluoroscopy.
- The sacral hiatus is identified and local anesthesia with 5 cc of 0.5% lidocaine is applied. An 18-gauge Touhy needle is then introduced with fluoroscopic confirmation of placement (▶ Fig. 24.3).[6]

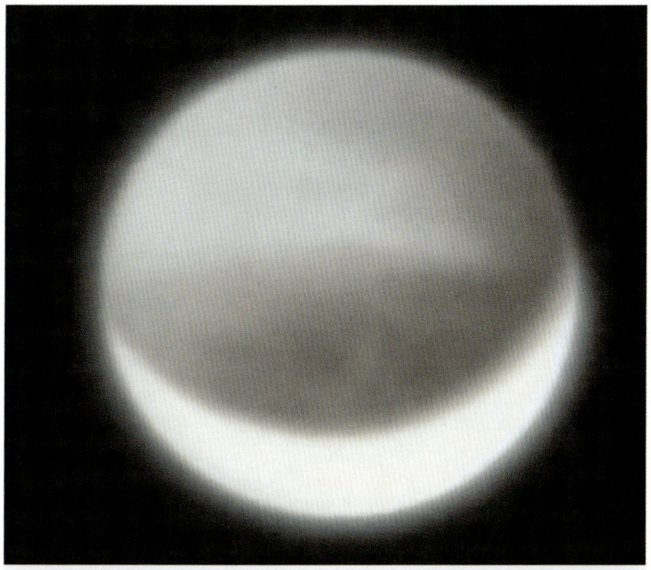

Fig. 24.2 Epiduroscopic view demonstrating adhesions in the epidural space.

- A 0.9-mm guidewire is inserted through the needle and advanced till L5–S1 level (▶ Fig. 24.4).[4]
- The needle is removed (▶ Fig. 24.5), a no. 11 scalpel is used to make a 3-mm incision around the wire (▶ Fig. 24.6),[5] and a 3.8 mm × 17.8 cm dilator is advanced over the guidewire (▶ Fig. 24.7).[6]
- The internal dilator is removed (▶ Fig. 24.8) and two-channel video-guided catheter (3.0 mm × 30 cm) is inserted with the fiberoptic microendoscope (0.8-mm spinal endoscope) (▶ Fig. 24.9).[6]
- With fluoroscopic guidance, the endoscope is directed to the level of suspected pathology (▶ Fig. 24.10).
- Saline irrigation is used for gentle distention of epidural space to allow for optimal visualization of nerve roots at multiple levels.
- After targeted nerve roots are identified, blunt dissection with hydrostatic distention allows for the creation of a pocket of space where the ozone can access the pathologic nerve roots (▶ Fig. 24.11).[4]
- Adhesiolysis is performed followed by administration of an oxygen–ozone gas mixture (20 mL of 30 µg/mL).[4]
- The catheter is then removed and the incision is sutured closed.
- The patient should remain in supine position for 2 to 4 hours after the procedure.[4]

Fig. 24.3 The 18-gauge Tuohy needle is inserted into the sacral hiatus and into the epidural space.

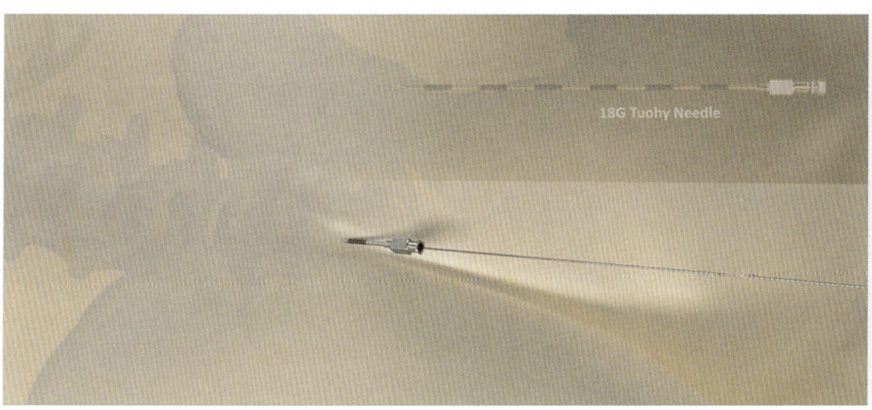

Fig. 24.4 The guidewire is inserted through the Tuohy needle into the epidural space.

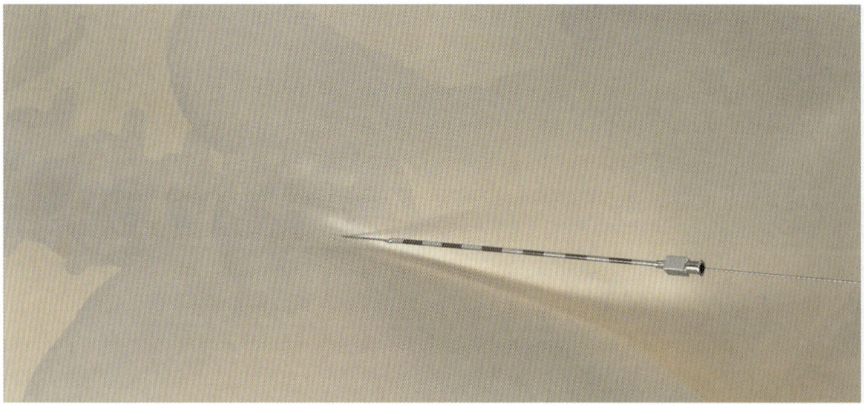

Fig. 24.5 The Tuohy needle is removed over the guidewire.

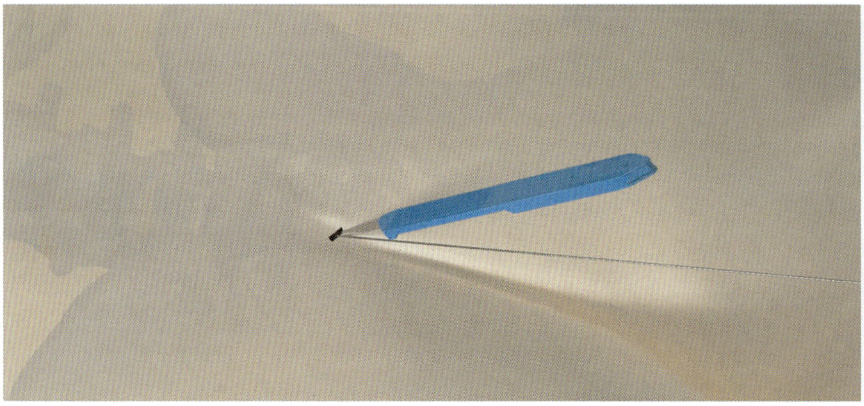

Fig. 24.6 The skin insertion site is incised with a blade to accommodate the dilator and sheath.

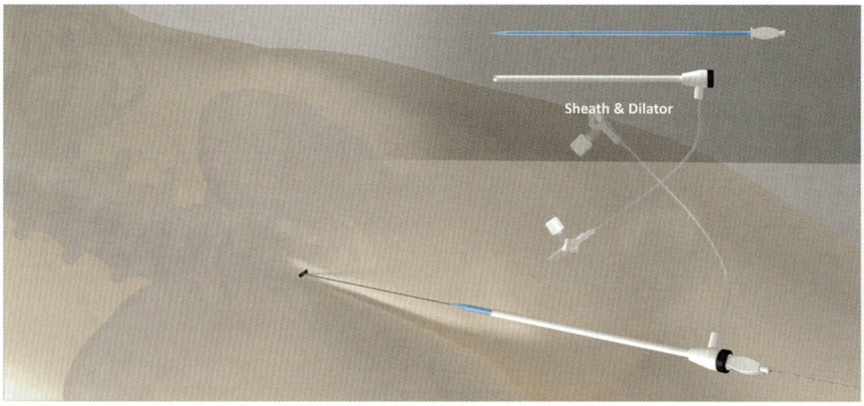

Fig. 24.7 The dilator and sheath are passed over the guidewire into the epidural space.

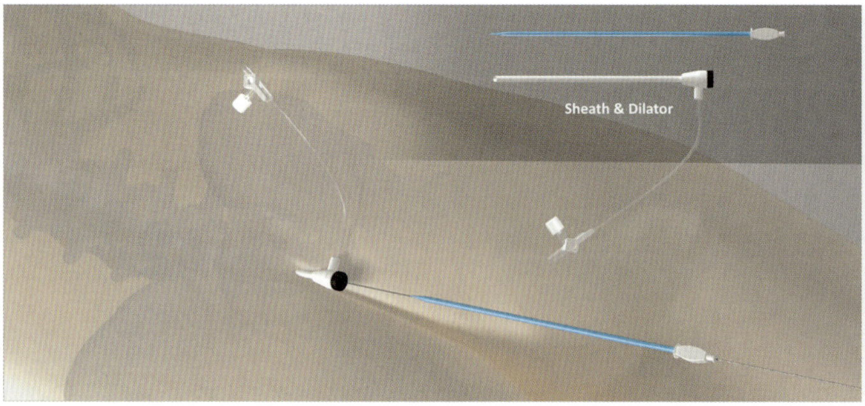

Fig. 24.8 The dilator and guidewire are removed, leaving the sheath in place.

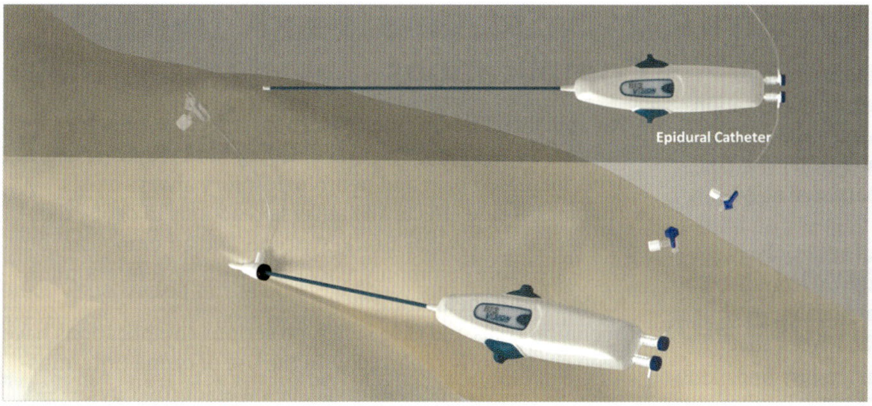

Fig. 24.9 The epiduroscope is inserted into the epidural space.

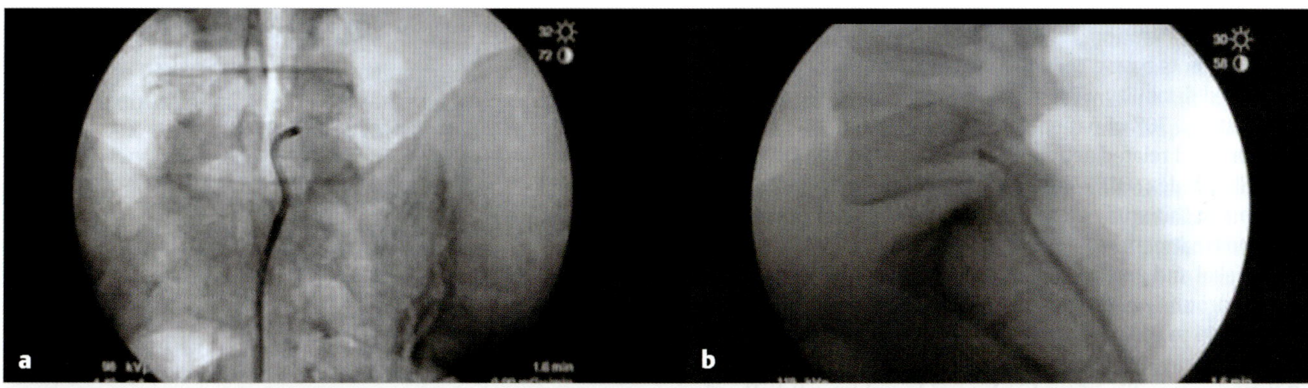

Fig. 24.10 **(a)** Anteroposterior fluoroscopy confirms epidural location and lumbar level. **(b)** Lateral fluoroscopy confirms epidural location and lumbar level.

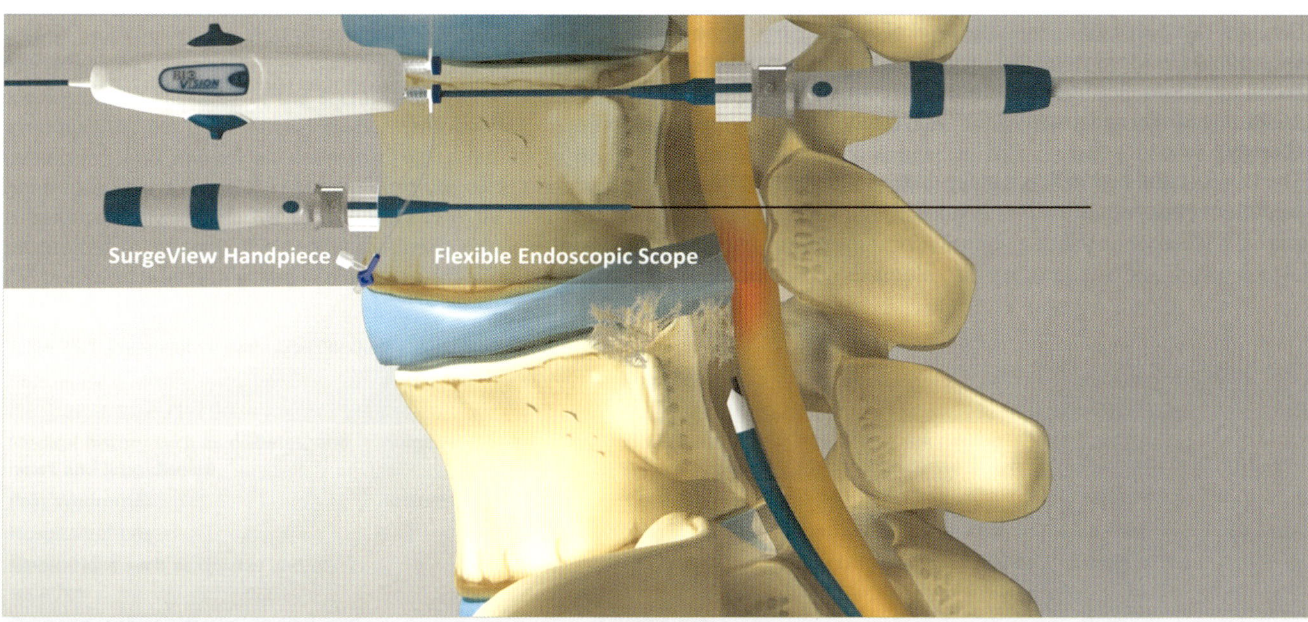

Fig. 24.11 The epiduroscope is steered to the desired ozone treatment location.

24.4 Efficacy and Outcomes

- Spinal endoscopy is a minimally invasive technique that allows for targeted diagnoses and treatment. It spares healthy tissue and serves as an alternative to surgery, thereby minimizing time off work, avoiding extended hospitalization, and reducing complications such as postsurgical infection and epidural fibrosis.[1,5]

Postoperative follow-up scheduling should be taken care of either before epiduroscopy or at the time of discharge, not only to evaluate for efficacy of treatment, but also to reassess neurologic function. Clearly, before the patient may be discharged, vital signs must be stable and within normal limits.

References

[1] Raffaeli W, Righetti D, Andruccioli J, Sarti D. How we can see and treat the epidural space: epiduroscopy. Eur J Pain Suppl. 2011; 5 S2:395–399

[2] Waldman SD. Atlas of Interventional Pain Management. 4th ed. Philadelphia, PA: W.B. Saunders; 2015

[3] Schütze G. Epiduroscopy – Spinal Endoscopy. Heidelberg: Springer; 2008

[4] Kitahata LM. Recent advances in epiduroscopy. J Anesth. 2002; 16(3):222–228

[5] Miller R, Pardo M. Basics of Anesthesia. 6th ed. London: Elsevier Health Sciences; 2011:165

[6] American Society of Anesthesiologists Task Force on Preanesthesia Evaluation. Practice advisory for preanesthesia evaluation: a report by the American Society of Anesthesiologists Task Force on Preanesthesia Evaluation. Anesthesiology. 2002; 96(2):485–496

[7] Jyrkkä J, Enlund H, Korhonen MJ, Sulkava R, Hartikainen S. Polypharmacy status as an indicator of mortality in an elderly population. Drugs Aging. 2009; 26(12):1039–1048

[8] Miller R, Pardo M. Basics of Anesthesia. 6th ed. London: Elsevier Health Sciences; 2011:181

[9] Tarhan S, Moffitt EA, Taylor WF, Giuliani ER. Myocardial infarction after general anesthesia. JAMA. 1972; 220(11):1451–1454

[10] Wallace AW, Au S, Cason BA. Association of the pattern of use of perioperative β-blockade and postoperative mortality. Anesthesiology. 2010; 113(4):794–805

[11] Foëx P, Sear JW. Challenges of β-blockade in surgical patients. Anesthesiology. 2010; 113(4):767–771

[12] Schouten O, Hoeks SE, Welten GM, et al. Effect of statin withdrawal on frequency of cardiac events after vascular surgery. Am J Cardiol. 2007; 100 (2):316–320

[13] Kertai MD, Bax JJ, Klein J, Poldermans D. Is there any reason to withhold beta blockers from high-risk patients with coronary artery disease during surgery? Anesthesiology. 2004; 100(1):4–7

[14] Kaye AD, Clarke R, Sabar R, et al. Herbal medicines: current trends in anesthesiology practice–a hospital survey. J Clin Anesth. 2000; 12:468–471

[15] Kaye AD, Sabar R, Clarke R, Vig S, Hofbauer R, Kaye AM. Herbal medications and anesthetics: a review on current concepts (part 1). Am J Anesth. 2000; 27 (7):405–407

[16] Manchikanti L, Falco FJ, Benyamin RM, et al. Assessment of bleeding risk of interventional techniques: a best evidence synthesis of practice patterns and perioperative management of anticoagulant and antithrombotic therapy. Pain Physician. 2013; 16(2) Suppl:SE261–SE318

[17] Jellin JM, Gregory PJ, eds. Natural Medicines Comprehensive Database. Stockton, CA: Therapeutic Research Center; 2013. Available online at: www.naturaldatabase.com. Accessed September 25, 2013

[18] Kaye AD, Baluch A, Kaye AJ, Frass M, Hofbauer R. Pharmacology of herbals and their impact in anesthesia. Curr Opin Anaesthesiol. 2007; 20(4):294–299

[19] Kaye AD, Kucera I, Sabar R. Perioperative anesthesia clinical considerations of alternative medicines. Anesthesiol Clin North America. 2004; 22(1):125–139

[20] Thomson AH. Introduction to clinical pharmacokinectics. Paed Perinat Drug Ther 2000;4:3–11.

[21] Benzon HT. Practical Management of Pain. Philadelphia, PA: Elsevier/Mosby; 2014

[22] Urman R, Kaye AD. Moderate and Deep Sedation in Clinical Practice. New York, NY: Cambridge University Press; 2012

[23] Barash PG, Cullen BF, Stoelting RK. Clinical Anesthesia. 7th ed. Philadelphia, PA: Lippincott Williams & Wilkins; 2013:583–611

[24] Trescot AM, Chopra P, Abdi S, Datta S, Schultz DM. Systematic review of effectiveness and complications of adhesiolysis in the management of chronic spinal pain: an update. Pain Physician. 2007; 10(1):129–146

[25] Chung F, Yuan H, Yin L, Vairavanathan S, Wong DT. Elimination of preoperative testing in ambulatory surgery. Anesth Analg. 2009; 108(2):467–475

[26] Fleisher LA, Beckman JA, Brown KA, et al. ACC/AHA 2007 guidelines on perioperative cardiovascular evaluation and care for noncardiac surgery. J Am Coll Cardiol. 2007; 50(17):1707–1732

[27] Gold BS, Young ML, Kinman JL, Kitz DS, Berlin J, Schwartz JS. The utility of preoperative electrocardiograms in the ambulatory surgical patient. Arch Intern Med. 1992; 152(2):301–305

[28] van Klei WA, Bryson GL, Yang H, Kalkman CJ, Wells GA, Beattie WS. The value of routine preoperative electrocardiography in predicting myocardial infarction after noncardiac surgery. Ann Surg. 2007; 246(2):165–170

[29] Sweitzer BJ. Preoperative evaluation and medication. In: Miller RD, Pardo MC, eds. Basics of Anesthesia. 6th ed. Philadelphia, PA: Elsevier Saunders; 2011:177–178

[30] Gill JB, Heavner JE. Visual impairment following epidural fluid injections and epiduroscopy: a review. Pain Med. 2005; 6(5):367–374

[31] Beltruti D, Groen GJ, Saberski L, Sandner-Kiesling A, Schutze G, Weber G. Epiduroscopy Consensus Decision March 2006. Pain Clin. 2007; 19(3):47–50

26 Epiduroscopy: Complications and Complication Avoidance

Siddarth Thakur and Salahadin Abdi

26.1 Introduction

Epiduroscopy is a minimally invasive technique useful in the diagnosis and treatment of a variety of chronic low back pain conditions. The procedure is generally well tolerated; however, as with any procedure, there are risks, and although rare, a number of epiduroscopic complications have been reported in the literature (▶ Table 26.1).

26.2 Visual Impairment

- Visual impairment is an uncommon but significant complication following epiduroscopy.
- It is believed to be caused by increased cerebrospinal fluid (CSF) pressure and subsequent retinal hemorrhage.
- The injection of fluid (usually saline) helps to open the epidural space during epiduroscopy, increasing epidural pressure and transiently increasing CSF pressure; the increase in pressure is directly proportional to the rate and volume of injection.
- The epidural injection of just 30 mL of saline over 1 minute can cause CSF pressure to rise to 60 mm Hg.[3]
- A sudden increase in CSF pressure is transmitted through the subarachnoid extension around the optic nerve, compressing the nerve and its vasculature, which results in the rupture of retinal blood vessels.
- Additional contribution to retinal hemorrhage may result from the increase in CSF pressure, decreasing cerebral blood flow with a reflex increase in ophthalmic artery pressure, resulting in venous collapse and the rupture of capillaries.
- In 2005, Gill and Heavner reported 12 cases of visual impairment after epiduroscopy. The volume of solution injected ranged from 20 to 120 mL. Approximately 58% of patient had bilateral retinal hemorrhages and 21% had residual vision loss or hemorrhage.[3]
- In order to minimize this complication, epidural injection should be administered at a low speed, 1 mL per 1 to 2 seconds, and the volume should not be exceed 100 mL per 60 minutes.[3]

26.3 Transient Neurologic Deficits and Seizure

- In 2015, Beyaz reported a case of generalized tonic–clonic seizure and transient complete motor paralysis and sensory loss after epiduroscopy.[2]
- Increased intracranial pressure (ICP) was hypothesized to be the etiology of the observed deficits.
- Physiologic ICP is 5 to 10 mm Hg at rest, and the epidural injection of just 10 mL has been shown to increase ICP to 11 to 63 mm Hg. Those with above-average ICP at rest can have a severe, devastating response to injection 10 mL of epidural

Table 26.1 Complications associated with epiduroscopy and prevention strategies

Complication	Presumed etiology	Prevention strategy
Visual impairment	Increased CSF pressure secondary to high rate and large volume epidural injection resulting in retinal hemorrhage[3]	Low injection volume with slow injection speed = 1 mL per 1–2 seconds
Transient neurologic deficits and seizure	Increased ICP secondary to large volume, bolus injection of saline in epidural space[2,3]	Slow injection speed and low injection volume = 100 mL per 60 minutes
		Avoid unnecessary epidural fluid administration.
Dural tear resulting in intradural cyst and orthostatic headache	Direct trauma to dura from endoscope[1,8]	Direct visualization of dura with epiduroscopy
		Aspiration of CSF when tear is suspected
		Contrast injection during fluoroscopy
		Monitor for pain exacerbation postprocedure
Dural tear resulting in encephalopathy and rhabdomyolysis	Direct trauma to dura from endoscope resulting in contrast media spread into CSF with resulting neurotoxicity and metabolic derangements[7]	Direct visualization of dura with epiduroscopy
		Aspiration of CSF when tear is suspected
		Contrast injection during fluoroscopy
Radiation exposure	Radiation emitted during fluoroscopy[5,6]	Utilize pulse mode fluoroscopy
		Judicious use of fluoroscopy with clear goals for every procedure
Intravascular injection	Tear in epidural vein	Contrast injection under fluoroscopy prior to administering other fluids
	Direct vascular injection[5]	

Abbreviations: CSF, cerebrospinal fluid; ICP, intracranial pressure.

solution, with ICP reaching around 300 mm Hg. The increase in ICP occurs within seconds of epidural injection and returns to baseline level in approximately 2 to 6 minutes.[2]

- In this case, 100-mL saline and 10-mL radiopaque substance was infused within 10 minutes, and the patient then became hypertensive, suffered respiratory arrest, and lost consciousness. This was followed by a generalized seizure. Upon regaining consciousness, the patient lost motor and sensory functions. Neuroimaging did not reveal abnormality, and the patient regained all motor and sensory functions and was fit for discharge the following day.[2]
- In 2008, Avellanal and Diaz-Reganon reported transient neurologic symptoms (< 30 seconds), headache, and hypoacusia, related to bolus injection of saline during epiduroscopy.[3]
- To prevent the increase in observed complications, it is important to precisely calculate the amount of liquids delivered to the epidural space, administering them slowly and carefully to minimize unnecessary fluid and drug administration.
- In 2006, at the World Initiative on Spinal Endoscopy, discussions regarding the amount of epidural fluid during epiduroscopy resulted in recommendations to limit the volume to 200 mL or less.[9]

26.4 Dural Tear

- Epidural tears during epiduroscopy can result in a number of complications, including encephalopathy, orthostatic headache, and intradural cyst.[1,7,8]
- Acute psychomotor disturbance and metabolic derangement after epiduroscopy has been reported, and an occult dural tear with leakage of radiographic contrast media into the CSF was the presumed causative factor.[7]
- In this case, soon after epiduroscopy was completed, the patient developed muscle weakness and hypoesthesia of bilateral lower extremities, followed by confusion, disorientation, agitation, tremors, and urinary incontinence. Computed tomography scan of the head performed 4.5 hours postprocedure identified diffuse contrast enhancement throughout the intracranial CSF. The patient also had markedly elevated creatinine kinase and myoglobin.
- The patient was treated conservatively with intravenous hydration, antibiotics, analgesics, and elevation of the head and trunk (to minimize further cephalad migration of the contrast agent). Fortunately, the psychomotor agitation and disorientation resolved within 13 hours postprocedure, lab abnormalities resolved, and the patient was discharged 7 days after the procedure.
- A dural tear can be diagnosed through direct visualization via an epiduroscope, aspiration of CSF from the epidural space, and contrast injection during fluoroscopy.
- The presence of dense scar and adhesions may make direct visualization challenging and the injection of saline into the epidural space may compromise the ability to aspirate CSF.
- With these diagnostic limitations in mind, all precautions must be taken to avoid inadvertent dural injury and intrathecal injection.
- In 2012, Ryu et al reported a case of an iatrogenic spinal intradural cyst after epiduroscopy.[8]
- The etiology was thought to be dural and arachnoid perforation, creating a ball-valve mechanism leading to the entrapment of CSF into an arachnoid cyst.
- After the patient underwent epiduroscopy, their back pain was exacerbated and did not respond to increased analgesics for 2 weeks. Follow-up magnetic resonance imaging of the lumbosacral spine demonstrated a large intradural cyst from L2–L3 to S1–S2 with the same signal intensity as CSF. The patient successfully underwent a partial L5 laminectomy with intradural exploration and fenestration of the cystic wall.
- The authors concluded that "the most frequent complication of epiduroscopy must be perforation of the dural sac."[8]
- In order to recognize this unforeseen complication, it is important to monitor for exacerbations of back pain after epiduroscopy and to exercise extreme caution when performing all epiduroscopic procedures, especially in patients who have a history of previous lumbar surgeries.[1]

26.5 Radiation Exposure

- As with any fluoroscopically guided procedure, there is a risk of radiation exposure to both patient and physician.
- In 2008, Komiya et al demonstrated average epiduroscopy procedure time of 50 minutes, with fluoroscopy times of 9 minutes and 26 seconds. Patients were assessed for radiation-associated skin injuries 1-month postprocedure, and no instances were detected.[6]
- The authors concluded that radiation exposure during epiduroscopy was well below the recommended limits.
- In 2009, Heavner and Bosscher described their experience with epiduroscopy, with average procedure times between 15 and 20 minutes and fluoroscope time of 45 seconds.[4]
- Strategies to mitigate radiation exposure during epiduroscopy include judicious use of pulse-mode fluoroscopy, using landmarks directly visualized during epiduroscopy for navigation, and clearly defined goals for each procedure.[4]

26.6 Intravascular Injection

- Intravascular injection is usually attributed to the entry of the needle or catheter into a blood vessel and is detectable using radiopaque contrast material with fluoroscopy.
- Although the direct vessel penetration is not necessary, in 2007, Heavner et al reported two cases of intravascular injection during epiduroscopy not due to direct entry into a vessel.[9]
- The most likely explanation of intravascular entry was related to the microenvironment of the epidural space of patients with failed back surgery with dense adhesions and fibrosis.
- During epidural administration of contrast an intravascular dye pattern was observed in both cases. The authors postulate that access to the intravascular space was likely through a tear in the wall of a vein, with the tear remaining patent secondary to fibrous attachments to the vessel wall.
- To minimize the risk of intravascular injection, the authors advocate always injecting radiopaque contrast material under fluoroscopy before injecting other fluids and avoiding blood vessels by continuously observing the epiduroscopic view.[5]

References

[1] Avellanal M, Diaz-Reganon G. Interlaminar approach for epiduroscopy in patients with failed back surgery syndrome. Br J Anaesth. 2008; 101(2):244–249

[2] Beyaz SG. Seizures and transient neurological deficits during epiduroscopy in a patient with failed back surgery syndrome. Pain Med. 2015; 16(4):825–827

[3] Gill JB, Heavner JE. Visual impairment following epidural fluid injections and epiduroscopy: a review. Pain Med. 2005; 6(5):367–374

[4] Heavner JE, Bosscher H. Epiduroscopy and radiation exposure. Reg Anesth Pain Med. 2009; 34(1):79

[5] Heavner JE, Wyatt DE, Bosscher HA. Lumbosacral epiduroscopy complicated by intravascular injection. Anesthesiology. 2007; 107(2):347–350

[6] Komiya K, Igarashi T, Suzuki H, Hirabayashi Y, Waechter J, Seo N. In vitro study of patient's and physician's radiation exposure in the performance of epiduroscopy. Reg Anesth Pain Med. 2008; 33(2):98–101

[7] Mizuno J, Gauss T, Suzuki M, Hayashida M, Arita H, Hanaoka K. Encephalopathy and rhabdomyolysis induced by iotrolan during epiduroscopy. Can J Anaesth. 2007; 54(1):49–53

[8] Ryu KS, Rathi NK, Kim G, Park CK. Iatrogenic intradural lumbosacral cyst following epiduroscopy. J Korean Neurosurg Soc. 2012; 52(5):491–494

[9] Sandner-Kiesling A, Weber G, Scheütze G, Beltrutti D, Groen GJ, Saberski L. Foundation of the World Initiative on Spinal Endoscopy(WISE). Pain Clin. 2007; 19(2):51–52

27 Clinical Outcomes of Epiduroscopy

Kent H. Nouri and Salahadin Abdi

27.1 Introduction

Numerous studies have shown the benefits of epiduroscopy on both quantifiable outcomes, including pain scores, functional ability, and activity rating scales, and qualitative measures, including sensory nerve function and contrast spread (▶ Table 27.1). By investigating the diagnostic and treatment value of epiduroscopy in addition to clinical complications reported in the current clinical and scientific literature, one can further understand its usage as it applies to the pain practitioner's therapeutic arsenal.

27.2 Diagnostic Utility

- A study in 2012 found that when compared with magnetic resonance imaging (MRI) or clinical evaluation, epiduroscopy was found to be a more reliable determinate of the vertebral level where significant spinal pathology occurs in patients with lower back and radiculopathic pain.[1] This was thought to be due to the fact that epiduroscopy provides a *functional* indicator of pathology versus an *observable* indicator found on imaging modalities.
- In correlating imaging and epiduroscopy, a study of 20 patients evaluated whether abnormalities identified on MRI scans could be confirmed with epiduroscopy and investigated whether targeted epidural injection of medication after adhesiolysis was able to reduce radicular pain. Epiduroscopy found adhesions in 19 of the 20 patients, 8 unobserved on previous MRI scans. Of the 20 patients, 11 showed a significant improvement in visual analog scale (VAS) scores after 3 and 12 months with epiduroscopy with adhesiolysis and steroid injection.[2]
- Epiduroscopy has been shown to successfully identify treatable pathology and may have a significant diagnostic and prognostic value. A study utilizing epiduroscopy for direct *visual* information (hyperemia, vascularity, fibrosis) and *mechanical* evidence (pain to touch, contrast spread, patency) was used to predict outcomes in 114 patients. By using epiduroscopy, a correct outcome was found in 78% of patients, with sensitivity in the prediction of a good or excellent outcome at 75% and specificity of no change or only a fair improvement with treatment at 82%. The study also found a sensitivity of epiduroscopy in the diagnosis of epidural pathology to be 91% but a specificity to exclude the presence of discrete epidural pathology using epiduroscopy of only 39%.[3]

27.3 Treatment Modality

- Nerve function testing has been performed pre- and postepiduroscopy to show whether mechanical alleviation of scar

Table 27.1 A summary of selected studies evaluating the outcome of epiduroscopy

Authors	Population	Treatment	Outcome
Sakai et al[4]	16 patients with chronic sciatica	Pre- and postepiduroscopy sensory nerve function testing	Improved CPT values of A-β and A-δ sensory nerve function, pain and disability scores
Manchikanti et al[6]	83 patients with lower back pain and radiculopathy	Interventional group received adhesiolysis with epiduroscopy and steroid injection, control epiduroscopy, and steroid injection alone	More than 50% pain relief at 3 mo (80%), 6 mo (57%), and 12 mo (48%) compared to 1 mo (33%) only in control group
Saberski[7]	35 patients with lower back pain secondary to herniated disc	Epiduroscopy with irrigation and steroid injection vs. laminectomy	Opioid continuation in 31.8% of epiduroscopy patients vs. 92.3% in postlaminectomy and return to work in 72 vs. 28%
Takeshima et al[8]	28 patients with failed back surgery syndrome	Epiduroscopy and adhesiolysis in patients with nerve root scarring vs. epidural space scarring only	Prolonged duration of pain relief in patients with nerve root scarring (6 mo) vs. epidural scarring (3 mo)
Murai et al[9]	124 patients with lower back and leg pain	Epiduroscopy with adhesiolysis in patients with prior lumbar surgery vs. no surgery	Significantly lower pain scores in patients at 1 and 3 mo in patients without prior lumbar surgery
Igarashi et al[10]	58 elderly patients with either monosegmental or multisegmental lumbar spinal stenosis	Epiduroscopy with adhesiolysis and steroid injection	Prolonged duration of pain relief in the monosegmental (12 mo) vs. mulitsegmental group (3 mo)
Raffaeli and Righetti[11]	14 patients who had short-term pain relief with prior epiduroscopy	Epiduroscopy and radiofrequency fibrolysis	More than 50% of patients noted 90% pain relief at 1 mo and 80% at 6 mo
Kim et al[12]	98 patients with chronic lower back and radicular pain	Epiduroscopy with laser adhesiolysis and steroid injection vs. epiduroscopy with steroid injection alone	Laser epiduroscopy had better VAS scores at 1 and 6 mo with a longer duration of pain relief
Donato et al[15]	350 patients with chronic lower back pain who failed conservative treatments	Epiduroscopy with ozone, hyaluronidase, and ciprofloxacin injection	At 2 years postprocedure, 69% of patients had significant pain relief and 78% continued to have a major decrease in disability indices

Abbreviations: CPT, current perception threshold: VAS, visual analog scale.

tissue can qualitatively improve sensory nerve dysfunction. Studies have shown that increased thresholds of A-β, A-δ and C fibers are found in affected dermatomes of patients with sciatica compared with unaffected dermatomes. By testing current perception threshold values of the affected lower extremity, it has been demonstrated that sensory A-β and A-δ fiber deficits improve by utilizing epiduroscopy to perform targeted adhesiolysis and epidural injection when compared with injection alone.[4] This may be due to an improvement in microcirculation once the ischemia caused by the constriction has improved with the epiduroscopy or remyelination.[5]

- Studies show the important role mechanical adhesiolysis and decompression of the epidural space may play in patients receiving epiduroscopy. A randomized, prospective, double-blind controlled study showed significant improvement in pain relief in 80% of patients at 3 months for those who received epiduroscopy and steroid injection with adhesiolysis, compared with no relief at 3 months in the group receiving epiduroscopy and steroid injection alone.[6] In the control group, the epiduroscope was introduced to the level of the sacral canal and steroid was injected but no adhesiolysis was performed. The interventional group underwent epiduroscopy with adhesiolysis and steroid injection. The outcome parameters were pain, functional status, and psychological and behavioral status. In the interventional group, 57% of patients showed significant improvements in pain scores and outcome measures including psychometric testing at 1, 3, and 6 months. The control group showed improvement in pain scores at 1 month only and none thereafter.
- In patients with lumbar disc herniation causing spinal pain, spinal endoscopy with adhesiolysis resulted in superior outcome in terms of improvement, returning to work (72 vs. 28%) and discontinuation of opioid medications (68.1 vs. 7.7%), when compared to laminectomy.[7]
- For radicular pain caused by adhesions surrounding nerve roots, epiduroscopy has been found to show a significant difference in diagnosis and outcomes. A study showed that when epiduroscopy was performed on patients with observable nerve root scarring versus those with epidural space scarring alone, the duration of pain relief was superior (6 vs. 3 months) if performed on patients with radiculopathic pain secondary to nerve root adhesions.[8]
- Epiduroscopy has also been used to evaluate specific subpopulations of patients with pain:
  - A study in 2007 evaluated postlumbar surgery patients versus nonoperated patients and found that while both groups improved with endoscopic adhesiolysis, the nonoperated group had significantly lower VAS scores at 1 and 3 months.[9]
  - A study of 58 elderly patients with a mean age of 71 years was conducted by dividing them into two groups based on the number of affected nerve roots.[10] A total of 34 patients in the monosegmental group and 24 in the multisegmental group underwent epiduroscopy, adhesiolysis, and steroid injection, and were evaluated in terms of VAS scores for back and leg pain. Pain relief was noted in both groups up to 12 months after the intervention; however, only the monosegmental group reported significant improvement at 12 months compared to a limit of 3 months of palliation in the multisegmental group.
- Novel therapies utilizing epiduroscopy have also been studied beyond traditional adhesiolysis and steroid injection treatments:
  - Radiofrequency using a ResAblator with 4-MHz radiofrequency output to disrupt adhesions was performed in 14 patients, with 57% stating more than 90% improvement in pain scores after 1 month.[11]
  - A study applying laser endoscopy to patients with refractory chronic low back and radicular pain and showed a longer duration of relief (6 months vs. 1 month) compared to epiduroscopy and steroid injection alone.[12] A holmium: yttrium-aluminum-garnet laser was utilized in two prospective case series of 34 and 68 patients, and positive results indicated by a decrease in VAS scores by 2 points at 8 weeks were obtained in 44 and 48.5% of patients, respectively.[13,14]
  - Epiduroscopy with adhesiolysis and the instillation of ozone, hyaluronidase, and ciprofloxacin was also performed in a 2013 study showing significant pain relief in 69% of patients at 2 years and a decrease is the Oswestry Low Back Pain Disability Index below 40% in 78% of patients.[15]
- In terms of cost effectiveness, a study found that when compared to nonendoscopic adhesiolysis, endoscopic treatment showed a 1-year quality of life improvement cost of $7,020 compared to $2,080.[16] This was found even though epiduroscopic patients had a longer duration of pain relief (20 vs. 12 weeks) and total number of procedures performed that year (3 vs. 1.3).

References

[1] Bosscher HA, Heavner JE. Diagnosis of the vertebral level from which low back or leg pain originates. A comparison of clinical evaluation, MRI and epiduroscopy. Pain Pract. 2012; 12(7):506–512

[2] Geurts JW, Kallewaard JW, Richardson J, Groen GJ. Targeted methylprednisolone acetate/hyaluronidase/clonidine injection after diagnostic epiduroscopy for chronic sciatica: a prospective, 1-year follow-up study. Reg Anesth Pain Med. 2002; 27(4):343–352

[3] Bosscher HA, Heavner JE. Lumbosacral epiduroscopy findings predict treatment outcome. Pain Pract. 2014; 14(6):506–514

[4] Sakai T, Aoki H, Hojo M, Takada M, Murata H, Sumikawa K. Adhesiolysis and targeted steroid/local anesthetic injection during epiduroscopy alleviates pain and reduces sensory nerve dysfunction in patients with chronic sciatica. J Anesth. 2008; 22(3):242–247

[5] Kayama S, Konno S, Olmarker K, Yabuki S, Kikuchi S. Incision of the anulus fibrosus induces nerve root morphologic, vascular, and functional changes. An experimental study. Spine. 1996; 21(22):2539–2543

[6] Manchikanti L, Boswell MV, Rivera JJ, et al. [ISRCTN 16558617] A randomized, controlled trial of spinal endoscopic adhesiolysis in chronic refractory low back and lower extremity pain. BMC Anesthesiol. 2005; 5(1):10

[7] Saberski LR. A retrospective analysis of spinal canal endoscopy and laminectomy outcomes data. Pain Physician. 2000; 3(2):193–196

[8] Takeshima N, Miyakawa H, Okuda K, et al. Evaluation of the therapeutic results of epiduroscopic adhesiolysis for failed back surgery syndrome. Br J Anaesth. 2009; 102(3):400–407

[9] Murai K, Suzuki H, Igarashi T, et al. Epiduroscopy for intractable low back pain or sciatica in operated and non-operated back patients: results from The Japan Society of Epiduroscopy. Pain Clin. 2007; 19(4):163–169

[10] Igarashi T, Hirabayashi Y, Seo N, Saitoh K, Fukuda H, Suzuki H. Lysis of adhesions and epidural injection of steroid/local anaesthetic during epiduroscopy potentially alleviate low back and leg pain in elderly patients with lumbar spinal stenosis. Br J Anaesth. 2004; 93(2):181–187

[11] Raffaeli W, Righetti D. Surgical radio-frequency epiduroscopy technique (R-ResAblator) and FBSS treatment: preliminary evaluations. Acta Neurochir Suppl (Wien). 2005; 92:121–125

[12] Kim JD, Jang JH, Jung GH, Kim JY, Jang SJ. Epiduroscopic laser disc and neural decompression. J Neurosurg Rev 2011;1 (S1):14-19. https://northamericanspine.com/assets/pdf/epiduroscopic-laser-disc-and-neural-decompression.pdf

[13] Ruetten S, Meyer O, Godolias G. Application of holmium:YAG laser in epiduroscopy: extended practicabilities in the treatment of chronic back pain syndrome. J Clin Laser Med Surg. 2002; 20(4):203–206

[14] Ruetten S, Meyer O, Godolias G. Endoscopic surgery of the lumbar epidural space (epiduroscopy): results of therapeutic intervention in 93 patients. Minim Invasive Neurosurg. 2003; 46(1):1–4

[15] Donato AD, Fontana C, Pinto R, Beltrutti D, Pinto G. The effectiveness of endoscopic epidurolysis in treatment of degenerative chronic low back pain: a prospective analysis and follow-up at 48 months. Acta Neurochir Suppl (Wien). 2011; 108:67–73

[16] Manchikanti L, Pampati V, Bakhit CE, Pakanati RR. Non-endoscopic and endoscopic adhesiolysis in post-lumbar laminectomy syndrome: a one-year outcome study and cost effectiveness analysis. Pain Physician. 1999; 2(3):52–58

28 Appendix: Intraoperative Endoscopic Images of Pathologic Anatomy of the Neuraxial Region

Günter Schütze

This appendix includes images taken from various clinical cases to demonstrate epiduroscopic techniques used to diagnose and treat spinal pathologies. These images also provide a library of intraoperative views of common and seldom seen spinal pathologies encountered during epiduroscopy to assist the clinician in identification of findings and diagnosis.

28.1 Pathology Images: Herniated Intervertebral Disk, Epidural Vasculature, Dura Mater Hemorrhage, Epidural Cyst

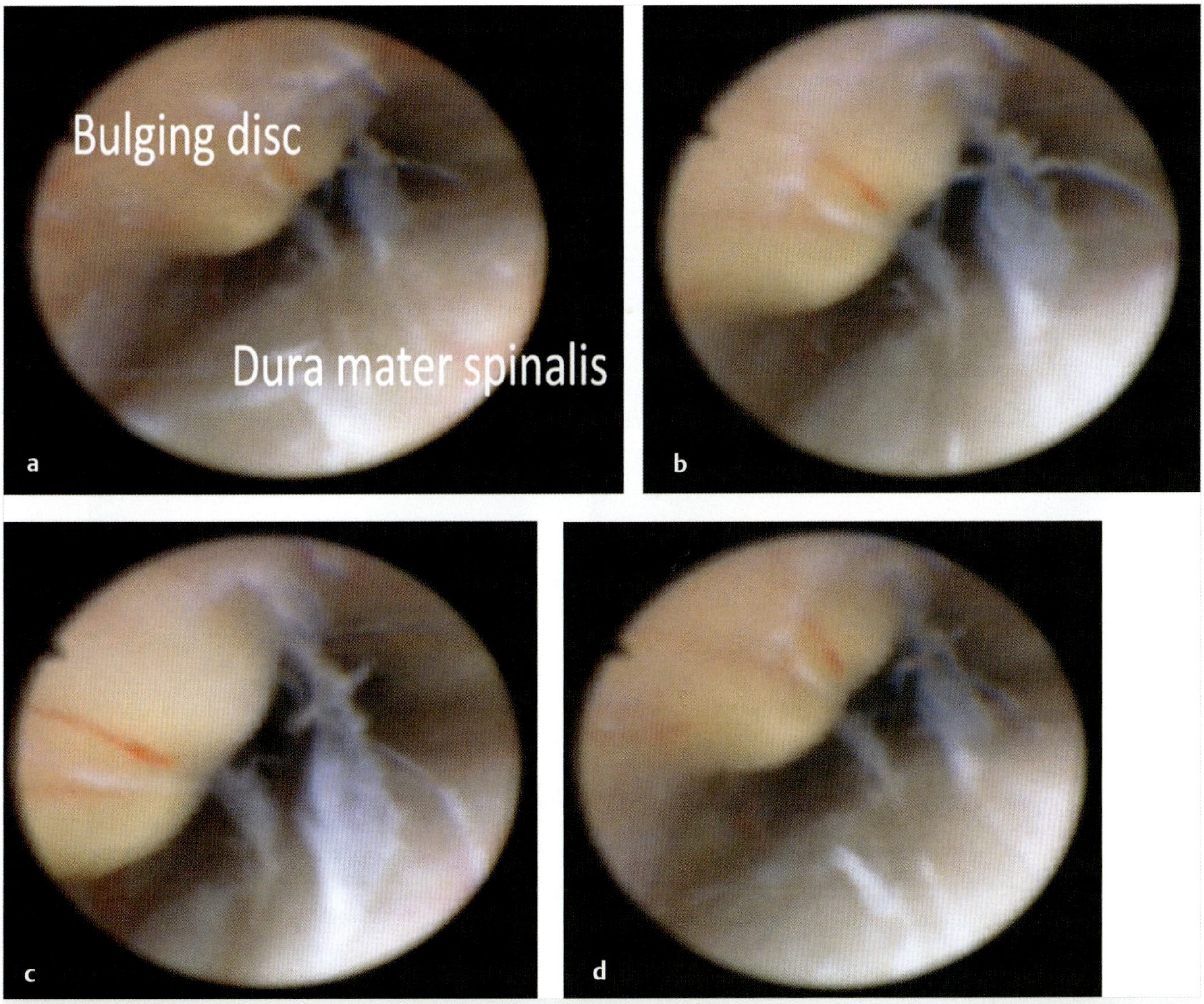

Fig. 28.1 Herniated intervertebral disk. (a–d) Epiduroscopic views of thoracic herniated disk. Note the subligamentous sequestration with soft epidural adhesions between the posterior longitudinal ligament and the spinal dura mater.

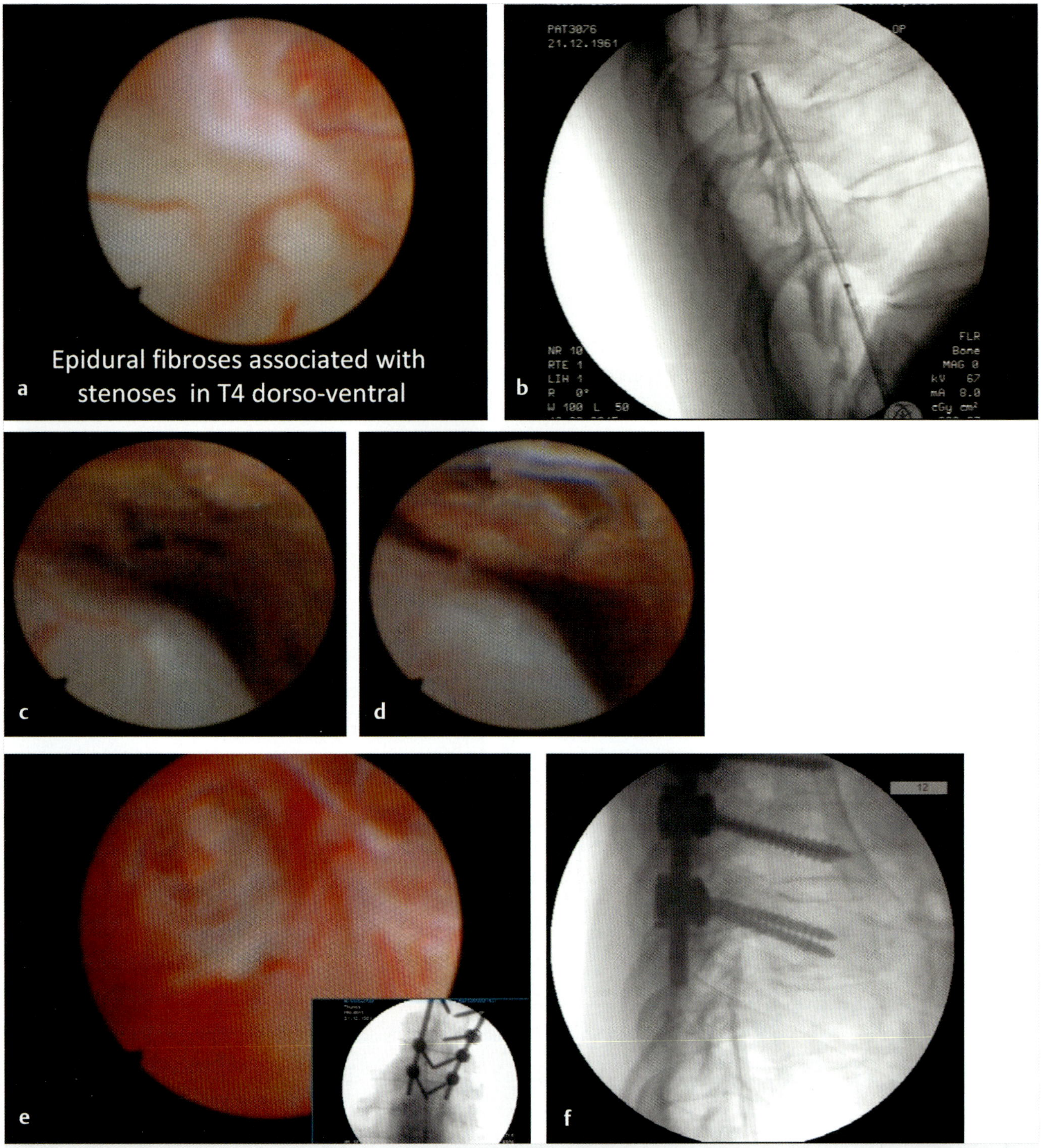

Fig. 28.2 Epidural vasculature. (a–e) Epiduroscopic view. (**f, g**) Epiduroscopic view where veins are constricted secondary to adhesions.

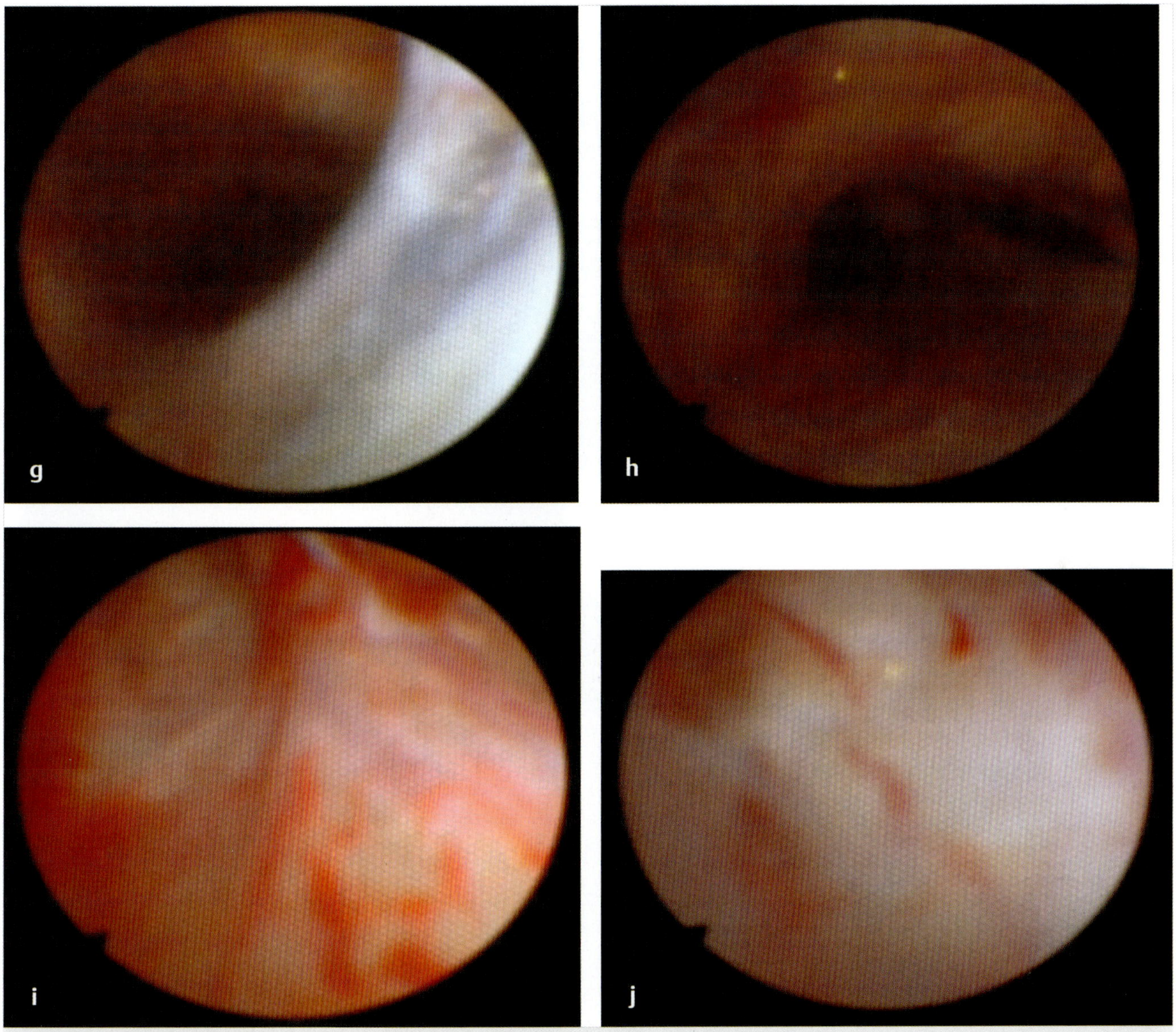

Fig. 28.2 (*continued*) (**f, g**) Epiduroscopic view where veins are constricted secondary to adhesions. (**h**) Epiduroscopic view where veins are surrounded by adhesions. (**i**) Epiduroscopic view with adhesions leading to vascular obstruction. (**j, h**) Epiduroscopic view with adhesions leading to vascular flow limitation.

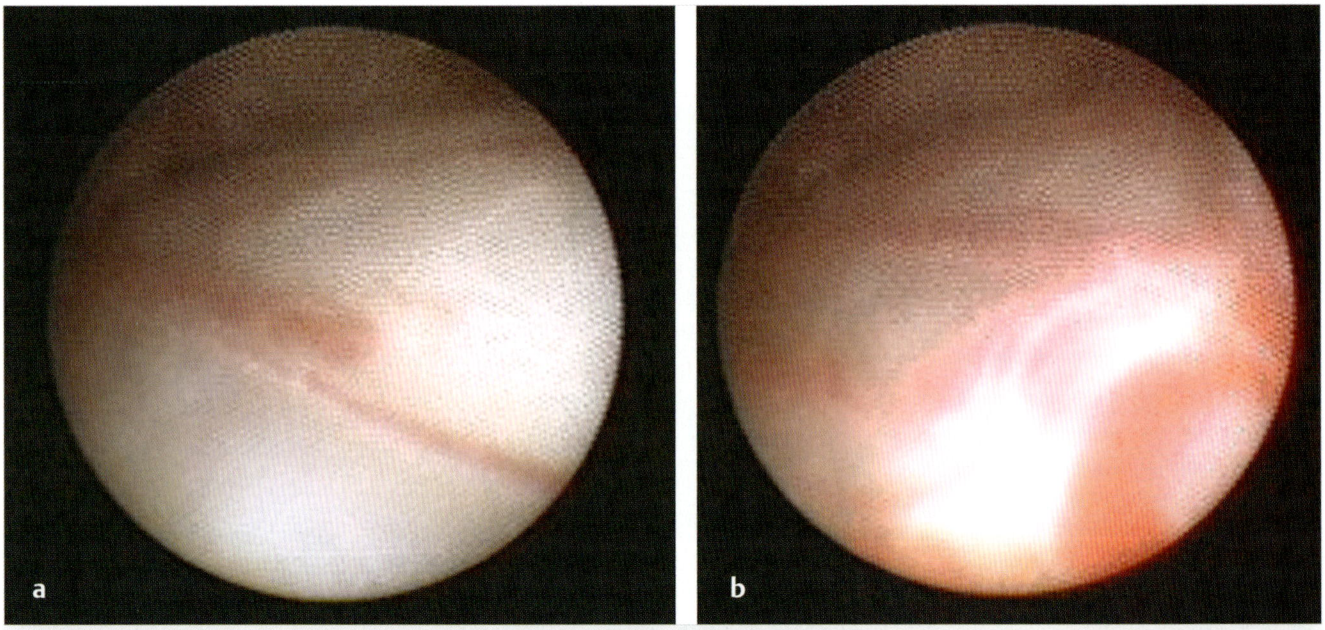

Fig. 28.3 **Dura mater hemorrhage.** (**a, b**) Epiduroscopic view of minimal hemorrhage tracking in the spinal dura mater.

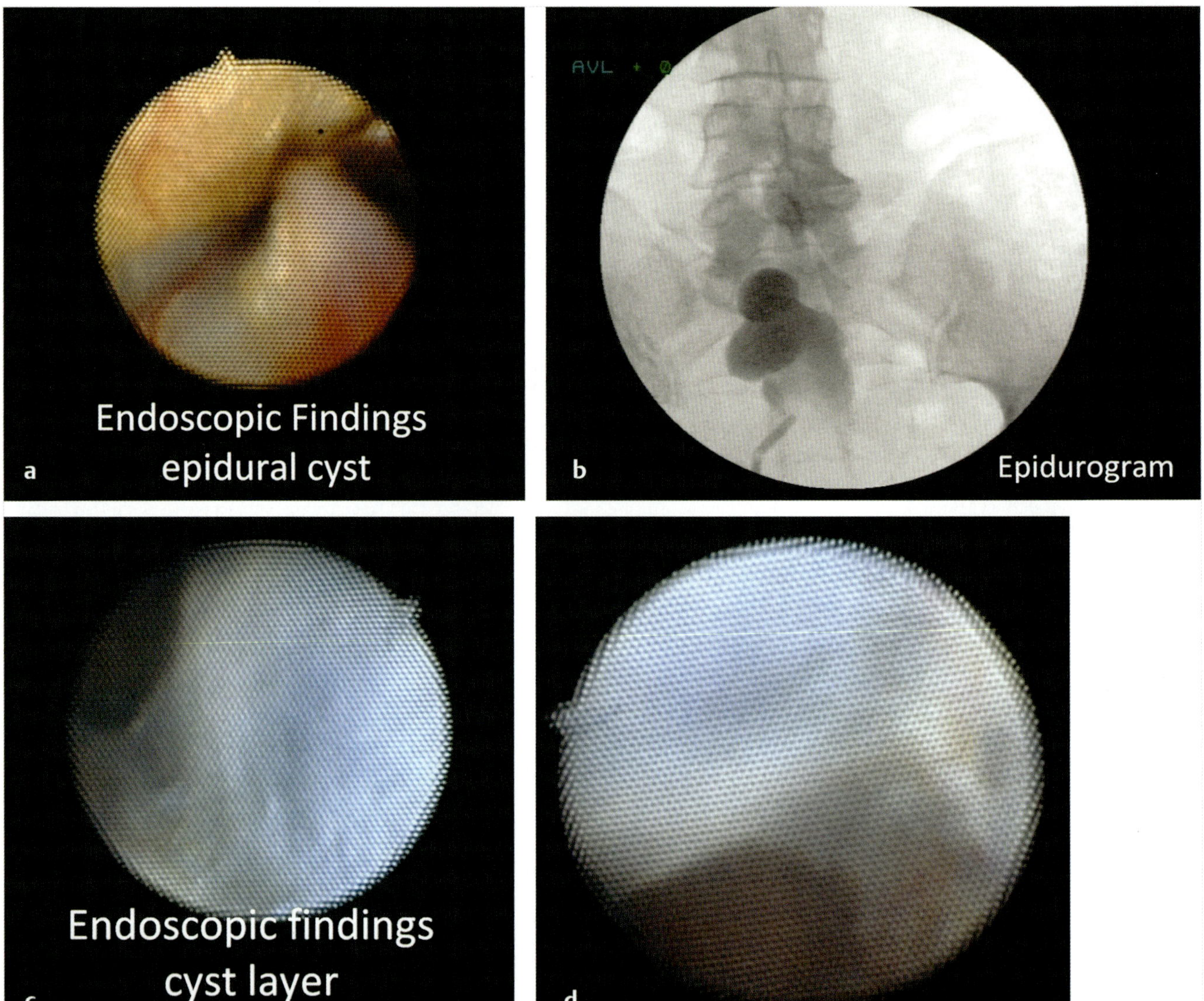

Fig. 28.4 **Epidural cyst.** (**a**) Epiduroscopic view. (**b**) Anteroposterior fluoroscopy with contrast injection demonstrating an epidural cyst. (**c, d**) Epiduroscopic view.

28.2 Pathology Images: Adhesions, Fibroses, and Chronic Inflammatory Processes

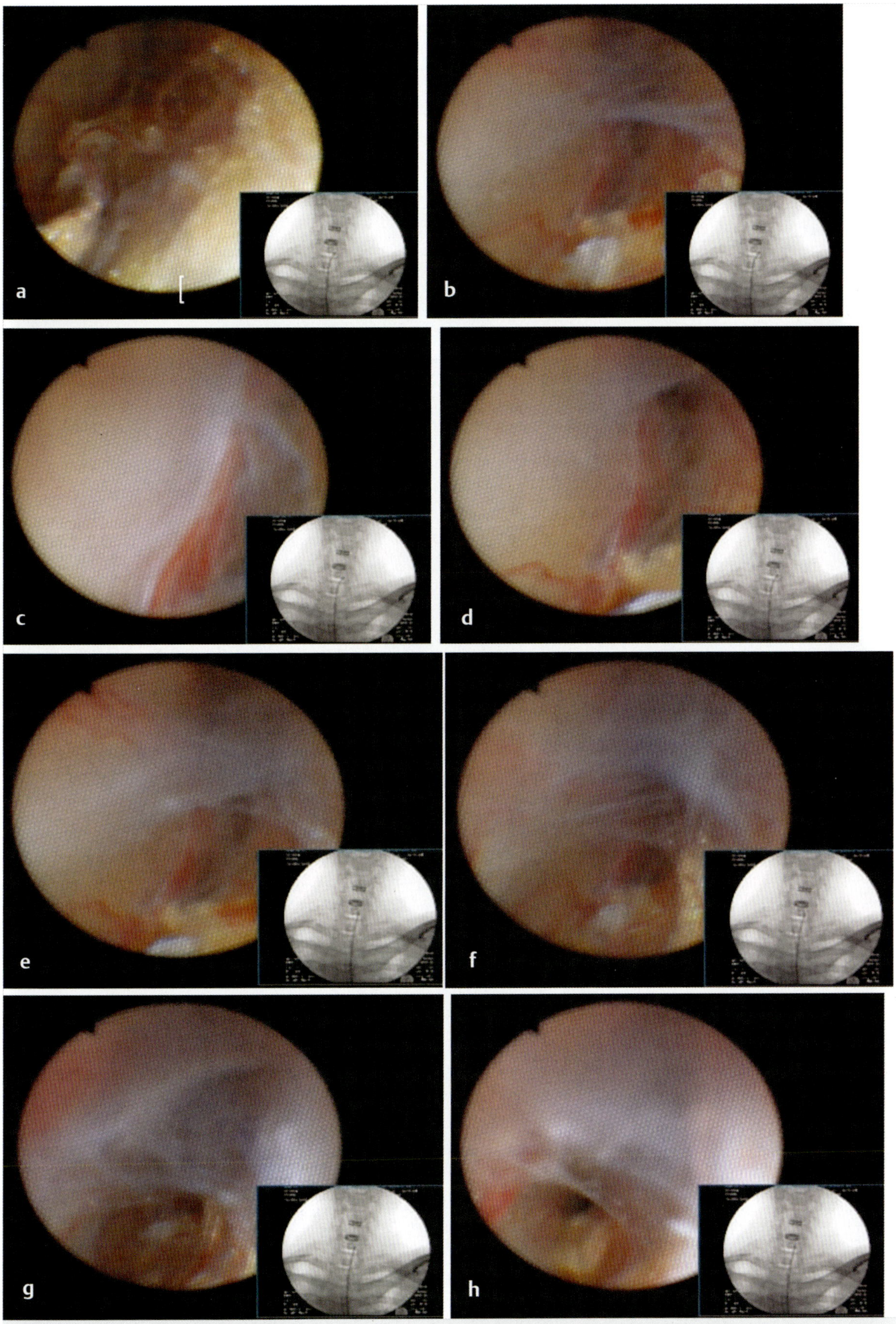

Fig. 28.5 (a–h) Cervical epidural adhesions.

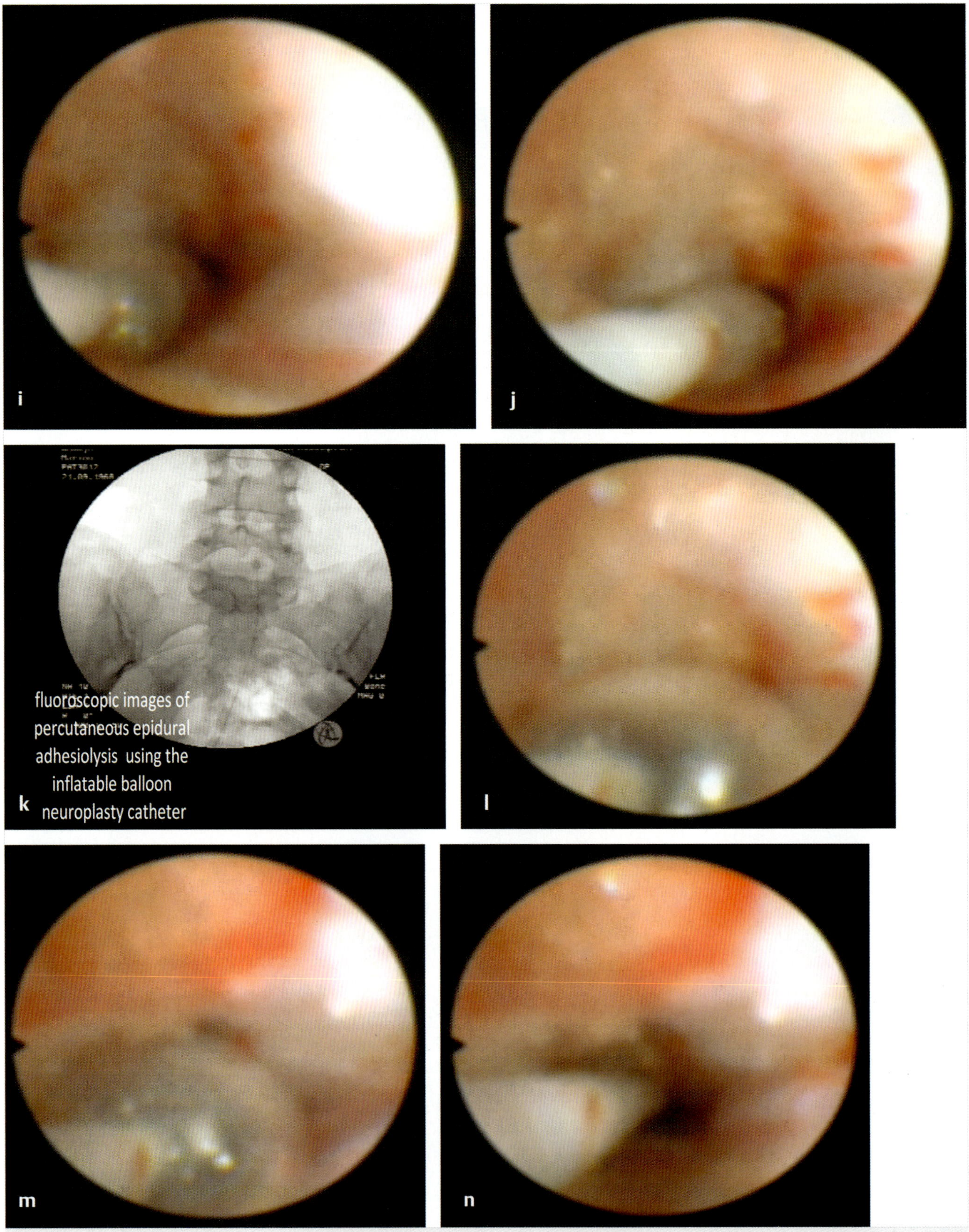
i
j
fluoroscopic images of percutaneous epidural adhesiolysis using the inflatable balloon neuroplasty catheter
k
l
m
n

Fig. 28.5 (m–n)

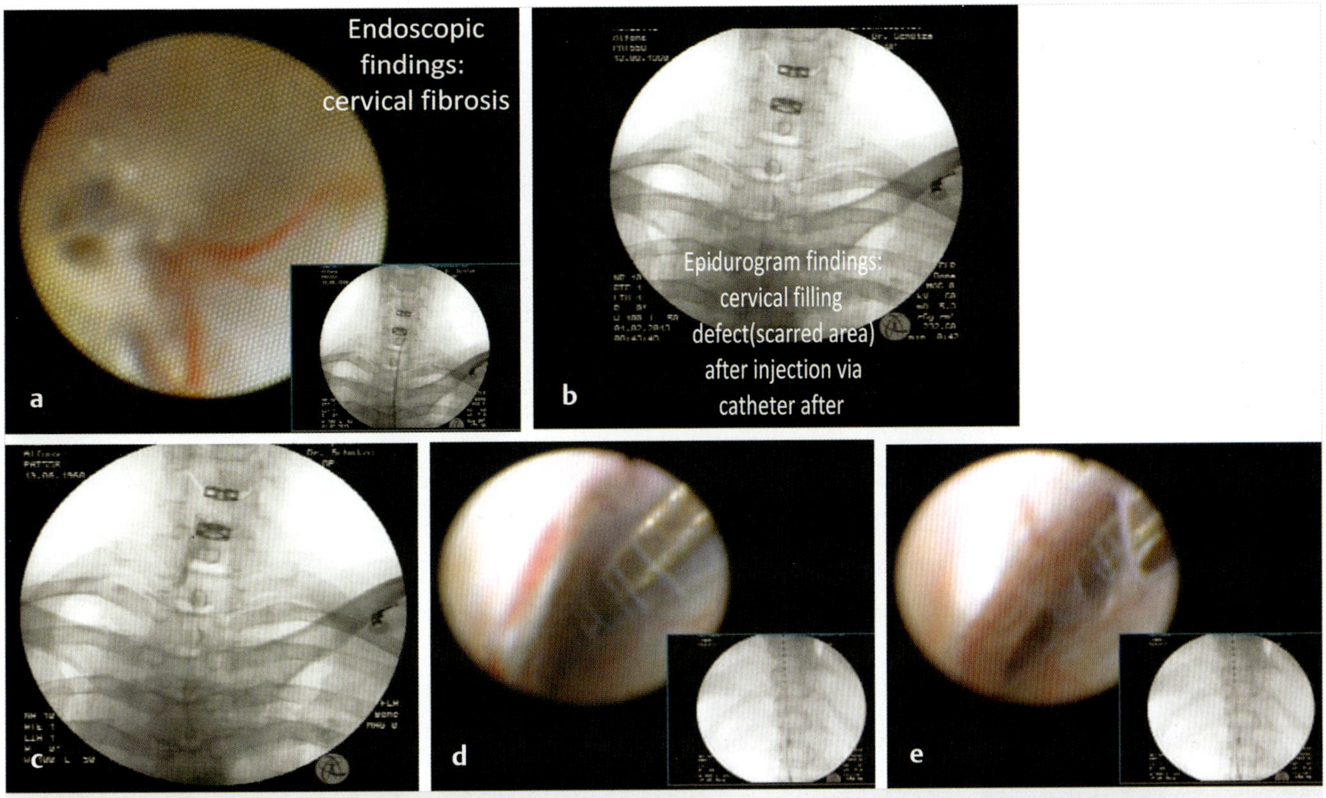

Fig. 28.6 Cervical epidural adhesions and fibrosis. (**a**) Cervical epidural fibrosis. (**b**) Anteroposterior fluoroscopy with contrast injection in the cervical epidural space demonstrates limited spread, indicating adhesions and fibrosis. (**c**) Contrast injection in the cervical epidural space demonstrates limited spread, indicating adhesions and fibrosis. Adhesiolysis can be performed to open up the epidural space.

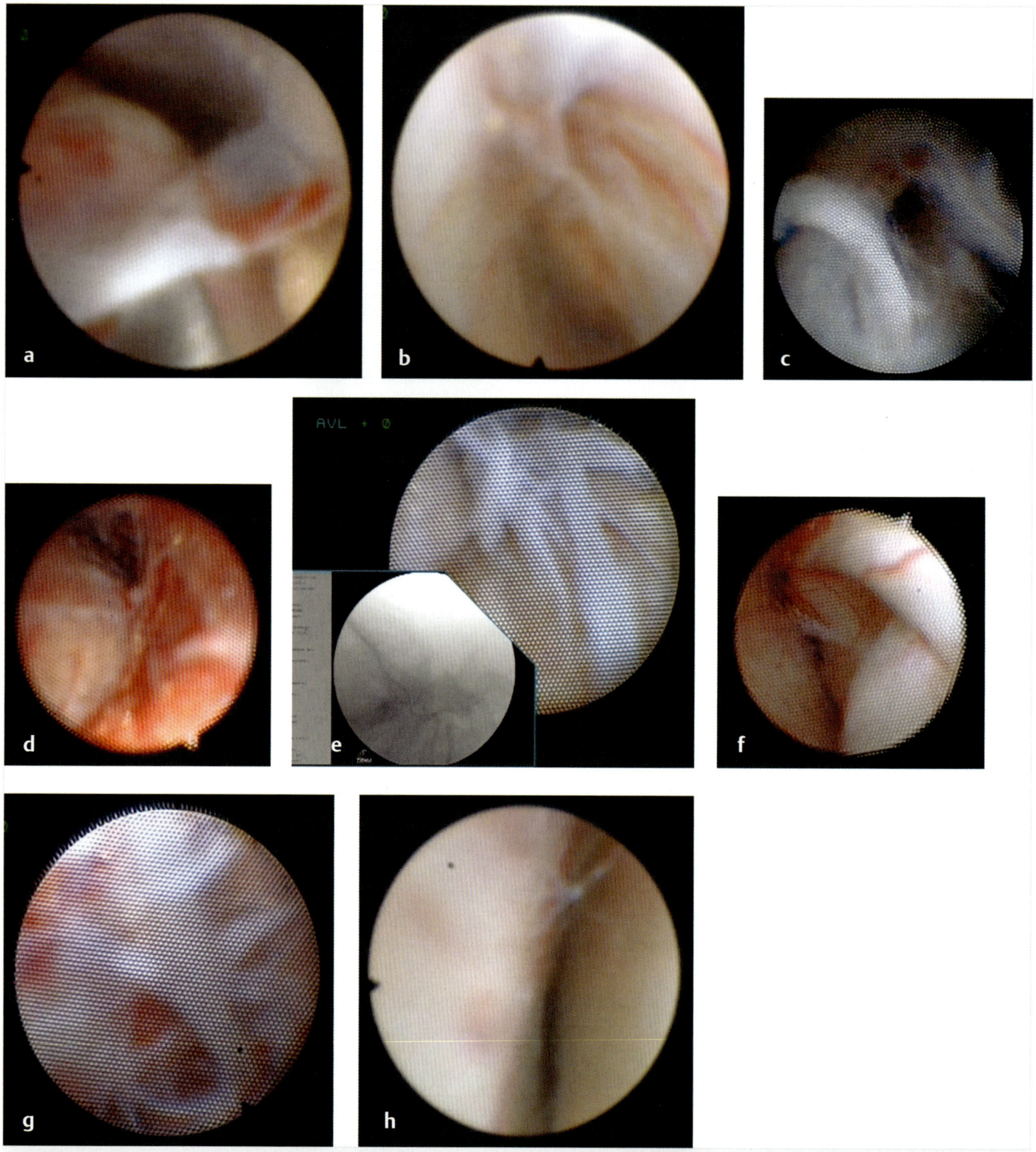

Fig. 28.7 Epidural proliferation of connective tissue. (a–b, f–i) Epiduroscopic view of epidural fibrosis. (c–e) Epidural fibrosis. (j) Anteroposterior fluoroscopic view of the epiduroscope in the cervical spine. (k–q) Epidural scars. (r) A forward-firing laser is used for adhesiolysis. (s–v) Epidural tissue fibrosis.

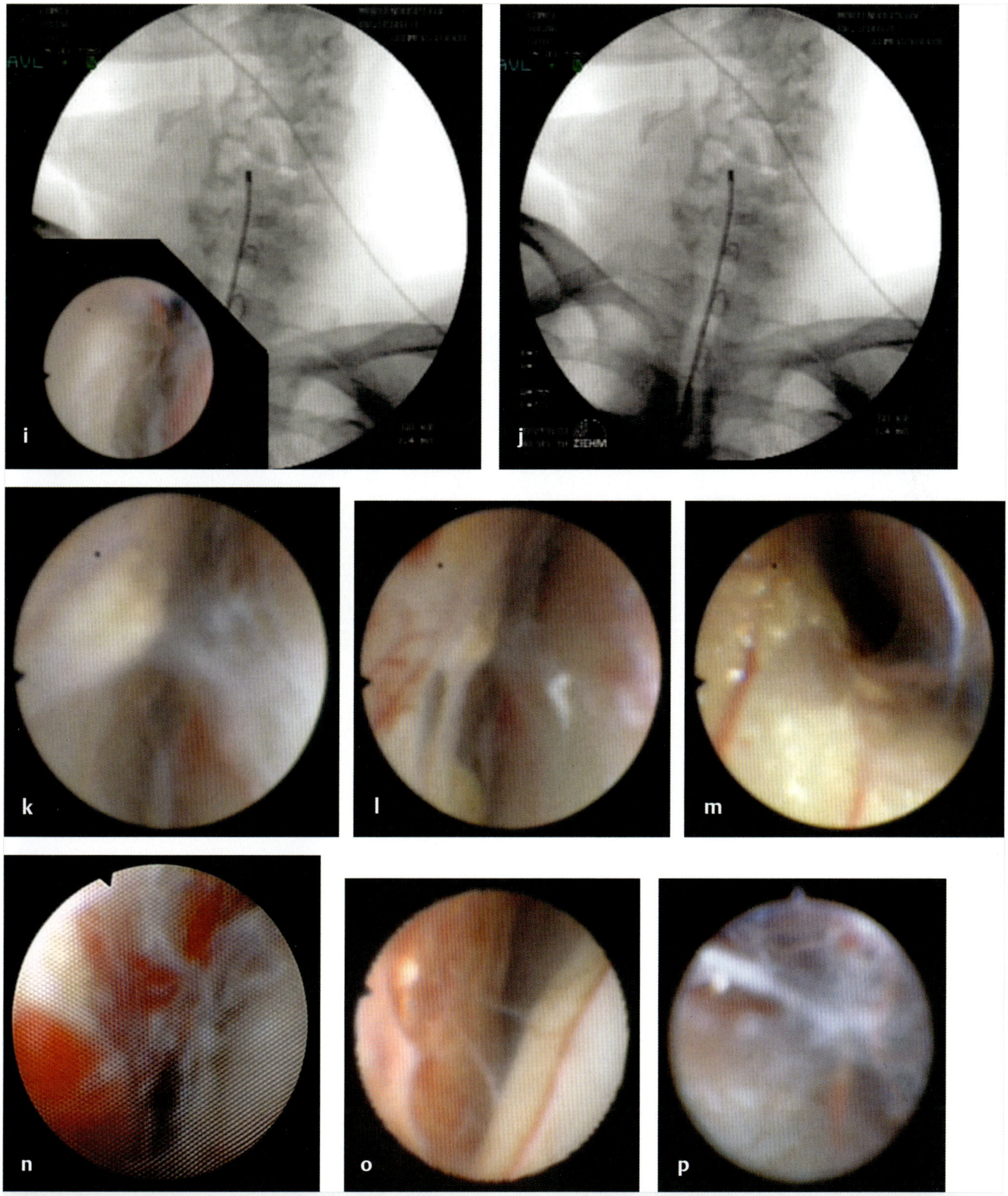

Fig. 28.7 (*Continued*) (i–p)

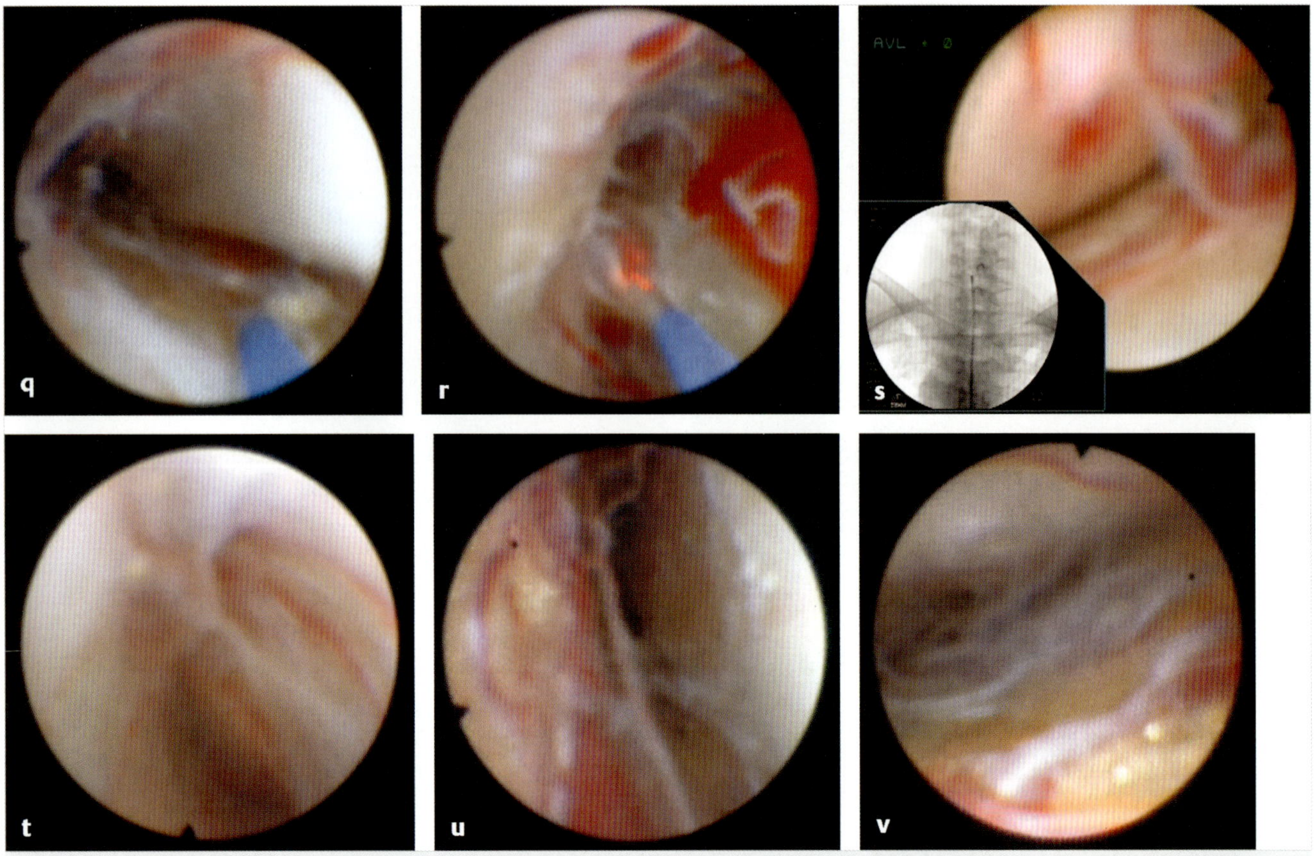

Fig. 28.7 (*Continued*) (q–v)

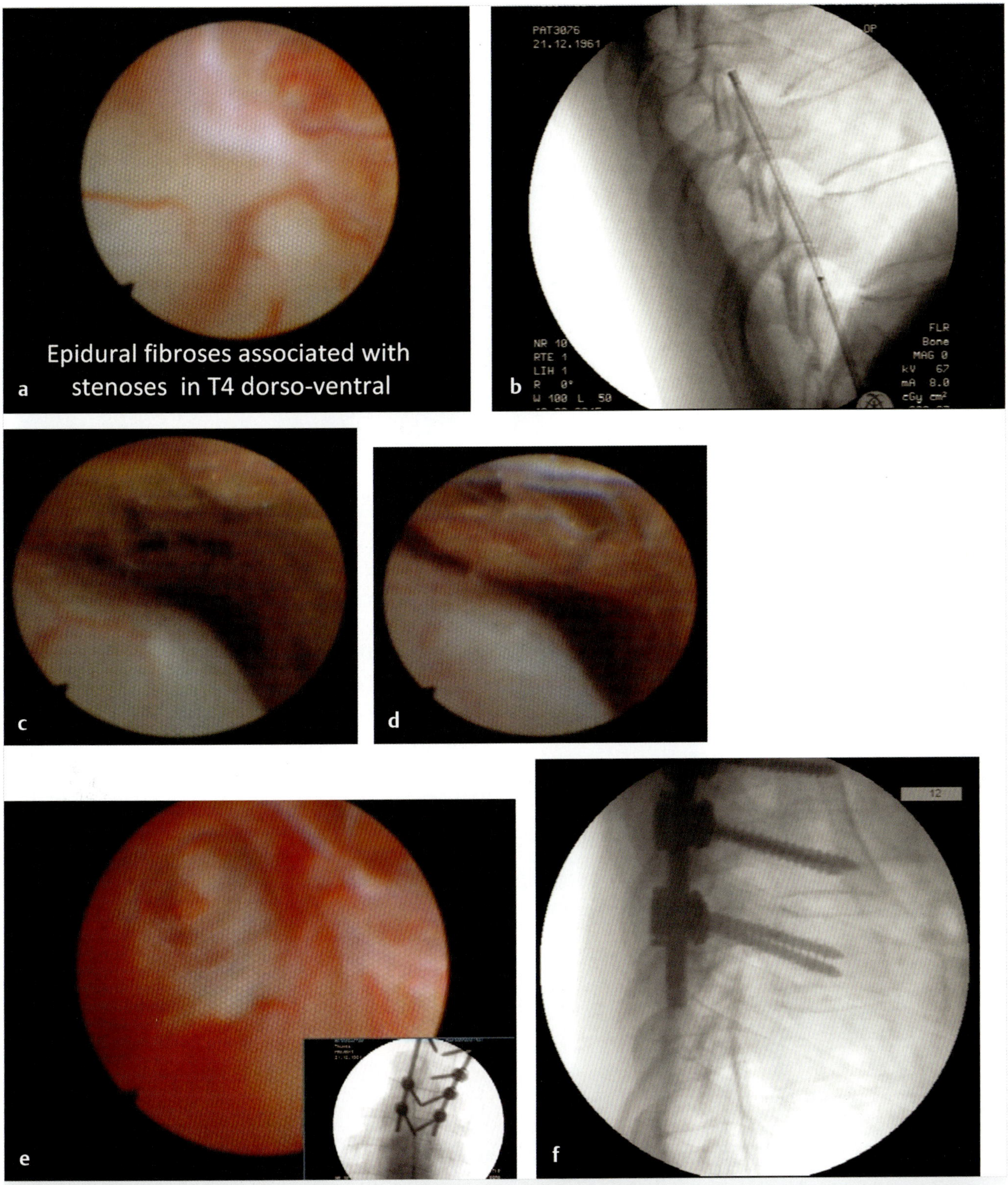

Fig. 28.8 Epidural fibrosis. (**a**) Fibrosis in the dorsal thoracic spine epidural space at T4. (**b**) Lateral fluoroscopy of the epiduroscope in the dorsal epidural space at T4. (**c, d**) Epiduroscopic view demonstrating epidural fibrosis. (**e**) Epiduroscopic view of the epidural space in a patient with previous spinal instrumentation demonstrating significant fibrosis. (**f**) Lateral fluoroscopic view of the epiduroscope in the dorsal epidural space in a patient with previous spinal instrumentation.

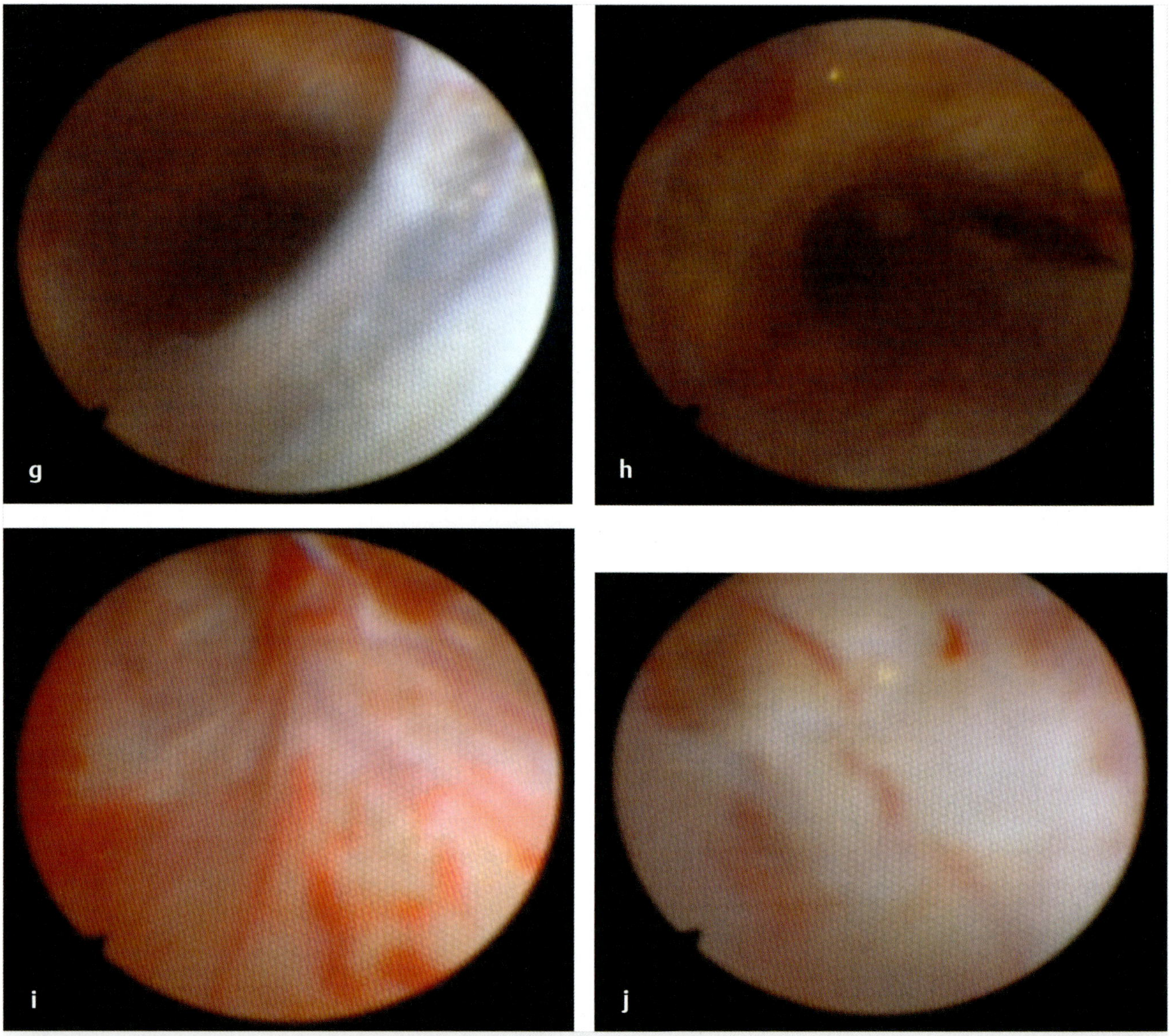

Fig. 28.8 (*Continued*) (**g–j**) Epiduroscopic view demonstrating epidural fibrosis.

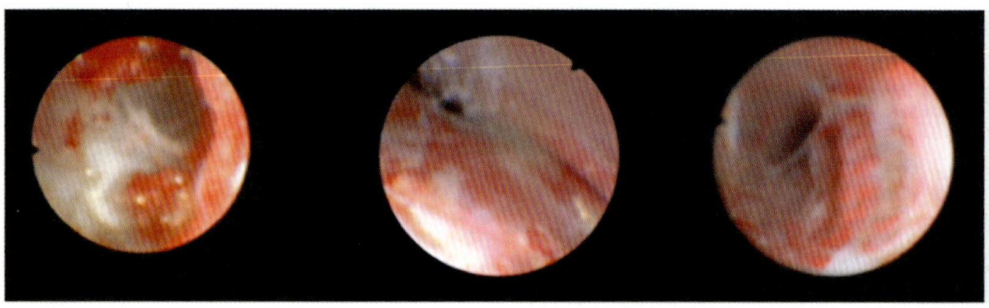

Fig. 28.9 Epiduritis. (**a–c**) Endoscopic view of chronic epidural inflammatory process. Both inflammation (seen as swelling redness) spreading from a point and wide-area inflammation are observed.

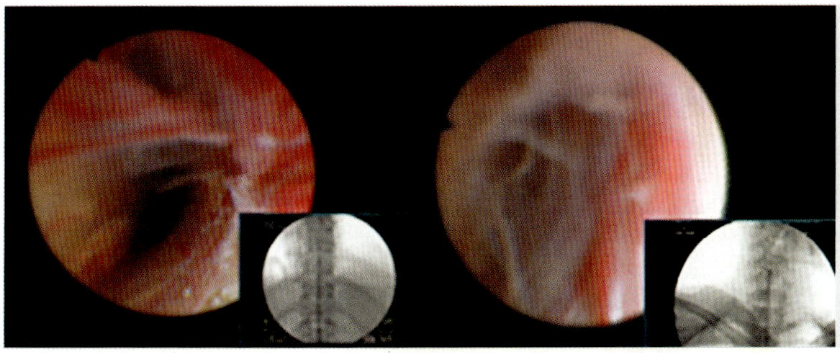

Fig. 28.10 Epiduritis. (**a**) Endoscopic view of thoracic epiduritis. (**b**) Endoscopic view of cervical epiduritis.

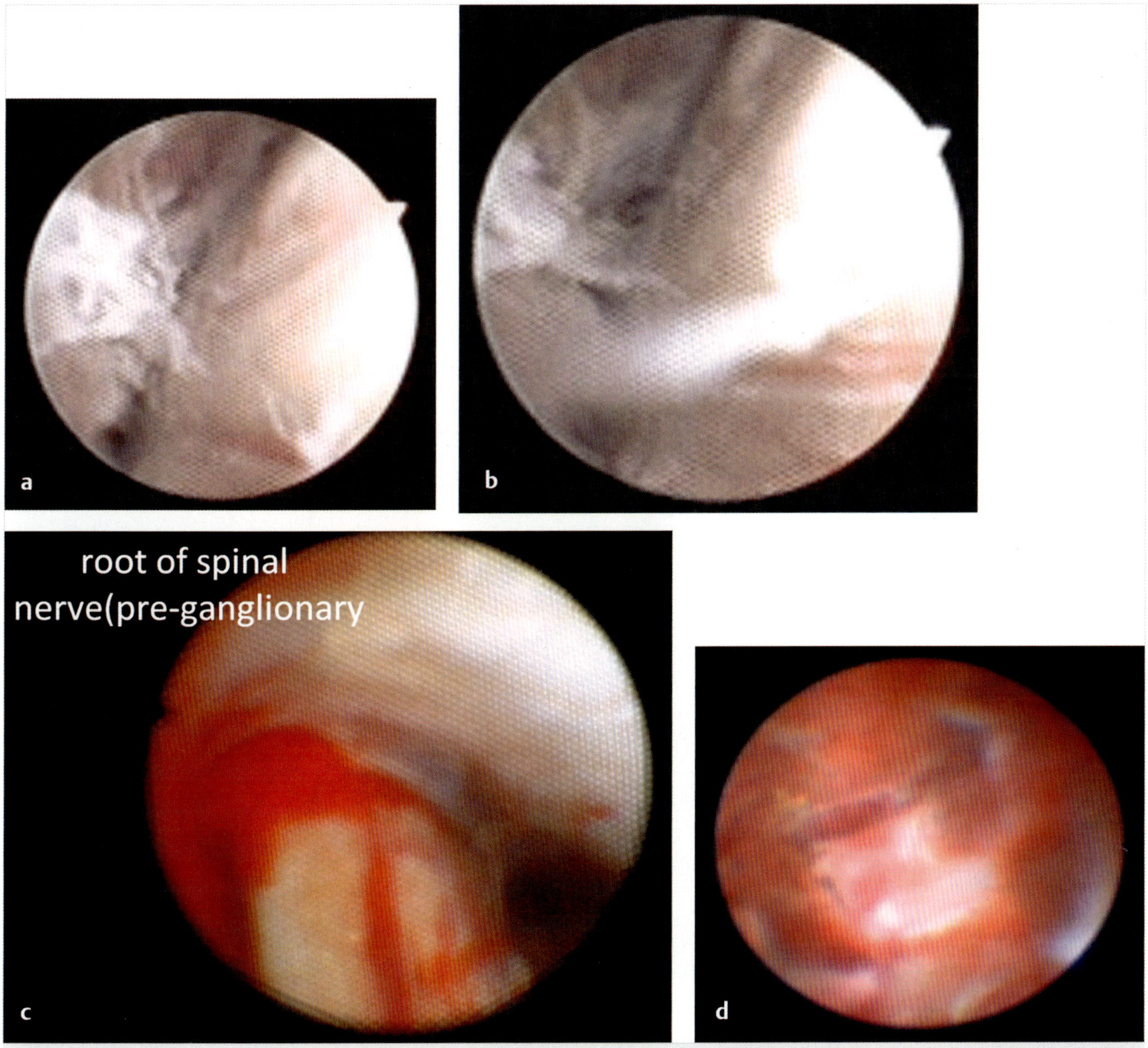

Fig. 28.11 Radiculitis. (**a,b**) Epiduroscopic view showing normal dorsal spinal nerve root. (**c,d**) Epiduroscopic view showing inflamed dorsal spinal nerve root.

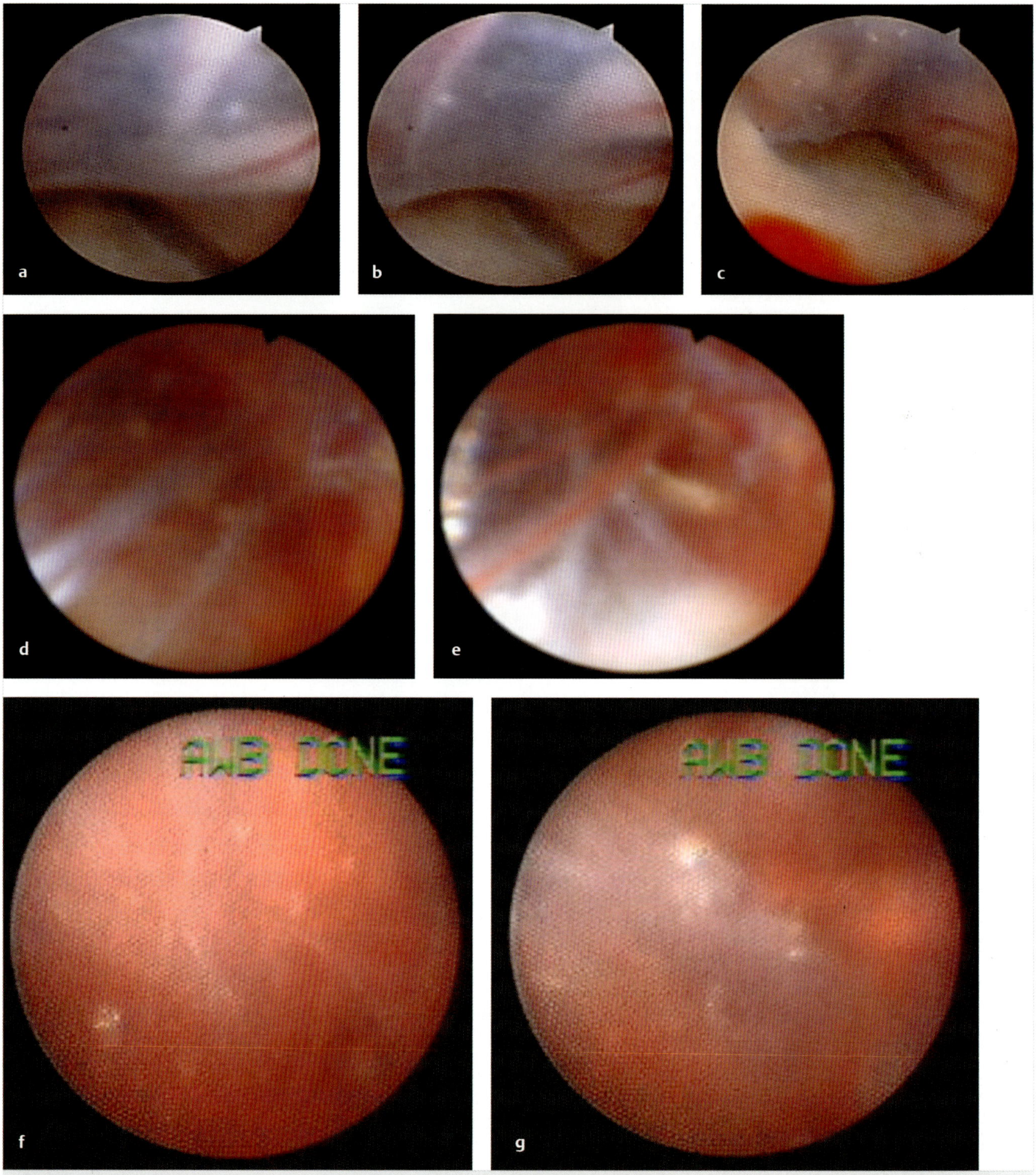

Fig. 28.12 Arachnoiditis. (**a–c**) Normal arachnoid. (**d**) Epiduroscopic view of adhesive arachnoiditis. (**e–g**) Epiduroscopic view of significant adhesions secondary to arachnoiditis .

28.3 Pathology Images: Implanted Spinal Cord Stimulation Electrode and Polymethylmethacrylate

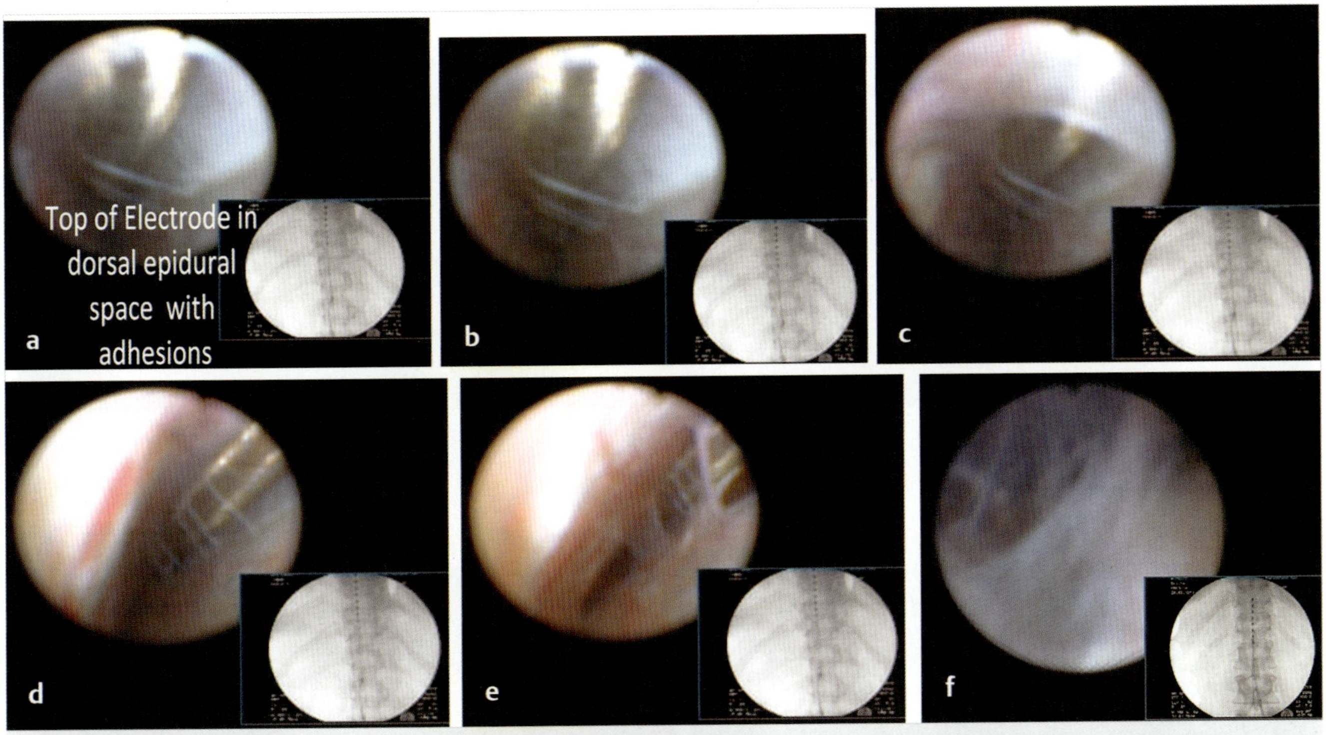

Fig. 28.13 Adhesions and scar tissue formation around spinal cord stimulator leads. (a–e) Epiduroscopic view of a fibrotic sheath formed around an implanted spinal cord stimulator lead. (f) Epiduroscopic view after intervening microsurgery around the tip of the spinal cord stimulator lead.

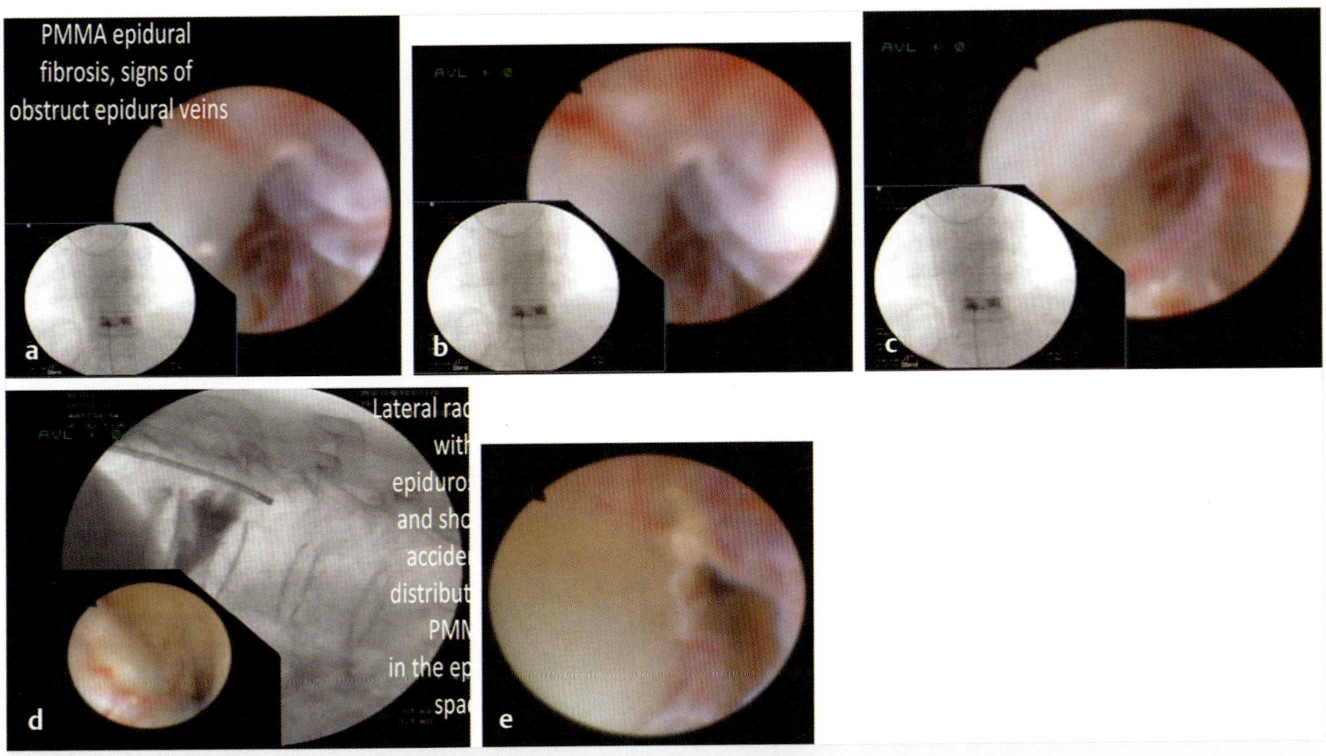

Fig. 28.14 Polymethacrylate. (a–e) Fibrosis and stenosis after a vertebroplasty with polymethylmethacrylate (PMMA).

28.4 Pathology Images: Special Findings in Epidurography, Spinal Ultrasound, and Thecaloscopy

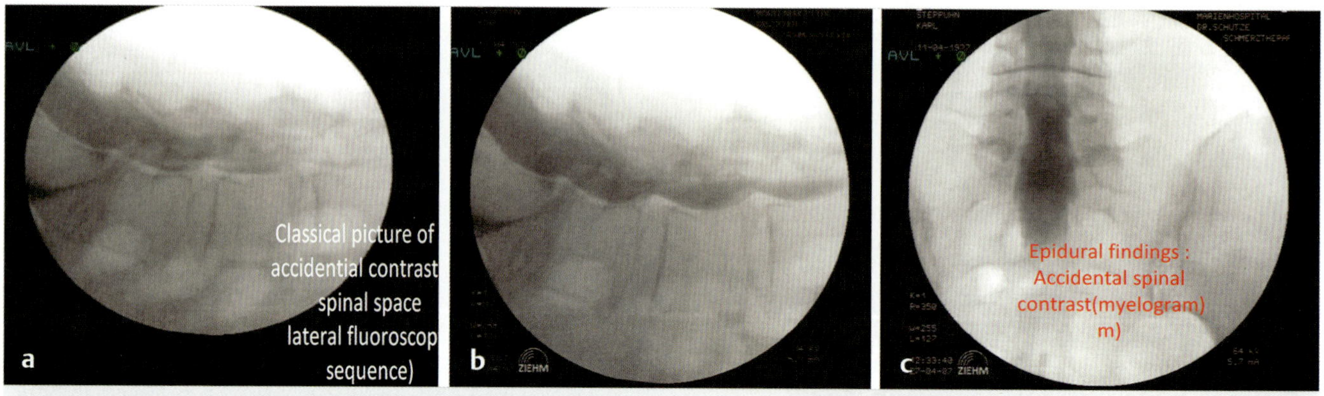

Fig. 28.15 Epidurogram findings of subdural contrast. (**a–b**) Lateral fluoroscopy demonstrating accidental contrast injection into the subarachnoid space. (**c**) Anteroposterior fluoroscopy demonstrating accidental contrast injection into the subarachnoid space.

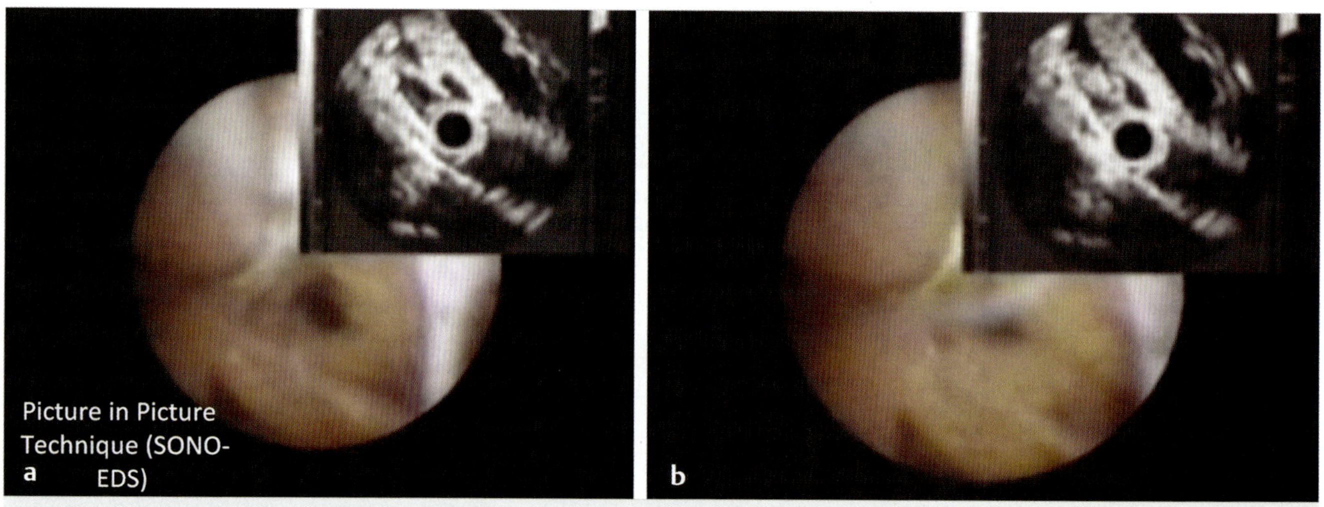

Fig. 28.16 Spinal ultrasound findings. (**a,b**) Ultrasonography used in conjunction with epiduroscopy.

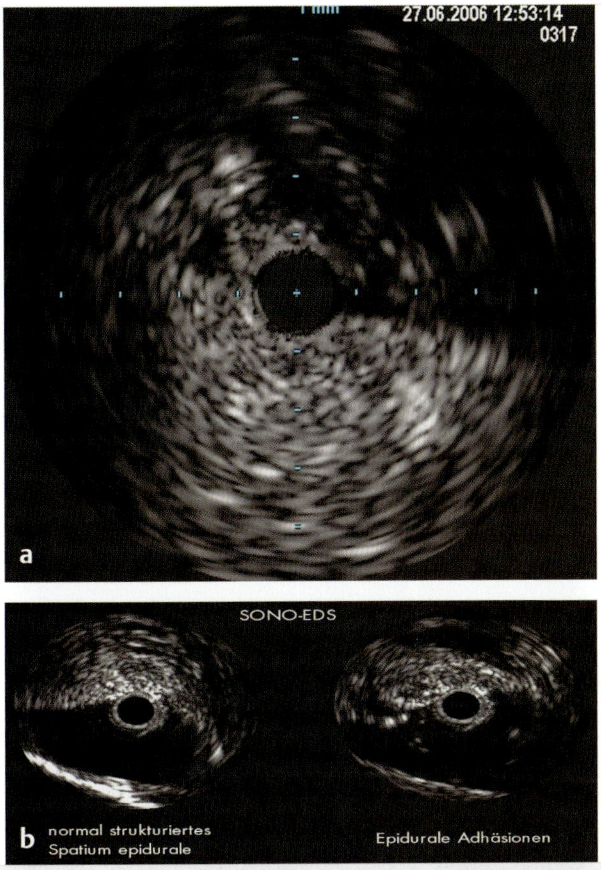

Fig. 28.17 **Spinal ultrasound findings.** (**a**) Sonogram in the epidural space, showing the distribution of fatty tissue. (**b**) Sonogram of dura mater, fatty tissue, and adhesions in the epidural space.

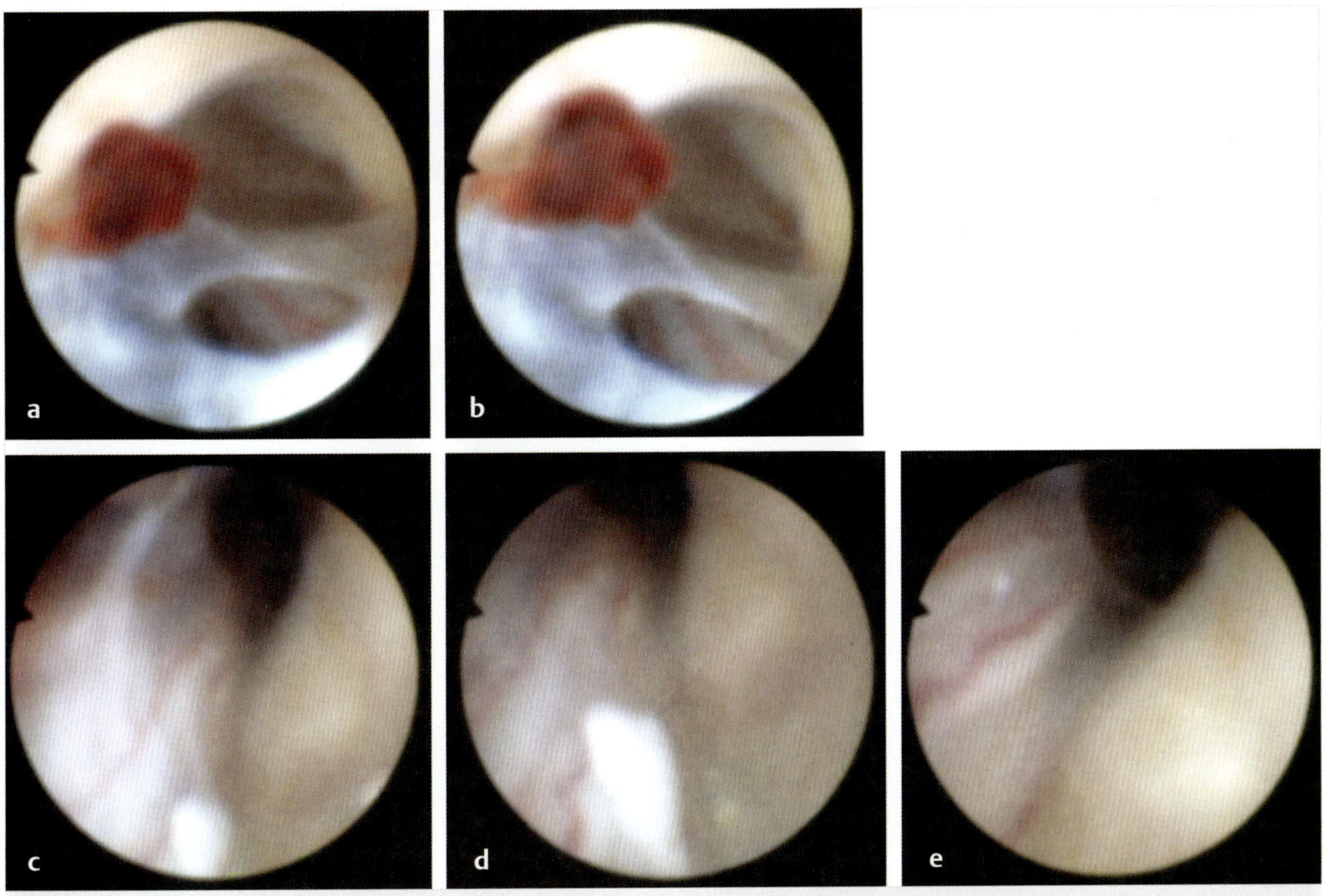

Fig. 28.18 **Endoscopic findings in the subarachnoid space.** (**a–b**) Epiduroscopic view of a durotomy. (**c–e**) Endoscopic view of the intrathecal space.

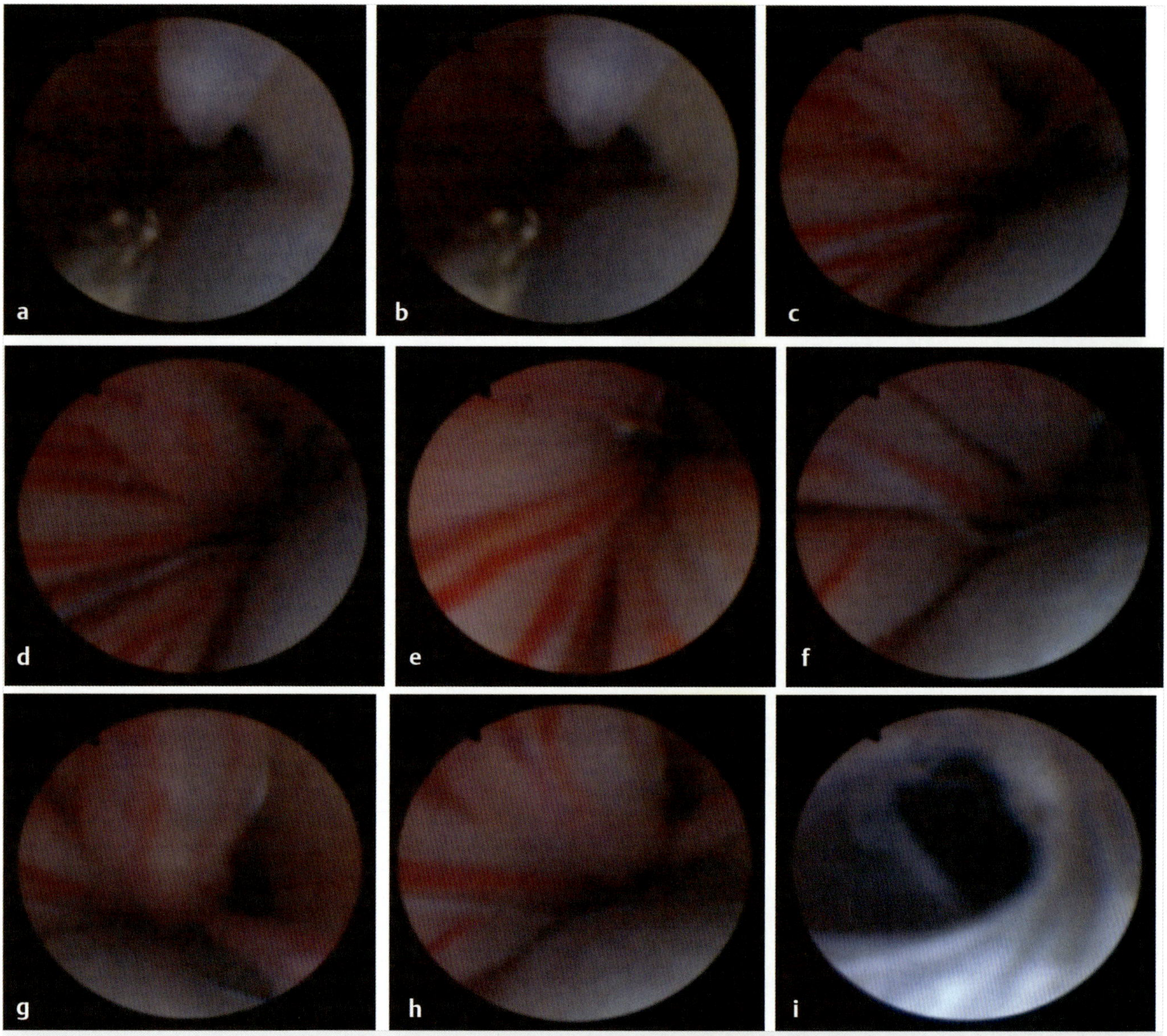

Fig. 28.19 Endoscopic findings in the intrathecal space. (**a**) Endoscopic intrathecal view of the spinal nerves. (**b**) Endoscopic intrathecal view of the spinal nerves. Flexible endoforceps can be used for biopsy. (**c–f**) Endoscopic intrathecal view of the spinal nerves. (**g**) Endoscopic intrathecal view of the spinal nerves from L2–L4. (**h**) Endoscopic intrathecal view of the spinal nerves. (**i**) Endoscopic view of the dural opening into the intrathecal space.

28.5 Images of Microinvasive Procedures: Catheter Placement, Biopsy, and Adhesiolysis

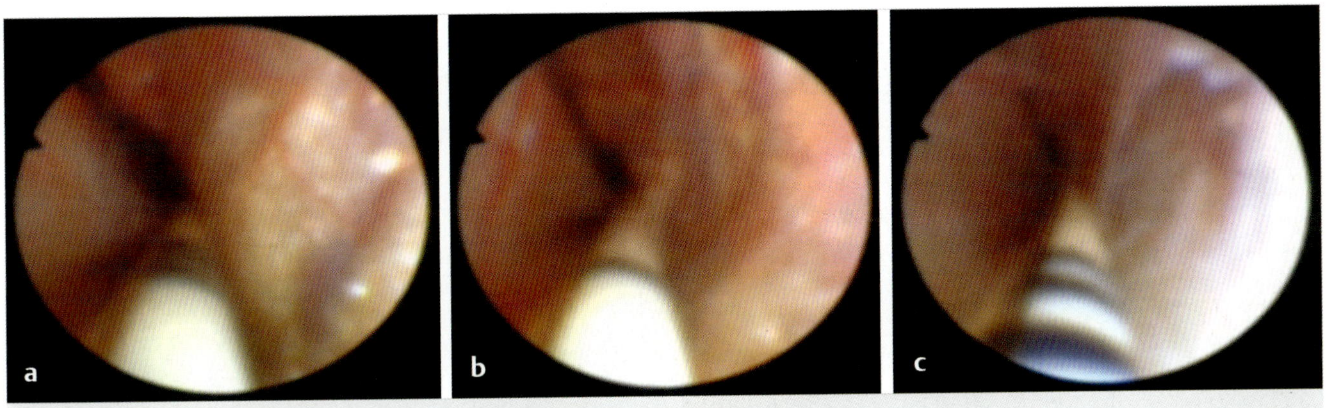

Fig. 28.20 Epidural catheter placement. (**a–c**) Epiduroscope-assisted catheter placement.

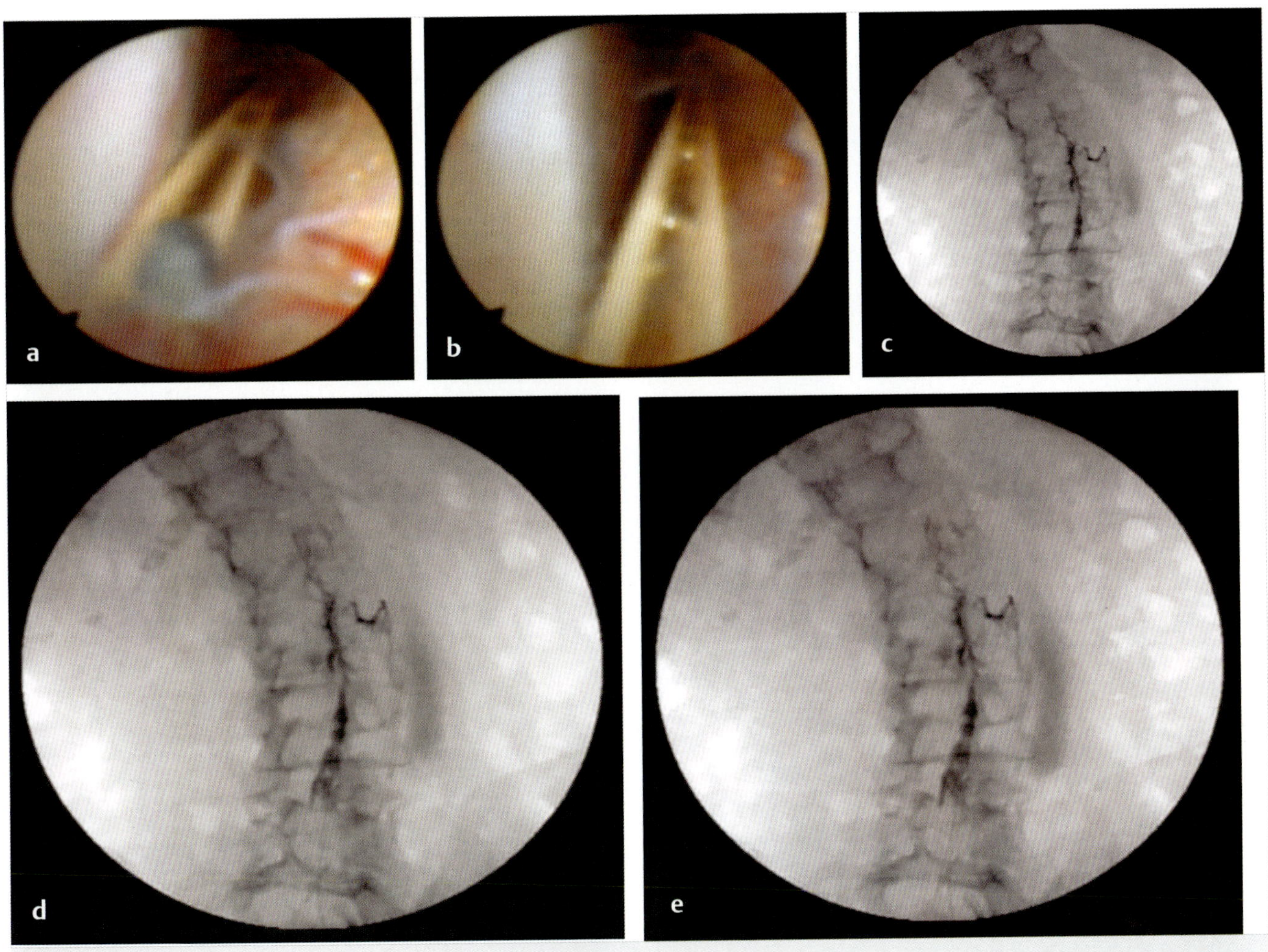

Fig. 28.21 Epidural catheter visualization and epidurogram findings: Contrast medium in the epidural vasculature. (**a,b**) Epiduroscope-assisted catheter placement. (**c**) Video still of epidural and intravascular contrast. (**d, e**) Epiduroscopic view of contrast injection into the epidural vasculature.

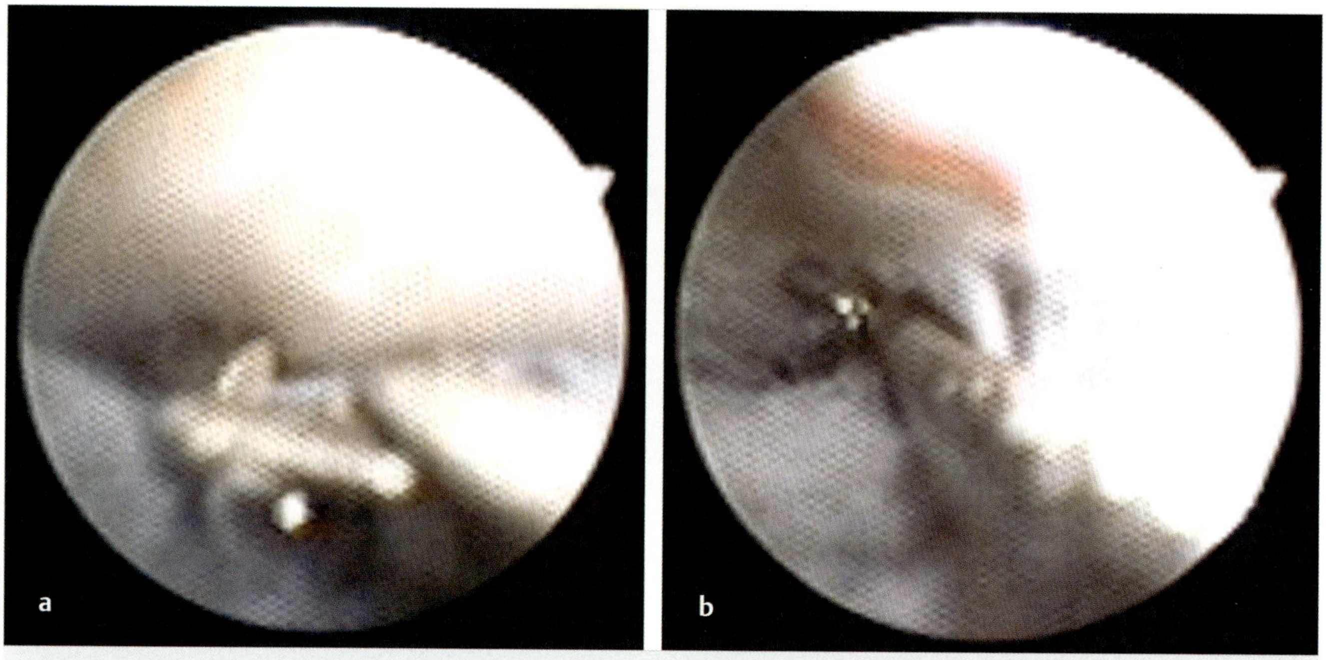

Fig. 28.22 Microinvasive procedure, biopsy. (a,b) Demonstration of endoscope-assisted biopsy with endoforceps.

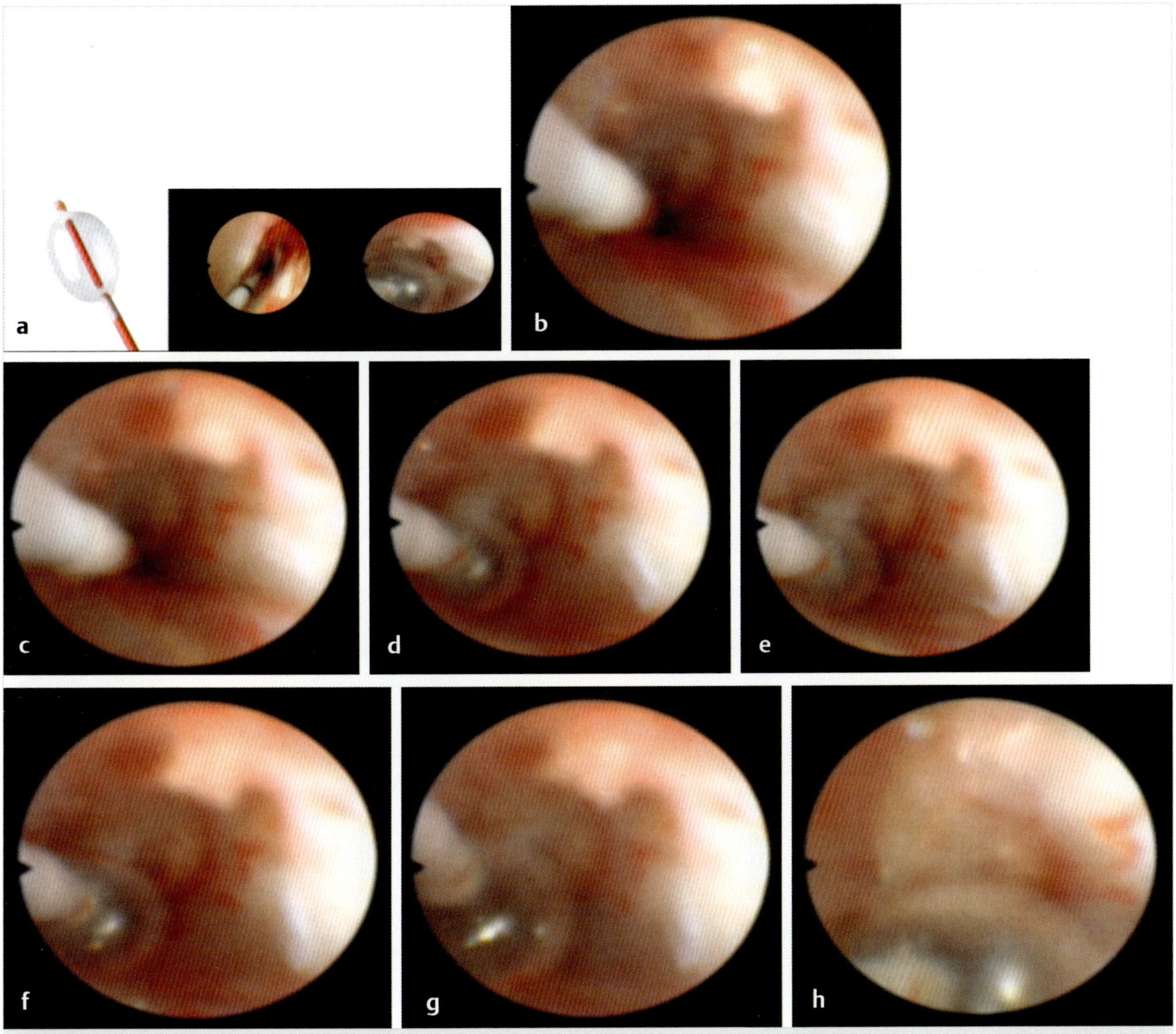

Fig. 28.23 Microinvasive procedure, epidural adhesiolysis via balloon catheter dilation. (**a**) Fogarty balloon catheter used for adhesiolysis. (**b, c**) The Fogarty balloon catheter is inserted in areas of adhesions. (**d, e**) The balloon catheter is inflated to perform adhesiolysis. (**f**) The Fogarty balloon catheter is inflated for adhesiolysis. (**g–i**) The balloon catheter is inflated to perform adhesiolysis. (**j**) After adhesiolysis, the balloon catheter is deflated. (**k**) Anteroposterior fluoroscopic view demonstrating adhesiolysis with the balloon catheter. (**l**) The balloon catheter is inflated to perform adhesiolysis. (**m**) View of the Fogarty balloon catheter inflated for adhesiolysis. (**n**) Increased spinal epidural space is created after deflation of the balloon catheter. This can facilitate placement of stimulator leads.

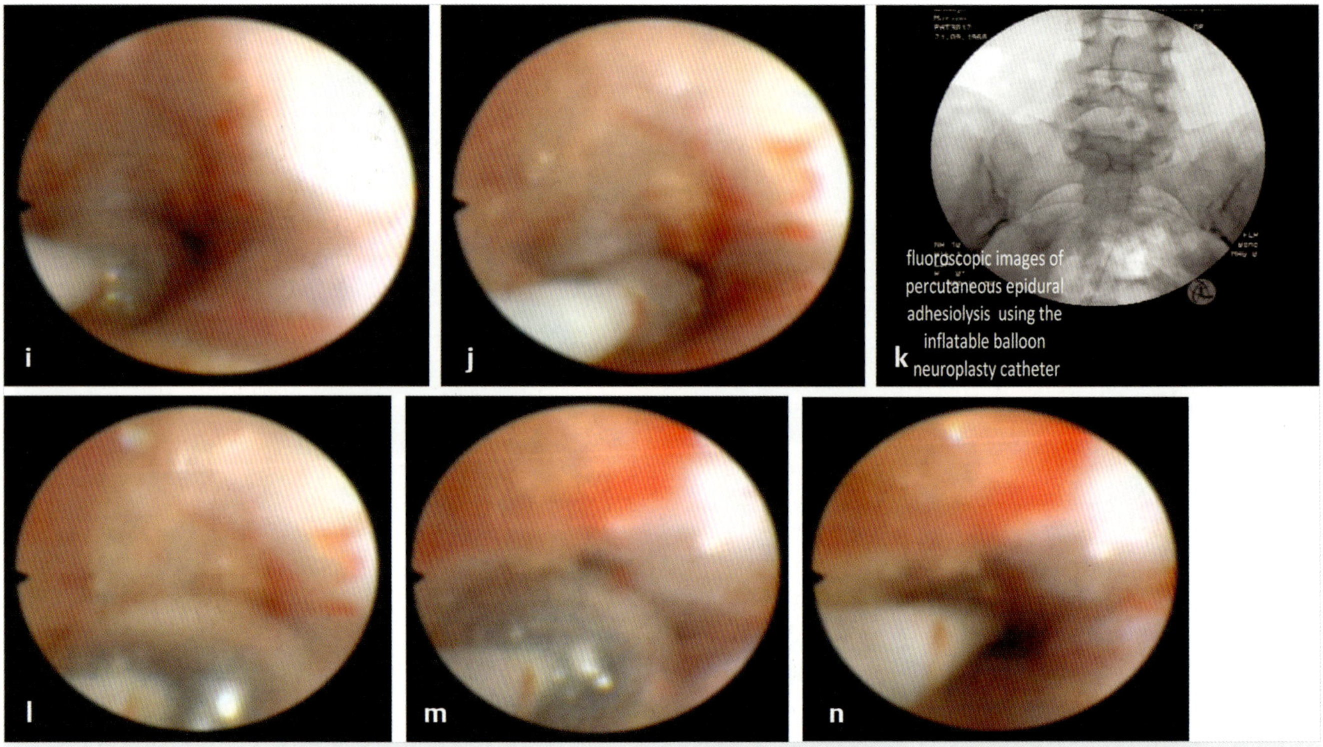

Fig. 28.23 (*Continued*)

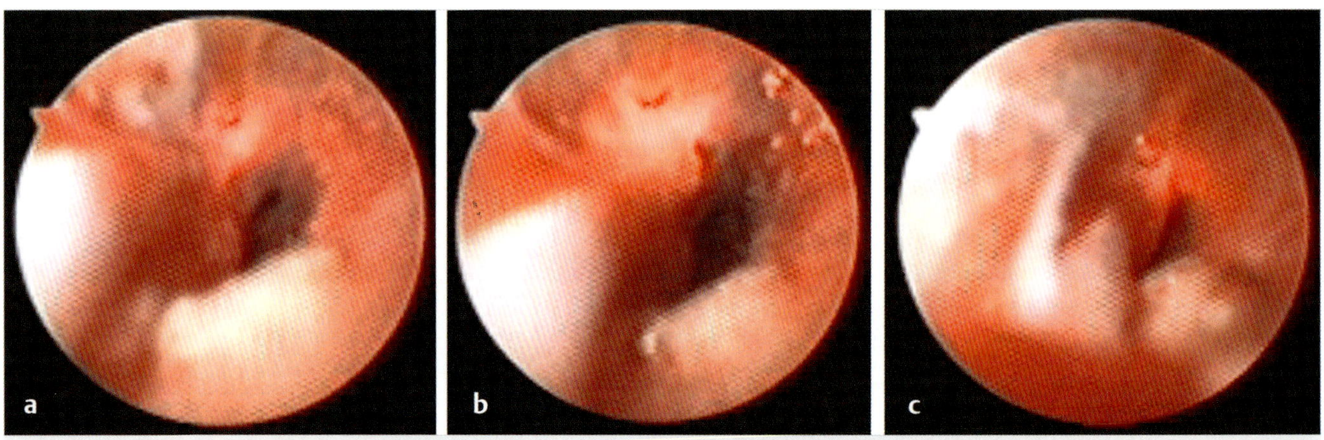

Fig. 28.24 Microinvasive procedure, epidural laser adhesiolysis. (a–c) Laser coagulation for adhesiolysis.

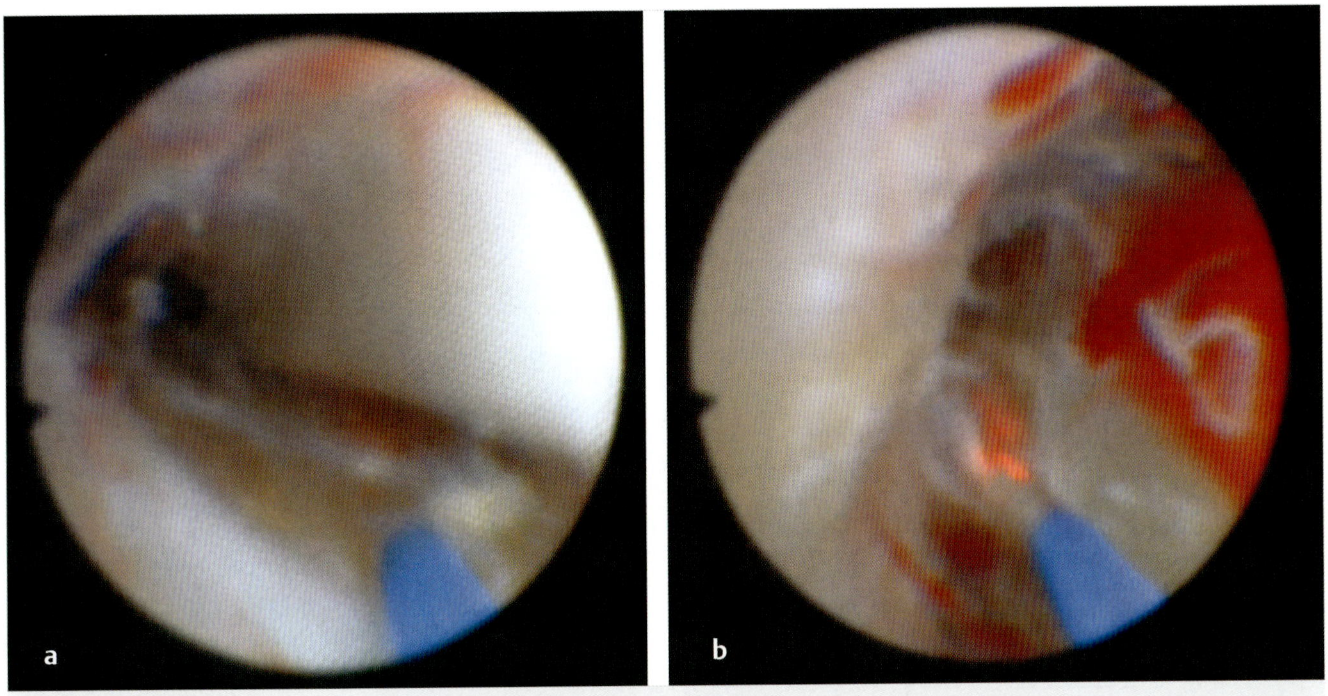

Fig. 28.25 Microinvasive procedure, epidural laser adhesiolysis. (a,b) A forward-firing laser is used for laser coagulation for adhesiolysis.

Index

Note: Page numbers set **bold** or *italic* indicate headings or figures, respectively.

A

B

C

D

E